Handbook of

Function and Generalized Function Transformations

Mathematical Sciences Reference Series

Series Editor
Daniel Zwillinger
Rensselaer Polytechnic Institute
Troy, New York

Titles in this series

CRC Standard Mathematical Tables and Formulae, 30th Edition
Daniel Zwillinger

Standard Mathematical Tables and Formulae (CD-ROM)
Daniel Zwillinger

Handbook of Function and Generalized Function Transformations
Ahmed I. Zayed

Partial Differential Equations and Mathematica
Prem K. Kythe, Pratap Puri, and Michael R. Schaferkotter

Advanced Engineering Mathematics
Dean Duffy

Handbook of *Function and Generalized Function Transformations*

Ahmed I. Zayed
Department of Mathematics
University of Central Florida
Orlando, Florida

CRC Press
Boca Raton New York London Tokyo

Library of Congress Cataloging-in-Publication Data

Zayed, Ahmed I.
Handbook of function and generalized function transformations / Ahmed I. Zayed.
p. cm.
Includes bibliographical references and index.
ISBN 0-8493-7851-6 (alk. paper)
1. Functions--Handbooks, manuals, etc. 2. Transformations (Mathematics)--Handbooks, manuals, etc. I. Title.
QA331.Z39 1996
515'.723--dc20 95-46198
CIP

International Standard Book Number 0-8493-7851-6
Library of Congress Card Number 95-46198
Printed in the United States of America 1 2 3 4 5 6 7 8 9 0
Printed on acid-free paper

Contents

List of Figures

Preface

For almost two centuries the method of function transformations has been used successfully in solving many problems in engineering, mathematical physics and applied mathematics. Function transformations include, but are not limited to, the well-known technique of linear integral transformations. A function transformation simply means a mathematical operation through which a real or complex-valued function f is transformed into another function F, or into a sequence of numbers, or more generally into a set of data. This is done for a variety of reasons, for example:

1. To transform a problem from one setting into another, where it can be easily solved.

2. The transform itself may have a physical meaning and ought to be studied in its own right.

3. The transform of f may be a set of data that is measured experimentally and from which f needs to be constructed.

Since its birth in the 1780s in the work of the great mathematician Laplace, on probability theory, the theory of function transformations has flourished and continues to do so. In the last few years, in particular, it has received a great impetus from the advent of wavelets. Not only is the wavelet transform an example of how practical function transformations can be, but it is also an example of a transformation that has gone beyond what it was designed to do as a technique. It has contributed to the development of modern mathematical analysis just as the Fourier transformation contributed to the advancement of classical analysis in the earliest years of the nineteenth century. Wavelet analysis has also revived interest in other related function transformations that were slowly fading away. Cases in point are the Bargmann, Gabor and Zak transformations.

Because modern science has put emphasis on specialization, the scientific literature is inundated every day with new publications, making it difficult for the experts to keep abreast of the new advances in their field, much less know what is transpiring in other fields. This is manifested in the fact that many function transformations were discovered in some discipline and then later rediscovered and even called different names by scholars in other disciplines. Interdisciplinary interaction is needed today more than ever before. This book is a humble step to fill that need. It collects and presents in an accessible style many function transformations used by engineers, physicists and mathematicians alike.

Some function transforms, such as the Fourier, Laplace and the Z-transforms, are taught at the undergraduate level to engineering, science and mathematics students; a few are taught at the graduate level, but the majority are left untouched on the presumption that only the experts need them. Yet they are cited in many technical articles in different fields, and in most cases the burden of learning them is left to the reader. There is clearly a need for a book to compile most of the useful function transformations and their important properties.

Granted, the classical books of R. Churchill [40], I. Sneddon [216] and E. C. Titchmarsh [225] provide an excellent source for a number of popular linear integral transformations. These books are still unsurpassed, but unfortunately outdated; since their last publication many function transformations have emerged on the scene.

On a parallel road, the classical theory of linear integral transformations has been extended to Schwartz distributions and generalized functions by many people. But the main credit indeed goes to Zemanian [276] who paved the way for that extension and called it the theory of generalized integral transformations. The classical and the generalized theories are usually treated separately, though their boundaries are vague and at best overlap.

The main objectives of this book are:

1. To complement the classical work of Churchill, Sneddon and Titchmarsh by presenting newly discovered transforms and in some instances old but useful transforms that have recently been rediscoverd.

2. To provide one text for both the classical and the generalized integral transformation theories. This is done in such a way that if the reader is not interested in the generalized theory, he/she can skip it without interrupting the coherence of the presentation.

3. To provide easy access to many function transformations and their basic properties without getting involved in many technical details.

When I accepted the challenge to write this book, my original idea was to compile as many function transformations as possible. I soon had to shun that idea as impractical: not because the number of transformations cited in the literature is overwhelming, but because most of them appeared to be trivial or obsolete. Or to put it more mildly, their existence was not justified by their creators, except for the mere fact that they satisfied the creators' mathematical curiosity.

Therefore, I had to use some filtering process for the inclusion in the book. The main criteria used are originality and applicability. Although most of the transformations presented in this book are linear integral transformations, some of them are *not* ! The Zak transform, for example, is a series transformation that maps a function of n variables into a function of $2n$ variables. Better yet, the Radon transformation is linear and defined by an integral; nevertheless, it does not fit very well in the theory of linear integral transformations because its kernel is very singular. In fact, its kernel is not even a function! The same goes for the X-ray and the k-plane transformations.

This book follows an unorthodox format. It is neither a textbook, though it can be used as such, nor a research monograph. But it is a handbook that is designed to give the reader easy access to the basic facts and properties of each transformation. It is not an intent of this book to delve into the frontiers of the subject, yet some new results that have never been published in a book form before are included. Non-technical proofs of important results are given. To keep the book at a manageable size, many results are presented without proofs, but references for them are provided. Sometimes proofs can be illustrative of the intricacies of the subject and revealing of the experts' tricks. Whenever we felt that that was the case, we included the proofs. This is particularly apparent in the chapters on the Bargmann, Gabor, Radon and wavelet transforms.

The book is tacitly divided into seven parts. To make it self-contained, I have included in the first part, which consists of Chapters 1 through 4, all the preliminary material. Since no advanced mathematics book can purport to be utterly self-contained, I have assumed that the reader has some knowledge of mathematical analysis at the level of beginning graduate courses in real and complex analysis.

Chapter 1 contains almost all that is needed from real, complex and functional analysis. Chapter 2 provides a summary of the basic facts about special functions and orthogonal polynomials. An introduction to generalized functions and Schwartz distributions is given in Chapter 3, and Chapter 4 discusses the general ideas of function and generalized function transformations.

Each of the other six parts comprises a collection of function transformations grouped together not alphabetically but by the proximity of their subjects. Part two, which consists of Chapters 5–11, deals with the Laplace and other related transformations. Chapters 12–18 make up part three and they include the Fourier transform and some of its relatives. The Abel and the Fractional integral transforms, Chapters 19 and 20, constitute part four, whereas the Hankel and the Sturm–Liouville transformations form the pivot for part five which consists of Chapters 21–27. Chapters 28–33, is made up of the more recent transformations, such as the Ambiguity, Radon and Wavelet transformations. The last part is Chapter 34 which, unlike the rest of the book, deals with some indispensable sequence transformations, such as the Fast Fourier, Z, and Walsh. Admittedly, the boundaries between these parts are ambiguous. But unfortunately this is inevitable because of the inter-relationships between the various transforms.

Each chapter, except for the first four, begins with a historical introduction, followed by definitions, existence theorems, elementary properties, relationships with other transforms, inversion formula(s), a Parseval-type relation, a convolution theorem (if any), an extension to generalized functions, and finally a section on applications.

The book is an outgrowth of lecture notes that I have used in graduate courses taught at the University of Central Florida on Transform Methods and Wavelets. Before I express my gratitude to those who have helped in this project, I would like to express my apology to those whose work I have either overlooked or not cited. My only excuse is that to make the book accessible to non-experts, I had to include only the basic facts.

Finally, thanks are due to the CRC Press for their support and cooperation. I am very thankful to my family, my wife Elena and my daughter Nora, for their patience and understanding and for putting up with my very long hours in front of the computer.

To my parents,
my wife Elena, and
my daughter Nora

1

Preliminaries

In this chapter we shall introduce some of the definitions and notations that will be used throughout the book, and collect certain results of a general nature that will be useful later on in studying several function transformations.

1.1 Notation

Most of the notations used in this book are standard, yet they may differ from those used in the physics and engineering literature.

We use $\mathcal{C}, \mathcal{R}$ to denote the sets of complex and real numbers, respectively, $\mathcal{Z}$ to denote the integers and $\mathcal{N}$ to denote the set of natural numbers. $\mathcal{C}^n$ and $\mathcal{R}^n$ denote, respectively, the complex and real n-dimensional Euclidean spaces. An arbitrary point x in $\mathcal{R}^n$ $(\mathcal{C}^n)$ is denoted by $x = (x_1, \cdots, x_n), x_i \in \mathcal{R}(\mathcal{C}), i = 1, \cdots, n$. Sometimes, we may use the vector notation $\mathbf{x} = (x_1, \cdots, x_n)$ to emphasize that x has n components. The differential operator

$$D^\alpha = \frac{\partial^{|\alpha|}}{\partial x_1^{\alpha_1} \cdots \partial x_n^{\alpha_n}}$$

has the usual meaning: $\alpha = (\alpha_1, \cdots, \alpha_n)$ is a multi-index with α_i $(i = 1, \cdots, n)$ being a non-negative integer and $| \alpha |= \sum_{i=1}^n \alpha_i$.

For $n - 1$, the kth derivative of a function $f(x)$ is denoted by $D^k f(x)$ or $f^{(k)}(x)$, where $D = \frac{d}{dx}$. By a function, we mean *a conventional function*, which is a mapping from a subset of $\mathcal{R}^n$ or $\mathcal{C}^n$ into an arbitrary set. This is emphasized to distinguish the notion of a function from that of *a generalized function* which will be explained later in Chapter 3. A function is said to be *real-valued or complex-valued* depending on whether its range is a subset of $\mathcal{R}$ or $\mathcal{C}$, respectively. The *support* of a function $f(x)$ defined on some open set I in $\mathcal{R}^n$, denoted by supp f, is the closure with respect to I of the set of points where $f(x) \neq 0$. The *characteristic function* of a set A is defined as

$$\chi_A(x) = \begin{cases} 1 \text{ if } x \in A \\ 0 \text{ if } x \notin A. \end{cases}$$

The Heaviside function, $H(x)$, is defined as $H(x) = \chi_{[0,\infty)}(x)$. For a real number $x \neq 0$, we define the sign of x, denoted by $\operatorname{sgn} x$, as $x/|x|$ or equivalently

$$\operatorname{sgn} x = \begin{cases} 1 & \text{if } x > 0, \\ -1 & \text{if } x < 0. \end{cases}$$

The *translation*, *modulation*, and *dilation* operators are defined by

$$T_a f(x) = f(x-a),$$

$$E_a f(x) = e^{iax} f(x),$$

$$D_a f(x) = \frac{1}{\sqrt{|\, a \,|}} f(x/a), a \neq 0,$$

respectively.

The *supremum, infimum, maximum* and *minimum* are abbreviated, sup, inf, max, and min, respectively. The Kronecker delta is defined as

$$\delta_{m,n} = \begin{cases} 0 \text{ if } m \neq n \\ 1 \text{ if } m = n. \end{cases}$$

The imaginary number, $\sqrt{-1}$, is denoted by i. The conjugate, $x - iy$, of a complex number $z = x + iy$ is denoted by $\bar{z}$, the real and imaginary parts of z are denoted by $\operatorname{Re} z$ and $\operatorname{Im} z$, respectively, the absolute value and the argument of z by $|z|$, $\arg z$, respectively, where $|z| = \sqrt{z\bar{z}} = \sqrt{x^2 + y^2}$ and $\arg z = \tan^{-1}(y/x)$.

1.2 Function Spaces

Let I denote an open set in $\mathcal{R}^n (n \geq 1)$ and $d\mu$ be an arbitrary Borel measure on I. We define $L^p(I, \mu)$, for $0 < p < \infty$, as the set of all measurable functions f on I such that $|f|^p$ is integrable on I with respect to $d\mu$, i.e.,

$$L^p(I,\mu) = \left\{ f : f \text{ is measurable on } I \text{ and } \int_I |f(\mathbf{x})|^p \, d\mu(\mathbf{x}) < \infty \right\}.$$

When $I = \mathcal{R}^n$ and $d\mu(\mathbf{x}) = d\mathbf{x}$ (the Lebesgue measure), the integral $\int_I f(\mathbf{x})\, d\mathbf{x}$ will mean

$$\int_{-\infty}^{\infty} \cdots \int_{-\infty}^{\infty} f(x_1, \cdots, x_n)\, dx_1 \cdots dx_n.$$

For $p = \infty$, we define $L^\infty(I, \mu)$ as the set of all functions that are essentially bounded on I with respect to μ. A function f is said to be *essentially bounded* on I with respect to μ if there exists a positive real number M such that $\mu\{x \in I : |f(x)| > M\} = 0$. In this case, we define the *essential supremum* , abbreviated ess. sup, by

$$\text{ess. sup}\, f = \inf\{M : \mu\{x \in I : |f(x)| > M\} = 0\}.$$

If ess. sup $f = M$, then $|f(x)| \leq M$ for almost all $x \in I$; the exception is a subset E of I with $\mu(E) = 0$.

We define $L^p_{loc}(I, \mu)(0 < p < \infty)$ as

$$L^p_{loc}(I, \mu) = \{f : f \text{ is measurable on } I \text{ and } \int_J |f(\mathbf{x})|^p d\mu(\mathbf{x}) < \infty,$$
$$\text{for every open set } J \text{ whose closure } \bar{J} \text{ is a compact subset of } I\}.$$

Clearly, $L^p(I, \mu)$ is a proper subset of $L^p_{loc}(I, \mu)$. An element of $L^1_{loc}(I, \mu)$ will be called a *μ-locally integrable* function on I, but it will be called a locally integrable function, for short, if μ is the Lebesgue measure. In most cases considered in this book, μ will be absolutely continuous with $d\mu(x) = W(x)dx$, for some non-negative weight function $W(x)$ on I, where dx indicates the Lebesgue measure. In this case, we write $L^p(I, W), L^p_{loc}(I, W)$ instead of $L^p(I, \mu), L^p_{loc}(I, \mu)$, respectively, and if $W(x) = 1$, we write $L^p(I), L^p_{loc}(I)$. Furthermore, when $I = \mathcal{R}^n$, we may write L^p and L^p_{loc} for short, while if $I = (a, b)$ is an open interval in $\mathcal{R}^1$, we may use the notation $L^p(a, b)$ and $L^p_{loc}(a, b)$.

For $p \geq 1$ and positive $W(x), L^p(I, W)$ is a Banach space (see Section 1.5) with norm

$$\|f\|_{L^p(I,W)} = \begin{cases} \left(\int_I |f(x)|^p W(x)\, dx\right)^{1/p}, & 1 \leq p < \infty \\ \underset{x \in I}{\text{ess. sup}}\, |f(x)|, & p = \infty \end{cases}.$$

When $I = \mathcal{R}^n$ and $W = 1$, we may write $\|f\|_p$ instead of $\|f\|_{L^p}$.

We call two positive real numbers p and q conjugate numbers if $(1/p) + (1/q) = 1$. Let I be a measurable set and f, g be measurable functions on I. Then

$$\left|\int_I f(x)g(x)\, d\mu(x)\right| \leq \left(\int_I |f(x)|^p\, d\mu(x)\right)^{1/p} \left(\int_I |g(x)|^q\, d\mu(x)\right)^{1/q}.$$

This inequality is known as *Hölder's inequality.* The case in which $p = q = 2$ is known as *Cauchy–Schwarz inequality.*

The set of all continuous functions on an open set I is denoted by $C(I)$ and the set of all functions having $k(k \geq 0)$ continuous derivatives on I is denoted by $C^k(I)$. We adopt the notation $C^o(I) = C(I)$. If I is an interval in $\mathcal{R}^1$, it is sometimes convenient to extend the definition of $C^k(I)$ to intervals that are not open. For example, if $I = [a, b)$ we say that $f \in C^k(I)$ if it is in $C^k(J)$, where $J = (a, b)$, and f has k right-hand derivatives at a, and they are continuous from the right at a. Similarly, one can extend the definition of $C^k(I)$ to the intervals $(a, b]$ and $[a, b]$.

The collection of all infinitely differentiable functions on some open set I in $\mathcal{R}^n$ will be denoted by $C^\infty(I)$, while $\mathcal{D}(I)$ will stand for the collection of all functions $f(x)$ that are infinitely differentiable on I such that the support of f is a compact subset of I. Finally, the set of all infinitely differentiable functions on $\mathcal{R}^n$ that, together with their derivatives of all orders, are rapidly decreasing on $\mathcal{R}^n$ will be denoted by S. The definition of a rapidly decreasing function is given below.

1.3 Analytic Functions

Let $f(z)$ be a complex-valued function of the complex variable $z \in \mathcal{C}$. We say that f is *analytic (holomorphic)* in a domain $\Omega \subset \mathcal{C}$ if it is differentiable at all points of Ω, and we say that f is *analytic at a point* z_0 if it is analytic in some open neighborhood of z_0. By a *domain* we mean an open and connected subset of $\mathcal{C}$. The set of all analytic (holomorphic) functions on a domain Ω will be denoted by $\mathcal{H}(\Omega)$.

Let f be analytic in a domain Ω and $z_0 \in \Omega$. Then f can be represented uniquely by a power series of the form

$$f(z) = \sum_{n=0}^{\infty} a_n (z - z_0)^n ,$$

in some neighborhood of z_0, where $a_n = f^{(n)}(z_0)/n!$. The series converges at least in the circle whose center is z_0 and whose radius is equal to the distance between z_0 and the boundary of Ω.

If f is single-valued and analytic in a domain Ω, then

$$f(z) = \frac{1}{2\pi i} \int_\gamma \frac{f(t)}{t - z} dt,$$

where γ is any simple, closed, positively oriented curve lying entirely in Ω and z is any point inside γ. This is *Cauchy's integral formula.* A more generalized version of it reads

$$f^{(n)}(z) = \frac{n!}{2\pi i} \int\limits_{\gamma} \frac{f(t)}{(t-z)^{n+1}}\, dt, \quad n = 0, 1, 2, \cdots,$$

where γ and z have the same meaning as before.

If f is analytic in a bounded domain Ω and continuous on the boundary $\partial\Omega$ of Ω, then the *maximum modulus principle* [208, p. 229] states that

$$| f(z) | \leq M(f, \partial\Omega), \text{ for all } z \in \Omega,$$

where $M(f, \partial\Omega) = \max_{z\in\partial\Omega} | f(z) |$.

THEOREM 1.1 [224, p. 99]
Let $f(z, w)$ be a continuous function of the complex variables z and w, where z ranges over a region D and w lies on a contour γ. If $f(z, w)$ is an analytic function of z in D for every w on γ and

$$F(z) = \int_{\gamma} f(z, w)\, dw\ ,$$

then $F(z)$ is an analytic function of z in D and

$$F'(z) = \int_{\gamma} \frac{\partial f(z, w)}{\partial z} dw.$$

The conclusion remains valid if γ is a contour going to infinity and the integral is uniformly convergent.

An *entire function* $f(z)$ is a function of complex variable analytic in the entire complex plane and can be represented by a power series

$$f(z) = \sum_{n=0}^{\infty} a_n z^n,$$

that converges everywhere, where $a_n = f^{(n)}(0)/n!$.

Let f be an entire function and

$$M_f(r) = \max_{|z|=r} | f(z) | .$$

Then the maximum modulus principle states that $M_f(r)$ is an increasing function of r. An entire function $f(z)$ is said to be of *finite order* if there exists a positive constant k such that the inequality

$$M_f(r) < e^{r^k}$$

is valid for all sufficiently large r.

The greatest lower bound (inf) of such numbers k is called the *order of the entire function.* If ρ is the order of $f(z)$, it can be shown [15, p. 8] that

$$\rho = \limsup_{r\to\infty} \frac{\ln\ln M_f(r)}{\ln r}.$$

By the *type of an entire* function $f(z)$ of order ρ we mean the greatest lower bound of positive numbers c such that

$$M_f(r) < e^{cr^\rho},$$

for sufficiently large r. Similarly, it can be shown that if the type of f is σ, then

$$\sigma = \lim_{r\to\infty} \frac{\ln M_f(r)}{r^\rho}.$$

An entire function of finite order is said to be of *maximum type* if $\sigma = \infty$ and of *minimum type* if $\sigma = 0$, otherwise it is said to be of normal type. An entire function of order 1 is called an entire function of *exponential type.* Polynomials are of exponential type zero, e^{at} is of exponential type $= a$, and the entire function $f(z) = \exp(\sigma z^m)$(m positive integer) is of order m and type σ, while $g(z) = \sin \pi z$ is of exponential type and type π.

Sometimes it is convenient to extend the definition of functions of exponential type to functions which are not even analytic; see Chapter 5 on the Laplace transform. For example, if $f(t)$ is a real-valued piecewise continuous function on $[a, \infty)$, we say that f is of exponential type (*real exponential type*) if there exists a real number c such that

$$\lim_{t\to\infty} f(t)e^{-ct} = 0.$$

The type σ of f is the greatest lower bound of such numbers c and can be given by

$$\sigma = \limsup_{t\to\infty} \frac{\ln \mid f(t) \mid}{t}.$$

Clearly, $-\infty \le \sigma < \infty$ and any function with compact support is of exponential type $= -\infty$. The function $f(t) = \cos t$ is of real exponential type zero, but as an entire function it is of type 1.

Like polynomials, entire functions of finite order can be factored out in terms of their zeros as can be seen from the following theorem which is known as the *Hadamard factorization theorem* for entire functions.

THEOREM 1.2 **(Hadamard's Factorization Theorem** [15, p. 22]) *Let $f(z)$ be an entire function of finite order ρ and $\{z_n\}_{n=1}^{\infty}$ be its non-zero zeros. Let p be the smallest integer for which the series*

$$\sum_{n=1}^{\infty} \frac{1}{|z_n|^{p+1}}$$

converges. Set

$$G(u;p) = (1-u)\exp\left(u + \frac{u^2}{2} + \cdots + \frac{u^p}{p}\right) \text{ and } G(u;0) = (1-u).$$

Then

$$f(z) = z^m e^{P(z)} \prod_{n=1}^{\infty} G\left(\frac{z}{z_n};p\right),$$

where m is the multiplicity of the zero at the origin and $P(z)$ is a polynomial whose degree does not exceed ρ.

An application of the Hadamard factorization theorem to the entire function $\sin \pi z$ leads to

$$\sin \pi z = \pi z \prod_{n=1}^{\infty}\left(1 - \frac{z^2}{n^2}\right).$$

For $\sigma \geq 0$ and $1 \leq p \leq \infty$, let B_σ^p denote the class of all entire functions f (on $\mathcal{C}$) of exponential type at most σ belonging to $L^p(\mathcal{R})$ when restricted to $\mathcal{R}$; that is $f \in B_\sigma^p$ if and only if

$$|f(z)| \leq C\exp\left(\sigma\,|z|\right), \ z \in \mathcal{C}$$

and

$$\int_{-\infty}^{\infty} |f(x)|^p dx < \infty \text{ if } 1 \leq p < \infty,$$

and

$$\text{ess. } \sup_{x\in\mathcal{R}} |f(x)| < \infty \text{ if } p = \infty.$$

It can be shown that in fact,

$$|f(z)| \leq \sup_{x\in\mathcal{R}} |f(x)| \exp(\sigma|y|), z \in \mathcal{C}$$

where $z = x + iy$.

By using the standard L^p-norm

$$\|f\|_p = \left(\int_{-\infty}^{\infty} |f(x)|^p dx\right)^{1/p} \quad \text{if } 1 \leq p < \infty,$$

and

$$\|f\|_\infty = \text{ess. } \sup_{x\in\mathcal{R}} |f(x)| \text{ if } p = \infty,$$

it can be shown that for $f \in B_\sigma^p$, we have

$$|f(x+iy)| \leq C\,\|f\|_p\, e^{\sigma|y|},$$

where C is a constant that depends only on p and σ.

The class $B_\sigma^p(\sigma > 0)$ is non-trivial, but the class B_0^p is trivial (it contains only the zero function). A well-known theorem of Paley and Wiener [182, p. 12] states that $f \in B_\sigma^2(\sigma > 0)$ if and only if

$$f(t) = \frac{1}{\sqrt{2\pi}} \int_{-\sigma}^{\sigma} F(w) e^{itw} dw,$$

for some $F \in L^2(-\sigma, \sigma)$.

The class B_σ^2 is called the *Paley–Wiener class of entire functions* and a member of this class is called a *band-limited function* or more precisely it is called a function band-limited to $[-\sigma, \sigma]$. Note that the function f itself is not limited, it is the Fourier transform of f, F, that is actually limited to $[-\sigma, \sigma]$.

A *meromorphic function* is a single-valued function having no singularities other than poles. Such a function has no more than a finite number of poles in every bounded region. Every meromorphic function is the quotient of two entire functions having no zeros in common. If f and g are meromorphic functions having exactly the same poles with their corresponding residues, then the difference, $f - g$, is an entire function. Let $\{z_n\}$ be a given set of points, finite or infinite, but having no finite limit point, and associate with every z_n a finite sequence of complex numbers $\{a_{n,i}\}_{i=1}^{k_n}$ and a rational function $h_n(z)$, called the *principle part* associated with z_n, defined by

$$h_n(z) = \frac{a_{n,1}}{(z - z_n)} + \frac{a_{n,2}}{(z - z_n)^2} + \cdots + \frac{a_{n,k_n}}{(z - z_n)^{k_n}},$$

where k_n is a positive integer. The *Mittag–Leffler theorem [208, p. 291]* asserts that one can construct a meromorphic function that has poles exactly at the points $\{z_n\}$ with the prescribed principle parts. If $f(z)$ is such a function then $f(z) + g(z)$ is the most general meromorphic function having the same poles with their corresponding principle parts as $f(z)$, where g is an entire function.

Since $\cot \pi z$ is a meromorphic function, being the quotient of two entire functions $\cos \pi z$ and $\sin \pi z$, it has a Mittag-Leffler expansion, namely,

$$\pi \cos \pi z = \frac{1}{z} + \sum_{n=1}^{\infty} \left(\frac{1}{z - n} + \frac{1}{z + n} \right).$$

1.4 Local Behavior of Functions and Integrals

Let $\Omega \subseteq \mathcal{R}^n$ and x_0 be a limit point of Ω, which may or may not belong to Ω. For two functions f and g defined on Ω we write $f = O(g)$ in Ω if there exists a constant C independent of x such that $|f(x)| \leq C|g(x)|$ for all $x \in \Omega$, and $f(x) = O(g(x))$ as $x \to x_0$ if there exists a neighborhood U of x_0 and a constant C such that $|f(x)| \leq C|g(x)|$ for all x in $U \cap \Omega$. But we shall write $f(x) = o(g(x))$ as $x \to x_0$ if for any $\dot{\varepsilon} > 0$, there exists a neighborhood U_ε of x_0 such that $|f(x)| \leq \epsilon|g(x)|$ for all $x \in U_\varepsilon \cap \Omega$. If $g \neq 0$ in Ω, then $f(x) = 0(g(x))$ in Ω (or as $x \to x_0$) if f/g is bounded in Ω (or as $x \to x_0$), and $f(x) = o(g(x))$ as $x \to x_0$ if $f/g \to 0$ as $x \to x_0$; moreover, if $f/g \to 1$ as $x \to x_0$, we may write $f \approx g$ or $f \sim g$ as $x \to x_0$.

If x_0 is the point at infinity and $g = 1$, then $f = 0(1)$ means that f is bounded as $|x| \to \infty$, while $f = o(1)$ means that f approaches zero as $|x| \to \infty$.

A function $f(x)$ is said to be *rapidly decreasing* (near infinity) if for every integer m, $f(x) = 0(|x|^{-m})$ as $|x| \to \infty$, and is said to be of *slow growth* or *polynomial growth* if there exists an integer m so that $f(x) = 0(|x|^m)$ as $|x| \to \infty$. Equivalently, f is rapidly decreasing if for every integer m,

$$(1 + |x|^2)^m f(x) = 0(1) \text{ as } |x| \to \infty.$$

The set of all infinitely differentiable functions f such that $f^{(k)}$, $k = 0, 1, 2, \ldots$, are rapidly decreasing is denoted by S, and when provided with an appropriate topology, it is called the *Schwartz space of functions.*

Let f be defined on a set Ω in $\mathcal{R}^n$. The *modulus of continuity* of f on Ω is defined by

$$\omega(\delta, f) = \omega(\delta) = \sup |f(x_1) - f(x_2)|,$$

where the sup is taken over all pairs $x_1, x_2 \in \Omega$ for which $|x_1 - x_2| \leq \delta$. If there exist two positive constants M and α such that

$$|f(x_1) - f(x_2)| \leq M|x_1 - x_2|^\alpha, \text{ for all } x_1, x_2 \in \Omega,$$

f is said to satisfy a *Lipschitz condition* of order α. The set of all such functions is denoted by $Lip_\alpha(\Omega)$.

A function $f(x)$ defined on an interval $I = [a, b]$ is said to be of *bounded variation* if there is a positive number M such that for every partition $x_0 = a < x_1 < x_2 < \cdots < x_n = b$,

$$V = \sum_{k=1}^{n} |f(x_k) - f(x_{k-1})| \leq M.$$

The least upper bound of the set of all possible Vs is called the *total variation* of f on $[a, b]$.

It is easy to show that if f is non-increasing or non-decreasing and bounded on $[a, b]$, then it is of bounded variation, and f is of bounded variation if and only if it is the difference of two non-decreasing functions.

Let f be an integrable function. If

$$\lim_{h\to 0}\frac{1}{h}\int_x^{x+h}|f(t)-f(x)|\,dt=0,$$

at the point x, we say that x is a *Lebesgue point* of f. The set of all Lebesgue points of f is called the *Lebesgue set* of f. If f is integrable on $[a,b]$, then almost every point of $[a,b]$ is a Lebesgue point of f. The equation defining Lebesgue points generalizes the fact that the derivative of the indefinite integral of f exists and equals f for almost all x.

For a finite interval I and $f\in L^1(I)$ we denote by $\Omega(\delta,f)$ the *integral modulus of continuity* of f, that is,

$$\Omega(\delta,f)=|I|^{-1}\left\|T_\delta f-f\right\|_{L^1(I)},$$

where T_δ is the translation operator defined in Section 1.1 and $|I|$ is the length of I. We clearly have

$$\Omega(\delta,f)\le\omega(\delta,f).$$

Let $I=[a,b]$ and $a<c<b$. If f is integrable on $[a,c-\delta]\cup[c+\delta,b]$ for all $\delta>0$, we define the *Cauchy principle value* of $\int_a^b f(x)dx$ as

$$(PV)\int_a^b f(x)dx=\lim_{\delta\to 0}\left(\int_a^{c-\delta}f(x)dx+\int_{c+\delta}^b f(x)dx\right).$$

If the integral $\int_a^b f(x)dx$ exists, then its Cauchy principle value also exists and is equal to it; however, the converse is not true. For example, if $a<c<b$, the integral $\int_a^b dx/(x-c)$ diverges, but

$$(PV)\int_a^b\frac{dx}{x-c}=\ln\left|\frac{b-c}{a-c}\right|.$$

Let $f(x,y)$ be an integrable function of x over the interval $[a,b]$ for all y in some closed set I and all b. Suppose that the integral

$$g(y)=\int_a^\infty f(x,y)dx$$

converges for all $y\in I$. Then the integral is said to be *uniformly convergent* if for any $\epsilon>0$ there exists b_0 that depends on ϵ but not on y such that

$$\left|g(y)-\int_a^b f(x,y)dx\right|<\epsilon$$

for all $b\ge b_0$.

1.5 Orthogonal Expansions in a Hilbert Space

Let X be a linear vector space. We say that X is *normed* if for each $x \in X$, there is a real number $\|x\|$, called the *norm* of x, such that

1. $\|x\| \geq 0$ and $\|x\| = 0$ if and only if $x = 0$,
2. $\|\alpha x\| = |\alpha| \|x\|$ for all $\alpha \in \mathcal{C}, x \in X$,
3. $\|x + y\| \leq \|x\| + \|y\|$.

A sequence $\{x_n\}_{n=1}^{\infty}$ in a normed vector space X is said to be a *Cauchy sequence* if for any $\epsilon > 0$, there exists n_0 such that $\|x_n - x_m\| < \epsilon$ for all $n, m > n_0$. A normed vector space is said to be *complete* if every Cauchy sequence converges to some element in X. A *complete normed* space is called a *Banach space.*

An *open ball* of radius r centered at $y \in X$ is defined as $B_r(y) = \{x \in X : \|x - y\| < r\}$ and the *open unit ball* $B_1(0)$ is the ball of radius 1 centered at the origin (the zero vector in X). A set $A \subset X$ is said to be *dense* in $B \subset X$ if for any $b \in B$ and any $\epsilon > 0$, there exists $a \in A$ such that $a \in B_\epsilon(b)$ or equivalently $B_\epsilon(b) \cap A \neq \emptyset$. A normed space X is said to be *separable* if it contains a dense set that is countable.

A *linear operator* L from a vector space X into another vector space Y is a mapping from X into $Y, L : X \to Y$ such that

$$L(\alpha x_1 + \beta x_2) = \alpha L(x_1) + \beta L(x_2) \text{ for all } \alpha, \beta \in \mathcal{C}, x_1, x_2 \in X.$$

If a linear operator L from X into Y is one-to-one, then there exists an operator M, called the inverse of L and denoted by L^{-1}, such that $ML = I_X$ and $LM = I_Y$, where I_X and I_Y are the identity operators on X and Y, respectively, i.e., $I_X(x) = x$ for all $x \in X$ and $I_Y(y) = y$ for all $y \in Y$. The operator L^{-1} is also linear. For, if $L(x_1) = y_1$ and $L(x_2) = y_2$, then

$$\alpha y_1 + \beta y_2 = \alpha L(x_1) + \beta L(x_2) = L(\alpha x_1 + \beta x_2),$$

hence

$$L^{-1}(\alpha y_1 + \beta y_2) = \alpha L^{-1}(y_1) + \beta L^{-1}(y_2).$$

A linear operator from a normed space X into another normed space Y is said to be *bounded* if there exists a constant M with the property that

$$\|L(x)\| \leq M \|x\| \text{ for all } x \in X.$$

The set of all bounded linear operators from X into Y will be denoted by $B(X,Y)$. If we define

$$(\alpha L_1 + \beta L_2)(x) = \alpha L_1(x) + \beta L_2(x),$$

for any $x \in X, \alpha, \beta \in \mathcal{C}, L_1, L_2 \in B(X,Y)$, the set $B(X,Y)$ becomes a vector space.

Let X be a Banach space and for any $L \in B(X,Y)$ define

$$\|L\| = \sup_{x \in X} \frac{\|L(x)\|}{\|x\|} = \sup_{\|x\|=1} \|L(x)\|.$$

It is easy to see that this defines a norm on $B(X,Y)$ making it a normed vector space. If Y is a Banach space, then so is $B(X,Y)$. Every bounded linear operator from X into Y is *continuous* in the sense that if $x_n \to x$ in X, then $L(x_n) \to L(x)$ in Y. The converse is also true, that is every continuous linear operator is bounded. If $Y = X$, we shall write $B(X)$ instead of $B(X,X)$. If $Y = \mathcal{R}(\mathcal{C})$, the space $B(X,Y)$ is called the space of all continuous linear functionals on X or the *dual space* of X. The dual space of X will be denoted by X^*, and $f \in X^*$ will be called a *continuous linear functional* on X.

If X is a Banach space, we shall say that an operator $L \in B(X)$ is invertible if there exists $M \in B(X)$ such that

$$LM = ML = I_X$$

If M exists, it will be denoted by L^{-1}.

The *spectrum* $\sigma(L)$ of $L \in B(X)$ is the set of all complex numbers λ such that $L - \lambda I$ is not invertible. If $\lambda \in \sigma(L)$, then either (i) $L - \lambda I$ is not one-to-one, or (ii) the range of $L - \lambda I$ is not all of X. If (i) holds, λ is said to be an *eigenvalue* of L and each x that satisfies the relation

$$Lx = \lambda x$$

is called an *eigenvector* corresponding to the eigenvalue λ.

Let X be a linear vector space. A mapping $\langle \cdot, \cdot \rangle$ from $X \times X$ into $\mathcal{C}, \langle \cdot, \cdot \rangle : X \times X \to \mathcal{C}$, is called an *inner product* on X if for any $x, y, z \in X, \alpha, \beta \in \mathcal{C}$ the following conditions are satisfied:

1. $\langle x, y \rangle = \overline{\langle y, x \rangle}$,
2. $\langle \alpha x + \beta y, z \rangle = \alpha \langle x, z \rangle + \beta \langle y, z \rangle$,
3. $\langle x, x \rangle \geq 0$ and $\langle x, x \rangle = 0$ if and only if $x = 0$.

A linear vector space with inner product is called an *inner product space*. It is easy to see that $\|x\| = \sqrt{\langle x, x\rangle}$ is a norm; hence, every inner product space is a normed space. Moreover,

$$|\langle x, y\rangle| \leq \|x\| \, \|y\| \, , \tag{1.1}$$

with $|\langle x, y\rangle| = \|x\| \, \|y\|$ if and only if x and y are linearly dependent. Inequality (1.1) is called the *Cauchy-Schwarz inequality*. An inner product space that is complete with respect to the norm induced by the inner product is called a *Hilbert space*; hence, every Hilbert space is a Banach space.

Let L be a bounded linear operator (transformation) $L : X \to X$. It can be shown [209, p. 93] that there exists a unique continuous linear operator $L^* : X \to X$ such that

$$\langle Lx, y\rangle = \langle x, L^* y\rangle$$

for all x and $y \in X$. The operator L^* is called the *adjoint* of L. If $L = L^*$, we say that L is *self-adjoint*. A bounded linear operator L is said to be *positive*, and we may write $L \geq 0$, if $\langle Lx, x\rangle \geq 0$ for every $x \in X$. If $L_1, L_2 \in B(X)$, we say that $L_1 \geq L_2$ if and only if $L_1 - L_2 \geq 0$. An operator $L \in B(X)$ is positive if and only if $L = L^*$ and $\sigma(L) \subset [0, \infty)$, see [209, p. 313]. If the spectrum of L is discrete, then there exists a sequence of numbers $\{\lambda_n\}$ and a sequence of vectors $\{\phi_n\} \subset X$ such that

$$L\phi_n = \lambda_n \phi_n, \quad \text{for all } n.$$

Let $\mathcal{H}$ be a Hilbert space with inner product $\langle \cdot, \cdot \rangle$ and norm $\|x\| = \sqrt{\langle x, x\rangle}$ for all $x \in \mathcal{H}$. A sequence of vectors $\{x_n\}_{n=1}^{\infty}$ in $\mathcal{H}$ is said to be *orthogonal* if

$$\langle x_n, x_m\rangle = 0, \text{ whenever } m \neq n,$$

and is said to be *orthonormal* if in addition

$$\langle x_n, x_n\rangle = 1, \text{ for all } n.$$

If S is a subset of $\mathcal{H}$, then $S^\perp$ will denote the set of all $y \in \mathcal{H}$ that are orthogonal to every $x \in S$, that is $y \in S^\perp$ if and only if $\langle y, x\rangle = 0$ for all $x \in S$. If S is a closed subspace of $\mathcal{H}$, then so is $S^\perp$ and $\mathcal{H}$ is the *direct sum* of S and $S^\perp$

$$\mathcal{H} = S \oplus S^\perp.$$

This means that every $x \in \mathcal{H}$ can be uniquely written in the form $x = x_1 + x_2$, where $x_1 \in S$ and $x_2 \in S^\perp$. The space $S^\perp$ is called the *orthogonal complement* of S.

When $\mathcal{H}$ is a space of functions defined on some set $\Omega \subseteq \mathcal{R}^n$ with inner product

$$\langle f, g\rangle = \int_{\Omega} f(x)\bar{g}(x) \, d\mu(x),$$

then $\{f_n\}_{n=1}^{\infty}$ is orthonormal if

$$\int_{\Omega} f_m(x)\bar{f}_n(x)d\mu(x) = \delta_{m,n}.$$

In particular, if $\mu(x)$ is a non-decreasing and non-constant function on $I = [a,b], -\infty < a < b < \infty$, then

$$\langle f, g \rangle = \int_a^b f(x)\bar{g}(x)\,d\mu(x), \tag{1.2}$$

defines an inner product on $L^2(I,\mu)$ making it an inner-product space.

The value $a = -\infty (b = \infty)$ is allowed under the requirement that $\mu(-\infty) = \lim_{x\to-\infty} \mu(x) \left(\mu(\infty) = \lim_{x\to\infty} \mu(x)\right)$ is finite. If $\mu(x)$ is absolutely continuous, i.e., $d\mu(x) = W(x)dx$, we shall call $W(x)$ a *weight function* if $\int_a^b W(x)dx > 0$. In this case,

$$\langle f, g \rangle = \int_a^b f(x)\bar{g}(x)W(x)\,dx.$$

If $\{x_n\}_{n=1}^{\infty}$ is an orthonormal sequence in $\mathcal{H}$, we associate with every $x \in \mathcal{H}$ the sequence of numbers $\{\langle x, x_n\rangle\}_{n=1}^{\infty}$ and the formal series expansion $\sum_{n=1}^{\infty} \langle x, x_n\rangle x_n$. The numbers $\langle x, x_n\rangle$ are called the *generalized Fourier coefficients* and the series $\sum_{n=1}^{\infty} \langle x, x_n\rangle x_n$ is called the *generalized Fourier series* of x with respect to the system $\{x_n\}_{n=1}^{\infty}$. Sometimes we may drop the word "generalized". An orthogonal sequence is linearly independent and the converse is, in a sense, true. This means that given a countable set of linearly independent vectors $\{y_n\}_{n=1}^{\infty}$ in an inner product space, we can construct an orthonormal family $\{x_n\}_{n=1}^{\infty}$ such that x_m is a linear combination of $y_1, \cdots, y_m$ for all m. This orthonormalization process is known as the *Gram-Schmidt process.*

A set $S \subset \mathcal{H}$ is said to be *complete* if no non-zero vector $y \in \mathcal{H}$ is orthogonal to S, that is if $\langle y, x\rangle = 0$ for all $x \in S$, then $y = 0$. For an orthonormal sequence $\{x_n\}_{n=1}^{\infty}$ in a Hilbert space the following are equivalent [88, p. 27]:

1. $\{x_n\}_{n=1}^{\infty}$ is complete.
2. The generalized Fourier series $\sum_{n=1}^{\infty} \langle x, x_n\rangle x_n$ converges to x for every $x \in \mathcal{H}$, i.e.,

$$x = \sum_{n=1}^{\infty} \langle x, x_n\rangle x_n.$$

3. The *generalized Parseval's relation*

$$\langle x, y\rangle = \sum_{n=1}^{\infty} \langle x, x_n\rangle \overline{\langle y, x_n\rangle}$$

   holds.

4. *Parseval's relation*

$$\|x\|^2 = \sum_{n=1}^{\infty} |\langle x, x_n\rangle|^2 ,$$

   holds.

Without the assumption of completeness, we only have

$$\sum_{n=1}^{\infty} | \langle x, x_n\rangle |^2 \leq \|x\|^2 ,$$

which is called the *Bessel inequality.* The expansion given in (2) for any $x \in \mathcal{H}$ in terms of the complete orthogonal set $\{x_n\}$ is also called an orthogonal (series) expansion.

Example A: Let $I = [\alpha , \beta]$ be an interval of length $1/b$, i.e., $\beta - \alpha = 1/b$. Then the set $\left\{\sqrt{b}\exp(2\pi inbx)\right\}_{n=-\infty}^{\infty}$ is a complete orthonormal set in $L^2(I)$. Hence, for any $f \in L^2(I)$, we have the Fourier series expansion

$$f(x) = \sqrt{b} \sum_{n=-\infty}^{\infty} \hat{f}_n e^{2\pi inbx} ,$$

with

$$\hat{f}_n = \sqrt{b} \int_{\alpha}^{\beta} f(x) e^{-2\pi inbx} dx ,$$

being the Fourier coefficients of f. Parseval's relation takes the form

$$\int_{\alpha}^{\beta} f(x)\overline{g}(x)\, dx = \sum_{n=-\infty}^{\infty} \hat{f}_n \overline{\hat{g}_n} \quad \text{for all } f, g \in L^2(I).$$

Now, let us describe a general procedure for generating orthogonal functions and hence generating orthogonal series expansions. Let us assume that we have a Hilbert space X, which in most cases we shall consider in this book, will be a space of functions. Let $L \in B(X)$ be self-adjoint and have a purely discrete spectrum, then its eigenvalues are real and its eigenvectors belonging to different eigenvalues are orthogonal. For,

$$\langle L\phi_n, \phi_m\rangle = \lambda_n \langle \phi_n, \phi_m\rangle = \langle \phi_n, L\phi_m\rangle = \bar{\lambda}_m \langle \phi_n, \phi_m\rangle ,$$

and by putting $n = m$, it follows that the eigenvalues are real. Hence,

$$(\lambda_n - \lambda_m)\langle\phi_n, \phi_m\rangle = 0,$$

and if $\lambda_n \neq \lambda_m$, then $\{\phi_n\}$ is an orthogonal sequence. Here we, of course, assumed that no ϕ_n satisfied $\langle\phi_n, \phi_n\rangle = 0$. When the set $\{\phi_n\}$ is complete, we then have for any $f \in X$

$$f = \sum_n \frac{\langle f, \phi_n\rangle}{\langle\phi_n, \phi_n\rangle}\phi_n.$$

This will happen, for example, if L is a self-adjoint compact operator as we shall see in Section 1.9. When X is a space of functions, we say that $\{\lambda_n\}$ are the eigenvalues of L and $\{\phi_n\}$ are the corresponding *eigenfunctions.*

1.6 Orthogonal Polynomials

Let $\mu(x)$ be a non-decreasing function with infinitely many points of increase in the finite or infinite interval $I = [a, b]$ and let $c_n = \int_a^b x^n d\mu(x)$ exists for $n = 0, 1, 2, \cdots$. Then $L^2(I, \mu)$ is a Hilbert space in which $\{x^n\}_{n=0}^{\infty}$ is a linearly independent set, provided that the inner product is defined by (1.2). By applying the Gram-Schmidt orthonormalization process, we obtain a unique set of orthonormal polynomials $\{p_n(x)\}_{n=0}^{\infty}$ determined by the condition that $p_n(x)$ is a polynomial of exact degree n in which the coefficient of x^n is positive. The orthonormality condition takes the form

$$\langle p_n(x), p_m(x)\rangle = \int_a^b p_n(x)p_m(x)d\mu(x) = \delta_{m,n}.$$

The numbers $\{c_n\}_{n=0}^{\infty}$ are called the *moments* and $\{p_n(x)\}_{n=0}^{\infty}$ are called the *orthogonal polynomials associated with* $d\mu$ in I. If I is an interval symmetric about the origin, i.e., $a = -b, b > 0$, and $d\mu(x) = W(x)dx$ with $W(x)$ being an even weight function, then $p_n(x)$ is even or odd according to n is even or odd. In fact, $p_n(-x) = (-1)^n p_n(x)$.

The set of orthogonal polynomials associated with $d\mu$ in a finite interval I is complete; hence for any $f \in L^2(I, \mu)$, the generalized Fourier series converges,

$$f(x) = \sum_{n=0}^{\infty} \langle f, p_n\rangle\, p_n(x).$$

The convergence is in the sense of $L^2(I, \mu)$. To get pointwise or uniform convergence, more conditions are usually needed.

For any three consecutive orthogonal polynomials, the following recurrence relation holds [220, p. 42]:

$$p_n(x) = (A_n x + B_n)\, p_{n-1}(x) - C_n p_{n-2}(x), \quad n = 1, 2, \cdots,$$

where A_n, B_n and C_n are constants with A_n and $C_n > 0, p_{-1}(x) = 0$, $p_0(x) = 1$. If the leading coefficient of $p_n(x)$ is denoted by k_n, we have

$$A_n = \frac{k_n}{k_{n-1}}, C_n = \frac{A_n}{A_{n-1}} = \frac{k_n k_{n-2}}{k_{n-1}^2}.$$

We also have the *Christoffel-Darboux formula*

$$p_0(x)p_0(y) + \cdots + p_n(x)p_n(y) = \frac{k_n}{k_{n+1}} \frac{p_{n+1}(x)p_n(y) - p_n(x)p_{n+1}(y)}{x - y}.$$

The *functions of the second kind* corresponding to a given system of orthogonal polynomials are defined by

$$q_n(z) = \int_a^b \frac{p_n(x)\, d\mu(x)}{x - z}, \quad n = 0, 1, 2, \cdots$$

$z \notin I$.

By combining the Christoffel-Darboux formula and the definition of q_n, we obtain

$$p_{n+1}(z)q_n(z) - p_n(z)q_{n+1}(z) = \frac{k_{n+1}}{k_n},$$

which, with the aid of the Christoffel-Darboux formula once more, yields

$$\sum_{k=0}^{n} q_k(y)q_k(z) = \frac{q_0(y) - q_0(z)}{(z - y)} + \frac{k_n}{k_{n+1}} \frac{q_{n+1}(z)q_n(y) - q_n(z)q_{n+1}(y)}{z - y}.$$

This is the analogue of the Christoffel-Darboux formula for the functions of the second kind.

For fixed $z \notin I, \{q_n(z)\}_{n=0}^{\infty}$ are the generalized Fourier coefficients of the function $1/(z - x)$ which is in $L^2(I, \mu)$. Therefore, the Bessel inequality yields

$$\sum_{n=0}^{\infty} |q_n(z)|^2 \leq \int_a^b \frac{d\mu(x)}{|z - x|^2}, \quad z \notin I.$$

An immediate consequence of this is $\lim_{n \to \infty} q_n(z) = 0, z \notin I$.

Examples:

1. For $a = -1, b = 1, W(x) = (1-x)^\alpha (1+x)^\beta; \alpha, \beta > -1$, the orthogonal polynomials, $p_n(x)$ are the Jacobi polynomials (up to a constant factor), which are usually denoted by $P_n^{(\alpha,\beta)}(x)$.
   Except for a constant factor, we have the following special cases: When $\alpha = \beta$, we obtain the ultraspherical (Gegenbauer) polynomials which in turn yield the following:

   (a) for $\alpha = 0 = \beta$, the Legendre polynomials $P_n(x)$,

   (b) for $\alpha = -\frac{1}{2} = \beta$, the Tchebichef polynomials of the first kind,

   $$T_n(x) = \cos n\theta, \text{ where } x = \cos\theta,$$

   (c) for $\alpha = \frac{1}{2} = \beta$, the Tchebichef polynomials of the second kind,

   $$U_n(x) = \sin(n+1)\theta/(\sin\theta), \quad x = \cos\theta.$$

2. For $a = 0, b = \infty, W(x) = x^\alpha e^{-x}, \dot{\alpha} > -1, p_n(x)$ are the generalized Laguerre polynomials (up to a constant factor) $L_n^{(\alpha)}(x)$.

3. For $a = -\infty, b = \infty, W(x) = e^{-x^2}, p_n(x)$, except for a constant factor, are the Hermite polynomials $H_n(x)$.

More properties of these classical orthogonal polynomials are given in forthcoming sections.

1.7 Orthogonal Functions and Series Expansions

In the previous section, we showed how to construct a system of orthonormal polynomials on a given interval with respect to a given weight function and obtained the Jacobi, Laguerre, and Hermite polynomials as a special case. In this section, we shall describe another approach for constructing more general systems of orthogonal functions and hence for obtaining more general types of generalized Fourier series expansions. These orthogonal functions are the eigenfunctions of certain boundary-value problems.

Consider the differential operator

$$L = p_0(x)\frac{d^n}{dx^n} + p_1(x)\frac{d^{n-1}}{dx^{n-1}} + \cdots + p_n(x), \tag{1.3}$$

where $p_0, p_1, \cdots, p_n$ are complex-valued functions defined on a finite interval $I = [a, b]$, with $p_k \in C^{n-k}(I), k = 0, 1, \ldots, n$, and $p_0(x) \neq 0$ for any $x \in I$. The *adjoint differential operator* L^* of L is defined by

$$L^* g = (-1)^n \frac{d^n}{dx^n}(\bar{p}_0 g) + (-1)^{n-1} \frac{d^{n-1}}{dx^{n-1}}(\bar{p}_1 g) + \cdots + \bar{p}_n g. \tag{1.4}$$

If $u, v \in C^n(I)$, then *Lagrange's formula*

$$\int_{x_1}^{x_2} \left\{ \bar{v}(Lu) - u\overline{(L^* v)} \right\} dx = [uv]_{x=x_2} - [uv]_{x=x_1} \tag{1.5}$$

holds for any $x_1, x_2 \in I$, where

$$[uv] = \Sigma_{m=1}^{n} \Sigma_{j+k=m-1, j\geq 0, k\geq 0} \, (-1)^j u^{(k)} \, (p_{n-m}\bar{v})^{(j)}.$$

The operator L is said to be *formally self-adjoint* if $L^* = L$. If p_k $(k = 0, 1, \ldots, n)$ are real-valued, then L is formally self-adjoint if and only if L has an even order $n = 2m (m \geq 1)$ and can be written in the form

$$Ly = (p_0 y^{(m)})^{(m)} + (p_1 y^{(m-1)})^{(m-1)} + \cdots + (p_{m-1} y^{(1)})^{(1)} + p_m y.$$

Let

$$U_j(y) = \sum_{k=1}^{n} \alpha_{j,k} y^{(k-1)}(a) + \beta_{j,k} y^{(k-1)}(b), \quad j = 1, 2, \ldots, n,$$

be linear forms in $y^{(k-1)}(a), y^{(k-1)}(b), k = 1, 2, \ldots, n$, and assume that the U_js are linearly independent.

It is known [43, p. 288] that, given any boundary forms $U_j, j = 1, \ldots, n$, we can find *complementary boundary forms* U_js, $j = n + 1, \ldots, 2n$, and a unique system of linearly independent boundary forms $V_1, \ldots, V_{2n}$, known as the *adjoint boundary forms*, such that

$$\int_a^b (Ly)\, \bar{z} dx = U_1 V_{2n} + \cdots + U_{m+1} V_{2n-m} + \cdots + U_{2n} V_1 + \int_a^b y\overline{(L^* z)} dx. \tag{1.6}$$

Now consider the boundary-value problem

$$\prod : \begin{cases} Ly = \lambda y, & x \in I \\ U_j(y) = 0, \; j = 1, 2, \ldots, n, \end{cases} \tag{1.7}$$

and its adjoint

$$\prod^* : \begin{cases} L^* y = \lambda y, & x \in I \\ V_j(y) = 0, \; j = 1, 2, \ldots, n. \end{cases} \tag{1.8}$$

It follows immediately from (1.6) that if y and z are two functions satisfying the boundary conditions (1.7) and (1.8), respectively, then

$$\int_a^b (Ly)\bar{z}dx = \int_a^b y\overline{(L^*z)}dx.$$

Since both problems Π and Π^* always have the trivial solution $y \equiv 0$, we shall be concerned only with the nontrivial solutions.

Let $y_1(x,\lambda),\ldots,y_n(x,\lambda)$ denote the fundamental system of solutions of the differential equation (1.7) that satisfy the initial conditions

$$y_j^{(v-1)}(a,\lambda) = \begin{cases} 0 \text{ for } j \neq v \\ 1 \text{ for } j = v; \end{cases} \quad j,v = 1,2,\ldots,n. \tag{1.9}$$

Let

$$U(\lambda) = \begin{bmatrix} U_1(y_1) \cdots U_1(y_n) \\ \cdots \qquad \cdots \\ U_n(y_1) \cdots U_n(y_n) \end{bmatrix},$$

and $\Delta(\lambda) = \det(U(\lambda))$. As a function of $\lambda, \Delta(\lambda)$ is an entire function; thus, unless identically zero, it can have at most countably many zeros with no finite limit point. The zeros of $\Delta(\lambda)$ are the eigenvalues of problem Π. The set of all eigenfunctions belonging to the same eigenvalue is a finite-dimensional vector space of dimension less than or equal to n. The dimension of this space is called the *multiplicity* of that eigenvalue. It is known that if λ_0 is a zero of $\Delta(\lambda)$ with multiplicity m, then its multiplicity as an eigenvalue is less than or equal to m. In particular, if λ_0 is a simple zero of $\Delta(\lambda)$, then there is one linearly independent eigenfunction corresponding to λ_0. In this case, we say that λ_0 is a *simple eigenvalue.* In general, the eigenvalues of problem Π are not simple.

Moreover, if λ_0 is an eigenvalue of problem Π with multiplicity m, then $\bar{\lambda}_0$ is also an eigenvalue of problem Π^* with the same multiplicity. The eigenfunctions of problem Π corresponding to the eigenvalue λ_0 are orthogonal to the eigenfunctions of problem Π^* corresponding to the eigenvalue μ_0, provided that $\lambda_0 \neq \bar{\mu}_0$. The orthogonality is considered with respect to the inner product

$$\langle f,g\rangle = \int_a^b f(x)\overline{g(x)}\,dx.$$

Finally, it is known that the *Green's function* $G(x,y,\lambda)$ of problem Π (or Π^*), which is a meromorphic function in λ, does not necessarily have simple poles; however, throughout the rest of this section we shall assume that it does. This will include the case in which Π is self-adjoint because it is known that for self-adjoint boundary-value problems, all the poles of the Green's function are simple.

Consider the boundary-value problem Π. Let $\{y_i(x,\lambda)\}_{i=1}^n$ be the fundamental system of solutions of the differential equation (1.7) that satisfy (1.9), and let W be their Wronskian

$$W(x) = \begin{vmatrix} y_1^{(n-1)}(x) & y_2^{(n-1)}(x) & \cdots & y_n^{(n-1)}(x) \\ y_1^{(n-2)}(x) & y_2^{(n-2)}(x) & \cdots & y_n^{(n-2)}(x) \\ \cdots & \cdots & & \cdots \\ y_1(x) & y_2(x) & \cdots & y_n(x) \end{vmatrix}.$$

Set

$$g(x,\xi) = \pm \frac{1}{W(\xi)} \begin{vmatrix} y_1(x) & y_2(x) & \cdots & y_n(x) \\ y_1^{(n-2)}(\xi) & y_2^{(n-2)}(\xi) & \cdots & y_n^{(n-2)}(\xi) \\ \cdots & \cdots & & \cdots \\ y_1(\xi) & y_2(\xi) & \cdots & y_n(\xi) \end{vmatrix},$$

where the positive sign is taken if $x > \xi$ and the negative sign if $x < \xi$; $x, \xi \in I$.

It can be shown [173, p. 37] that the *Green's function* $G(x,\xi,\lambda)$ of problem Π is given by

$$G(x,\xi,\lambda) = \frac{(-1)^n}{\Delta(\lambda)} H(x,\xi,\lambda),$$

where

$$\Delta(\lambda) = \begin{vmatrix} U_1(y_1) & U_1(y_2) & \cdots & U_1(y_n) \\ U_2(y_1) & U_2(y_2) & \cdots & U_2(y_n) \\ \cdots & \cdots & \cdots & \cdots \\ U_n(y_1) & U_n(y_2) & \cdots & U_n(y_n) \end{vmatrix},$$

and

$$H(x,\xi,\lambda) = \begin{vmatrix} y_1(x) & y_2(x) & \cdots & y_n(x) & g(x,\xi) \\ U_1(y_1) & U_1(y_2) & \cdots & U_1(y_n) & U_1(g) \\ U_2(y_1) & U_2(y_2) & \cdots & U_2(y_n) & U_2(g) \\ \cdots & \cdots & \cdots & \cdots & \cdots \\ U_n(y_1) & U_n(y_2) & \cdots & U_n(y_n) & U_n(g) \end{vmatrix}.$$

Let us denote the poles of G, or equivalently the eigenvalues of Π, by $\{\lambda_k\}_{k=0}^{\infty}$. Although all are assumed to be simple poles of G, some or all of the λ_ks may not be simple zeros of $\Delta(\lambda)$; hence, they may not be simple eigenvalues of Π, i.e., their multiplicities as eigenvalues may be greater than 1.

We now obtain series expansions for a class of functions in terms of the eigenfunctions of problems $\prod$ and $\prod^*$.

THEOREM 1.3 [173, p. 129]
Assume that the boundary-value problem Π is such that all the eigenvalues are simple zeros of the function $\Delta(\lambda)$. Then any function $f \in C^n(I)$

satisfying the boundary conditions (1.7) *can be expanded in terms of the eigenfunctions of* Π *in a series of the form*

$$f(x) = \sum_{n=1}^{\infty} \hat{f}(n)\phi_n(x),$$

with

$$\hat{f}(n) = \int_a^b f(x)\overline{\psi_n(x)}dx,$$

where ϕ_n *and* ψ_n *are the eigenfunctions of problems* Π *and* Π^* *corresponding to the eigenvalues* λ_n *and* $\bar{\lambda}_n$*, respectively; these eigenfunctions can be normalized so that*

$$\int_a^b \phi_n(x)\overline{\psi_n(x)}dx = 1.$$

The series converges uniformly at least for one particular order of its terms.

COROLLARY 1.1

If Π *is self-adjoint, i.e.,* $\Pi = \Pi^*$*, then any function* $f \in C^n(I)$ *satisfying the boundary-condition (1.7) can be expanded in terms of the eigenfunctions of* Π *in a uniformly convergent series*

$$f(x) = \sum_{n=1}^{\infty} \hat{f}(n)\phi_n(x),$$

with

$$\hat{f}(n) = \int_a^b f(x)\overline{\phi_n}(x)dx,$$

where ϕ_n *is the eigenfunction corresponding to the eigenvalue* λ_n.

It should be noted that while the series expansion of f given in Corollary 1.1 is an orthogonal expansion, the one given in Theorem 1.3 is not, since the eigenfunctions of problem Π are not orthogonal. Such an expansion will generally be called a non-orthogonal expansion or, more accurately, a biorthogonal expansion. This type of expansion will be discussed in more detail in Section 1.10.

1.8 Sturm-Liouville Series Expansions

Solving the boundary-value problem $\prod$ explicitly is, in general, very difficult, but if we restrict ourselves to some special cases, more specific results can be obtained. As a special case of the results presented in the previous section, we consider the case in which $n = 2$. The problem Π reduces to

$$p_0(x)y'' + p_1(x)y' + p_2(x)y = \lambda y, \qquad a \leq x \leq b$$

$$\alpha_{11}y(a) + \alpha_{12}y'(a) + \beta_{11}y(b) + \beta_{12}y'(b) = 0,$$

$$\alpha_{21}y(a) + \alpha_{22}y'(a) + \beta_{21}y(b) + \beta_{22}y'(b) = 0.$$

We shall consider only the self-adjoint case. For the differential equation to be formally self-adjoint, we must have $p_0' = p_1$. Hence, the differential equation can be reduced to

$$(p_0(x)y')' + (-\lambda + p_2(x))\, y = 0.$$

More generally, let us consider

$$(p(x)y')' + (\lambda r(x) + s(x))\, y = 0,$$

with $p(x), r(x), s(x)$ being continuous on $[a, b]$ and $p(x), r(x) > 0$. Note that because of the assumption that p_0 is continuous and $p_0 \neq 0$ in $[a, b]$, without loss of generality, we may further assume that $p_0(x) > 0$.

The self-adjoint boundary-value problem

$$(p(x)y')' + (\lambda r(x) + s(x))\, y = 0, \qquad a \leq x \leq b,$$

$$a_1 y(a) + a_2 y'(a) = 0,$$

$$b_1 y(b) + b_2 y'(b) = 0,$$

is called a *regular Sturm-Liouville boundary-value problem* and the differential equation is called a *Sturm-Liouville differential equation.*

From the results of the previous section, it follows that all the eigenvalues of a regular Sturm-Liouville boundary-value problem are simple, eigenfunctions belonging to different eigenvalues are orthogonal and the generalized Fourier expansions of functions satisfying the boundary conditions are still valid; see Corollary 1.1. If we denote the eigenfunctions by ϕ_n, then the orthogonality relation takes the form

$$\int_a^b \phi_n(x)\phi_m(x)r(x)dx = 0, \qquad \text{if } n \neq m.$$

By means of the transformation

$$z = \int \sqrt{\frac{r(x)}{p(x)}}dx, \qquad u = [r(x)p(x)]^{\frac{1}{4}}\, y,$$

the Sturm-Liouville differential equation can be transformed into

$$u'' - Q(z)u = \lambda u,$$

in which

$$Q(z) = \frac{\theta''(z)}{\theta(z)} + \frac{q(x)}{r(x)} \text{ and } \theta(z) = [r(x)p(x)]^{\frac{1}{4}}.$$

Hence, we have a second version of the regular Sturm-Liouville boundary-value problems:

$$y'' - q(x)y = \lambda y, \qquad a \le x \le b,$$
$$\cos\alpha\ y(a) + \sin\alpha\ y'(a) = 0$$
$$\cos\beta\ y(b) + \sin\beta\ y'(b) = 0,$$

where $\cos\alpha = a_1/\sqrt{a_1^2 + a_2^2}$, $\sin\alpha = a_2/\sqrt{a_1^2 + a_2^2}$ and similarly for $\cos\beta$ and $\sin\beta$.

In this version, the orthogonality relation for the eigenfunctions takes the form

$$\int_a^b \phi_n(x)\phi_m(x)dx = 0, \qquad \text{if } n \ne m.$$

Corollary 1.1 asserts that if $f(x)$ has a continuous second derivative and satisfies the Sturm-Liouville boundary conditions, then it can be expanded in a uniformly convergent generalized Fourier series with respect to the (normalized) eigenfunctions $\{\phi_n\}_{n=1}^{\infty}$ of the Sturm-Liouville problem, i.e.,

$$f(x) = \sum_{n=1}^{\infty} \hat{f}(n)\phi_n(x),$$

where

$$\hat{f}(n) = \int_a^b f(x)\phi_n(x)dx.$$

Such an expansion is called a *Sturm-Liouville series expansion.* If f is only a square integrable function on $I = [a, b]$, i.e., $f \in L^2(I)$, then the Sturm-Liouville series expansion is guaranteed to converge to f only in the sense of $L^2(I)$.

A Sturm-Liouville boundary-value problem is said to be *singular* if $p(x) = 0$ at one of the end points of the interval $[a, b]$, or if any of the functions p, r or s becomes infinite at either $x = a$ or $x = b$ or both, or if the interval $[a, b]$ becomes infinite. In the second version of the Sturm-Liouville problem, the problem is singular if either Q has a singularity at at least one of the end points of the interval, or if the interval is infinite.

A singular Sturm-Liouville problem does not, in general, have a discrete spectrum (eigenvalues); however, if further restrictions are imposed, a singular problem can have similar properties to a regular one. For example, a singular problem may have a purely discrete spectrum and a complete set of eigenfunctions.

We conclude this section by showing that the orthogonality relations of the Jacobi, Laguerre and Hermite polynomials follow from the procedure described in this section because these polynomials are eigenfunctions of Sturm-Liouville problems.

1. **The Jacobi Polynomials:** The functions

$$(\sin\frac{\theta}{2})^{\alpha+1/2}(\cos\frac{\theta}{2})^{\beta+1/2}\ P_n^{(\alpha,\beta)}(\cos\theta);\alpha,\beta>1$$

are the eigenfunctions of the Sturm-Liouville problem

$$y''-q(\theta)y=\lambda y, \qquad 0\le\theta\le\pi,$$

with $y(0)$ and $y(\pi)$ being finite, where

$$q(\theta)=\frac{1}{4}\left[\frac{\alpha^2-\frac{1}{4}}{\sin^2\frac{\theta}{2}}+\frac{\beta^2-\frac{1}{4}}{\cos^2\frac{\theta}{2}}\right].$$

The eigenvalues are

$$-\left(n+\frac{1}{2}(\alpha+\beta+1)\right)^2,$$

and the orthogonality relation is

$$\int_0^\pi\left(\sin\frac{\theta}{2}\right)^{2\alpha+1}\left(\cos\frac{\theta}{2}\right)^{2\beta+1}P_m^{(\alpha,\beta)}(\cos\theta)P_n^{(\alpha,\beta)}(\cos\theta)d\theta$$

$$=\frac{1}{2^{\alpha+\beta+1}}\int_{-1}^1(1-x)^\alpha(1+x)^\beta P_m^{(\alpha,\beta)}(x)P_n^{(\alpha,\beta)}(x)\,dx=0 \text{ if } m\neq n.$$

2. **The Laguerre Polynomials:** The functions

$$\phi_n(x)=e^{-x^2/2}x^{\alpha+\frac{1}{2}}L_n^\alpha(x^2)$$

are the eigenfunctions of the Sturm-Liouville problem

$$y''-\left(x^2+\frac{\alpha^2-\frac{1}{4}}{x^2}\right)y=\lambda y, \quad 0\le x<\infty,$$

$$y(0) \text{ and } y(\infty)=\lim_{x\to\infty}y(x) \text{ are finite,}$$

with $-(4n+2\alpha+2)$ being the eigenvalues. The orthogonality relation takes the form

$$2\int_0^\infty e^{-x^2}x^{2\alpha+1}L_m^\alpha(x^2)L_n^\alpha(x^2)dx=\int_0^\infty e^{-x}x^\alpha L_m^\alpha(x)L_n^\alpha(x)dx=0$$

if $m\neq n$.

3. **The Hermite Polynomials:** The eigenfunctions of the Sturm-Liouville problem

$$y'' - x^2 y = \lambda y, \quad -\infty < x < \infty$$

$$y(-\infty) \text{ and } y(\infty) \text{ are finite,}$$

are $\phi_n(x) = e^{-x^2/2} H_n(x)$ with $-(2n+1)$ being the corresponding eigenvalues. The orthogonality relation is

$$\int_{-\infty}^{\infty} e^{-x^2} H_m(x) H_n(x) dx = 0 \text{ if } m \neq n.$$

In addition to the orthogonality relation of these classical polynomials, we obtain the well-known orthogonality relation for the Bessel functions.

4. **The Bessel Functions:** The eigenfunctions of the Sturm-Liouville problem

$$y'' - \left(\frac{\nu^2 - \frac{1}{4}}{x^2} \right) y = \lambda y, \quad 0 \leq x \leq b, \quad \nu > -\frac{1}{2},$$

$$y(0) = 0 = y(b),$$

are $\phi_n(x) = \sqrt{x} J_\nu(\sqrt{\lambda_n} x)$ and the eigenvalues, λ_n, are the zeros of $J_\nu(b\sqrt{\lambda})$. The orthogonality relation becomes

$$\int_0^b x J_\nu \left(\sqrt{\lambda_m} x \right) J_\nu \left(\sqrt{\lambda_n} x \right) dx = 0 \text{ if } m \neq n.$$

The orthogonal expansions of functions in terms of these classical orthogonal polynomials follow from the general Sturm-Liouville series expansion discussed above.

1.9 More General Orthogonal Expansions

A more general approach to generate orthogonal functions than the one presented in the previous section is based on integral equations rather than differential equations and boundary-value problems. In this section, we shall describe briefly how this can be achieved.

DEFINITION 1.1 *A linear operator L from a Banach space X into itself is said to be compact (or completely continuous) if it maps every* *bounded set into a relatively compact set, i.e., into a set whose closure is compact.*

Equivalently, a linear operator is compact if and only if every bounded sequence $\{x_n\}$ in X contains a subsequence $\{x_{n_i}\}$ such that $\{Lx_{n_i}\}$ converges to a point of X. If X is finite dimension, then every linear operator mapping X into itself is compact. More generally, if X is infinite dimension, but $L(X) = \{Lx : x \in X\}$ is finite dimension, then L is compact. Every compact operator is readily seen to be bounded (continuous), but the converse is not true. Compact operators have the following properties [209, pp. 97-103]:

1. The set of all compact operators on a Banach space X is a vector space,
2. If L is compact and A is bounded, then LA, AL are compact,
3. If X is infinite dimension and L is compact, then the inverse L^{-1} of L cannot be bounded; otherwise $I = LL^{-1}$ would be compact.
4. If $\{L_n\}$ is a sequence of compact operators on X such that $L_n \to L$ in the norm, i.e., $\|L_n - L\| \to 0$, then L is compact. The norm of an operator is defined as usual by
$$\|L\| = \sup_{\|x\|=1} \|Lx\| .$$
5. The adjoint L^* of a compact operator L is also compact.
6. If L is compact, then $\sigma(L)$ is compact, at most countable, and has at most one limit point, namely, 0.

If we further assume that X is a Hilbert space, we can obtain more interesting properties of compact operators on X. One such property can be summarized in the following theorem.

THEOREM 1.4 **(Hilbert-Schmidt** [127, p. 248]**)**
Let L be a self-adjoint compact operator mapping a Hilbert space X into itself. Then, there is an orthonormal system of eigenvectors $\{\phi_n\}$ corresponding to the non-zero eigenvalues $\{\lambda_n\}$ of L such that every element $f \in X$ has a unique representation of the form
$$f = \sum_n \alpha_n \phi_n + f',$$
where f' satisfies the condition $Lf' = 0$. Moreover
$$Lf = \sum_n \lambda_n \alpha_n \phi_n,$$

and in the case where there are infinitely many non-zero eigenvalues

$$\lim_{n\to\infty} \lambda_n = 0\,.$$

Let S denote the span of $\{\phi_n\}$ and $S^\perp$ denote its orthogonal complement (see Section 1.5). Because every vector $x \in S^\perp$ is an eigenvector of L corresponding to the eigenvalue $\lambda = 0$, we can add an arbitrary orthonormal basis of $S^\perp$ to the system $\{\phi_n\}$ described above to obtain a complete orthonormal basis $\{\psi_n\}$ of X consisting entirely of eigenvectors of L. The following result can now be easily verified.

THEOREM 1.5

Let L be a self-adjoint compact operator on a Hilbert space X. Then there is a complete orthonormal system $\{\psi_n\}$ of eigenvectors of L such that every $f \in X$ has a unique representation of the form

$$f = \sum_n \langle f, \psi_n\rangle\, \psi_n$$

with

$$Lf = \sum_n \lambda_n \langle f, \psi_n\rangle\, \psi_n,$$

where $\lambda_1, \lambda_2, \ldots,$ are the eigenvalues corresponding to $\psi_1, \psi_2, \ldots$.

We conclude this section with an important application of compact operators on a Hilbert space.

Let $K(x,\xi)$ be a bounded, piecewise continuous function on the square $Q = I \times I$, where $I = [a,b]$. Define the operator L on $L^2(I)$ by

$$(Ly)(x) = \int_a^b K(x,\xi)y(\xi)d\xi, \quad y \in L^2(I).$$

It can be easily shown that L is a compact operator on $L^2(I)$ and it is self-adjoint if K is real and symmetric, i.e., $K(x,\xi) = K(\xi,x)$.

Now consider the *homogeneous Fredholm integral equation of the second kind*

$$Ly = \int_a^b K(x,\xi)y(\xi)d\xi = \frac{1}{\lambda}y(x), \tag{1.10}$$

where λ is a constant and K is a symmetric real-valued, bounded and piecewise continuous function on Q. Eq. (1.10) gives rise to an eigenvalue problem for the compact operator L, and hence, the values of λ for which this integral equation has a non-trivial solution are called the eigenvalues and the corresponding solutions are called the eigenfunctions. From the above discussion on compact operators, the following properties readily follow:

1. Unless K is separable, i.e., it can be written in the form

$$K(x, \xi) = \sum_{i=1}^{n} \alpha_i(x)\beta_i(\xi),$$

the set $\{\lambda_n\}$ of eigenvalues is infinite, and can be ordered in such a way that

$$0 < |\lambda_1| < |\lambda_2| < \cdots$$

and

$$\lim_{n\to 0} |\lambda_n| = \infty.$$

2. To each eigenvalue, λ_n, there corresponds a finite set of independent eigenfunctions, thus every eigenvalue has finite multiplicity $\nu(n)$.

3. The eigenfunctions corresponding to distinct eigenvalues are orthogonal. That is, if ϕ_n and ϕ_m are eigenfunctions corresponding to λ_n and λ_m, respectively, with $\lambda_n \neq \lambda_m$, then

$$\int_a^b \phi_n(x)\ \phi_m(x)\ dx = 0, \quad n \neq m.$$

4. Since each eigenvalue has a finite number of independent eigenfunctions, $\left\{\tilde{\phi}_{n,k}\right\}_{k=1}^{\nu(n)}$, we can use the Gram-Schmidt orthogonalization process to orthogonalize $\left\{\tilde{\phi}_{n,k}\right\}_{k=1}^{\nu(n)}$ and obtain a complete orthonormal set of eigenfunctions $\{\phi_{n,k}\}_{k=1,\,n=1}^{\nu(n),\infty}$. Since this set is countable, we can reorder it and write it as $\{\psi_m\}_{m=1}^{\infty}$. Moreover, for any $f \in L^2(I)$ in the form $f(x) = \int_a^b K(x,\xi)g(\xi)d\xi$, for some $g \in L^2(I)$, we have the generalized Fourier series expansion

$$f = \sum_{m=1}^{\infty} \langle f, \psi_m \rangle\, \psi_m.$$

This procedure of generating orthogonal eigenfunctions is more general than the one described in Section 1.7 because the boundary-value problem Π is equivalent to an integral equation of type (1.10), in which the kernel function K is the Green's function of the problem Π. However, the converse in general is not true.

1.10 Non-Orthogonal Expansions

By a *non-orthogonal expansion* in a Hilbert space X we mean a series expansion of any $x \in X$ in the form $x = \sum_n a_n x_n$, where $\{x_n\}$ is a complete set of vectors that is not necessarily orthogonal. An example of such expansions can be seen in Theorem 1.3. Although non-orthogonal expansions are more difficult to work with than orthogonal expansions, in some problems they are unavoidable and one is led to deal with them directly. This section is devoted to the study of such expansions.

1.10.1 Frames and Non-Orthogonal Basis

Let $\mathcal{H}$ be a Hilbert space with inner product $\langle \cdot, \cdot \rangle$ and norm $\|x\| = \sqrt{\langle x, x \rangle}$, for $x \in \mathcal{H}$.

DEFINITION 1.2 *A sequence of vectors $\{x_n\}_{n=1}^{\infty}$ in a Hilbert space $\mathcal{H}$ is said to be a basis (Schauder basis) of $\mathcal{H}$ if to each $x \in \mathcal{H}$, there corresponds a unique sequence of scalars $\{c_n\}_{n=1}^{\infty}$ such that*

$$x = \sum_{n=1}^{\infty} c_n x_n, \tag{1.11}$$

where the convergence is understood to be in the norm, that is

$$\left\| x - \sum_{n=1}^{N} c_n x_n \right\| \to 0 \text{ as } N \to \infty.$$

Recall that a basis $\{x_n\}_{n=1}^{\infty}$ of $\mathcal{H}$ is said to be orthogonal if $\langle x_m, x_n \rangle = 0$, whenever $m \neq n$, and an orthogonal basis is said to be orthonormal if, in addition, $\langle x_n, x_n \rangle = 1$ for all n.

An orthogonal basis is complete in the sense that if $\langle x, x_n \rangle = 0$ for all n, then $x = 0$. Every orthogonal set in a separable Hilbert space is countable and every separable Hilbert space has an orthonormal basis. For orthonormal basis, the expansion (1.11) is given by

$$x = \sum_{n=1}^{\infty} \langle x, x_n \rangle x_n, \tag{1.12}$$

with

$$\|x\|^2 = \sum_{n=1}^{\infty} |\langle x, x_n \rangle|^2 . \tag{1.13}$$

More generally, for any $x, y \in \mathcal{H}$

$$\langle x, y\rangle = \sum_{n=1}^{\infty} \langle x, x_n\rangle \overline{\langle y, x_n\rangle}. \tag{1.14}$$

It can be shown [251, p. 29] that every basis $\{x_n\}_{n=1}^{\infty}$ of a Hilbert space possesses a unique *biorthonormal basis* $\{x_n^*\}_{n=1}^{\infty}$, which means that

$$\langle x_m, x_n^*\rangle = \delta_{m,n},$$

and $\{x_n^*\}_{n=1}^{\infty}$, is also a basis of $\mathcal{H}$. Moreover, for every $x \in \mathcal{H}$ we have

$$x = \sum_{n=1}^{\infty} \langle x, x_n^*\rangle x_n = \sum_{n=1}^{\infty} \langle x, x_n\rangle x_n^*.$$

Such a non-orthogonal expansion will be called a *biorthogonal expansion.* If $\langle x_m, x_n^*\rangle = 0$ whenever $m \neq n$, but $\langle x_n, x_n^*\rangle$ is not necessarily equal to one, $\{x_n^*\}_{n=1}^{\infty}$ will be called a biorthogonal basis of $\{x_n\}_{n=1}^{\infty}$. In this case we have for any $x \in \mathcal{H}$

$$x = \sum_{n=1}^{\infty} (\bar{d}_n)^{-1} \langle x, x_n^*\rangle x_n = \sum_{n=1}^{\infty} (d_n)^{-1} \langle x, x_n\rangle x_n^*, \tag{1.15}$$

where $d_n = \langle x_n^*, x_n\rangle \neq 0$. It can be easily verified that for all $n, d_n \neq 0$.

DEFINITION 1.3 *Let $\mathcal{G} = \{g_n\}$ be a basis in a separable Hilbert space $\mathcal{H}$. Then*

1. *$\mathcal{G}$ is called unconditional if*

$$\sum c_n g_n \in \mathcal{H} \text{ implies that } \sum |c_n|\, g_n \in \mathcal{H}.$$

2. *$\mathcal{G}$ is said to be bounded if there exist two non-negative numbers A and B such that for all n*

$$A \leq \|g_n\| \leq B.$$

3. *$\mathcal{G}$ is said to be a Riesz basis if there exist a topological isomorphism $T : \mathcal{H} \to \mathcal{H}$ and an orthonormal basis $\{u_n\}$ of $\mathcal{H}$ such that $Tg_n = u_n$ for each n.*

DEFINITION 1.4 *Let $\mathcal{G} = \{g_n\}$ be a sequence in $\mathcal{H}$ (not necessarily a basis of $\mathcal{H}$). Then $\mathcal{G}$ is called a frame if there exist two numbers $A, B > 0$ such that for any $f \in \mathcal{H}$, we have*

$$A\|f\|^2 \leq \sum_n |\langle f, g_n\rangle|^2 \leq B\|f\|^2.$$

The numbers A and B are called the frame bounds. The frame is said to be tight if $A = B$ and is exact if it ceases to be a frame whenever any single element is deleted from the frame. To every frame $\mathcal{G}$ there corresponds an operator S, known as the frame operator, which maps $\mathcal{H}$ into itself and is defined by

$$Sf = \sum_n \langle f, g_n \rangle g_n, \text{ for all } f \in \mathcal{H}.$$

In view of the definition, frames are complete, since if $\langle f, g_n \rangle = 0$ for all n, then

$$A \|f\|^2 \leq \sum_n |\langle f, g_n \rangle|^2 = 0,$$

which implies that $f = 0$.

The proof of the following theorem can be found in [11, pp. 97-162].

THEOREM 1.6

In a separable Hilbert space $\mathcal{H}$ the following conditions are equivalent:

1. *$\{g_n\}$ is an exact frame,*
2. *$\{g_n\}$ is a bounded unconditional basis,*
3. *$\{g_n\}$ is a Riesz basis.*

If $\{g_n\}$ is an orthonormal basis of $\mathcal{H}$, then the Parseval equality

$$\|f\|^2 = \sum_n |\langle f, g_n \rangle|^2, \qquad f \in \mathcal{H}$$

shows that an orthonormal basis is a tight exact frame with frame bounds $A = B = 1$. However, tight frames are not necessarily orthonormal bases. For example,

$$\left\{ \left(\frac{1}{\sqrt{2}}, \frac{1}{\sqrt{2}} \right), \left(-\frac{1}{\sqrt{2}}, \frac{1}{\sqrt{2}} \right), \left(\frac{1}{\sqrt{2}}, -\frac{1}{\sqrt{2}} \right), \left(-\frac{1}{\sqrt{2}}, -\frac{1}{\sqrt{2}} \right) \right\}$$

is a tight frame in $\mathcal{R}^2$ with frame bounds $A = 2 = B$, but it is not even orthonormal since these four vectors are linearly dependent.

If $\{g_n\}$ is a tight frame with frame bound $A = 1 = B$, and if $\|g_n\| = 1$ for all n, then $\{g_n\}$ is an orthonormal basis. For, if $\{g_n\}$ is a tight frame with frame bound $= 1$, then

$$\|g_k\|^2 = \sum_n |\langle g_k, g_n \rangle|^2 = \|g_k\|^4 + \sum_{n \neq k} |\langle g_k, g_n \rangle|^2.$$

But since $\|g_k\| = 1$, it follows that $\langle g_k, g_n \rangle = 0$ for all $n \neq k$. The completeness of $\{g_n\}$ is a consequence of the fact that frames are complete.

Therefore, we can say that for a tight frame $\{g_n\}$, the frame bound is a measurement of how close the frame to being an orthonormal basis, the closer the frame bound to 1, the closer the frame to being orthonormal, provided that $\|g_n\| = 1$ for all n. The last restriction cannot be dropped since if $\{g_n\}_{n=1}^{\infty}$ is an orthonormal basis of $\mathcal{H}$, then

$$\left\{g_1, g_2/\sqrt{2}, g_2/\sqrt{2}, g_3/\sqrt{3}, g_3/\sqrt{3}, g_3/\sqrt{3}, \cdots\right\}$$

is a tight frame with frame bound $A = 1$, but clearly it is not an orthonormal basis.

Tightness and exactness are not related, as illustrated by the following example. If $\{g_n\}$ is an orthonormal basis of $\mathcal{H}$, then $\{g_1, g_1, g_2, g_2, \cdots\}$ is a tight frame with frame bounds $A = 2 = B$, but it is not exact, whereas $\{\sqrt{2}g_1, g_2, g_3, \cdots\}$ is exact but not tight since the frame bounds are easily seen to be $A = 1$ and $B = 2$. For more details, see [95].

The following theorem summarizes some of the main properties of frames and frame operators that will be needed later.

THEOREM 1.7 [11, p. 101]
Let $\mathcal{H}$ be a separable Hilbert space and $\{g_n\}_{n=-\infty}^{\infty} \subset \mathcal{H}$ be a frame with frame bounds A and B. Then

1. *The frame operator $Sf = \sum_n \langle f, g_n \rangle g_n$ is a bounded linear operator on $\mathcal{H}$ with $AI \leq S \leq BI$, where I is the identity operator.*

2. *S is invertible with $B^{-1}I \leq S^{-1} \leq A^{-1}I$. Moreover, S^{-1} is a positive operator; hence it is self-adjoint.*

3. *$\{S^{-1}g_n\}$ is a frame with frame bounds B^{-1}, A^{-1}.*

4. *Every $f \in \mathcal{H}$ can be written in the form*

$$f = \sum_n \langle f, S^{-1}g_n \rangle g_n = \sum_n \langle f, g_n \rangle S^{-1}g_n. \tag{1.16}$$

5. *If there exists a sequence of scalars $\{c_n\}$ such that $f = \sum_n c_n g_n$, then*

$$\sum_n |c_n|^2 = \sum_n |a_n|^2 + \sum_n |a_n - c_n|^2,$$

where $a_n = \langle f, S^{-1}g_n \rangle$.

6. *In addition, if* $\{g_n\}$ *is an exact frame, then* $\{g_n\}$ *and* $\{S^{-1}g_n\}$ *are biorthonormal, i.e.,* $\langle g_m, S^{-1}g_n\rangle = \delta_{m,n}$.

The frame $\{S^{-1}g_n\}$ is called the *dual frame* of the frame $\{g_n\}$. Equation (1.16) is the analogue of (1.15) for frames and the two coincide if any of the conditions of Theorem 1.6 is satisfied, and in this case $S^{-1}g_n = g_n^*$.

Part (5) shows that although the expansion of f in terms of the frame $\{g_n\}$ is not necessarily unique, using the expansion with coefficients $f_n = \langle f, S^{-1}g_n\rangle$ is the most economical in the sense that for any other coefficients $\{c_n\}$ with $f = \sum_n c_n g_n$, we have $\sum_n |c_n|^2 \geq \sum_n |f_n|^2$.

1.10.2 Reproducing-Kernel Hilbert Spaces

We conclude this section with an important concept that is closely related to orthogonal and non-orthogonal expansions in a Hilbert space, namely, the concept of reproducing kernels.

Throughout the rest of this section $\mathcal{H}$ will denote a Hilbert space consisting of functions defined on some set X. A function $K(x,t), x, t \in X$ is called a reproducing kernel of $\mathcal{H}$ if

1. For every $t, K(x,t)$ as a function of x belongs to $\mathcal{H}$.

2. For every $x, t \in X$ and f$\in \mathcal{H}$,

$$f(t) = \langle f(x), K(x,t)\rangle_x, \tag{1.17}$$

   where the subscript x indicates that the scalar product applies to functions of x.

DEFINITION 1.5 *A Hilbert space* $\mathcal{H}$ *is called a reproducing-kernel Hilbert space if it possesses a reproducing kernel.*

It is readily seen that if a reproducing kernel exists, it is unique. As for its existence, it is known [3] that $\mathcal{H}$ has a reproducing kernel if and only if for every $x \in X$, the evaluation functional $f(x), f \in \mathcal{H}$ is continuous.

If $\{e_n\}_{n=1}^{\infty}$ is an orthonormal basis of $\mathcal{H}$ and if K is its reproducing kernel, then

$$K(t,x) = \sum_{n=1}^{\infty} e_n(t)\overline{e_n(x)}. \tag{1.18}$$

For, if we fix $t \in X$ and set

$$\langle f, K_t\rangle = \langle f(y), K(y,t)\rangle_y = f(t),$$

then by (1.12) and (1.17)

$$K_t = \sum_{n=1}^{\infty} \langle K_t, e_n \rangle e_n = \sum_{n=1}^{\infty} \overline{e_n(t)} e_n,$$

which, in view of (1.17) once more and (1.14), yields

$$K(t,x) = \langle K_x, K_t \rangle = \sum_{n=1}^{\infty} e_n(t)\overline{e_n(x)}.$$

Similarly, we can verify that if $\{e_n\}_{n=1}^{\infty}$ is a basis (not necessarily orthogonal), then the reproducing kernel may be given by

$$K(t,x) = \sum_{n=1}^{\infty} \left(\bar{d}_n\right)^{-1} e_n(t)\overline{e_n^*(x)} = \sum_{n=1}^{\infty} (d_n)^{-1} e_n^*(t)\overline{e_n(x)}, \tag{1.19}$$

where $d_n = \langle e_n^*, e_n \rangle$ and $\{e_n^*\}_{n=1}^{\infty}$ is a biorthogonal basis for $\{e_n\}_{n=1}^{\infty}$. If $\{e_n\}_{n=1}^{\infty}$ is an orthonormal basis, $e_n = e_n^*, d_n = 1$, and (1.19) reduces to (1.18).

If $\lim_{n\to\infty} \|f_n - f\| = 0$ in a reproducing-kernel Hilbert space, then $\lim_{n\to\infty} f_n(x) = f(x)$ for each $x \in X$. The convergence is uniform on every subset of X in which $K(x,x)$ is uniformly bounded as can be seen from the relation

$$\begin{aligned} |f_n(y) - f(y)| &= |\langle f_n(x) - f(x), K(x,y)\rangle_x| \le \|f_n - f\| \, \|K(x,y)\|_x \\ &= \|f_n - f\| \sqrt{K(y,y)}, \end{aligned}$$

where

$$\|K(x,y)\|_x^2 = \langle K(x,y), K(x,y)\rangle_x .$$

1.11 Summability of Series and Integrals

In the preceding sections, we discussed orthogonal and non-orthogonal series expansions. We have been concerned mainly with series that converge in some sense, either in some Hilbert space or pointwise or uniformly. Divergent series and integrals arise very often in many problems. At first, they were a source of difficulties and controversies, but nowadays not many mathematicians object to them. There are several methods available to deal with and manipulate divergent series and integrals, the two most important of which are the method of summability and the method of weak convergence.

In the summability method if a series or an integral is divergent, it is replaced by a transformation thereof that would have a limit when an appropriate limit procedure is applied. But for this method to provide a true generalization of the notion of convergence, it is required that it have the property that the limit of the transformation of a series (integral) must exist whenever the original series (integral) converges, and that the limit of the transformation and the sum of the convergent series (integral) must be equal.

The other method, which is newer and more elegant, takes a more philosophical approach: the trouble with divergent series (integrals) is not due to an intrinsic defect of the series (integral), but due to our notion of convergence. If we relax our notion of convergence, many divergent series (integrals) will converge. It is no exaggeration to say that any series (integral) can be forced to converge. For example, any trigonometric series with arbitrary coefficients is known to converge in certain spaces of generalized functions [78, 240]. For the definition and properties of generalized functions, see Chapter 3.

In this section, we shall describe briefly some of the different summability techniques used for series and integrals, but we shall relegate the discussion of the method of weak convergence to Chapter 3.

1.11.1 Summability of Series

Let $\{a_n\}_{n=0}^{\infty}$ be a sequence of numbers and $s_n = \sum_{k=0}^{n} a_k$. If $\lim\limits_{n\to\infty} s_n$ exists and is equal to s, we say that the series $\sum_{k=0}^{\infty} a_k$ converges to s and we write $s = \sum_{k=0}^{\infty} a_k$. If a series is not convergent, it is said to be divergent. The sequence $\{s_n\}_{n=0}^{\infty}$ is called the *sequence of partial sums* of the series $\sum_{k=0}^{\infty} a_k$. If the series $\sum_{k=0}^{\infty} |a_k|$ is convergent, then $\sum_{k=0}^{\infty} a_k$ is said to be *absolutely convergent.*

Let

$$M = (a_{ij})_{i,j=0}^{\infty} = \begin{pmatrix} a_{00} & a_{01} & a_{02} & \cdots \\ a_{10} & a_{11} & a_{12} & \cdots \\ \vdots & \vdots & \vdots & \\ a_{n0} & a_{n1} & a_{n2} & \cdots \\ \vdots & \vdots & \vdots & \end{pmatrix}$$

be a doubly infinite matrix of numbers. Let

$$S = \begin{pmatrix} s_0 \\ s_1 \\ s_2 \\ \vdots \end{pmatrix} \text{ and } \sum = \begin{pmatrix} \sigma_0 \\ \sigma_1 \\ \sigma_2 \\ \vdots \end{pmatrix},$$

where $\sum = MS$, or

$$\sigma_n = a_{n0}s_0 + a_{n_1}s_1 + \cdots + a_{n_k}s_k + \cdots, \quad n = 0, 1, 2, \cdots.$$

If $\lim_{n\to\infty} \sigma_n = s$, we shall say that the sequence of partial sums $\{s_n\}_{n=0}^{\infty}$ or the series $\sum_{k=0}^{\infty} a_k$ is *M-summable* and that the M-sum of the series $\sum_{k=0}^{\infty} a_k$ is s.

Assume that $A_n^+ = \sum_{k=0}^{\infty} |a_{nk}|$ and $A_n = \sum_{k=0}^{\infty} a_{nk}$ exist and are finite for all n.

DEFINITION 1.6 *The matrix M is said to be regular if $\lim_{n\to\infty} s_n = s$ implies that $\lim_{n\to\infty} \sigma_n = s$. In this case we say that the M-summability method is regular.*

The following theorem, which is known as the Silverman-Toeplitz theorem, gives sufficient conditions for M to be regular.

THEOREM 1.8 [196, p. 23]
The matrix M is regular if the following conditions are satisfied:

1. $\lim_{n\to\infty} a_{nk} = 0$, *for* $k = 0, 1, 2, \ldots,$
2. *the sequence* $\{A_n^+\}_{n=0}^{\infty}$ *is bounded,*
3. $\lim_{n\to\infty} A_n = 1$.

I. Cesàro Summability (C-summability)

Let $A_n^\alpha, n = 0, 1, 2, \ldots$ and $\alpha > -1$, be a sequence of numbers defined by

$$\sum_{n=0}^{\infty} A_n^\alpha x^n = (1-x)^{-\alpha-1}, \quad |x| < 1.$$

One can easily verify that if α is a non-negative integer, then

$$A_n^0 = 1,\ A_n^1 = (n+1),\ \text{and } A_n^k = A_0^{k-1} + A_1^{k-1} + \cdots + A_n^{k-1},\ k = 1, 2, \cdots.$$

Similarly, let $s_n^\alpha, n = 0, 1, 2, \ldots, \alpha > -1$, be a sequence of numbers defined by

$$\sum_{n=0}^{\infty} s_n^\alpha x^n = (1-x)^{-\alpha} \sum_{n=0}^{\infty} s_n x^n = (1-x)^{-\alpha-1} \sum_{n=0}^{\infty} a_n x^n, |x| < 1, \quad (1.20)$$

where $\{s_n^0 = s_n\}_{n=0}^{\infty}$ is the sequence of partial sums of the series $\sum_{k=0}^{\infty} a_k$.

DEFINITION 1.7 *The series $\sum_{k=0}^{\infty} a_k$ is said to be α-Cesàro summable or (C, α)-summable to s if*

$$\lim_{n\to\infty} \sigma_n^\alpha = s,$$

where

$$\sigma_n^\alpha = s_n^\alpha / A_n^\alpha.$$

We may write $(C, \alpha) \sum_{k=0}^{\infty} a_k = s$.

For example, when $\alpha = 1$, we have $A_n^1 = (n+1), s_n^1 = s_0 + \cdots + s_n$, and

$$\sigma_n^1 = \frac{s_0 + \cdots + s_n}{(n+1)} = \sum_{k=0}^{n} \left(1 - \frac{k}{n+1}\right) a_k.$$

Thus, $\sum_{k=0}^{\infty} a_k$ is $(C, 1)$-summable to s if

$$\lim_{n\to\infty} \sum_{k=0}^{n} (1 - k/(n+1))\, a_k = s.$$

σ_n^1 is the arithmetic mean of $s_0, \cdots, s_n$ and the $(C, 1)$ summability method is readily seen to be regular since the corresponding matrix M is given by

$$M = \begin{pmatrix} 1 & 0 & 0 & 0 \cdots\cdots \\ 1/2 & 1/2 & 0 & 0 \cdots\cdots \\ 1/3 & 1/3 & 1/3 & 0 \cdots\cdots \\ \vdots & \vdots & \vdots & \end{pmatrix}.$$

In fact, more can be said about the (C, α) summability.

THEOREM 1.9
The (C, α) summability method is regular and if a series is (C, α) summable to s, then it is also $(C, \alpha + h)$ summable to s for any $h > 0$. Moreover, if $\sum_{k=0}^{\infty} a_k$ is (C, α) summable, then $a_k = 0(k^\alpha)$ as $k \to \infty$.

The Cesàro summability method gives a genuine generalization of the notion of convergence since, for example, if we let $a_k = (-1)^k, k = 0, 1, 2, \cdots$, then $s_{2n} = 1, s_{2n+1} = 0$, and $\lim_{n\to\infty} \sigma_n^1 = 0$. Thus, $\sum_{k=0}^{\infty} a_k$ is $(C, 1)$ summable to 0 although $\sum_{k=0}^{\infty} a_k$ is divergent.

A sufficient condition for the series $\sum_{k=0}^{\infty} a_k$ to converge to s whenever it is $(C, 1)$ summable to s is that $T_n = \sum_{k=0}^{n} k a_k = o(n)$. In fact, if $\sum_{k=0}^{\infty} a_k$ is $(C, 1)$ summable to s, then $\sum_{k=0}^{\infty} a_k$ converges to s if and only if $T_n = o(n)$. For, if we define

$$\Delta_n = s_n - \sigma_n^1,$$

then

$$\Delta_n = \frac{1}{(n+1)} T_n.$$

If $\sum_{k=0}^{\infty} a_k$ converges to s, then $\sum_{k=0}^{\infty} a_k$ is $(C,1)$ summable to s since the $(C,1)$ summability method is regular. Thus, $\lim_{n\to\infty} s_n = \lim_{n\to\infty} \sigma_n^1 = s$ and it follows that $\lim_{n\to\infty} \Delta_n = 0$; thus, $T_n = o(n)$. Conversely, if $T_n = o(n)$, then $\lim_{n\to\infty} s_n = \lim_{n\to\infty} \sigma_n^1 = s$. In particular, if $\sum_{k=0}^{\infty} a_k$ is $(C,1)$ summable to s and $a_k = o(1/k)$, then $\sum_{k=0}^{\infty} a_k$ converges to s. This immediately follows from the observation that if $a_k = o(1/k)$, then $T_n = o(n)$.

II. Abel Summability

DEFINITION 1.8 *The series $\sum_{k=0}^{\infty} a_k$ is said to be Abel summable (or A-summable) to s if the series $\sum_{k=0}^{\infty} a_k x^k$ is convergent for $|x| < 1$ and*

$$\lim_{x\to 1^-} \sum_{k=0}^{\infty} a_k x^k = s,$$

where x tends to 1 from the left along the real axis.

Equivalently, we have, in view of (1.20), that $\sum_{k=0}^{\infty} a_k$ is Abel summable to s if $\sum_{k=0}^{\infty} a_k x^k$ is convergent for $|x| < 1$ and

$$\lim_{x\to 1^-} (1-x)^{\alpha+1} \sum_{k=0}^{\infty} s_k^{\alpha} x^k = \lim_{x\to 1^-} (1-x) \sum_{k=0}^{\infty} s_k x^k = s.$$

This can be used to prove the following theorem which, among other things, shows that the Abel summability method is regular.

THEOREM 1.10 [279, p. 80]
If $\sum_{k=0}^{\infty} a_k$ is (C,α) summable to s, then it is also Abel summable to s. In fact,

$$\lim_{\substack{z\to 1\\ \gamma}} \sum_{k=0}^{\infty} a_k z^k = s,$$

as z approaches 1 along any non-tangential path γ. If $\sum_{k=0}^{\infty} a_k$ is Abel summable to s, then $\sum_{k=0}^{\infty} a_k$ converges to s if and only if

$$T_n = \sum_{k=0}^{n} k a_k = o(n).$$

In particular, if $\sum_{k=0}^{\infty} a_k$ is Abel summable to s and $a_k = o(1/k)$, then $\sum_{k=0}^{\infty} a_k$ converges to s.

A *non-tangential path* γ terminating at 1 is any path lying between two chords of the unit circle passing through the point $z = 1$. Such a path is characterized by the inequality

$$|1 - z| \,/\, (1 - |z|) \leq \text{ constant.}$$

Another important type of summability method that will be used later is the Riesz summability method.

III. Riesz Summability

DEFINITION 1.9 *Let $\lambda = \{\lambda_k\}_{k=0}^{\infty}$ be a sequence of real numbers and $\alpha \geq 0$. The series $\sum_{k=0}^{\infty} a_k$ is said to be Riesz summable to s with respect to λ and α, or (R, λ, α)-summable if*

$$\lim_{\lambda\to\infty} \sum_{\lambda_k < \lambda} \left(1 - \frac{\lambda_k}{\lambda}\right)^{\alpha} a_k = s.$$

This type of summability is particularly useful in studying summability properties of Sturm-Liouville eigenfunction expansions; see Section 1.8.

1.11.2 Summability of Integrals

Most of the summability methods for series have their counterparts for integrals.

I. Cesàro Summability (C-summability)

DEFINITION 1.10 *We say that the integral $\int_0^\infty f(x)dx$ is α-Cesàro summable to s or (C, α) summable to s $(\alpha \geq 0)$ if*

$$\lim_{\lambda\to\infty} \int_0^\lambda \left(1 - \frac{x}{\lambda}\right)^{\alpha} f(x)dx = s.$$

Similarly, the integral $\int_{-\infty}^\infty f(x)dx$ is α-Cesàro summable to s or (C, α) summable to s $(\alpha \geq 0)$ if

$$\lim_{\lambda\to\infty} \int_{-\lambda}^\lambda \left(1 - \frac{|x|}{\lambda}\right) f(x)\,dx = s.$$

The $(C, 0)$ summability is equivalent to ordinary convergent. If $\alpha = 1$, we put $g(x) = \int_0^x f(y)dy$ and obtain by integration by parts

$$\frac{1}{\lambda}\int_0^\lambda g(x)dx = g(\lambda) - \frac{1}{\lambda}\int_0^\lambda xf(x)dx = \int_0^\lambda \left(1 - \frac{x}{\lambda}\right) f(x)dx,$$

hence

$$\int_0^\lambda \left(1 - \frac{x}{\lambda}\right) f(x)\, dx = \frac{1}{\lambda} \int_0^\lambda dx \int_0^x f(y) dy,$$

which is analogous to the $(C, 1)$ sum

$$\sigma_n^1 = \frac{\sum_{k=0}^n s_k}{(n+1)}$$

of series. The process of (C, α) summability of integrals has similar properties to those of series. For example, if an integral is (C, α) summable to s, it is also (C, β) summable to s for any $\beta \geq \alpha$. This implies that the (C, α) summability is regular. Moreover, the Cesàro summability method provides a more general concept of convergence than ordinary convergence of integrals. For example, the integral $\int_0^\infty \sin ax\, dx$ $(a > 0)$ does not exist; however, it is $(C, 1)$ summable to $(1/a)$ because

$$\lim_{\lambda \to \infty} \int_0^\lambda \left(1 - \frac{x}{\lambda}\right) \sin ax\, dx = \lim_{\lambda \to \infty} \left(\frac{1}{a} - \frac{\sin a\lambda}{a^2 \lambda}\right) = \frac{1}{a}.$$

II. Abel, Weierstrass and Riesz Summabilities

As with series, there are many summability techniques available for integrals, many of which fall into the following category. Let ϕ be a non-negative locally integrable function with $\phi(0) = 1$. We say that the integral $\int_{-\infty}^\infty f(x)dx$ is ϕ-summable to s if the integral $\int_{-\infty}^\infty \phi(\epsilon x) f(x)dx$ exists for every ϵ in some neighborhood of 0 and

$$\lim_{\epsilon \to 0} \int_{-\infty}^\infty \phi(\epsilon x) f(x) dx = s.$$

The function ϕ is called the summability kernel and ϵ the summability variable.

As a special case, we obtain the following definition.

DEFINITION 1.11 *The integral $\int_{-\infty}^\infty f(x)dx$ is said to be Abel summable to s if*

$$\lim_{\epsilon \to 0} \int_{-\infty}^\infty e^{-\epsilon x} f(x) dx = s,$$

and Weierstrass summable to s if

$$\lim_{\epsilon \to 0} \int_{-\infty}^\infty e^{-\epsilon^2 x^2} f(x) dx = s.$$

We say that the integral $\int_b^\infty f(x)dx$ is Riesz α-summable to s or (R, α) summable $(\alpha \geq 0)$ to s if

$$\lim_{\lambda \to \infty} \frac{1}{\Gamma(\alpha+1)} \int_b^\lambda \left(1 - \frac{x^2}{\lambda^2}\right)^\alpha f(x) dx = s.$$

All the above summability methods are regular in the sense that if an integral converges absolutely to s, then it is summable to s. Again, there are divergent integrals that are summable. For example, the integral $\int_0^\infty \sin(at)\,dt$ is divergent, yet it is Abel summable to $(1/a)$ since

$$\lim_{\epsilon \to 0} \int_0^\infty e^{-\epsilon x} \sin(ax)dx = \frac{1}{a},$$

while the divergent integral $\int_0^\infty x \sin(bx)dx$ is Weierstrass summable to 0 since

$$\lim_{\epsilon \to 0} \int_0^\infty e^{-\epsilon^2 x^2} x \sin(bx)dx = \frac{1}{2} \lim_{p \to 0} \int_0^\infty e^{-pt} \sin\left(b\sqrt{t}\right) dt = 0,$$

and $\int_0^\infty \cos(ax)dx$ is divergent, yet Riesz summable to 0 because

$$\lim_{\lambda \to \infty} \frac{1}{\Gamma(\alpha+1)} \int_0^\lambda \left(1 - \frac{x^2}{\lambda^2}\right)^\alpha \cos(ax)dx$$

$$= \frac{1}{\Gamma(\alpha+1)} \lim_{\lambda \to \infty} \int_0^1 \lambda \left(1 - y^2\right)^\alpha \cos(a\lambda y)dy$$

$$= 2^{\alpha-1/2}\sqrt{\pi} \lim_{\lambda \to \infty} \frac{\lambda J_{\alpha+1/2}(a\lambda)}{(a\lambda)^{\alpha+1/2}} = 0, \quad \alpha > 0,$$

where J_ν is the Bessel function of the first kind and order ν.

2

Special Functions

Many linear integral transformations are either defined in terms of or have properties related to special functions, such as the Bessel, the zeta and the hypergeometric functions. To make this book as self-contained as possible, we shall introduce in this chapter those special functions that will be used in the sequel. Most of the material presented here can be found in any text or reference book on special functions, in particular, in [60, 61, 79, 220].

2.1 The Gamma and Beta Functions

1. **Definition and Properties:** The Γ-function is defined as

$$\Gamma(z) = \int_0^\infty t^{z-1} e^{-t} dt, \qquad \operatorname{Re} z > 0.$$

Hence,

$$\Gamma(z) = \int_0^1 t^{z-1} e^{-t} dt + \int_1^\infty t^{z-1} e^{-t} dt = P(z) + F(z),$$

where

$$P(z) = \int_0^1 t^{z-1} e^{-t} dt = \int_0^1 t^{z-1} \left(\sum_{n=0}^\infty \frac{(-t)^n}{n!} \right) dt = \sum_{n=0}^\infty \frac{(-1)^n}{n!(z+n)},$$

and $F(z) = \int_1^\infty t^{z-1} e^{-t} dt$ is an entire function.

This shows that the Γ-function is a meromorphic function with simple poles at $z = -n$, $n = 0, 1, 2, \cdots$. The residue at the pole $z = -n$ is $(-1)^n/n!$ (cf. Sec. 1.3). It also follows from the definition of the Γ-function that $\Gamma(1) = 1$ and

$$\Gamma(z) = \frac{1}{z}\int_0^\infty t^z e^{-t} dt = \frac{1}{z}\Gamma(z+1)$$

or

$$\Gamma(z+1) = z\Gamma(z). \tag{2.1}$$

Thus, if n is a positive integer, we have

$$\Gamma(z+n) = z(z+1)(z+2)\cdots(z+n-1)\Gamma(z),$$

which implies that

$$\Gamma(n+1) = 1.2.3.\cdots n = n!.$$

Since $1/\Gamma(z)$ is an entire function, it can be written in terms of the canonical product of its zeros (the poles of Γ) as suggested by the Hadamard factorization theorem for entire functions (cf. Theorem 1.2). In fact, we have

$$\frac{1}{\Gamma(z)} = ze^{\gamma z}\prod_{n=1}^{\infty}\left[\left(1+\frac{z}{n}\right)e^{-z/n}\right], \tag{2.2}$$

where γ is *Euler's constant* defined by

$$\gamma = \lim_{n\to\infty}\left(\sum_{k=1}^{n}\frac{1}{k} - \ln n\right) = 0.577215\cdots. \tag{2.3}$$

From (2.2), it follows easily that

$$\frac{1}{\Gamma(z)\Gamma(-z)} = -z^2\prod_{n=1}^{\infty}\left(1-\frac{z^2}{n^2}\right) = -\pi^{-1}z\sin(\pi z),$$

or

$$\Gamma(z)\Gamma(-z) = -\frac{\pi}{z\sin(\pi z)} \qquad (z \text{ non-integer}),$$

which gives

$$\Gamma(z)\Gamma(1-z) = \frac{\pi}{\sin(\pi z)} \qquad (z \text{ non-integer}). \tag{2.4}$$

Upon replacing z by $z + 1/2$, we obtain

$$\Gamma\left(\frac{1}{2}+z\right)\Gamma\left(\frac{1}{2}-z\right) = \pi\sec(\pi z),$$

in particular, $\Gamma\left(\frac{1}{2}\right) = \sqrt{\pi}$.

The Γ-function is known to satisfy the *multiplication formula of Gauss and Legendre*:

$$\prod_{k=0}^{m-1} \Gamma\left(z+\frac{k}{m}\right) = (2\pi)^{(m-1)/2} m^{(1/2)-mz} \Gamma(mz); \; m = 2, 3, 4, \cdots,$$

from which we obtain, for $m = 2$, *Legendre's duplication formula*

$$\Gamma(2z) = 2^{2z-1}\pi^{-1/2}\Gamma(z)\Gamma\left(z+\frac{1}{2}\right). \tag{2.5}$$

Other formulas satisfied by the Γ-function are

$$\Gamma(x)/\Gamma(x+y) = e^{\gamma y} \prod_{k=0}^{\infty} \left[1+\frac{y}{(x+k)}\right] e^{-y/(k+1)}, \tag{2.6}$$

and

$$e^{y\psi(x)}\Gamma(x)/\Gamma(x+y) = \prod_{k=0}^{\infty} \left[1+\frac{y}{(x+k)}\right] e^{-y/(x+k)}, \tag{2.7}$$

where γ is given by (2.3) and $\psi(x)$ is the *logarithmic derivative* of the *Γ-function*, i.e.,

$$\psi(x) = \frac{d\log\Gamma(x)}{dx} = \frac{\Gamma'(x)}{\Gamma(x)}.$$

The ψ function is a meromorphic function with simple poles at $z = 0, -1, -2, \cdots$, and whose Mittag–Leffler expansion (cf. Section 1.3) can be given explicitly by

$$\psi(z) = -\gamma + (z-1)\sum_{k=0}^{\infty} \frac{1}{(k+1)(z+k)}.$$

In addition, we have

$$e^{y\psi(x)} = e^{-\gamma y - y/x} \prod_{k=1}^{\infty} e^{xy/[k(k+x)]}. \tag{2.8}$$

Another function related to the gamma function is the *beta function* , which is defined by

$$B(x,y) = \int_0^1 t^{x-1}(1-t)^{y-1}dt; \quad \text{Re } x, \; \text{Re } y > 0.$$

In the above integral if we make the change of variable, $t = u - 1$, we find that

$$B(x,y) = B(y,x),$$

and if we set $t = u/(u+1)$, we obtain

$$B(x,y) = \int_0^\infty \frac{u^{x-1}}{(1+u)^{x+y}} du; \qquad \text{Re } x, \text{ Re } y > 0.$$

By multiplying the relation

$$\int_0^\infty t^{x+y-1} e^{-(1+v)t} dt = \frac{\Gamma(x+y)}{(1+v)^{x+y}},$$

by v^{x-1}, integrating with respect to v between 0 and ∞, and interchanging the integrals, we obtain the main relation between the gamma and the beta functions

$$B(x,y) = \frac{\Gamma(x)\Gamma(y)}{\Gamma(x+y)}.$$

2. **Asymptotic Expansions:** The following asymptotic expansions of the gamma function as $|z| \to \infty$ are useful:

$$\Gamma(z) = e^{-z} e^{(z-1/2)\log z} (2\pi)^{1/2} \left[1 + \frac{z^{-1}}{12} + O(z^{-2})\right], |\arg z| < \pi,$$

and

$$\frac{\Gamma(z+a)}{\Gamma(z+b)} = z^{a-b} \left[1 + \frac{1}{2}\frac{(a-b)(a+b-1)}{z} + O(z^{-2})\right].$$

The first asymptotic expansion is known as *Stirling's formula.*

2.2 The Zeta and Mittag–Leffler Functions

2.2.1 The Zeta-Function

1) **The Generalized ζ-Function:** The generalized zeta-function is defined by

$$\zeta(s,\nu) = \sum_{n=0}^\infty \frac{1}{(\nu+n)^s}, \quad \nu \neq 0, -1, -2, \cdots, \quad \text{Re } s > 1.$$

It can also be defined as

$$\zeta(s,\nu) = \frac{1}{\Gamma(s)} \int_0^\infty t^{s-1} e^{-\nu t} (1-e^{-t})^{-1} dt \quad \text{Re } s > 1, \quad \text{Re } \nu > 0.$$

2) **Riemann's ζ-Function:** When $\nu = 1$, the generalized ζ-function reduces to Riemann's zeta-function

$$\zeta(s) = \zeta(s,1) = \sum_{n=1}^{\infty} \frac{1}{n^s}, \qquad \text{Re } s > 1,$$

which is also called the zeta-function for short. From the definition, it easily follows that

$$\begin{aligned}\zeta(s)(1-2^{-s}) &= 1 + \frac{1}{2^s} + \frac{1}{3^s} + \cdots - (\frac{1}{2^s} + \frac{1}{4^s} \cdots) \\ &= \sum_{n=1}^{\infty} \frac{1}{(2n-1)^s}.\end{aligned}$$

Therefore, multiplying the ζ-function by $(1-2^{-s})$ results in the elimination of all the terms in which n is a multiple of 2 from the original series of the zeta-function. Similarly,

$$\zeta(s)(1-2^{-s})(1-3^{-s}) = 1 + \frac{1}{3^s} + \frac{1}{5^s} + \cdots - (\frac{1}{3^s} + \frac{1}{9^s} \cdots),$$

where all the terms in which n is a multiple of 2 or 3 have been eliminated from the original series of the zeta-function. Proceeding inductively, we find that

$$\zeta(s) = \prod_{p=2}^{\infty} (1-p^{-s})^{-1}, \quad p \text{ is prime.}$$

An integral representation of the zeta-function is given by

$$\zeta(s) = \frac{1}{\Gamma(s)} \int_0^{\infty} \frac{t^{s-1}}{e^t - 1} dt, \qquad \text{Re } s > 1.$$

An important relation satisfied by the zeta-function is

$$\zeta(1-s) = 2^{1-s}\pi^{-s}\cos(\pi s/2)\Gamma(s)\zeta(s).$$

Let n be a positive integer. Then $n = p_1^{a_1} \cdots p_k^{a_k}$, where the p_i's are distinct primes and the a_k's are positive integers. Möbius' function, $\mu(n)$, is defined by

$$\mu(n) = \begin{cases} 1 & \text{if } n = 1, \\ (-1)^k & \text{if } a_1 = a_2 = \cdots = a_k = 1, \\ 0 & \text{otherwise .} \end{cases}$$

Then

$$\frac{1}{\zeta(s)} = \sum_{n=1}^{\infty} \mu(n) n^{-s}.$$

If $d(n)$ denotes the number of divisors of n (including 1 and n) and $\sigma_k(n)$ denotes the sum of the kth powers of the divisors of n, i.e., $\sigma_k(n) = \sum_{d|n} d^k$, then

$$\zeta(s)\zeta(s-k) = \sum_{n=1}^{\infty} \sigma_k(n) n^{-s}, \qquad \text{Re } s > \max(1, k+1),$$

where $d|n$ means d divides n.

2.2.2 The Mittag–Leffler Function

The *Mittag–Leffler function* $E_a(z)$ is defined by

$$E_a(z) = \sum_{k=0}^{\infty} \frac{z^k}{\Gamma(ak+1)}, \qquad a > 0.$$

It is an entire function of order $1/a$ that reduces to e^z when $a = 1$. In fact,

$$E_1(z) = e^z, \qquad E_2(z^2) = \cosh z.$$

The *generalized Mittag–Leffler function* $E_{a,b}(z)$ is defined by

$$E_{a,b}(z) = \sum_{k=0}^{\infty} \frac{z^k}{\Gamma(ak+b)}, \qquad a, b > 0,$$

and it reduces to the Mittag–Leffler function when $b = 1$. It can also be defined by the integral

$$E_{a,b}(z) = \frac{1}{2\pi i} \int_c \frac{t^{a-b} e^t}{t^a - z} dt,$$

where the integration path, c, is a loop that starts and ends at $-\infty$ and encircles the disc $|t| \le |z|^{1/a}$ in the positive sense. Moreover, it satisfies the relations

$$E_{a,b}(z) = \frac{1}{\Gamma(b)} + z E_{a,a+b}(z),$$

$$\int_0^{\infty} e^{-t} t^{b-1} E_{a,b}(t^a z)\, dt = \frac{1}{1-z},$$

and

$$\left(\frac{d}{dz}\right)^m [z^{b-1} E_{a,b}(z^a)] = z^{b-m-1} E_{a,b-m}(z^a).$$

An asymptotic expansion for $E_{a,b}(z)$ as $z \to \infty$ is

$$E_{a,b}(z) = -\sum_{n=1}^{N-1} \frac{z^{-n}}{\Gamma(b-an)} + O\left(|z|^{-N}\right), \qquad |\arg(-z)| < (1-a/2)\pi.$$

For $0 < a < 2$ and $|\arg(z)| < a\pi/2$, we have

$$E_a(z) = a^{-1} e^{z^{1/a}} + O\left(|z|^{-1}\right).$$

2.3 The Hypergeometric Functions

Some of the formulas presented in this section are somewhat difficult to typeset. One way to simplify the notation is to introduce the *Pochhammer symbol*

$$(a)_0 = 1;\ (a)_n = a(a+1)\dots(a+n-1),\ n = 1, 2, 3, \cdots.$$

In terms of the Γ-function, we can write

$$(a)_n = \frac{\Gamma(a+n)}{\Gamma(a)}, \qquad n = 0, 1, 2, \cdots.$$

The following relations can be easily verified:

1. $(1)_n = n!$,
2. $(a)_{n+k} = (a)_k(a+k)_n = (a)_n(a+n)_k$,
3. $(-k)_n = \begin{cases} \frac{(-1)^n k!}{(k-n)!}, & 0 \le n \le k \\ 0, & n > k \end{cases}.$

Another notation that will be used frequently is the *binomial coefficient symbol*

$$\binom{\alpha}{k} = \frac{\Gamma(\alpha+1)}{\Gamma(k+1)\Gamma(\alpha-k+1)}.$$

2.3.1 The Generalized Hypergeometric Functions

1. **Definition:** The generalized hypergeometric series

$${}_pF_q\left[\begin{matrix} a_1, \cdots, a_p \\ b_1, \cdots, b_q \end{matrix}; z\right] = {}_pF_q\left(a_r; b_t; z\right),$$

where p and q are non-negative integers and z is a complex number, is defined as

$${}_pF_q\left[\begin{matrix} a_1, \cdots, a_p \\ b_1, \cdots, b_q \end{matrix}; z\right] = {}_pF_q(a_r; b_t; z) = \sum_{n=0}^{\infty} \frac{(a_1)_n \cdots (a_p)_n}{(b_1)_n \cdots (b_q)_n} \frac{z^n}{n!}.$$

The series terminates if one of the parameters a_i is a non-positive integer and is not defined if one of the parameters b_i is a non-positive integer. Therefore, we shall exclude these values of b_i from our definition. With the exception of those special values of the parameters that make the series terminate, the series converges for all z if $p \le q$, converges in the region $|z| < 1$ if $p = q+1$, and diverges for all $z \ne 0$ if $p > q+1$. The series converges for $|z| = 1$ if $p = q+1$ and $\operatorname{Re}\left[\sum_1^q b_i - \sum_1^p a_i\right] > 0$.

2. **Differential Equation:** The generalized hypergeometric function ${}_pF_q$ formally satisfies a differential equation of order $q+1$, namely,

$$\left\{z\frac{d}{dz}\prod_{i=1}^{q}\left(z\frac{d}{dz}+b_i-1\right)-z\prod_{i=1}^{p}\left(z\frac{d}{dz}+a_i\right)\right\}u=0,$$

which can be rewritten as

$$\sum_{n=1}^{q} z^{n-1}(\alpha_n z-\beta_n)\,u^{(n)}+\alpha_0 u+z^q u^{(q+1)}=0 \text{ when } q\ge p,$$

and

$$\sum_{n=1}^{q} z^{n-1}\,(\alpha_n z-\beta_n)\,u^{(n)}+\alpha_0 u+z^q(1-z)u^{(q+1)}=0 \text{ when } p=q+1,$$

where α_n and β_n are constants.
We have the following important relation for $p\le q$.

3. **The Laplace Transform:**

$$\int_0^\infty e^{-st}\,{}_pF_q\left[\begin{matrix}a_1,\cdots,a_p\\ b_1,\cdots,b_q\end{matrix};t\right]dt=\frac{1}{s}\,{}_{p+1,q}F\left[\begin{matrix}1,a_1,\cdots,a_p\\ b_1,\cdots,b_q\end{matrix};\frac{1}{s}\right].$$

2.3.2 The Gauss Hypergeometric Function

The most widely used hypergeometric functions are ${}_2F_1$ and ${}_1F_1$, known as the *Gauss and confluent (Kummer) hypergeometric functions*, respectively. Their importance is due to the fact that many of the classical special functions and orthogonal polynomials can be written in terms of them. Because of the important role that these two hypergeometric functions play in integral transforms, we shall collect some of their properties that will be needed later on.

If there is no other type of hypergeometric functions appearing in a formula, we shall simplify the notation and denote the Gauss hypergeometric function ${}_2F_1$ by F for short. The following notation will be used

$${}_2F_1\left(\begin{matrix}a,b\\ c\end{matrix};z\right)={}_2F_1\,(a,b;c;z)=F(a,b;c;z).$$

Here is a list of some of the main properties of F.

1. **Differential Equation:** F satisfies the second-order differential equation

$$z(1-z)\frac{d^2u}{dz^2}+[c-(a+b-1)z]\frac{du}{dz}-abu=0$$

2. **Differential Relation:**

$$\frac{d^n}{dz^n}F(a,b;c;z) = (a)_n(b)_n\left[(c)_n\right]^{-1}F(a+n,\ b+n;\ c+n;z),$$

3. **Integral Representation (Euler's Formula):**

$$F(a,b;c;z) = \frac{\Gamma(c)}{\Gamma(b)\Gamma(c-b)}\int_0^1 t^{b-1}(1-t)^{c-b-1}(1-tz)^{-a}dt,$$

for $\operatorname{Re} c > \operatorname{Re} b > 0$ and $|\arg(1-z)| < \pi$. We also have

$$\frac{\Gamma(a)\Gamma(b)}{\Gamma(c)}F(a,b;c;z) = \frac{1}{2\pi i}\int_{-i\infty}^{i\infty}\frac{\Gamma(a+t)\Gamma(b+t)\Gamma(-t)}{\Gamma(c+t)}(-z)^t dt,$$

$|\arg(-z)| < \pi$, where the path of integration is chosen in such a way (if necessary) as to separate the poles at $t = 0, 1, 2, \cdots$, from those at $t = -a-n$, and $t = -b-n, n = 0, 1, 2, \cdots$.

4. **Analytic Continuation:** The formula

$$F(a,b;c;z) = (1-z)^{-a}F\left(a, c-b; c; z/(z-1)\right)$$

may be used to continue F analytically outside its original domain of analyticity, which is the domain of convergence of its defining power series. In fact, it is used to continue F analytically to the domain $\operatorname{Re} z < 1/2$ since the left-hand side is valid for $|z| < 1$ and the right-hand side is valid for $|z/(z-1)| < 1$.

The integral representations given above are also used to continue F analytically. It can be shown that F is analytic in the complex z-plane cut along the segment $[1, \infty]$. As for the point $z = 1$, we have $\lim_{z\to 1^-}(1-z)F(a,b;c;z) = 0, \quad Re\ (c-a-b) > 0.$

5. **The Hypergeometric Function as a Function of the Parameters:** For fixed z, $F(a,b;c;z)/\Gamma(c)$ is an entire function of the parameters a, b, and c.

6. **Asymptotic Expansions:**

   (a) $F(a,b;c;z) = c_1 z^{-a} + c_2 z^{-b} + 0\left(z^{-a-1}\right) + 0\left(z^{-b-1}\right)$ as $|z| \to \infty$, provided that $a-b$ is not an integer, where c_1 and c_2 are constants. If $a-b$ is an integer, z^{-a} or z^{-b} has to be multiplied by $\ln z$.

(b) For a, b, z fixed,

$$F(a,b;c;z) = 1 + \frac{ab}{c}z + \cdots + \frac{(a)_n(b)_n}{(c)_n}\frac{z^n}{n!} + 0\left(|c|^{-n-1}\right)$$

as $|c| \to \infty$.

7. **Relationships with Other Special Functions:**

(a) **The Jacobi functions** $P_t^{(\alpha,\beta)}(x)(\alpha, \beta > -1, t \in C)$:

$$P_t^{(\alpha,\beta)}(x) = \frac{\Gamma(t+\alpha+1)}{\Gamma(1+\alpha)\Gamma(1+t)}F\left(-t, t+2\gamma; 1+\alpha; \frac{1-x}{2}\right),$$

where $2\gamma = \alpha + \beta + 1$. The Jacobi function reduces to the Jacobi polynomials when $t = n$, i.e.,

$$P_n^{(\alpha,\beta)}(x) = \binom{n+\alpha}{n}F\left(-n, n+2\gamma; \alpha+1; \frac{1-x}{2}\right).$$

Because of the poles of the Γ-function, it is more convenient to define the Jacobi function $R_t^{(\alpha,\beta)}(x)$ as

$$R_t^{(\alpha,\beta)}(x) = F\left(-t, t+2\gamma; 1+\alpha; \frac{1-x}{2}\right).$$

Thus, the Jacobi function $R_t^{(\alpha,\beta)}$ is an entire function in t that reduces to a multiple of the Jacobi polynomial of degree n when $t = n$.

(b) **The Legendre Functions:** As a special case of the Jacobi function, we obtain for $\alpha = 0 = \beta$ the more known Legendre function $P_t(x) = F\left(-t, t+1; 1; (1-x)/2\right)$. The Legendre function of the second kind $Q_t(x)$ can also be expressed in terms of F

$$Q_t(x) = \frac{\sqrt{\pi}\Gamma(t+1)}{\Gamma\left(t+\frac{3}{2}\right)(2x)^{t+1}}F\left(\frac{t}{2}+1, \frac{t}{2}+\frac{1}{2}; t+\frac{3}{2}; \frac{1}{z^2}\right),$$

where $|z| > 1$ and $t \neq -1, -2, \cdots$.

(c) **The Associated Legendre Functions:** The associated Legendre functions of the first and second kinds $P_t^\mu(x)$ and $Q_t^\mu(x)$ are solutions of the differential equation

$$(1-x^2)y'' - 2xy' + \left[\nu(\nu+1) - \frac{\mu^2}{1-x^2}\right]y = 0,$$

and can be expressed in terms of F as

$$P_t^\mu(z) = \frac{1}{\Gamma(1-\mu)} \left(\frac{z+1}{z-1}\right)^{\mu/2} F\left(-t, t+1; 1-\mu; \frac{1-z}{2}\right),$$

$$|1-z| < 2,$$

and

$$Q_t^\mu(z) = e^{\mu\pi i} \frac{\sqrt{\pi}\Gamma(t+\mu+1)}{2^{t+1}\Gamma\left(t+\frac{3}{2}\right) z^{\mu+t+1}} \left(z^2-1\right)^{\mu/2} \times$$

$$F\left(\frac{t+\mu+2}{2}, \frac{t+\mu+1}{2}; t+\frac{3}{2}; \frac{1}{z^2}\right), \quad |z| > 1.$$

When μ is a positive integer $m (m = 1, 2, 3, \ldots)$ P_t^μ reduces to

$$P_t^m(z) = \frac{\Gamma(t+m+1)}{(-2)^m\Gamma(t-m+1)\Gamma(m+1)} \left(1-z^2\right)^{m/2} \times$$

$$F\left(1+m+t, m-t; 1+m; \frac{1-z}{2}\right).$$

Moreover, we have

$$P_\nu^m(z) = \frac{\Gamma(\nu+m+1)}{\Gamma(\nu-m+1)} P_\nu^{-m}(z).$$

The following is an important integral representation for $P_\nu^\mu(z)$, known as the *Mehler–Dirichlet formula*

$$P_\nu^\mu(\cos\theta) = \frac{\sqrt{2}(\sin\theta)^\mu}{\sqrt{\pi}\Gamma(1/2-\mu)} \int_0^\theta (\cos\phi - \cos\theta)^{-\mu-1/2} \cos[(\nu+1/2)\phi]\, d\phi\,, \tag{2.9}$$

where $0 < \theta, \pi$ and Re $\mu < 1/2$.

We also have the following asymptotic relations: As $|z| \to \infty$

$$P_\nu^\mu(z) = \begin{cases} O\left(z^\nu\right) & \text{if Re } \nu > -1/2, \\ O\left(z^{-\nu-1}\right) & \text{if Re } \nu < -1/2, \end{cases}$$

$$P_\nu^\mu(z) = O\left((z-1)^{-\mu/2}\right) \text{ as } z \to 1\,, \mu \neq 1, 2, \cdots,$$

and

$$P_\nu^m(z) = \nu^{m-1/2} e^{(\nu+1/2)z}\,, \text{ as } |\nu| \to \infty.$$

2.3.3 The Confluent Hypergeometric Function

Now we turn our attention to the confluent (Kummer) hypergeometric function ${}_1F_1$. It is customary to use *Humbert's notation* $\Phi(a;c;z)$ to denote ${}_1F_1(a;c;z)$.

1. **Differential Equation:** Φ satisfies the differential equation

$$z\frac{d^2u}{dz^2} + (c-z)\frac{du}{dz} - az = 0.$$

2. **Differential Relation:**

$$\frac{d^n}{dz^n}\,\Phi(a;c;z) = \frac{(a)_n}{(c)_n}\,\Phi(a+n;c+n;z)$$

3. **Integral Representation:**

$$\Phi(a;c;z) = \frac{\Gamma(c)}{\Gamma(a)\Gamma(c-a)}\int_0^1 e^{zu}u^{a-1}(1-u)^{c-a-1}du,\ \mathrm{Re}\,c > \mathrm{Re}\,a > 0.$$

Another independent solution of the differential equation satisfied by Φ is

$$\Psi(a;c;z) = \frac{1}{\Gamma(a)}\int_0^{\infty} e^{-zu}u^{a-1}(1+u)^{c-a-1}du,\ \mathrm{Re}\,a > 0.$$

The two solutions are related by

$$\Psi(a;c;z) = \frac{\Gamma(1-c)}{\Gamma(a-c+1)}\,\Phi(a;c;z) + \frac{\Gamma(c-1)}{\Gamma(a)}\,z^{1-c}\,\Phi(a-c+1;2-c;z).$$

4. **Asymptotic Expansions:**

$$\Phi(a;c;z) = \frac{\Gamma(c)}{\Gamma(a)}\,e^z z^{a-c}\left[1+0\left(|z|^{-1}\right)\right],\ \text{as}\ \ \mathrm{Re}\,z \to \infty,$$

and

$$\Phi(a;c;z) = \frac{\Gamma(c)}{\Gamma(c-a)}\,(-z)^{-a}\left[1+0\left(|z|^{-1}\right)\right],\ \text{as}\ \mathrm{Re}\,z \to -\infty.$$

For a, z bounded,

$$\Phi(a;c;z) = 1 + 0\left(|c|^{-1}\right)\ \text{as}\ c \to \infty,$$

while if $c - a$ and z are bounded

$$\Phi(a; c; z) = e^z \left[1 + 0\left(|c|^{-1}\right)\right] \quad \text{as } c \to \infty.$$

As for Ψ, we have

$$\Psi(a; c; z) = z^{-a}\left[\sum_{k=0}^{n} \frac{(-1)^k (a)_k (1+a-c)_k}{k!} z^{-k} + 0\left(|z|^{-n-1}\right)\right]$$

as $|z| \to \infty$ for $|\arg z| < \frac{3\pi}{2}$.

If a and z are bounded, then

$$\begin{aligned}\Psi(a; c; z) &= (-c)^{-a}\left[1 + 0\left(|c|^{-1}\right)\right] \\ &+ \sqrt{2\pi}\, z^{1-c} \exp\left(z - c + \left(c - \frac{3}{2}\right)\log c\right)\left[1 + 0\left(|c|^{-1}\right)\right]\end{aligned}$$

as $c \to \infty$ and $|\arg c| < \pi - \epsilon$.

5. **Relationships with Other Special Functions:**

   (a) **The Laguerre Function and Polynomials:**
   The generalized Laguerre polynomial of degree n, $L_n^\alpha(z)$, is given by

$$L_n^\alpha(z) = \frac{(a+1)_n}{n!}\, \Phi(-n; \alpha+1; z) = \frac{(-1)^n}{n!} \Psi(-n; \alpha+1; z).$$

   As with the Jacobi functions and polynomials, we may define the *Laguerre function* as

$$L_t^\alpha(z) = \frac{1}{\Gamma(t+1)} \Phi(-t; \alpha+1; z), \qquad t \in C$$

   so that when t is replaced by n, the Laguerre function reduces to a multiple of the Laguerre polynomial of degree n.

   (b) **The Hermite Polynomials:**
   The Hermite polynomial $H_n(z)$ of degree n is given by

$$H_{2n}(z) = (-1)^n \frac{(2n)!}{n!}\, \Phi\left(-n; \frac{1}{2}; z^2\right),$$

$$H_{2n+1}(z) = (-1)^n \frac{(2n+1)!}{n!}\, (2z)\Phi\left(-n; \frac{3}{2}; z^2\right).$$

(c) **The Bessel and Modified Bessel Functions:**
The Bessel function $J_\nu(z)$ can be defined as

$$J_\nu(z) = \frac{1}{2^\nu \Gamma(\nu+1)}\, z^\nu e^{-iz} \Phi\left(\nu + \frac{1}{2}; 2\nu+1; 2iz\right);$$

hence, the modified Bessel function of the first kind, $I_\nu(z) = e^{-i\nu\frac{\pi}{2}} J_\nu\left(z e^{i\frac{\pi}{2}}\right)$, can be written as

$$I_\nu(z) = \frac{z^\nu}{2^\nu \Gamma(\nu+1)}\, e^{-z} \Phi\left(\nu + \frac{1}{2}; 2\nu+1; 2z\right).$$

The modified Bessel function of the third kind, which is also called the *MacDonald function* or the *Basset function*, is defined by

$$K_\nu(z) = \frac{\pi}{2\ \sin\ \nu\pi}\, [I_{-\nu}(z) - I_\nu(z)], \quad \nu \neq 0, \pm 1, \pm 2, \ldots.$$

Thus, with some easy calculations it can be expressed in terms of Ψ as

$$K_\nu(z) = \sqrt{\pi} e^{-z} (2z)^\nu\ \Psi\left(\nu + \frac{1}{2}; 2\nu+1; 2z\right).$$

(d) **The Whittaker Functions:**
The Whittaker functions $M_{\nu,\mu}$ and $W_{\nu,\mu}$ are defined as

$$M_{\nu,\mu}(z) = e^{-z/2} z^{c/2}\ \Phi(a; c; z),$$

and

$$W_{\nu,\mu}(z) = e^{-z/2} z^{c/2}\ \Psi(a; c; z),$$

where $a = \frac{1}{2} - v + \mu$ and $c = 2\mu + 1$.

(e) **The Incomplete Gamma Functions:**
From the relation $\Phi(a; a; z) = e^z$, it follows that the *incomplete gamma function*

$$\gamma(a, x) = \int_0^x e^{-t} t^{a-1} dt, \quad a > 0,$$

and its complementary function, called the *complementary incomplete gamma function*,

$$\Gamma(a, x) = \int_x^\infty e^{-t} t^{a-1} dt, \quad a > 0,$$

can both be expressed in terms of Φ and Ψ as

$$\gamma(a, x) = \Gamma(a) - \Gamma(a, x) = a^{-1}x^a\Phi(a; a+1; -x),$$

and

$$\Gamma(a, x) = e^{-x}\Psi(1-a; 1-a; x).$$

(f) **The Error Functions:**
The error function and the *complementary error function* defined, respectively, by

$$\text{Erf}(x) = \int_0^x e^{-t^2}\,dt = \frac{1}{2}\gamma\left(\frac{1}{2}, x^2\right),$$

and

$$\text{Erfc}(x) = \int_x^\infty e^{-t^2}\,dt = \frac{1}{2}\Gamma\left(\frac{1}{2}, x^2\right),$$

can be written as

$$\text{Erf}(x) = x\Phi\left(\frac{1}{2}; \frac{3}{2}; -x^2\right),$$

and

$$\text{Erfc}(x) = e^{-x^2}\Psi\left(\frac{1}{2}; \frac{1}{2}; x^2\right).$$

(g) **Fresnel Integrals:**
Closely related to the error functions are the *Fresnel integrals*

$$C(x) = \frac{1}{\sqrt{2\pi}}\int_0^x \frac{\cos t}{\sqrt{t}}\,dt,$$

and

$$S(x) = \frac{1}{\sqrt{2\pi}}\int_0^x \frac{\sin t}{\sqrt{t}}\,dt.$$

It can be easily verified that

$$C(x) = \frac{1}{\sqrt{2\pi}}\sqrt{x}\left[\Phi\left(\frac{1}{2}; \frac{3}{2}; -xe^{i\pi/2}\right) + \Phi\left(\frac{1}{2}; \frac{3}{2}; -xe^{-i\pi/2}\right)\right],$$

and

$$S(x) = \frac{i}{\sqrt{2\pi}}\sqrt{x}\left[\Phi\left(\frac{1}{2}; \frac{3}{2}; -xe^{i\pi/2}\right) - \Phi\left(\frac{1}{2}; \frac{3}{2}; -xe^{-i\pi/2}\right)\right].$$

(h) **The Parabolic Cylinder Function:**
The solutions of the differential equation

$$y'' + \left(\nu + \frac{1}{2} - \frac{1}{4}z^2\right) y = 0,$$

are called parabolic cylinder functions. If we define

$$\begin{aligned} \mathrm{D}_\nu(z) &= 2^{(\nu-1)/2} e^{-z^2/4} z \Psi\left(\frac{1-\nu}{2}; \frac{3}{2}; \frac{z^2}{2}\right) \\ &= 2^{(\nu+1)/2} z^{-1/2} W_{\frac{1}{2}(\nu+\frac{1}{2}), -\frac{1}{4}}(z^2/2), \end{aligned}$$

we find that $\mathrm{D}_\nu(z), \mathrm{D}_\nu(-z), \mathrm{D}_{-\nu-1}(iz)$ and $\mathrm{D}_{-\nu-1}(-iz)$ satisfy the same differential equation. The parabolic cylinder function D_ν reduces to a multiple of the Hermite functions when ν is an integer. More precisely,

$$\mathrm{D}_n(z) = 2^{-n/2} e^{-z^2/4} H_n\left(z/\sqrt{2}\right).$$

(i) **Other Functions Defined by Integrals:**
Here we list some common functions defined by integrals and their relations with the confluent hypergeometric functions:

The Exponential Integral

$$\mathrm{Ei}(x) = \int_{-\infty}^{x} \frac{e^t}{t} dt = -e^x \Psi(1; 1; -x), \quad x \neq 0$$

where the integral is taken in the Cauchy principle value, i.e.,

$$\int_{-\infty}^{x} \frac{e^t}{t} dt = \lim_{\epsilon \to 0} \left[\int_{-\infty}^{-\epsilon} \frac{e^t}{t} dt + \int_{\epsilon}^{x} \frac{e^t}{t} dt\right] \text{ for } x > 0.$$

The Logarithmic Integral

$$\mathrm{li}(x) = \int_0^x \frac{dt}{\log t} = \mathrm{Ei}(\ln x) = -x\Psi(1; 1; -\ln x).$$

The Sine and Cosine Integrals

$$\mathrm{Si}(x) = \int_0^x \frac{\sin t}{t} dt = \frac{\pi}{2} - \frac{i}{2} e^{-ix} \Psi(1; 1; ix) + \frac{i}{2} e^{ix} \Psi(1; 1; -ix),$$

$$\mathrm{Ci}(x) = -\int_x^\infty \frac{\cos t}{t} dt = -\frac{1}{2} e^{-ix} \Psi(1; 1; ix) - \frac{1}{2} e^{ix} \Psi(1; 1; -ix);$$

$$si(x) = -\int_x^\infty \frac{\sin t}{t} dt = \mathrm{Si}(x) - \frac{\pi}{2}.$$

2.4 Classical Orthogonal Polynomials

In this section, we collect some of the main properties of the Jacobi, Laguerre and Hermite polynomials, namely, their differential equations, orthogonality and recurrence relations, their generating functions, Rodrigues' formulas and asymptotic behavior.

2.4.1 The Jacobi Polynomials $P_n^{(\alpha,\beta)}(x)(\alpha, \beta > -1)$

1. **Differential Equation:**

$$(1-x^2)y'' + [\beta - \alpha - (\alpha+\beta+2)x]\, y' + n(n+\alpha+\beta+1)y = 0.$$

2. **Orthogonality Relation:**

$$\int_{-1}^{1} (1-x)^{\alpha}(1+x)^{\beta}\, P_n^{(\alpha,\beta)}(x)\, P_m^{(\alpha,\beta)}(x)\, dx =$$

$$\frac{2^{\alpha+\beta+1}}{2n+\alpha+\beta+1}\,\frac{\Gamma(n+\alpha+1)\Gamma(n+\beta+1)}{\Gamma(n+1)\Gamma(n+\alpha+\beta+1)}\,\delta_{m,n}.$$

3. **Recurrence Relations:**

   (a) **Three-Term Recurrence Relations:**

$$2n(n+\alpha+\beta)(2n+\alpha+\beta-2)P_n^{(\alpha,\beta)}(x) = (2n+\alpha+\beta-1)$$
$$\left\{(2n+\alpha+\beta)(2n+\alpha+\beta-2)x + \alpha^2 - \beta^2\right\} P_{n-1}^{(\alpha,\beta)}(x)$$
$$-2(n+\alpha-1)(n+\beta-1)(2n+\alpha+\beta)P_{n-2}^{(\alpha,\beta)}(x), \quad n = 2, 3, 4, \cdots,$$
$$P_0^{(\alpha,\beta)}(x) = 1 \text{ and } P_1^{(\alpha,\beta)}(x) = \frac{1}{2}(\alpha+\beta+2)x + \frac{1}{2}(\alpha-\beta).$$

   (b) **Recurrence Relation for the Derivative:**

$$\frac{d}{dx}\left\{P_n^{(\alpha,\beta)}(x)\right\} = \frac{1}{2}(n+\alpha+\beta+1)P_{n-1}^{(\alpha+1,\beta+1)}(x),$$

and

$$(1-x^2)\frac{d}{dx}\left\{P_n^{(\alpha,\beta)}(x)\right\} = A_n P_{n-1}^{(\alpha,\beta)}(x) + B_n P_n^{(\alpha,\beta)}(x) + C_n P_{n+1}^{(\alpha,\beta)}(x),$$

where

$$A_n = \frac{2(n+\alpha)(n+\beta)(n+\alpha+\beta+1)}{(2n+\alpha+\beta)(2n+\alpha+\beta+1)},\ B_n = \frac{(\alpha-\beta)(2n)(n+\alpha+\beta+1)}{(2n+\alpha+\beta)(2n+\alpha+\beta+2)},$$

$$C_n = \frac{-2n(n+1)(n+\alpha+\beta+1)}{(2n+\alpha+\beta+1)(2n+\alpha+\beta+2)}.$$

4. **Generating Function:**

$$\sum_{n=0}^{\infty} P(x)w^n = \frac{2^{\alpha+\beta}}{\sqrt{1-2xw+w^2}} \left[1-w+\sqrt{1-2xw+w^2}\right]^{-\alpha}$$

$$\times \left[1+w+\sqrt{1-2xw+w^2}\right]^{-\beta},$$

for w sufficiently small.

5. **Rodrigues' Formula:**

$$P_n^{(\alpha,\beta)}(x) = \frac{(-1)^n}{2^n n!}(1-x)^{-\alpha}(1+x)^{-\beta}\frac{d^n}{dx^n}\left[(1-x)^{n+\alpha}(1+x)^{n+\beta}\right].$$

6. **Inequalities and Asymptotic Properties:**
Let $q = \max(\alpha, \beta)$. Then

$$\max_{-1\leq x\leq 1}\left|P_n^{(\alpha,\beta)}(x)\right| = \max P_n^{(\alpha,\beta)}(\pm 1) = \binom{n+q}{n} \approx n^q/\Gamma(q+1)$$

as $n \to \infty$, where $\alpha > -1, \beta > -1, q \geq -\frac{1}{2}$.

If $-1 < \alpha, \beta < -\frac{1}{2}$, the largest maximum of $\left|P_n^{(\alpha,\beta)}(x)\right|$ is of order $n^{-1/2}$ as $n \to \infty$. We also have

$$\frac{d^m}{dx^m}P_n^{(\alpha,\beta)}(x) = 0(n^q) \text{ as } n \to \infty.$$

where

$$q = \max\left(2m+\alpha, 2m+\beta, m-\frac{1}{2}\right).$$

For x in the interior of the interval $[-1,1]$, we have a sharper estimate

$$\left(\sin\frac{\theta}{2}\right)^{\alpha+1/2}\left(\cos\frac{\theta}{2}\right)^{\beta+1/2} P_n^{(\alpha,\beta)}(\cos\theta) = \frac{1}{\sqrt{n\pi}}\cos(N\theta+\gamma)+0\left(n^{-3/2}\right),$$

uniformly in $0 < \epsilon \leq \theta \leq \pi - \epsilon$, where $x = \cos\theta$ and $N = n + (\alpha + \beta + 1)/2, \gamma = -(\alpha+1/2)\pi/2$.

If z is in the complex plane cut along the interval $[-1,1]$, then

$$\left|P_n^{(\alpha,\beta)}(z)\right| \approx \left|z+\sqrt{z^2-1}\right|^n, \text{ as } n \to \infty,$$

where the right-hand member is > 1 and represents the sum of the semi-axes of the ellipse with foci at ± 1 and passing through z.

2.4.2 The Laguerre Polynomials $L_n^\alpha(x)(\alpha > -1)$

1. **Differential Equation:**

$$xy'' + (\alpha + 1 - x)y' + ny = 0.$$

2. **Orthogonality Relation:**

$$\int_0^\infty e^{-x} x^\alpha L_n^\alpha(x) L_m^\alpha(x) dx = \frac{\Gamma(n+\alpha+1)}{\Gamma(n+1)} \delta_{m,n}.$$

3. **Recurrence Relations:**

   (a) **Three-Term Recurrence Relation:**

$$nL_n^\alpha(x) = (2n+\alpha-1-x)L_{n-1}^\alpha(x) - (n+\alpha-1)L_{n-2}^\alpha(x), \quad n = 2, 3, 4, \cdots$$

$$L_0^\alpha(x) = 1 \text{ and } L_1^\alpha(x) = \alpha + 1 - x.$$

   (b) **Recurrence Relation for the Derivative:**

$$\frac{d}{dx}\left\{L_n^\alpha(x)\right\} = -L_{n-1}^{\alpha+1}(x) = x^{-1}\left\{nL_n^\alpha(x) - (n+\alpha)L_{n-1}^\alpha(x)\right\}.$$

4. **Generating Function:**

$$\sum_{n=0}^\infty \frac{L_n^\alpha(x)}{\Gamma(n+\alpha+1)} w^n = e^w (xw)^{-\alpha/2} J_\alpha\left[2\sqrt{xw}\right].$$

5. **Rodrigues' Formula:**

$$L_n^\alpha(x) = \frac{x^{-\alpha} e^x}{n!} \frac{d^n}{dx^n}\left(e^{-x} x^{n+\alpha}\right).$$

6. **Inequalities and Asymptotic Properties:**
   Let c and w be fixed positive numbers. Then as $n \to \infty$

$$L_n^\alpha(x) = \begin{cases} x^{-\alpha/2-1/4}\, 0\left(n^{\alpha/2-1/4}\right) & \text{if } c/n \le x \le w \\ 0(n^\alpha) & \text{if } 0 \le x \le c/n \end{cases},$$

   and for $x > 0$

$$e^{-x/2} x^{\alpha/2} L_n^\alpha(x) = \frac{n^{\alpha/2-1/4}}{\sqrt{\pi} x^{1/4}} \cos\left[2\sqrt{nx} - \frac{\alpha\pi}{2} - \frac{\pi}{4}\right] + 0\left(n^{\alpha/2-3/4}\right)$$

   as $n \to \infty$ with the bound for the remainder holding uniformly in $[c, w]$.
   If z is in the complex plane cut along the non-negative real axis, then

$$\lim_{n\to\infty} \frac{\ln\left|L_n^\alpha(z)\right|}{n^{1/2}} = 2\, Re\left[(-z)^{1/2}\right],$$

   with $(-z)^{1/2}$ taken as real and positive if $z < 0$.

2.4.3 The Hermite Polynomials $H_n(x)$

1. **Differential Equation:**

$$y'' - 2xy' + 2ny = 0.$$

2. **Orthogonality Relation:**

$$\int_{-\infty}^{\infty} e^{-x^2} H_n(x) H_m(x) dx = \sqrt{\pi} 2^n n!\, \delta_{m,n}.$$

3. **Recurrence Relations:**

   (a) **Three-Term Recurrence Relation:**

$$H_n(x) = 2xH_{n-1}(x) - 2(n-1)H_{n-2}(x), \quad n = 2, 3, 4, \cdots$$

$$H_0(x) = 1 \text{ and } H_1(x) = 2x.$$

   (b) **Recurrence Relation for the Derivative:**

$$\frac{d}{dx}\{H_n(x)\} = 2nH_{n-1}(x),$$

   and

$$\frac{d}{dx}\{H_n(x)\} = 2xH_n(x) - H_{n+1}(x).$$

4. **Generating Function:**

$$\sum_{n=0}^{\infty} \frac{H_n(x)}{n!} w^n = \exp(2xw - w^2).$$

5. **Rodrigues' Formula:**

$$H_n(x) = (-1)^n e^{x^2} \frac{d^n}{dx^n}\left(e^{-x^2}\right).$$

6. **Inequalities and Asymptotic Properties:**

Let $N = 2n + 1$. Then for x real

$$e^{-x^2/2} H_n(x) = \frac{\Gamma(n+1)}{\Gamma\left(\frac{n}{2}+1\right)} \left\{ \cos\left(\sqrt{N}x - \frac{n\pi}{2}\right) + \frac{x^3}{6\sqrt{N}} \sin\left(\sqrt{N}x - \frac{n\pi}{2}\right) + 0\left(\frac{1}{n}\right)\right.$$

as $n \to \infty$. We also have

$$e^{-x^2/2} |H_{2n}(x)| \leq 2^{2n} n!\, (2 - g_n),$$

$$x^{-1}e^{-x^2/2}\left|H_{2n+1}(x)\right| \le 2^{2n+2}(n+1)!\; g_{n+1},$$

where

$$g_n = \frac{\left(\frac{1}{2}\right)_n}{n!} = \frac{1}{\sqrt{\pi n}} + 0\left(\frac{1}{n^{3/2}}\right).$$

For z in the complex plane cut along the real axis, we have

$$(2n)^{-1/2}\ln\left\{\frac{\Gamma\left(\frac{n}{2}+1\right)}{\Gamma(n+1)}\left|H_n(x)\right|\right\} \to \left|\mathrm{Im}\, z\right| \text{ as } n \to \infty.$$

2.5 The Bessel Functions

Using Frobenius' method to solve the differential equation

$$z^2y'' + zy' + (z^2 - \nu^2)y = 0, \qquad \nu \in \mathcal{R}$$

in a neighborhood of the regular singularity $z = 0$, leads to the solution

$$J_\nu(z) = \sum_{k=0}^{\infty} \frac{(-1)^k z^{\nu+2k}}{2^{\nu+2k}k!\Gamma(\nu+k+1)},$$

and $J_{-\nu}$.

1. **Definition:** The first solution $J_\nu(z)$ is called the Bessel function of the first kind and order ν.
   If ν is not an integer J_ν and $J_{-\nu}$ are independent. If ν is a negative integer, the factor $1/\Gamma(\nu+k+1)$ must be replaced by 0 whenever $\nu+\alpha+1 \le 0$. For all values of ν, $z^{-\nu}J_\nu(z)$ is an even entire function in z.
   We introduce other functions that are intimately related to the Bessel function

$$Y_\nu(z) = \frac{1}{(\sin\nu\pi)}\left\{J_\nu(z)\cos\nu\pi - J_{-\nu}(z)\right\}, \quad \nu \ne -1, -2, \cdots,$$

$$H_\nu^{(1)}(z) = J_\nu(z) + iY_\nu(z) = \frac{1}{(i\sin\pi\nu)}\left\{J_{-\nu}(z) - J_\nu(z)e^{-i\nu\pi}\right\},$$

$$H_\nu^{(2)}(z) = J_\nu(z) - iY_\nu(z) = \frac{1}{(i\sin\pi\nu)}\left\{J_\nu(z)e^{i\nu\pi} - J_{-\nu}(z)\right\},$$

$$I_\nu(z) = e^{-i\pi\nu/2}J_\nu\left(e^{i\pi/2}z\right) = \sum_{k=0}^{\infty}\frac{z^{\nu+2k}}{2^{\nu+2k}k!\Gamma(\nu+k+1)},$$

and

$$K_\nu(z) = \frac{\pi}{(2\sin\pi\nu)}\left\{I_{-\nu}(z) - I_\nu(z)\right\}.$$

Y_ν is called the *Bessel function of the second kind* or *Neumann's function*, $H_\nu^{(1)}$ and $H_\nu^{(2)}$ are the *Bessel functions of the third kind* or *the Hankel functions*, I_ν and $I_{-\nu}$ are the *modified Bessel functions of the first kind*, and K_ν is the *modified Bessel function of the third kind* or *MacDonald's function.*

2. **Differential Equation:**

The functions $y_1 = x^a J_\nu(bx^c)$ and $y_2 = x^a Y_\nu(bx^c)$ are solutions of the differential equation

$$y'' - (\frac{2a-1}{x})y' + \left(b^2c^2x^{2c-2} + \frac{a^2-\nu^2c^2}{x^2}\right)y = 0\,,\ bx^c > 0.$$

3. **Differentiation and Recurrence Relations:**

We list some differentiation formulas and recurrence relations for the Bessel function of the first kind; similar formulas and relations for the other types of the Bessel functions can be obtained:

$$\frac{d}{dz}\left[z^\nu J_\nu(z)\right] = z^\nu J_{\nu-1}(z), \tag{2.10}$$

$$\frac{d}{dz}\left[z^{-\nu} J_\nu(z)\right] = -z^{-\nu} J_{\nu+1}(z), \tag{2.11}$$

and hence by repeated differentiation, we have

$$\left(\frac{1}{z}\frac{d}{dz}\right)^m \left[z^\nu J_\nu(z)\right] = z^{\nu-m} J_{\nu-m}(z), \tag{2.12}$$

and

$$\left(\frac{1}{z}\frac{d}{dz}\right)^m \left[z^{-\nu} J_\nu(z)\right] = (-1)^m z^{-\nu-m} J_{\nu+m}(z). \tag{2.13}$$

Moreover, we have

$$2J'_\nu(z) = J_{\nu-1}(z) - J_{\nu+1}(z), \tag{2.14}$$

$$J'_\nu(z) = J_{\nu-1}(z) - \frac{\nu}{z}J_\nu(z) \tag{2.15}$$

and

$$J'_\nu(z)J_{-\nu}(z) - J'_{-\nu}(z)J_\nu(z) = \frac{2\sin\nu\pi}{\pi z},\ (\nu \text{ is non-integer}).$$

The three-term recurrence relation is

$$2\nu J_\nu(z) = z\left[J_{\nu-1}(z) + J_{\nu+1}(z)\right]. \tag{2.16}$$

4. **Integral Representations (Poisson's Integral Representation):**

$$J_\nu(z) = \frac{z^\nu}{2^\nu\sqrt{\pi}\Gamma\left(\nu+\frac{1}{2}\right)}\int_{-1}^{1}(1-t^2)^{\nu-1/2}e^{izt}dt$$
$$= \frac{z^\nu}{2^{\nu-1}\sqrt{\pi}\Gamma(\nu+1/2)}\int_0^{\pi/2}\cos(z\sin\theta)(\cos\theta)^{2\nu}d\theta, \quad (2.17)$$

for $\operatorname{Re} v > -\frac{1}{2}$. Moreover, we have

$$J_n(z) = \frac{1}{2\pi}\int_\alpha^{\alpha+2\pi} e^{i(n\theta - z\sin\theta)}d\theta\,, \quad [242, \text{p. } 20] \quad (2.18)$$

and

$$K_\mu(z) = \frac{\sqrt{\pi}(z/2)^\mu}{\Gamma(\mu+1/2)}\int_0^\infty e^{-z\cosh t}(\sinh t)^{2\mu}\,dt. \quad [242, \text{p. } 172] \quad (2.19)$$

5. **Generating Function:**
There is also a generating function for the Bessel function of the first kind of integer order, namely,

$$\sum_{k=-\infty}^{\infty} J_k(z)t^k = \exp\left[\frac{z}{2}\left(t-\frac{1}{t}\right)\right].$$

6. **Rodrigues' Formula:**
A Rodrigues-type formula for MacDonald's function of order $n+\frac{1}{2}$ is

$$K_{n+\frac{1}{2}}(z) = (-1)^n\sqrt{\frac{\pi}{2z}}z^{n+1}\left(\frac{1}{z}\frac{d}{dz}\right)^n\frac{e^{-z}}{z}.$$

7. **Relations with Other Special Functions:**
Below is a list of some important relations between the Bessel functions of different types and other special functions:

   (a) $J_{-\frac{1}{2}}(z) = \sqrt{2/(\pi z)}\cos z$ and $J_{\frac{1}{2}}(z) = \sqrt{2/(\pi z)}\sin z$,

   (b) $J_\nu(z) = \left(2^\nu\Gamma(\nu+1)\right)^{-1}z^\nu e^{-iz}\,{}_1F_1\left(\nu+\frac{1}{2};2\nu+1;2iz\right)$
   $= \left(2^\nu\Gamma(\nu+1)\right)^{-1}z^\nu\,{}_0F_1\left(\nu+1;-\frac{z^2}{4}\right).$

   (c) $K_\nu(z) = \sqrt{\pi/(2z)}W_{0,\nu}(2z)$, in particular, $K_{\frac{1}{2}}(z) = \sqrt{\pi/(2z)}e^{-z}$.

   (d) $\operatorname{Ai}(z) = (1/\pi)\sqrt{z/3}K_{\frac{1}{3}}\left(2z^{3/2}/3\right), |\arg z| < \frac{2\pi}{3}$, where $\operatorname{Ai}(z)$ is the Airy function, which is a solution of the differential equation

$$y'' - zy = 0.$$

This leads to the integral representation

$$\sqrt{\frac{z}{3}}K_{\frac{1}{3}}\left(2z^{\frac{3}{2}}\right) = \int_0^\infty \cos(t^3 + 3tz)dt, \qquad Re\ z > 0$$

(e) **Product Formula:**

$$J_\nu(z)J_\mu(z) = \frac{(z/2)^{\nu+\mu}}{\Gamma(\nu+1)\Gamma(\mu+1)} \times$$
$${}_2F_3\left(\frac{\nu+\mu+1}{2}, \frac{\nu+\mu+2}{2}; 1+\nu, 1+\mu, 1+\nu+\mu; -z^2\right)$$

8. **Asymptotic Expansions:**
For ν real and z small, we have the following estimates

$$J_\nu(z) \sim \frac{z^\nu}{2^\nu\Gamma(\nu+1)}, \text{ as } z \to 0^+, \quad \nu \neq -1, -2, \cdots,$$

$$Y_\nu(z) \sim -\frac{2^\nu\Gamma(\nu)}{\pi}\frac{1}{z^\nu} \text{ as } z \to 0^+, \quad \nu > 0,$$

$$Y_0(z) \sim \frac{2}{\pi}\ln z \text{ as } z \to 0^+,$$

$$I_\nu(z) \sim \frac{z^\nu}{2^\nu\Gamma(1+\nu)} \text{ as } z \to 0^+, \quad \nu > 0,$$

$$K_\nu(z) \sim \frac{\Gamma(\nu)}{2^{1-\nu}}\frac{1}{z^\nu} \text{ as } z \to 0^+, \quad \nu > 0,$$

$$K_0(z) \sim -\ln z \text{ as } z \to 0^+,$$

while for large z we have

$$J_\nu(z) \sim \sqrt{\frac{2}{\pi z}}\cos\left[z - \frac{\left(\nu+\frac{1}{2}\right)\pi}{2}\right], \qquad z \to \infty,$$

$$Y_\nu(z) \sim \sqrt{\frac{2}{\pi z}}\sin\left[z - \frac{\left(\nu+\frac{1}{2}\right)\pi}{2}\right], \qquad z \to \infty,$$

$$I_\nu(z) \sim \frac{e^z}{\sqrt{2\pi z}}, \quad z \to \infty, \quad \nu \geq 0$$

and

$$K_\nu(z) \sim \sqrt{\frac{\pi}{2z}}e^{-z}, \quad z \to \infty, \quad \nu \geq 0.$$

9. **Results Related to the Zeros of the Bessel Function:**
The zeros of the Bessel function $J_\nu(z)$ (ν real), i.e., the solutions of the equation $J_\nu(z) = 0$, are all simple, except possibly the zero at the origin. If ν is an integer n, J_n has no complex zeros and has an infinite number of real zeros which are symmetrically located about the origin. The zero at the origin is of order n. But if ν is an arbitrary real number and $|\arg z| < \pi$, then J_ν has an infinite number of positive real zeros, which will be denoted by $\alpha_{\nu,n}(n = 1, 2, 3, \cdots)$ with $\alpha_{\nu,1} < \alpha_{\nu,2} < \cdots$, and a finite number of conjugate complex zeros.

Since $z^{-\nu}J_\nu(z)$ is an entire function in z, it can be expressed in terms of the canonical product of its zeros as suggested by the Hadamard factorization theorem for entire functions (Theorem 1.2). In fact, we have

$$J_\nu(z) = \frac{z^\nu}{2^\nu\Gamma(\nu+1)} \prod_{k=1}^{\infty}\left(1 - \frac{z^2}{\alpha_{\nu,k}^2}\right).$$

A very useful property of the Bessel function is the orthogonality relation

$$\int_0^b xJ_\nu(\alpha_{\nu,m}x/b)J_\nu(\alpha_{\nu,n}x/b)dx = \frac{b^2}{2}\left[J_{\nu+1}(\alpha_{\nu,n})\right]^2 \delta_{m,n};\ m, n = 1, 2, 3, \cdots,$$

where $b > 0,\ \nu > -1$.

10. **Other Related Functions:**

(a) **Lommel's Function:**
Lommel's function, $s_{\mu,\nu}(z)$ is a solution of the inhomogeneous Bessel differential equation

$$z^2W'' + zW' + (z^2 - \nu^2)W = z^{2\mu+1}.$$

It is defined by

$$s_{\mu,\nu}(z) = \frac{z^{\mu+1}}{(\mu-\nu+1)(\mu+\nu+1)}\, {}_1F_2(1; \frac{\mu-\nu+3}{2}, \frac{\mu+\nu+3}{2}; -\frac{z^2}{4}).$$

(b) **Struve's function:**
Struve's function, $\mathbf{H}_\nu(z)$ is defined by

$$\mathbf{H}_\nu(z) = \frac{2}{\Gamma(\nu+3/2)\sqrt{\pi}}(z/2)^{\nu+1}\, {}_1F_2(1\,; 3/2+\nu\,, 3/2\,; -\frac{z^2}{4}).$$

It follows from the definition that $(z/2)^{-\nu}\mathbf{H}_\nu(z)$ is an entire function of ν and z. Moreover,

$$\frac{d}{dz}z^\nu\mathbf{H}_\nu(z) = z^\nu\mathbf{H}_{\nu-1}(z),$$

and

$$\mathbf{H}_{1/2}(z) = (\pi z/2)^{-1/2}(1 - \cos z).$$

It is related to Lommel's function by

$$\mathbf{H}_\nu(z) = \frac{2^{1-\nu}}{\sqrt{\pi}\Gamma(\nu + 1/2)} s_{\nu,\nu}(z).$$

An integral representation for Struve's function is

$$\mathbf{H}_\nu(z) = \frac{2}{\Gamma(\nu + 1/2)\sqrt{\pi}}(z/2)^\nu \int_0^1 (1 - t^2)^{\nu-1/2} \sin(zt)\, dt,$$

where Re $\nu > -1/2$.

2.6 The E-Function and Meijer's G-Function

2.6.1 MacRobert's E-Function

We have seen in Section 2.3.1 that the generalized hypergeometric function ${}_pF_q$ is not well defined when $p > q + 1$. An attempt to extend the definition of ${}_pF_q$ to the case where $p > q+1$ was made by T.M. MacRobert in the late 1930s. He introduced the following definition of the E-function, which is usually called *MacRobert's E-function.*

1. **Definition:** For $p \leq q + 1$, we have

$$E(p; a_r : q; b_s : z) = E(a_1, \cdots, a_p; b_1, \cdots, b_q; z)$$

$$= \frac{\prod_{i=1}^{p} \Gamma(a_i)}{\prod_{i=1}^{q} \Gamma(b_i)} \, {}_pF_q\left(a_1, \cdots, a_p; b_1, \cdots, b_q; -\frac{1}{z}\right)$$

where $z \neq 0$ if $p < q$ and $|z| > 1$ if $p = q + 1$. For $p \geq q + 1$, we have

$$E(p; a_r : q; b_s : z) = \sum_{k=1}^{p} \frac{\prod_{i=1}^{p\prime} \Gamma(a_i - a_k)}{\prod_{i=1}^{q} \Gamma(b_i - a_k)} \Gamma(a_k) z^{a_k} \times {}_{q+1}F_{p-1}$$

$$\left[a_k, a_k - b_1 + 1, \cdots, a_k - b_q + 1; a_k - a_1, \cdots, *, \cdots, a_k - a_p + 1; (-1)^{p+q} z\right]$$

where $|z| < 1$ if $p = q + 1$.

The prime on the product sign indicates the omission of the factor $\Gamma(a_k - a_k)$ and the asterisk, $*$, in F indicates the omission of the term $a_k - a_k + 1$. An empty product is replaced by 1 and non-positive integer values of the a_k parameters are excluded.

We are not going to discuss the E-function any further since it is a special case of Meijer's G-function that will be introduced in the next section. In fact

$$E(a_1, \cdots, a_p; b_1, \cdots, b_q; z) = G^{p,1}_{q+1,p}\left(z \,\middle|\, \begin{matrix} 1, b_1, \cdots, b_q \\ a_1, \cdots, a_p \end{matrix}\right).$$

2.6.2 Meijer's G-Function

Another extension of the hypergeometric function ${}_pF_q$ to the case where $p > q + 1$ is provided by *Meijer's G-function*, which we shall call the G-function for short.

1. **Definition:** We define

$$G^{m,n}_{p,q}\left(z \,\middle|\, \begin{matrix} a_1, \cdots, a_p \\ b_1, \cdots, b_q \end{matrix}\right) =$$

$$= \frac{1}{2\pi i}\int_\gamma \frac{\prod_{j=1}^{m}\Gamma(b_j - s)\prod_{j=1}^{n}\Gamma(1 - a_j + s)}{\prod_{j=m+1}^{q}\Gamma(1 - b_j + s)\prod_{j=n+1}^{p}\Gamma(a_j - s)} z^s ds,$$

where $a_1, \cdots, a_p, b_1, \cdots, b_q$ are constants and $0 \le m \le q$, $0 \le n \le p$. In the above formula, an empty product is interpreted as 1, and the parameters are such that no pole of $\Gamma(b_j - s), j = 1, \ldots, m$, coincides with any pole of $\Gamma(1 - a_k + s), k = 1, \ldots, n$ and γ is the integration path. We shall often denote this function by

$$G^{m,n}_{p,q}\left(z \,\middle|\, \begin{matrix} a_r \\ b_s \end{matrix}\right) \text{ or simply } G^{m,n}_{p,q}(z)$$

for short. By choosing the path of integration appropriately, one can show that if no two $b_j (j = 1, \ldots, m)$ differ by an integer

$$G^{m,n}_{p,q}\left(z \,\middle|\, \begin{matrix} a_r \\ b_s \end{matrix}\right) = \sum_{k=1}^{m} \frac{\prod_{i=1}^{m\prime}\Gamma(b_i - b_k)\prod_{i=1}^{n}\Gamma(1 + b_k - a_i)}{\prod_{i=m+1}^{q}\Gamma(1 + b_k - b_i)\prod_{i=n+1}^{p}\Gamma(a_i - b_k)} z^{b_k} \times$$

$${}_pF_{q-1}\left[1 + b_k - a_1, \cdots, 1 + b_k - a_p;\right.$$
$$\left.1 + b_k - b_1, \cdots, *, 1 + b_k - b_q; (-1)^{p-m-n} z\right],$$

where either $p < q$ or $p = q$ and $|z| < 1$. The prime on the product sign denotes the omission of the term when $i = k$, and the $*$ in the parameter set of ${}_pF_{q-1}$ denotes the omission of the term $1 + b_k - b_k$. Similarly, if no two $a_k, k = 1, \cdots, n$, differ by an integer, we have

$$G_{p,q}^{m,n}\left(z \middle| \begin{matrix} a_r \\ b_s \end{matrix}\right) = \sum_{k=1}^{n} \frac{\prod\limits_{i=1}^{n\prime} \Gamma(a_k - a_i) \prod\limits_{i=1}^{m} \Gamma(b_i - a_k + 1)}{\prod\limits_{i=n+1}^{p} \Gamma(a_i - a_k + 1) \prod\limits_{i=m+1}^{q} \Gamma(a_k - b_i)} z^{a_k - 1} \times$$

$${}_qF_{p-1}\,[1 + b_1 - a_k, \cdots, 1 + b_q - a_k\ ;$$

$$1 + a_1 - a_k, \cdots, *, \cdots, 1 + a_p - a_k; (-1)^{q-m-n} z^{-1},]$$

where $q < p$ or $q = p$ and $|z| > 1$.

2. **Properties and Identities:**
The G-function is analytic in z and symmetric in the parameters $a_1, \cdots, a_n$, likewise in $a_{n+1}, \cdots, a_p$, in $b_1, \cdots, b_m$ and in $b_{m+1}, \cdots, b_q$. If one of the parameters a_i(for $i = 1, 2, \ldots, n$) coincides with one of the parameters b_i(for $i = m+1, m+2, \cdots, q$), or one of the b_i(for $i = 1, 2, \cdots, m$) coincides with one of the a_i(for $i = n+1, \cdots, p$), then the order of the G-function decreases. For example

$$G_{p,q}^{m,n}\left(z \middle| \begin{matrix} a_1, & \cdots, & a_p \\ b_1, & \cdots, & b_{q-1}, a_1 \end{matrix}\right) = G_{p-1,q-1}^{m,n-1}\left(z \middle| \begin{matrix} a_2, & \cdots, & a_p \\ b_1, & \cdots, & b_q \end{matrix}\right),$$

$n, p, q \geq 1$.
The G-function with $p > q$ can be transformed into the G-function with $p < q$ by means of the relation

$$G_{p,q}^{m,n}\left(\frac{1}{z} \middle| \begin{matrix} a_r \\ b_s \end{matrix}\right) = G_{q,p}^{n,m}\left(z \middle| \begin{matrix} 1 - b_s \\ 1 - a_r \end{matrix}\right).$$

Multiplication of the G-function by powers of z leads to the relation

$$z^\sigma G_{p,q}^{m,n}\left(z \middle| \begin{matrix} a_r \\ b_s \end{matrix}\right) = G_{p,q}^{m,n}\left(z \middle| \begin{matrix} a_r + \sigma \\ b_s + \sigma \end{matrix}\right),$$

and differentiation yields

$$z\frac{d}{dz} G_{p,q}^{m,n}(z|_{b_s}^{a_r}) = G_{p,q}^{m,n}\left(z|_{b_1,\ldots,b_q}^{a_1-1,a_2,\ldots,a_p}\right) + (a_1 - 1) G_{p,q}^{m,n}(z|_{b_s}^{a_r}); n \geq 1.$$

Other properties of the G-function are listed below:

3. **Differential Equation:**
$G_{p,q}^{m,n}$ satisfies the differential equation

$$\left\{(-1)^{p-m-n} z \prod_{j=1}^{p}\left(z\frac{d}{dz} - a_j + 1\right) - \prod_{j=1}^{q}\left(z\frac{d}{dz} - b_j\right)\right\} y = 0$$

$(p \leq q)$.

4. **Relations with Other Special Functions:**

(a) ${}_pF_q(a_1, \cdots, a_p; b_1, \cdots, b_q; z)$

$$= \frac{\prod_{i=1}^{q} \Gamma(b_i)}{\prod_{i=1}^{p} \Gamma(a_i)} G_{p,q+1}^{1,p}\left(-z \,\middle|\, \begin{matrix} 1-a_1, & \cdots, & 1-a_p & \\ 0, & 1-b_1, & \cdots, & 1-b_q \end{matrix}\right).$$

(b) $z^\mu J_\nu(z) = 2^\mu G_{0,2}^{1,0}\left(\frac{z^2}{4} \,\middle|\, \frac{\nu+\mu}{2}, \frac{\mu-\nu}{2}\right)$.

(c) $z^\mu K_\nu(z) = 2^{\mu-1} G_{0,2}^{2,0}\left(\frac{z^2}{4} \,\middle|\, \frac{\nu+\mu}{2}, \frac{\mu-\nu}{2}\right)$.

(d) $K_\nu(z) = \sqrt{\pi} e^z G_{1,2}^{2,0}\left(2z \,\middle|\, {}^{\frac{1}{2}}_{\nu,-\nu}\right)$.

(e) $W_{k,m}(z) = z^{(1-b-c)/2}\, e^{z/2}\, G_{1,2}^{2,0}\left(z \,\middle|\, {}^{a}_{b,c}\right)$,
where $k = \frac{1}{2}(1+b+c) - a,\ m = \frac{1}{2}(b-c)$.

5. **Asymptotic Expansions:**
For $p \leq q$,

$$G_{p,q}^{m,n}(x) = 0\left(|x|^\beta\right) \text{ as } x \to 0,$$

where $\beta = \max\{\operatorname{Re} b_k\}$ for $k = 1, 2, \ldots, m,$.
The asymptotic expansion of the G-function as $z \to \infty$ is complicated; therefore, we give only some important special cases.
As $z \to \infty$, the G-function vanishes exponentially if $p < q$, $m > \frac{1}{2}(p+q), n = 0, |\arg z| < \left(m - \frac{1}{2}(p+q)\right)\pi$, and is of the order of some power of z if $p < q, n \geq 1, m+n > \frac{1}{2}(p+q)$ and $|\arg z| < (m+n-\frac{1}{2}(p+q))\pi$.

3

Generalized Functions

In Chapter 1, we emphasized the distinction between conventional functions and generalized functions. By a conventional function, we mean a function whose domain is a subset of $\mathcal{R}^n(\mathcal{C}^n)$ and whose range is an arbitrary set, but for all practical purposes it is usually taken as a subset of $\mathcal{R}^m$ or $\mathcal{C}^m$. Generalized functions, on the other hand, are not functions in the classical sense, but rather they are mathematical objects. The nature of these objects depends on the approach one uses to construct them.

There are different approaches to introduce generalized functions. In one of them, due to G. Temple [222, 223], M. Lighthill [148], and D. Jones [117], generalized functions are simply sequences of conventional functions satisfying certain conditions; in another approach, due to M. Sato [211, 212] and J. Colombeau [44], generalized functions are elements of a quotient module of rings of very smooth functions. In a third approach, generalized functions are constructed from continuous functions in a similar way to that in which the real numbers are constructed from the rationals. This approach was first introduced by J. Mikusinski [163] and later refined by T. Boehme [20], J. Mikusinski and P. Mikusinski [164]. In the approach we shall introduce in this chapter, which is due to L. Schwartz [214], I. Gelfand and G. Shilov [70], and A. Zemanian [276, 278], generalized functions are continuous linear functionals acting on some locally-convex space of smooth functions, i.e., generalized functions are elements of a dual space. Therefore, before we introduce generalized functions, we need to introduce the underlying locally-convex spaces, which will be called testing-function spaces.

3.1 Countably-Normed Spaces and Their Duals

3.1.1 Countably-Normed Spaces

To introduce the concept of a testing-function space, first we need to introduce the definition of a seminorm.

DEFINITION 3.1 (**seminorm**) *A function γ from a linear vector space X into $\mathcal{R}, \gamma : X \to \mathcal{R}$ is called a seminorm if*

1. $\gamma(\alpha x) = |\alpha|\,\gamma(x)$, for $\alpha \in \mathcal{C}, x \in X$,

2. $\gamma(x + y) \leq \gamma(x) + \gamma(y)$, for all $x, y \in X$.

Note that since $\gamma(0) = 0$, it follows that $\gamma(x) \geq 0$ for all x. For, $0 = \gamma(x - x) \leq \gamma(x) + \gamma(-x) = 2\gamma(x)$. Every norm is clearly a seminorm, but the converse is not true, since it is possible that $\gamma(x) = 0$ for some $x \neq 0$. If $\gamma_1, \cdots, \gamma_n$ are seminorms, then so are $a\gamma_i (a > 0), i = 1, \ldots, n; \gamma_1 + \cdots + \gamma_n$, and $\max\{\gamma_1, \ldots, \gamma_n\}$. Two seminorms γ_1 and γ_2 are *equivalent* if there exist two positive numbers a and b such that $a\gamma_1(x) \leq \gamma_2(x) \leq b\gamma_1(x)$ for all $x \in X$.

DEFINITION 3.2 *Let $S = \{\gamma_\alpha\}_{\alpha \in A}$ be a family of seminorms defined on a linear vector space X, where the index α runs over a finite or an infinite set A. The family S is said to be separating if for any $x \in X, x \neq 0$, there exists at least one seminorm γ_{α_0} such that $\gamma_{\alpha_0}(x) \neq 0$.*

Equivalently, S is separating if and only if $\gamma_\alpha(x_0) = 0$ for all $\alpha \in A$ implies that $x_0 = 0$. A sufficient condition for S to be separating is that one of the seminorms is actually a norm.

We have assumed that the reader is familiar with the notion of a linear vector space, and therefore we did not make any notational distinction between the zero vector, $x = 0 \in X$, and the number $\gamma(x) = 0 \in \mathcal{R}$.

DEFINITION 3.3 *Let $S = \{\gamma_\alpha\}_{\alpha \in A}$ be an infinite set of seminorms defined on a linear vector space X. If S is separating, we say that X is an infinitely normed space and we may denote it by (X, S) to stress the dependence on S. If the index set A is countable, i.e., $S = \{\gamma_n\}_{n=1}^{\infty}$, we shall say that X is a countably normed space.*

Since the seminorms are assumed to be separating, they, as a collection, behave more or less like norms; therefore, it is not misleading to call a space an infinitely normed space instead of an infinitely seminormed space.

DEFINITION 3.4 *A sequence* $\{x_n\}_{n=1}^{\infty}$ *in an infinitely normed space* (X, S) *is said to be a Cauchy sequence if for any* $\gamma \in S$, *we have*

$$\lim_{m,n \to \infty} \gamma(x_m - x_n) = 0,$$

and is said to converge to x_0 *if* $x_0 \in X$ *and for any* $\gamma \in S$, *we have*

$$\lim_{n \to \infty} \gamma(x_n - x_0) = 0.$$

Because S is separating, the limit of any convergent sequence is unique.

DEFINITION 3.5 *An infinitely normed space* (X, S) *is said to be complete if every Cauchy sequence in* X *is convergent and a complete countably normed space is called a Fréchet space.*

Every countably normed space is a locally convex topological vector space. The reader who is familiar with the theory of locally convex topological vector spaces has probably recognized by now that we have only defined sequential convergence and completeness. Admittedly, these are not the most general notions of convergence and completeness in a locally convex topology. Nevertheless, they are sufficient for our forthcoming introduction of generalized functions.

It is worth noting that the mere definition of convergence of sequences in a vector space gives rise to a topology on that space by declaring, for example, that a set A is *closed* if and only if every convergent sequence $\{x_n\}_{n=1}^{\infty}$ in A converges to an element in A, i.e., if $\lim_{n \to \infty} x_n = x_0$, then $x_0 \in A$. In other words, A is closed if and only if it contains the limits of all its convergent sequences. A set A is *compact* if and only if every sequence in A contains a convergent subsequence. *Open sets* are defined as the complements of closed sets, etc.

Let S and T be two families of separating seminorms defined on a linear vector space X. We say that the topology generated by S is *stronger* (more precisely, not weaker) than the topology generated by T if and only if every sequence convergent with respect to S is convergent with respect to T. In this case, we write $T \leq S$. We also say that the two topologies generated by S and T are *equivalent*, or simply, S and T are equivalent if and only if $S \leq T$ and $T \leq S$. In other words, S and T are equivalent if and only if they have exactly the same convergent sequences. The following theorem gives a characterization of the statement that $T \leq S$.

THEOREM 3.1
Let $S = \{\gamma_\alpha\}_{\alpha \in A}$ and $T = \{\rho_\beta\}_{\beta \in B}$ be two families of seminorms defined on a linear vector space X. S is stronger than T if and only if for every $\rho_\beta \in T$ there exists a finite set of seminorms $\gamma_1, \ldots, \gamma_n \in S$ such that for all $x \in X$,

$$\rho_\beta(x) \leq C_1 \max\{\gamma_1(x), \cdots, \gamma_n(x)\},$$

or alternatively

$$\rho_\beta(x) \leq C_2 [\gamma_1(x) + \cdots + \gamma_n(x)],$$

where C is a positive constant. Both C and n depend on ρ_β.

3.1.2 The Dual Spaces of Countably-Normed Spaces

Now we turn our attention to continuous linear operators and transformations on countably normed spaces.

Let (X, S) and (Y, T) be two countably normed spaces with $S = \{\gamma_n\}_{n=0}^{\infty}$ and $T = \{\rho_n\}_{n=0}^{\infty}$. By a linear operator $\mathcal{L}$ from (X, S) into (Y, T) we mean a mapping $\mathcal{L} : X \to Y$ such that

$$\mathcal{L}(\alpha x_1 + \beta x_2) = \alpha \mathcal{L}(x_1) + \beta \mathcal{L}(x_2); \alpha, \beta \in \mathcal{C},\ x_1, x_2 \in X.$$

If $Y = X, \mathcal{L}$ is called a *linear transformation* on X.

In general, the domain of $\mathcal{L}$ may not be all of X and likewise the range of $\mathcal{L}$ may not be all of Y. If any such case arises and makes a difference in the discussion, we shall make a note of it, but for the time being, we assume that the domain of $\mathcal{L}$, denoted by $Dom\mathcal{L} = D(\mathcal{L})$, is equal to X.

DEFINITION 3.6 *Let (X, S) and (Y, T) be two countably normed vector spaces, and let $\mathcal{L}$ be a linear operator from X into Y. We say that $\mathcal{L}$ is continuous at $x_0 \in X$ if $\lim_{n\to\infty} x_n = x_0$ in X implies that*

$$\lim_{n\to\infty} \mathcal{L}(x_n) = \mathcal{L}(x_0) \text{ in } Y.$$

The operator $\mathcal{L}$ is said to be continuous if it is continuous at every point in X. The vector space of all continuous linear operators from X into Y will be denoted by $C(X, Y)$ and if $Y = X, C(X, X)$ will be called the space of all continuous linear transformations on X and will be denoted by $C(X)$.

Since $x_n \to x_0$ is equivalent to $x_n - x_0 \to 0$, for $\mathcal{L}$ to be continuous it is necessary and sufficient that $\mathcal{L}$ be continuous at the origin. It is easy to obtain an analogue of Theorem 3.1 for linear operators from X into Y. In fact, a necessary and sufficient condition for $\mathcal{L}$ to be continuous from (X, S)

into (Y, T) is that for each $\rho_n \in T$, there exists a positive constant C and a positive integer m such that for all $x \in X$

$$\rho_n(\mathcal{L}(x)) \leq C \max\{\gamma_1(x), \ldots, \gamma_m(x)\}$$

or alternatively

$$\rho_n(\mathcal{L}(x)) \leq C [\gamma_1(x) + \cdots + \gamma_m(x)],$$

where C and m depend on n and not on x.

If $\mathcal{L}$ is one-to-one, then its inverse, $\mathcal{L}^{-1}$, exists and is linear; see Section 1.5. In addition, if $\mathcal{L}$ is onto, i.e., $\mathcal{L}(X) = Y$, then $\mathcal{L}^{-1}$ is a linear transformation from Y into X. Two countably normed spaces (X, S) and (Y, T) are called *isomorphic* if there exists a one-to-one continuous linear operator $\mathcal{L}$ from X onto Y such that $\mathcal{L}^{-1}$ is a continuous linear operator from Y into X. In this case, we say that $\mathcal{L}$ is an isomorphism from X onto Y, and if $X = Y$, we say that $\mathcal{L}$ is an *automorphism* on X.

DEFINITION 3.7 *A continuous linear operator from (X, S) into $\mathcal{R}$ is called a continuous linear functional on X and the vector space of all continuous linear functionals on $X, C(X, \mathcal{R})$, is called the dual space of X. The dual space of X will be denoted by X^*.*

Since the structure of the dual space depends on the seminorms S, it may be more accurate to designate the dual space by $(X, S)^*$; however, to simplify the matter we shall adhere to the notation X^*. This will cause no confusion because in most cases we shall be dealing with only one fixed family of seminorms on a given space X. If two non-equivalent families of seminorms on a space X are used, we shall then utilize the notation $(X, S)^*$ to distinguish between the corresponding dual spaces. Another word about the notation: although we have used the notation $f(x)$ to denote the number that the functional $f \in X^*$ assigns to the vector $x \in X$, it is customary to denote that number by $\langle f, x \rangle$ instead. This notation will be adopted throughout this book.

The next theorem gives a characterization for members of the dual space of a countably normed space.

THEOREM 3.2 [276, p. 12]
Let (X, S) be a countably normed space with $S = \{\gamma_n\}_{n=1}^{\infty}$ and at least one of the seminorms, say γ_1, is a norm. Let the countable family of seminorms $T = \{\rho_m\}_{m=1}^{\infty}$ be defined by

$$\rho_m = \max\{\gamma_1, \ldots, \gamma_m\}.$$

If $f \in X^*$, then there exist a positive number C and a non-negative integer k such that for every $x \in X$

$$|\langle f, x\rangle| \leq C\rho_k(x),$$

where C and k depend on f but not on x.

Let (X_1, S_1) be a countably normed space and $X_2 \subset X_1$. As a subset of X_1, X_2 inherits a topology from (X_1, S_1), namely, the topology generated by S_1. This makes (X_2, S_1) a countably normed space, or more precisely, a countably normed subspace of (X_1, S_1). If X_2 has its own topology that is generated by another countable family of separating semi-norms S_2 such that $S_1 \leq S_2$, then (X_2, S_2) is also a subspace of (X_1, S_1). If $\{x_n\}_{n=1}^{\infty}$ and x_0 belong to X_2 with $x_n \to x_0$ in (X_2, S_2) as $n \to \infty$, then $x_n \to x_0$ in (X_1, S_1) as n$\to \infty$.

DEFINITION 3.8 *Let (X_1, S_1) be a countably normed space with $X_2 \subset X_1$. By the restriction of $f \in (X_1, S_1)^*$ to X_2 we mean that unique linear functional g defined by*

$$\langle g, x\rangle = \langle f, x\rangle, \qquad \text{for every } x \in X_2.$$

The restriction of f to X_2 will be denoted by $f|_{X_2}$.

The restriction of $f \in (X_1, S_1)^*$ to X_2 gives rise to an element in $(X_2, S_1)^*$, or loosely speaking, we may say that the restriction of $f \in (X_1, S_1)^*$ is in $(X_2, S_1)^*$. But, if X_2 has its own topology which is generated by a family of seminorms S_2, then the restriction of $f \in (X_1, S_1)^*$ to X_2 is not in general in $(X_2, S_2)^*$. If the topology of X_2 generated by S_2 is stronger than that inherited from S_1, i.e., $S_1 \leq S_2$, then the restriction of $f \in (X_1, S_1)^*$ to X_2 is in $(X_2, S_2)^*$. For if $\{x_n\}_{n=1}^{\infty}$ is a sequence in X_2 converging to x_0 in the S_2 topology, it also converges to x_0 in the topology inherited from S_1 and since $f \in (X_1, S_1)^*$, it follows that $\{f(x_n)\}_{n=1}^{\infty}$ converges to $f(x_0)$.

However, even if we identify $f \in (X_1, S_1)^*$ with its restriction to X_2 when X_2 is provided with a stronger topology S_2 than the one it inherits from (X_1, S_1), still we cannot regard $(X_1, S_1)^*$ as a subspace of $(X_2, S_2)^*$. The reason is there may not be a one-to-one correspondence between $(X_1, S_1)^*$ and any subspace of $(X_2, S_2)^*$. This will occur, for example, if there exist two functionals f_1 and f_2 in $(X_1, S_1)^*$ with $f_1 \neq f_2$, but $f_1(x) = f_2(x)$ for all $x \in X_2$. On the other hand, if (X_2, S_2) is dense in (X_1, S_1), then any two functionals that agree on X_2 must agree on X_1. For if $x_0 \in X_1$, then there exists a sequence $\{x_n\}_{n=1}^{\infty}$ in X_2 such that $x_n \to x_0$ in S_2 and hence in S_1, and by continuity $f_1(x_n) = f_2(x_n) \to f_1(x_0) = f_2(x_0)$. We summarize these results in the following theorem.

THEOREM 3.3 [276, p. 21]
Let (X_1, S_1) and (X_2, S_2) be two countably normed spaces with $X_2 \subset X_1$ and $S_1 \leq S_2$. If (X_2, S_2) is dense in (X_1, S_1), then there is a one-to-one correspondence between $(X_1, S_1)^*$ and a subspace of $(X_2, S_2)^*$. This correspondence is the one that assigns to every $f \in (X_1, S_1)^*$ its restriction, $f|_{X_2}$, to X_2. If we identify f with its restriction $f|_{X_2}$, then $(X_1, S_1)^*$ is a subspace of $(X_2, S_2)^*$.

Evidently, if X is a countably normed space, then X^* is a linear vector space (cf. Section 1.5). Now we provide X^* with a family of seminorms so that X^* becomes an infinitely normed space. For any fixed $x \in X$, define the seminorm

$$\rho_x(f) = |\langle f, x\rangle|, \qquad f \in X^*.$$

The family $\{\rho_x\}_{x\in X}$ is a separating family of seminorms on X^*. For, if $f \neq 0$, there exists $x \in X$ such that $\langle f, x\rangle \neq 0$; hence $\rho_x(f) \neq 0$. It now follows from the definition that a sequence $\{f_n\}_{n=1}^{\infty}$ in X^* is a Cauchy sequence if

$$\lim_{m,n\to\infty} \langle f_m - f_n, x\rangle = 0, \qquad \text{for every } x \in X,$$

and $\{f_n\}_{n=1}^{\infty}$ converges to f_0 in X^* if

$$\lim_{n\to\infty} \langle f_n, x\rangle = \langle f_0, x\rangle \text{ for every } x \in X.$$

Since every $x \in X$ gives rise to a seminorm ρ_x on X^*, we may denote this family of seminorms on X^* by X, and obtain that (X^*, X) is an infinitely normed space. In general, (X^*, X) is not a countably normed space. The topology generated by X on X^* is called the *weak topology*. Unless otherwise stated, hereafter the topology of X^* will be the weak topology.

THEOREM 3.4
Let X be a complete countably normed space. Then its dual X^* (provided with the weak topology) is also complete.

3.1.3 Examples of Countably Normed Spaces

1. Let I be an open set in $\mathcal{R}^n$ and X be the set of all C^∞-functions on I that vanish outside some fixed compact subset K of I. That is $\phi \in X$ if and only if $\phi \in C^\infty(I)$ and $\operatorname{supp}\phi \subseteq K \subset I$. Define the seminorms

$$\gamma_m(\phi) = \sup_{\mathbf{x}\in K} |D^p\phi(\mathbf{x})|, \ |p| \leq m, \ m = 0, 1, 2, \ldots,$$

where $p = (p_1, \ldots, p_n)$ is a multi-index, $D^p = \frac{\partial^{|p|}}{\partial x_1^{p_1} \cdots \partial x_n^{p_n}}$, $|p| = p_1 + p_2 + \cdots + p_n$, and $\phi \in X$. It is easy to show that γ_0 is a norm

and $(X, \{\gamma_m\}_{m=0}^{\infty})$ is a complete countably normed space. This space will be denoted by $\mathcal{D}_K(I)$ and if $I = \mathcal{R}^n$, it will be simply denoted by $\mathcal{D}_K$.
The set X is not empty. For example, if K contains the closed unit ball $\bar{B}_1(0) = \{\mathbf{x} \in \mathcal{R}^n : |\mathbf{x}| \leq 1\}$ then

$$\phi(\mathbf{x}) = \begin{cases} \exp\left(1/(|\mathbf{x}|^2 - 1)\right), & \text{if } |\mathbf{x}| < 1 \\ 0, & \text{if } |\mathbf{x}| \geq 1 \end{cases},$$

is in X.

2. Let I be an open set in $\mathcal{R}^n$ and $\{K_m\}_{m=1}^{\infty}$ be a sequence of compact sets such that $K_1 \subset K_2 \subset \cdots$, and $I = \bigcup_{p=1}^{\infty} K_p$. Let X be the set of all C^∞-functions on I. Define the seminorms

$$\gamma_{m,p}(\phi) = \sup_{\mathbf{x} \in K_p} \left|D^k \phi(\mathbf{x})\right|, \ |k| \leq m, \quad m = 0, 1, 2, \cdots.$$

Then $\left(X, \{\gamma_{m,p}\}_{m=0,p=1}^{\infty,\infty}\right)$ is a complete countably normed space. This space is usually denoted by $\mathcal{E}(I)$ and when $I = \mathcal{R}^n$, we may write $\mathcal{E}$ instead of $\mathcal{E}(\mathcal{R}^n)$. It should be noted that the definition of $\mathcal{E}(I)$ is independent of the sequence $\{K_p\}_{p=1}^{\infty}$. In other words, if $\left\{\tilde{K}_p\right\}_{p=1}^{\infty}$ is another sequence of compact sets such that $\tilde{K}_1 \subset \tilde{K}_2 \subset \cdots$, $I = \bigcup_{p=1}^{\infty} \tilde{K}_p$, with

$$\tilde{\gamma}_{m,p}(\phi) = \sup_{\mathbf{x} \in \tilde{K}_p} \left|D^k \phi(\mathbf{x})\right|, |k| \leq m, \ m = 0, 1, 2, \cdots,$$

then the two families of seminorms $\{\gamma_{m,p}\}$ and $\{\tilde{\gamma}_{m,p}\}$ are equivalent.

3. Let $n = 1$ and $\mathcal{P}_t$ be that subspace of $\mathcal{E}$ consisting of all periodic functions ϕ with period T, i.e., $\phi(x + T) = \phi(x)$ for all x. It follows that a sequence $\{\phi_n\}_{n=1}^{\infty}$ in $\mathcal{P}_T$ converges to $\phi_0 \in \mathcal{P}_T$ if and only if $\lim_{n\to\infty} \phi_n^{(k)}(x) = \phi_0^{(k)}(x)$ uniformly on any compact set for $k = 0, 1, 2, \ldots$.

4. Let X be the set of all C^∞-functions ϕ on $\mathcal{R}^n$ such that $D^k\phi$, $|k| = 0, 1, 2, \cdots$, are rapidly decreasing. Let

$$\gamma_{m,|k|}(\phi) = \sup_{\mathbf{x} \in \mathcal{R}^n} \left|\left(1 + |\mathbf{x}|^2\right)^m D^k \phi(\mathbf{x})\right|, m, |k| = 0, 1, 2, \cdots.$$

Since $D^k\phi$ is rapidly decreasing for each $|k| = 0, 1, 2, \cdots, \gamma_{m,|k|}(\phi)$ is finite for each $m = 0, 1, 2, \cdots$, and because $\gamma_{0,0}$ is a norm, it follows that $S = \left\{\gamma_{m,|k|}\right\}_{m,|k|=0}^{\infty}$ is a countable set of separating seminorms.

Hence (X, S) is a countably normed space. This space, which will be denoted by $\mathcal{S}$, is called *the Schwartz space of rapidly decreasing functions* or *the Schwartz space of functions* for short.

5. Let G be an open subset of the Riemann sphere $\mathcal{O}$ and $\mathcal{H}(G)$ be the ring of locally holomorphic (analytic) functions on G. When provided with the topology of uniform convergence on compact subsets of G, i.e., the topology generated by the norms

$$\gamma_{K_n}(\phi) = \sup_{z \in K_n} |\phi(z)|, \quad \phi \in \mathcal{H}(G),$$

where $\{K_n\}_{n=1}^{\infty}$ is a sequence of compact subsets of G such that $K_1 \subset K_2 \subset \cdots$, and $G = \bigcup_{n=1}^{\infty} K_n, \mathcal{H}(G)$ becomes a Fréchet space which will also be denoted by $\mathcal{H}(G)$.

6. Let I be an interval, finite or infinite, and let $\eta(x, a, b)$ be a continuous function on I for some real parameters a and b such that $\eta \neq 0$ for any $x \in I$. Let

$$R = \theta_0 D^{n_1} \theta_1 D^{n_2} \theta_2 \cdots D^{n_p} \theta_{n_p},$$

where $D = \frac{d}{dx}, n_k$ is a non-negative integer and θ_k is an infinitely differentiable function on I with the property that θ_k and its derivatives of all orders are never zero on I, where $k = 0, 1, \cdots, n_p$.

We define the space $\mathcal{L}_{a,b}$ as follows: The space $\mathcal{L}_{a,b}$ consists of all complex-valued C^{∞}-functions $\phi(x)$ on I for which the seminorms

$$\gamma_{k,a,b}(\phi) = \sup_{x \in I} \left| \eta(x, a, b) R^k \phi(x) \right|, \quad k = 0, 1, 2, \cdots$$

assume finite values, where $R^0 = I$ (the identity), $R^k = R^{k-1} R$. For fixed a and b, we write $\gamma_k(\phi)$ instead of $\gamma_{k,a,b}(\phi)$. If η depends only on one parameter (say a), we write $\mathcal{L}_a$ for the underlying space. Clearly, $\mathcal{L}_{a,b}$ is a linear space under the pointwise addition of functions and multiplication by complex numbers. The collection $\{\gamma_k\}_{k=0}^{\infty}$ is a separating family of seminorms since each $\gamma_k (k = 1, 2, \ldots)$ is a seminorm and γ_0 is a norm.

3.2 Countable-Union Spaces and Their Duals.

3.2.1 Countable-Union Spaces

Let $\{(X_n, S_n)\}_{n=1}^{\infty}$ be a sequence of countably normed spaces such that

1. $X_1 \subset X_2 \subset X_3 \subset \cdots$,
2. the topology of X_n is stronger than the topology it inherits from X_{n+1}, that is $S_{n+1} \leq S_n$ for all n.

Let $X = \bigcup_{n=1}^{\infty} X_n$. Then X is easily seen to be a linear vector space. We now provide X with a topology. It is the topology generated by the following notion of convergence. A sequence $\{x_n\}_{n=1}^{\infty}$ is said to converge to x_0 in X if there exists a positive integer m such that x_0 and $x_n (n = 1, 2, \cdots)$ are all in X_m and $x_n \to x_0$ in X_m as $n \to \infty$. Since $S_{m+i} \leq S_m, i = 1, 2, \cdots$, it follows that $x_n \to x_0$ in X_{m+i} for all $i = 1, 2, \cdots$.

DEFINITION 3.9 *Let $\{(X_n, S_n)\}_{n=1}^{\infty}$ be a sequence of countably normed spaces satisfying the above two conditions. Then $X = \bigcup_{n=1}^{\infty} X_n$, provided with the aforementioned topology, is called a countable-union space.*

A sequence $\{x_n\}_{n=1}^{\infty}$ in X is said to be *Cauchy* if there exists a positive integer m such that $\{x_n\}_{n=1}^{\infty}$ is a Cauchy sequence in X_m. If every Cauchy sequence in X converges to an element in X, we say that X is *complete.* Clearly, if every X_m is complete, then X itself is complete. Moreover, $\{x_n\}_{n=1}^{\infty}$ converges in X to x_0 if and only if x_0 and $x_n (n = 1, 2, \cdots)$ are all contained in one of the X_m's and for each $\gamma \in S_m, \gamma(x_n - x_0) \to 0$ as $n \to \infty$.

A subset Y of a countable-union space X is said to be *dense* in X if for any $x_0 \in X$ there exists a sequence $\{x_n\}_{n=1}^{\infty} \subset Y$ such that $x_n \to x_0$ in X as $n \to \infty$. A countable-union space will be called a *strict countable-union space* if the topology of X_m is identical to the topology it inherits from X_{m+1},i.e., $S_m = S_{m+1}$ for all m. Hence, $S_m = S_{m+i}, i = 1, 2, 3, \cdots$. This is the case, for example, if $\gamma_{m,k}(x) = \gamma_{m+1,k}(x)$ for all $x \in X_m$, all k and m, where $S_m = \{\gamma_{m,k}\}_{k=1}^{\infty}$. It follows that any countably normed space (X, S) is a strict countable-union space since we may set $(X_m, S_m) = (X, S)$ for all m. However, the converse is not true even if every S_m contains only a single norm $\{\gamma_m\}$.

3.2.2 The Dual Spaces of Countable-Union Spaces

DEFINITION 3.10 *Let $X = \bigcup_{n=1}^{\infty} X_n$ be a countable-union space. A linear functional f on X is said to be continuous if and only if it is continuous on every X_n. The set of all continuous linear functionals on X is called the dual space of X and will be denoted by X^*.*

It is readily seen that X^* is a vector space under the standard addition and scalar multiplication operations. It follows from the definition that if

$\{x_n\}_{n=1}^{\infty}$ is a sequence converging to x_0 in X and if $f \in X^*$, then $\langle f, x_n \rangle \to \langle f, x_0 \rangle$ as $n \to \infty$. We provide X^* with the following notion of convergence: A sequence $\{f_n\}_{n=1}^{\infty}$ in X^* is said to converge to $f_0 \in X^*$ if and only if for every $x \in X, \langle f_n, x \rangle \to \langle f_0, x \rangle$ as n$\to \infty$. A sequence $\{f_n\}_{n=1}^{\infty}$ in X^* is said to be Cauchy if for every $x \in X, \langle f_m - f_n, x \rangle \to 0$ as $m, n \to \infty$, and X^* is said to be complete if and only if every Cauchy sequence converges to an element of X^*.

Let X and Y be two countable-union spaces with $X \subset Y$ and assume that the convergence in X is stronger than that in Y, that is if $\{x_n\}_{n=1}^{\infty}$ is a sequence in X such that $x_n \to x_0 \in X$ as $n \to \infty$, then $x_n \to x_0$ in Y. In this case, we may also say that the convergence in Y is weaker than that in X.

If the convergence in X is both stronger and weaker than that in Y, we shall say that the convergence in X and Y are *equivalent*. As in the case with countably normed spaces, the restriction of a continuous linear functional on Y does not always give rise to a continuous linear functional on X. That is we cannot identify Y^* with a subspace of X^*. But if the convergence in X is stronger than that in Y, and if X is dense in Y, then Y^* is a subspace (can be identified with a subspace) of X^*.

If $X = \bigcup_{n=1}^{\infty} X_n$ is a countable-union space with each X_n $(n = 1, 2, \ldots,)$ being a complete countably normed space, then X^* is also complete.

3.2.3 Examples of Countable-Union Spaces

1. Let $\mathcal{D}_K$ be the countably normed space discussed in Example 1, Section 3.1.3. As a set $\mathcal{D}_K = \{\phi \in C^{\infty} : \phi(\mathbf{x}) = 0 \text{ for } \mathbf{x} \notin K\}$, but when provided with the topology generated by the seminorms

$$\gamma_{|k|}(\phi) = \sup_{\mathbf{x} \in K} \left| D^k \phi(\mathbf{x}) \right|, \quad |k| = 0, 1, 2, \cdots,$$

$(\mathcal{D}_K, S_K)$ is a countably normed space, where $S_K = \left\{\gamma_{|k|}\right\}_{|k|=0}^{\infty}$. Let $K_1 \subset K_2 \subset K_3 \cdots$ be a sequence of compact sets such that $\mathcal{R}^n = \bigcup_{m=1}^{\infty} K_m$. For example, we may choose

$$K_m = \{\mathbf{x} \in \mathcal{R}^n : |\mathbf{x}| \leq m\}.$$

Set $\mathcal{D} = \bigcup_{m=1}^{\infty} \mathcal{D}_{K_m}$. Then $\phi \in \mathcal{D}$ if and only if ϕ is infinitely differentiable on $\mathcal{R}^n$ and vanishes outside some compact set. A sequence $\{\phi_k\}_{k=1}^{\infty}$ in $\mathcal{D}$ converges to $\phi_0 \in \mathcal{D}$ if and only if there exists a compact set K_m such that $\operatorname{supp} \phi_k \subset K_m$ for all $k = 1, 2, \cdots$, and $\phi_k \to \phi_0$ in $\mathcal{D}_{K_m}$. Since each $\mathcal{D}_{K_m}$ is complete, then so is $\mathcal{D}$. It is almost trivial to show that a sequence $\{\phi_m\}_{m=1}^{\infty}$ in $\mathcal{D}$ converges to $\phi_0 \in \mathcal{D}$ if and only if there exists some fixed compact subset K of $\mathcal{R}^n$ such that $\operatorname{supp} \phi_m \subset K$ for all m and $D^k \phi_m \to D^k \phi_0$, uniformly on $\mathcal{R}^n$ as

$m \to \infty$, for all $|k| = 0, 1, 2, \cdots$.
It can be easily shown that the definition of $\mathcal{D}$ is independent of the sequence $\{K_m\}_{m=1}^{\infty}$. In fact, if $\left\{\tilde{K}_m\right\}_{m=1}^{\infty}$ is another sequence of compact sets such that $\tilde{K}_1 \subset \tilde{K}_2 \subset \tilde{K}_3 \subset \cdots$, and $\mathcal{R}^n = \bigcup_{m=1}^{\infty} \tilde{K}_m$, then the countable-union space $\tilde{\mathcal{D}}$ formed by using $\left\{\tilde{K}_m\right\}_{m=1}^{\infty}$ is exactly the same as $\mathcal{D}$ and the convergent sequences in both spaces are also the same.

2. Let I be an open set in $\mathcal{R}^n$ and $\{K_m\}_{m=1}^{\infty}$ be a sequence of compact sets such that $K_1 \subset K_2 \subset K_3 \subset \cdots$, and $I = \bigcup_{m=1}^{\infty} K_m$. Define $\mathcal{D}(I) = \bigcup_{m=1}^{\infty} \mathcal{D}_{K_m}$. Then $\mathcal{D}(I)$ consists of all C^{∞}-functions on I whose supports are contained in I. Analogous to $\mathcal{D}, \mathcal{D}(I)$ is a complete countable-union space. If $I = \mathcal{R}^n$, then $\mathcal{D}(\mathcal{R}^n) = \mathcal{D}$.
Although the space $\mathcal{D}$ is a special case of the space $\mathcal{D}(I)$, historically $\mathcal{D}$ was introduced first and has played a more important role in the development of the theory of generalized functions.

3. Let $\mathcal{Z}_a$ $(a > 0)$ denote the set of all entire functions $\phi(z)$ satisfying

$$\left|z^k \phi(z)\right| \le C_k e^{a|y|}, \quad k = 0, 1, 2, \cdots, \quad z = x + iy$$

for some constant C_k. Define

$$\gamma_k(\phi) = \sup \left|z^k e^{-a|y|} \phi(z)\right|, \quad k = 0, 1, 2, \cdots.$$

Note that if $a < b$, then $\mathcal{Z}_a \subset \mathcal{Z}_b$. Choose a sequence of rational numbers $\{r_n\}_{n=1}^{\infty}$ such that $0 = r_1 < r_2 < \cdots$, $\lim_{n\to\infty} r_n = \infty$, and let $\mathcal{Z} = \bigcup_{n=1}^{\infty} \mathcal{Z}_{r_n}$. Then $\mathcal{Z}$ is a complete countable-union space in which a sequence $\{\phi_m\}_{m=1}^{\infty}$ converges to ϕ_0 if and only if there exist two constants a and C_k such that

$$\left|z^k e^{-a|y|} \phi_m(z)\right| \le C_k, \quad k = 0, 1, 2, \cdots, \text{ for all } m = 1, 2, \cdots,$$

and $\lim_{m\to\infty} \phi_m = \phi_0$ uniformly on every compact subset of the complex plane. It can be shown that $\mathcal{Z}$ is independent of the sequence $\{r_n\}_{n=1}^{\infty}$. The space $\mathcal{Z}$ can also be characterized as being the image of $\mathcal{D}$ under the Fourier transformation, i.e., $\phi \in \mathcal{Z}$ if and only if there exists $\psi \in \mathcal{D}$ such that $\hat{\psi} = \phi$, where $\hat{\psi}$ is the Fourier transform of ψ.

4. Let K be a compact subset of the complex plane $\mathcal{C}$ and consider the system of open neighborhoods $\{G_n\}_{n=1}^{\infty}$ of K defined by

$$G_n = \left\{z : \, dist(z, K) < \frac{1}{n}\right\}.$$

Hence, $G_1 \supset G_2 \supset \cdots$, and consequently, $\mathcal{H}(G_1) \subset \mathcal{H}(G_2) \subset \cdots$, where $\mathcal{H}(G_i)$ is the space of all holomorphic functions on G_i (cf. Example 5, Section 3.1.3). Any $f(z) \in \mathcal{H}(G_n)$ defines an element in $\mathcal{H}(G_{n+1})$ by restriction, i.e., $f(z)|_{G_{n+1}}$. We define $\mathcal{H}(K)$ by

$$\mathcal{H}(K) = \bigcup_{n=1}^{\infty} \mathcal{H}(G_n).$$

Thus, $f(z) \in \mathcal{H}(K)$ if and only if it is locally holomorphic in some neighborhood of K. We endow $\mathcal{H}(K)$ with the inductive limit topology, i.e., the topology induced by the topologies of $\{\mathcal{H}(G_n)\}_{n=1}^{\infty}$. It is known that the dual space $\mathcal{H}^*(K)$ of $\mathcal{H}(K)$ is isomorphic to $\mathcal{H}_0(C\backslash K) = \{f : f \text{ is holomorphic on } C\backslash K \text{ and } f(\infty) = 0\}$.

5. Let η and R be as defined in Example 6, Section 3.1.3 and assume that for $a \le c$ and $d \le b$, we have $0 < \eta(x, a, b) \le \eta(x, c, d)$ on I. Then it follows that $\mathcal{L}_{c,d} \subset \mathcal{L}_{a,b}$ and that the convergence in $\mathcal{L}_{c,d}$ implies the convergence in $\mathcal{L}_{a,b}$. This can be seen from the relation

$$\gamma_{k,a,b}(\phi) = \sup_{x \in I} \left|\eta(x, a, b) R^k \phi(x)\right| \le \sup_{x \in I} \left|\eta(x, c, d) R^k \phi(x)\right| = \gamma_{k,c,d}(\phi).$$

Let w and z be such that $-\infty < w < z < \infty$. Choose two monotonic sequences of real numbers $\{a_\nu\}_{\nu=1}^{\infty}$ and $\{b_\nu\}_{\nu=1}^{\infty}$ such that $a_\nu \to w^+$ and $b_\nu \to z^-$. Thus, we can define a countable-union space $\mathcal{L}(w, z) = \bigcup_{\nu=1}^{\infty} \mathcal{L}_{a_\nu, b_\nu}$.

One can easily verify the following:

(i) The definition of $\mathcal{L}(w, z)$ is independent of the choices of $\{a_\nu\}_{\nu=1}^{\infty}$ and $\{b_\nu\}_{\nu=1}^{\infty}$, and if $w \le u < v \le z$, then $\mathcal{L}(u, v) \subset \mathcal{L}(w, z)$. Moreover, $\mathcal{L}(w, z) \subseteq \mathcal{L}_{w,z}$.

(ii) A sequence $\{\phi_\nu\}_{\nu=1}^{\infty}$ converges in $\mathcal{L}(w, z)$ if and only if it converges in one of the spaces $\mathcal{L}_{a_\nu, b_\nu}$. In addition to that, $\mathcal{L}(w, z)$ is complete and so is its dual space $\mathcal{L}^*(w, z)$.

(iii) The operator R is continuous on $\mathcal{L}_{a,b}$ since $\gamma_{k,a,b}(R\phi) = \gamma_{k+1,a,b}(\phi)$.

3.3 Generalized Functions

3.3.1 Testing-Function Spaces and Their Duals

As indicated at the beginning of this chapter, in our approach a generalized function is defined as an element of the dual space of some testing-function space. Therefore, to introduce generalized functions, we first need to introduce the concept of a testing-function space.

DEFINITION 3.11 *Let I be an open subset of $\mathcal{R}^n$. A set of functions $V(I)$ is said to be a testing-function space (on I) if:*

1. *all the elements of $V(I)$ are infinitely differentiable functions on I, i.e., they belong to $C^\infty(I)$,*
2. *$V(I)$ is either a complete countably normed or a complete countable-union space,*
3. *the convergence of a sequence $\{\phi_m\}_{m=1}^\infty$ to ϕ_0 in $V(I)$ implies that $\lim_{m\to\infty} D^k\phi_m = D^k\phi_0$ uniformly on every compact subset of I, for all $|k| = 0, 1, 2, \cdots$.*

DEFINITION 3.12 *The dual space $V^*(I)$ of $V(I)$ is called a space of generalized functions on I and any element f of $V^*(I)$ is called a generalized function on I.*

From the definition and the above discussion, it follows that the spaces of generalized functions are complete. If f is a conventional function that is Lebesgue measurable on I such that for any $\phi \in V(I)$, the integral

$$\int_I f(\mathbf{x})\phi(\mathbf{x})d\mathbf{x}$$

exists in the Lebesgue sense, and if for any sequence $\{\phi_m\}_{m=1}^\infty$ converging to ϕ_0 in $V(I)$ we have

$$\int_I f(\mathbf{x})\phi_m(\mathbf{x})d\mathbf{x} \to \int_I f(\mathbf{x})\phi_0(\mathbf{x})d\mathbf{x}$$

as $m \to \infty$, we say that f is a regular generalized function belonging to $V^*(I)$. More precisely, f can be identified with an element $\tilde{f}$ in $V^*(I)$ as follows. Let us define $\tilde{f}$ by

$$\left\langle \tilde{f}, \phi \right\rangle = \int_I f(\mathbf{x})\phi(\mathbf{x})d\mathbf{x},$$

for every $\phi \in V(I)$. In view of the linearity and the continuity of the integral, it readily follows that $\tilde{f} \in V^*(I)$. We identify f with $\tilde{f}$ and write

$$\langle f, \phi \rangle = \int_I f(\mathbf{x})\phi(\mathbf{x})d\mathbf{x};$$

hence $f \in V^*(I)$. A generalized function $f \in V^*(I)$ that is not regular is called singular.

DEFINITION 3.13 *A generalized function $f \in V^*(I)$ is said to vanish in an open set $J \subset I$ if for any $\phi \in V(I)$ with supp $\phi \subset J$, we have $\langle f, \phi \rangle = 0$. The null set of f is the union of all open subsets J of I on which f vanishes. The complement of the null set of f with respect to I is called the support of f. f is said to be concentrated in a subset A of I if A contains the support of f.*

Thus, the support of f, written supp$f$, is the smallest closed set B (with respect to I) such that f is zero on $I - B$.

DEFINITION 3.14 *Two generalized functions f and g in $V^*(I)$ are said to be equal on an open set $J \subset I$ if and only if $f - g$ vanishes on J. A generalized function is said to be regular on J if it coincides with a regular generalized function concentrated in J.*

The limit of a sequence of regular generalized functions may not be regular. Take for example, $f_\epsilon(x) = (1/x)\,\Phi_\epsilon(x), \epsilon > 0$, where $\Phi_\epsilon(x)$ is the characteristic function of the set $(-\infty, -\epsilon) \cup (\epsilon, \infty)$. Then, for any $\phi \in \mathcal{D}$

$$\langle f_\epsilon, \phi \rangle = \int_{|x|>\epsilon} \frac{\phi(x)}{x} dx,$$

and it follows that $\lim_{\epsilon \to 0} f_\epsilon = f$, where $f(x) = 1/x$ if $x \neq 0$, yet f is not locally integrable in any neighborhood of the origin; hence it is not regular.

3.3.2 A Special Type of Generalized Functions

In this section we introduce the δ-function and some of its generalizations because of the important role they play in function transformations.

1. The *Dirac-delta function*, $\delta(\mathbf{x})$, is defined by

$$\langle \delta, \phi \rangle = \phi(\mathbf{0}), \;\; \phi \in \mathcal{D}.$$

   It is a prototype of singular generalized functions. Showing that it is indeed singular is easy. Assume to the contrary that it is regular. Hence, there exists a locally integrable function f such that

$$\int_{\mathcal{R}^n} f(\mathbf{x})\phi(\mathbf{x})\, d\mathbf{x} = \phi(\mathbf{0}), \;\; \text{for all } \phi \in \mathcal{D}.$$

For $a > 0$, let

$$\phi_a(\mathbf{x}) = \begin{cases} \exp(a^2/\left(|\mathbf{x}|^2 - a^2\right), & |\mathbf{x}| < a \\ 0, & |\mathbf{x}| \geq a \end{cases}.$$

Then $\phi_a \in \mathcal{D}$ and

$$\int\limits_{\mathcal{R}^n} f(\mathbf{x})\phi_a(\mathbf{x})d\mathbf{x} = \int\limits_{\bar{B}_a(\mathbf{0})} f(\mathbf{x})\phi_a(\mathbf{x})d\mathbf{x} = \frac{1}{e},$$

where $\bar{B}_a(\mathbf{0})$ is the closed ball with center at $\mathbf{0} = (0, \cdots, 0)$ and radius a, i.e., $\bar{B}_a(\mathbf{0}) = \{\mathbf{x} \in \mathcal{R}^n, |\mathbf{x}| \leq a\}$. But if f were a locally integrable function, Lebesgue's theorem would imply that

$$\lim_{a \to 0} \int\limits_{\bar{B}_a(\mathbf{0})} f(\mathbf{x})\phi_a(\mathbf{x})d\mathbf{x} = 0,$$

which is a contradiction.

2. Let S be a manifold in $\mathcal{R}^n$of dimension less than $n, d\sigma$ be the induced Lebesgue measure on S and f be a locally integrable function. For any $\phi \in \mathcal{D}$, let

$$\langle f, \phi \rangle = \int\limits_S f(\mathbf{x})\phi(\mathbf{x})d\sigma.$$

It is easy to see that this equation defines a continuous linear functional or a generalized function that is concentrated in S.

3. For $n > 1$, let S be an $(n-1)$-dimensional manifold (hypersurface) whose equation is given by $P(x_1, \cdots, x_n) = 0$. Let S be smooth so that $gradP \neq 0$. For any point of the hypersurface $P = 0$ choose a sufficiently small neighborhood U so that a new coordinate system $\{u_1, \cdots, u_n\}$ can be chosen in which $P = 0$ is one of the coordinate hypersurfaces. Let $u_1 = P$ and choose the remaining u_i coordinates $(i = 2, \cdots, n)$ arbitrarily but in such a way that the Jacobian $J\left(\frac{\partial \mathbf{x}}{\partial \mathbf{u}}\right)$of the x_i with respect to the u_i is never zero. This is always possible since $gradP \neq 0$ on $P = 0$.

For any test function ϕ, let $\phi_1(u_1, \cdots, u_n) = \phi(x_1, \cdots, x_n)$ and $\psi(u_1, \cdots, u_n) = \phi_1(u_1, \cdots, u_n)J\left(\frac{\partial \mathbf{x}}{\partial \mathbf{u}}\right)$, or for short $\phi_1(\mathbf{u}) = \phi(\mathbf{x})$ and $\psi(\mathbf{u}) = \phi_1(\mathbf{u})J\left(\frac{\partial \mathbf{x}}{\partial \mathbf{u}}\right)$.

The generalized function $\delta(P)$ is defined as

$$\langle \delta(P), \phi \rangle = \int \psi(0, u_2, \cdots, u_n)du_2 \cdots du_n,$$

where the integral is taken over $\mathcal{R}^n$.

Formally, we may write

$$\langle\delta(P),\phi\rangle = \int \delta(P)\phi(\mathbf{x})d\mathbf{x} = \int \delta(P)\phi_1(\mathbf{u})J\left(\frac{\partial\mathbf{x}}{\partial\mathbf{u}}\right)d\mathbf{u}$$

$$= \int \delta(u_1)\psi(\mathbf{u})d\mathbf{u} = \int \psi(0,u_2,\cdots,u_n)du_2\cdots du_n.$$

The last integral is taken over the hypersurface $u_1 = P = 0$; thus, $\delta(P)$ is concentrated in that surface.

In particular, if S is the hyperplane whose equation is $x_1 = 0$, we obtain

$$\langle\delta(x_1),\phi(\mathbf{x})\rangle = \int \delta(x_1)\phi(\mathbf{x})d\mathbf{x} = \int\left(\int \delta(x_1)\phi(x_1,\cdots,x_n)dx_1\right)dx_2\cdots dx_n$$

$$= \int \phi(0,x_2,\cdots,x_n)dx_2\cdots dx_n.$$

Similarly,

$$\left\langle\delta^{(k)}(x_1),\phi(\mathbf{x})\right\rangle = \int \delta^{(k)}(x_1)\phi(\mathbf{x})d\mathbf{x} = (-1)^k\int \left.\frac{\partial^k\phi}{\partial x_1^k}\right|_{x_1=0} dx_2\cdots dx_n.$$

3.3.3 Spaces of Generalized Functions

Now we give examples of spaces of generalized functions, but first let us observe that the spaces $\mathcal{E},\mathcal{P}_T,\mathcal{D},\mathcal{S},\mathcal{Z},\mathcal{E}(I),\mathcal{D}(I),\mathcal{H}(K),\mathcal{L}_{a,b}$ and $\mathcal{L}(a,b)$ are all testing-function spaces; therefore, their dual spaces are spaces of generalized functions. Moreover, since $\mathcal{D}\subseteq\mathcal{S}\subseteq\mathcal{E}$ and $\mathcal{D}$ is dense in $\mathcal{E}$, it follows that $\mathcal{D}^*\supseteq\mathcal{S}^*\supseteq\mathcal{E}^*$, and similarly $\mathcal{E}^*(I)\subset\mathcal{D}^*(I)$.

1. The space $\mathcal{D}^*$ is known as the space of *Schwartz distributions* and $\mathcal{D}^*(I)$ as the space of *Schwartz distributions* on I.

2. The space $\mathcal{E}^*$ is the space of all Schwartz distributions with compact support, i.e., $f\in\mathcal{E}^*$ if and only if $f\in\mathcal{D}^*$ and f has compact support.

3. The space $\mathcal{P}_T^*$ is called the space of periodic distributions with period T.

4. The space $\mathcal{S}^*$ is the space of *tempered distributions.* It contains all locally integrable functions with polynomial growth at $\pm\infty$ (cf. Section 1.4), which are sometimes called *distributions with slow growth.*

5. The space $\mathcal{Z}^*$ is the space of *ultra-distributions.* In view of the inclusion $\mathcal{Z}\subset\mathcal{S}$, the fact that the convergence in $\mathcal{Z}$ is stronger than that in $\mathcal{S}$, and that $\mathcal{Z}$ is dense in S, it follows that $\mathcal{S}^*\subset\mathcal{Z}^*$. In fact,

$\mathcal{Z} \subset \mathcal{S} \subset \mathcal{S}^* \subset \mathcal{Z}^*$. It is worth noting that although both $\mathcal{D}$ and $\mathcal{Z}$ are subspaces of S, neither one of the inclusions $\mathcal{Z} \subset \mathcal{D}$ and $\mathcal{D} \subset \mathcal{Z}$ is true. In fact, $\mathcal{Z} \cap \mathcal{D}$ contains only the zero function since the only entire function that can vanish on an interval is the zero function.

6. The space $\mathcal{H}^*(K)$ is the space of *analytic hyperfunctions* on K.

7. The spaces $\mathcal{L}^*_{a,b}$ and $\mathcal{L}^*(a,b)$ are very useful in extending linear integral transformations to generalized functions. It is worth noting that if f is a locally integrable function on I such that $f(x)/\eta(x,a,b)$ is absolutely integrable on I, then $f \in \mathcal{L}^*_{a,b}$. For,

$$|\langle f, \phi\rangle| = \left|\int_I \frac{f(x)}{\eta(x,a,b)}\eta(x,a,b)\phi(x)dx\right| \leq \gamma_{0,a,b}(\phi) \int_I \left|\frac{f(x)}{\eta(x,a,b)}\right| dx < \infty,$$

for any $\phi \in \mathcal{L}_{a,b}$.

3.4 Operations with Generalized Functions

3.4.1 Operations and Transformations

By definition, the space of generalized functions $V^*(I)$ is a linear vector space that is complete with respect to the predescribed notion of convergence (weak or sequential convergence); hence, if $\alpha, \beta \in \mathcal{C}$, $f, g \in V^*(I)$

1. $\alpha f + \beta g \in V^*(I)$,

2. For any Cauchy sequence $\{f_m\}_{m=1}^{\infty}$ in $V^*(I)$, $\lim_{m\to\infty} f_m = f_0 \in V^*(I)$. In particular, if $S_m = \sum_{k=1}^{m} f_k$ is a Cauchy sequence, then the series $\sum_{k=1}^{\infty} f_k$ converges to some element $f_0 \in V^*(I)$ and we write $\sum_{k=1}^{\infty} f_k = f_0$.

3. If L is a continuous linear transformation on $V(I)$, then we can define the adjoint transformation L^* on $V^*(I)$ as

$$\langle L^* f, \phi\rangle = \langle f, L\phi\rangle, \text{ for all } \phi \in V(I) \text{ and } f \in V^*(I).$$

L^* is easily seen to be a continuous linear transformation on $V^*(I)$. As a special case, we obtain

(a) if θ is a C^∞-function on I such that $\theta\phi \in V(I)$ for any $\phi \in V(I)$ and $\lim_{m\to\infty} (\theta\phi_m) = \theta\phi_0$ whenever $\lim_{m\to\infty} \phi_m = \phi_0$ in $V(I)$, then $\theta f \in V^*(I)$ for any $f \in V^*(I)$ and

$$\langle \theta f, \phi\rangle = \langle f, \theta\phi\rangle, \quad \phi \in V(I),\ f \in V^*(I).$$

(b) Let $\theta_0, \cdots, \theta_m$ be C^∞-functions on I such that for any $\phi \in V(I)$, $D^k(\theta_i\phi) \in V(I)$ for $i = 0, 1, \cdots, m$ and $|k| = 0, 1, 2, \cdots$. Set $L = \theta_0 D^{k_1}\theta_1 D^{k_2}\theta_2 \cdots D^{k_m}\theta_m$. Hence, L is a linear transformation on $V(I)$. If it is also continuous, then we can define L^* by

$$\langle L^* f, \phi\rangle = \langle f, L\phi\rangle, \ \phi \in V(I), \ f \in V^*(I).$$

As a special case if $L = D$, we obtain that every generalized function can be differentiated. If f is a regular generalized function, we can write the right-hand side, $\langle f, L\phi\rangle$, as an integral, and then formally integrate by parts to obtain a representation of L^* in terms of the adjoint of the differential operator L. But it should be emphasized that the adjoint of L as a continuous linear transformation on $V^*(I)$ does not in general coincide with the adjoint of L as a differential operator.

(c) Let L be T_a (the translation operator), or D_a (the dilation operator), or E_a (the modulation operator). If L is a continuous linear transformation on $V(I)$, then we can define the adjoint of the translation, dilation, and modulation operators on $V^*(I)$.

4. If $V(I) \subset L^2(I)$ and the convergence in $V(I)$ is stronger than that of $L^2(I)$, then $V(I)$ is a subspace of $V^*(I)$ consisting of regular generalized functions, i.e., every $\tilde{\phi} \in V(I)$ is a regular generalized function in $V^*(I)$. For, given any fixed $\tilde{\phi} \in V(I)$,

$$\left|\left\langle \tilde{\phi}, \phi\right\rangle\right| = \left|\int_I \tilde{\phi}(\mathbf{x})\phi(\mathbf{x})d\mathbf{x}\right| \le \left\|\tilde{\phi}\right\|_2 \|\phi\|_2 < \infty,$$

for any $\phi \in V(I)$. Moreover, if $\lim_{m\to\infty} \phi_m = \phi_0$ in $V(I)$, then

$$\lim_{m\to\infty} \left\langle \tilde{\phi}, \phi_m\right\rangle = \left\langle \tilde{\phi}, \phi_0\right\rangle$$

since

$$\left|\left\langle \tilde{\phi}, \phi_m\right\rangle - \left\langle \tilde{\phi}, \phi_0\right\rangle\right| \le \left\|\tilde{\phi}\right\|_2 \|\phi_m - \phi_0\|_2 \to 0$$

as $m \to \infty$.

3.4.2 Examples of Operations on Generalized Functions

Since multiplication by C^∞-functions, translation, dilation, modulation and differentiation are continuous linear transformations on $\mathcal{D}$, we have for any $f \in \mathcal{D}^*$ and all $\phi \in \mathcal{D}$

1.
$$\langle \theta f, \phi\rangle = \langle f, \theta\phi\rangle,$$

where $\theta \in C^\infty$,

2.

$$\langle T_a f, \phi \rangle = \langle f, T_{-a}\phi \rangle$$

which is formally written in the form

$$\langle f(x-a), \phi(x) \rangle = \langle f(x), \phi(x+a) \rangle .$$

A generalized function f is said to be *periodic* with period a if $T_a f = f$. This means that for every $\phi \in \mathcal{D}$,

$$\langle f(x-a), \phi(x) \rangle = \langle f(x), \phi(x) \rangle .$$

The number a is called a period. The set of all distributions that are periodic with period T can be identified with the space $\mathcal{P}_T^*$, (cf. item 3, Section 3.3.3).

3.

$$\langle D_a f, \phi \rangle = \left\langle f, D_{1/a}\phi \right\rangle , \quad a \neq 0$$

or formally

$$\left\langle f\left(\frac{x}{a}\right), \phi(x) \right\rangle = |a| \left\langle f, \phi(ax) \right\rangle$$

4.

$$\langle E_a f, \phi \rangle = \langle f, E_a \phi \rangle$$

5.

$$\left\langle \frac{\partial}{\partial x_i} f, \phi \right\rangle = \left\langle f, -\frac{\partial \phi}{\partial x_i} \right\rangle , \quad i = 1, \cdots, n,$$

or more generally if $p = (p_1, \cdots, p_n)$ is a multi-index and

$$L = \sum_{|p|=0}^{m} a_p(\mathbf{x}) D^p = \sum_{|p|=0}^{m} a_p(\mathbf{x}) \frac{\partial^{|p|}}{\partial x_1^{p_1} \cdots \partial x_n^{p_n}} ,$$

where the $a_p(\mathbf{x})$'s are C^∞-functions, then we have

$$\langle Lf, \phi \rangle = \langle f, L^* \phi \rangle ,$$

in which

$$L^* \phi = \sum_{|p|=0}^{m} (-1)^{|p|} \frac{\partial^{|p|} \left(a_p(\mathbf{x})\phi(\mathbf{x})\right)}{\partial x_1^{p_1} \cdots \partial x_n^{p_n}} .$$

6. Let $\{f_m\}_{m=1}^{\infty}$ be a sequence of locally integrable functions such that $\lim_{m\to\infty} f_m = f_0$ uniformly on any compact subset of $\mathcal{R}^n$, then f_0 is

a regular Schwartz distribution. For, given any $\phi \in \mathcal{D}$, choose a compact set K such that $\operatorname{supp}\phi \subset K$. We then have

$$|\langle f_m, \phi\rangle - \langle f_0, \phi\rangle| = \left| \int_K (f_m(\mathbf{x}) - f_0(\mathbf{x}))\, \phi(\mathbf{x}) d\mathbf{x} \right|$$

$$\leq \sup_{\mathbf{x}\in K} |\phi(\mathbf{x})| \int_K |f_m(\mathbf{x}) - f_0(\mathbf{x})|\, d\mathbf{x} \to 0 \quad \text{as } m \to \infty.$$

3.4.3 Ordinary and Generalized Derivatives

We have just seen that any $f \in \mathcal{D}^*$ can be differentiated in the sense of generalized functions to yield another generalized function $(f')_g \in \mathcal{D}^*$. If f is also a differentiable conventional function with ordinary derivative $(f')_0$, it is natural to ask if $(f')_g = (f')_0$. The answer to this question is the focus of this section.

The δ-function is concentrated at the origin with $\operatorname{supp}\delta = \{\mathbf{0}\}$, and it is a regular generalized function in any open set not containing the origin. In fact, all the derivatives of δ are concentrated at the origin. Moreover, if we restrict ourselves to $\mathcal{R}$, and set $H(x) = \chi_{(0,\infty)}(x)$, the *Heaviside function*, we find that for any $\phi \in \mathcal{D}$

$$\langle H', \phi\rangle = -\langle H, \phi'\rangle = -\int_0^\infty \phi'(x)dx = \phi(0) = \langle \delta, \phi\rangle .$$

Hence, $H' = \delta$, where the derivative is taken in the sense of generalized functions. The ordinary (conventional) derivative of H, which unfortunately is usually denoted also by H', is given by

$$H'(x) = \begin{cases} 0 & \text{if } x \neq 0 \\ \text{undefined} & \text{if } x = 0 \end{cases} .$$

In general, if f is a locally integrable function whose derivative f' is not continuous, then the derivatives of f in the generalized function sense and in the ordinary sense do not coincide. In other words, the generalized function that arises from the conventional (classical) derivative of f and the derivative of the generalized function that arises from f are not always the same. However, the following is true.

THEOREM 3.5 [70, Vol. 1, p. 22]
Let f be a piecewise continuous function with discontinuities only of the first kind (jump discontinuities) $h_1, h_2, \cdots, h_n$ at $x_1, x_2, \cdots, x_n$ and let f have piecewise continuous derivative f'. Then

$$f' = f_1' + \sum_{k=1}^{n} h_k \delta(x - x_k),$$

where f_1 *is a continuous function whose derivative* f_1' *coincides with* f' *everywhere except at* $x_1, \cdots, x_n$*. The derivative of the regular generalized function generated by* f_1 *coincides with the regular generalized function generated by the ordinary derivative of* f*.*

3.5 Local Behavior of Generalized Functions

3.5.1 The Order of a Generalized Function

We have seen that the δ-function and all its derivatives are concentrated at a single point. It turns out that the converse is also true; that is, every generalized function concentrated at a single point x_0 can be represented as a finite linear combination of $\delta(x - x_0)$ and its derivatives. Moreover, the fact that the δ-function is the second derivative of the continuous function

$$H^{(-1)}(x) = \begin{cases} x - x_0, & x \geq x_0 \\ 0, & x \leq x_0 \end{cases},$$

suggests that any generalized function concentrated at a point is the mth derivative of some continuous function. In fact, not only is this true, but a more general result is also true: any generalized function concentrated on a compact set is the mth derivative of some continuous function. This is a special case of the following theorem.

THEOREM 3.6 [278, p. 92]
Let I *be a fixed finite closed interval in* $\mathcal{R}$ *and let* f *be a generalized function in* $\mathcal{D}^*$ *(Schwartz distribution) defined over a neighborhood of* I*. Then there exist a nonnegative integer* m *and a continuous function* F *such that*

$$\langle f, \phi \rangle = \left\langle F^{(m)}, \phi \right\rangle$$

for every $\phi \in \mathcal{D}(I)$*.*

COROLLARY 3.1
If $f \in \mathcal{D}^*$ *has compact support, then there exist a non-negative integer* m *and a continuous function* F *such that* $f(x) = F^{(m)}(x)$*.*

Theorem 3.6 leads to the following definition.

DEFINITION 3.15 *Let* f *be a generalized function in* $\mathcal{D}^*$ *and* I *be a fixed interval in* $\mathcal{R}$*. The smallest nonnegative integer* m *such that for all*

$\phi \in \mathcal{D}(I)$

$$\langle f, \phi \rangle = \left\langle F^{(m)}, \phi \right\rangle$$

is called the order of f on I, where F is some continuous function.

Theorem 3.6 states that every generalized function is locally of finite order. A generalized function may have different orders on different intervals, for example, on any open interval not containing the origin, the kth derivative of the δ-function is of order zero, whereas on any open interval containing the origin the order is $k+2$.

Globally, a generalized function is not necessarily of finite order. The *global order* of a generalized function f is defined as the smallest nonnegative integer m such that

$$\langle f, \phi \rangle = \left\langle F^{(m)}, \phi \right\rangle$$

for all $\phi \in \mathcal{D}$. If no such m exists, we say that the generalized function has *infinite order*. Any continuous function is a generalized function of order zero and every generalized function with compact support is of finite order (globally). An example of an infinite-order generalized function is $f(x) = \sum_{n=0}^{\infty} \delta^{(n)}(x-n)$.

3.5.2 The Value of a Generalized Function at a Point

We now turn our attention to another local behavior of generalized functions. Let f be a Schwartz distribution, i.e., a generalized function in $\mathcal{D}^*$. If $\lim\limits_{\alpha \to 0} f(\alpha x + x_0)$ exists, i.e.,

$$\lim_{\alpha \to 0} \langle f(\alpha x + x_0), \phi(x) \rangle = \lim_{\alpha \to 0} \frac{1}{|\alpha|} \left\langle f(x), \phi\left(\frac{x - x_0}{\alpha}\right) \right\rangle$$

exists for every $\phi \in \mathcal{D}$, then it must be a constant [149]. This leads to the following definition.

DEFINITION 3.16 *Let f be a generalized function in $\mathcal{D}^*$. If the limit $\lim\limits_{\alpha \to 0} f(\alpha x + x_0)$ exists, it will be called the value of f at x_0, and x_0 will be called a regular point of f, otherwise it is called a singular point.*

If f is a locally integrable function that is continuous at x_0, then x_0 is a regular point of the generalized function f and the value of that generalized function at x_0 is the same as the ordinary value $f(x_0)$.

If a locally integrable function f has an ordinary derivative at x_0, then $f'(x_0)$ is the value of the generalized function f' at x_0. In particular, if

f is a locally integrable function and if $F(x) = \int_a^x f(t)dt$ has an ordinary derivative at x_0, then this derivative is the value of the generalized function f at x_0. It is possible that a function f does not have a derivative at x_0, but its derivative in the sense of generalized functions has a value at x_0. For example, $f(x) = 3x^2 \sin(1/x) - x\cos(1/x)$ does not have an ordinary derivative at $x_0 = 0$, but its derivative in the sense of generalized functions has the value zero at $x_0 = 0$. To this end, we need to show that $\lim_{a\to 0} f'(ax) = 0$ in the sense of generalized functions. This will follow by differentiation if we can show that $\lim_{a\to 0}(f(ax)/a) = 0$ in the sense of generalized functions. Let $G_a(x) = ax^3 \sin(1/ax)$. It is easy to see that $G'_a(x) = (f(ax)/a)$ and $\lim_{a\to 0} G_a(x) = 0$ in the sense of generalized functions; hence,

$$\lim_{a\to 0} G'_a(x) = \lim_{a\to 0} \frac{f(ax)}{a} = 0.$$

Every $x_0 \neq 0$ is a regular point of the Heaviside function $H(x)$ and the value of H at x_0 in the ordinary sense is the same as the value in the generalized function sense. The point $x_0 = 0$ is a singular point of H since

$$\lim_{a\to 0} H(ax) = \lim_{a\to 0}\left[\frac{a}{|a|}\left(H(x) - \frac{1}{2}\right) + \frac{1}{2}\right]$$

does not exist.

Consequently, the Dirac-delta function $\delta(x)$ has the value 0 at every point $x_0 \neq 0$ and since $\lim_{a\to 0} \delta(ax)$ does not exist it has no value at $x_0 = 0$. On the other hand, the point $x_0 = 0$ is a regular point for the generalized function $f(x) = \sin(1/x)$, which as a conventional function has no value at $x = 0$. This follows from the second mean value theorem and the observation that

$$F'(0) = \lim_{x\to 0} \frac{F(x)}{x} = \lim_{x\to 0} \frac{1}{x}\int_0^x \sin\frac{1}{t}dt = \lim_{x\to 0}\frac{1}{x}\int_{1/x}^{\infty} \frac{1}{t}\frac{\sin t}{t}dt,$$

$$= \lim_{x\to 0} \int_{1/x}^{\xi} \frac{\sin t}{t}dt = 0,$$

where $F(x) = \int_0^x \sin(1/t)\,dt$, and $1/x < \xi < \infty$.

The following theorem gives a necessary and sufficient condition for a Schwartz distribution to have a value at a point.

THEOREM 3.7 (Lojasiewcz [149])
A generalized function $f \in \mathcal{D}^*$ has the value c at x_0 if and only if there exist an integer $k \geq 0$ and a continuous function F such that $F^{(k)} = f$ and

$$\lim_{x\to x_0} \frac{F(x)}{(x - x_0)^k} = \frac{c}{k!}.$$

As a corollary it follows that if the Schwartz distribution f' has a value c at x_0, then the distribution f also has the same value c at x_0.

We close this section with the following definition.

DEFINITION 3.17 *A generalized function $f \in \mathcal{D}^*(\mathcal{R}^n)$ is said to have the asymptotic bound $g(\mathbf{x}) \geq 0$ in a neighborhood of a point x_0 if there exist a constant c and a neighborhood U of x_0 such that for any $\phi \in \mathcal{D}(\mathcal{R}^n)$ with support in U,*

$$|\langle f, \phi\rangle| \leq c \int_{\mathcal{R}^n} g(\mathbf{x})\, |\phi(\mathbf{x})|\, d\mathbf{x}.$$

We write $f = 0(g)$ as $x \to x_0$.

If f is generated by a locally integrable function, then $f = 0(g)$ if and only if $f = 0(g)$ in the generalized function sense.

3.6 Analytic Representations of Generalized Functions

Let $V^*(I)$ be a space of generalized functions on some interval $I \subset \mathcal{R}$. For $z \in \mathcal{C} - \bar{I}$, consider the *Cauchy kernel function* $g(x) = 1/(x - z)$, which is a C^∞-function in x for $x \in I$.

DEFINITION 3.18 *Let $g \in V(I)$. We define the analytic representation , $\tilde{f}(z)$, of any $f \in V^*(I)$ as*

$$\tilde{f}(z) = \frac{1}{2\pi i}\left\langle f(x), \frac{1}{x-z}\right\rangle, \quad z \in \mathcal{C} - \bar{I}.$$

The function $\tilde{f}(z)$ is analytic in the complex-plane cut along the closed interval $\bar{I}$. For example, since for any $z \in \mathcal{C} - \mathcal{R}$, the function g belongs to $\mathcal{E}$, the space of all infinitely differentiable functions, then for any distribution $f \in \mathcal{E}^*$, we have

$$\tilde{f}(z) = \frac{1}{2\pi i}\left\langle f(x), \frac{1}{x-z}\right\rangle, \quad Im\ z \neq 0$$

which defines a pair of analytic functions, one in the upper and one in the lower half planes. In fact, more can be said. These two functions can be continued analytically across the real axis to form one analytic function.

THEOREM 3.8 (**Bremermann** [27, pp. 44–48])
Let $f \in \mathcal{E}^*$. *Then* $\tilde{f}(z)$ *is an analytic function in the complex plane cut along the support of* $\tilde{f}$ *with*

$$\frac{d^n}{dz^n}\tilde{f}(z) = \frac{n!}{(2\pi i)}\left\langle f(x), \frac{1}{(x-z)^{n+1}}\right\rangle, \quad z \notin supp\ f,$$

and

$$\tilde{f}(z) = 0\left(|z|^{-1}\right) \quad as\ z \to \infty.$$

Moreover, for any test function $\phi \in \mathcal{E}$ *such that* $\phi^{(k)}, k = 0, 1, \cdots,$ *are bounded, we have*

$$\lim_{\epsilon \to 0} \int_{-\infty}^{\infty} \left[\tilde{f}(x+i\epsilon) - \tilde{f}(x-i\epsilon)\right] \phi(x)dx = \langle f, \phi \rangle .$$

Examples:

1. The analytic representation of the Dirac-delta function is readily seen to be

$$\tilde{\delta}(z) = \frac{1}{2\pi i}\left\langle \delta(x), \frac{1}{x-z}\right\rangle = -\frac{1}{2\pi i z}, \quad z \neq 0,$$

   and

$$\frac{d^n}{dz^n}\tilde{\delta}(z) = \frac{1}{2\pi i}(-1)^{n+1}\frac{n!}{z^{n+1}}.$$

2. The analytic representation of the Heaviside function is given by

$$\tilde{H}(z) = -\frac{1}{(2\pi i)}\log(-z), \ Imz \neq 0.$$

Other spaces of generalized functions on which we can define the concept of analytic representation may be found in [27]. For example, let I be a bounded open interval, $x \in I, z \in \mathcal{C} - \bar{I}$ and $V(I)$ be the space of all analytic functions on I. Then $g(x) = 1/(x-z) \in V(I)$, and consequently the analytic representation $\tilde{f}$ of any $f \in V^*(I)$ is well defined. The same is true if I is a compact subset of $\mathcal{R}$.

4

Function Transformations

4.1 Introduction

Function transformations have been used for over a century to facilitate the study of various problems in applied mathematics, physics and engineering. Some function transformations were introduced to solve specific problems, but they were later found to be useful in solving problems other than those they were designed for. A few were introduced for no obvious practical reason other than to satisfy mathematical curiosity.

Notwithstanding the vast scope of applications of function transformations, there does not seem to be a general consensus, at least among mathematicians, on the definition of a function transformation. In the next section, we will attempt to fill this gap in the literature and give one such definition. Function transformations will be classified as continuous or discrete depending on the nature of their ranges. Our definition will encompass known classes of function transformations, the most important of which is the class of linear integral transformations.

Regardless of how one defines a particular function transformation, the definition should entail three equally important properties: existence, uniqueness and invertibility. The first two are self-evident and the third, though not essential, is dictated by applications; a transformation that is not invertible is of no practical value.

A function transformation simply means a mathematical operation through which a real or complex-valued function f is transformed into another function F, or into a sequence of numbers, or more generally into a set of data. This is done for a variety of reasons. For example:

1. In solving some difficult mathematical problems, it may be useful to transform this problem by means of some function transformation to a new setting where the problem can be easily solved. Having solved the problem, one has to transform the solution back to the original setting, and this is where the inverse transformation comes into play.

A prototype situation in which this method is used is when one tries to solve certain initial-value problems involving ordinary or partial differential equations. Many of the classical linear integral transformations, such as the Fourier, Laplace, Hankel and Mellin, are good examples of function transformations used for that particular purpose.

2. Sometimes the transform function F itself has some physical meaning and ought to be studied in its own right. For example, in electrical engineering the original function $f(t)$ may represent a signal that is a function of time, e.g., a radar or radio signal. The Fourier transform $F(\omega)$ of $f(t)$ in this case represents the frequency content of the signal f, and it is as valuable as the time representation of the signal. In fact, it is often more practical to work with F rather than with f.

   Another example is the Gabor transformation which transforms a signal $f(t)$ into a doubly-indexed sequence of complex numbers $c_{mn}(f)$, where $m, n = 0, \pm 1, \pm 2, \ldots$. This sequence conveys essential information about the structure of the signal f in the time-frequency domain (the t-ω plane). By using this sequence, one can study both the time and frequency behavior of the signal simultaneously.

3. In some applications, the transform function F, is a set of data that can be measured experimentally and from which the original function f needs to be reconstructed.
   An example of such a transformation is the Radon transformation, which enables one to reconstruct a function f of $N (N > 1)$ variables from the values of its integrals over all hyperplanes of dimension $N-1$. As a special case, if $N = 2$, one can use the Radon transform to reconstruct a function $f(x, y)$ from the values of its integrals over all straight lines in the x-y plane.

   The original function f may represent the internal distribution, perhaps density of some physical quantity, of a hard-to-reach object like a human brain. In practice, f needs to be reconstructed from its Radon transform F, which can be measured physically. One of the most important applications of the Radon transform is in the field of computer aided tomography, commonly known to the public as CAT scan.

Some function transformations have been discovered by engineers, such as the Gabor and the Windowed (sliding window) Fourier transforms; some by physicists, such as the wavelets and the coherent state representation; and others by mathematicians, such as the Laplace, Legendre and Jacobi transforms. A number of transformations were discovered by scientists in one discipline and later were rediscovered by other scientists in a different discipline. The Radon transform is a good example of that. Therefore, it is generally fair to say that major contributions to the field of function transformations are shared equally by mathematicians, physicists and engineers.

In the next sections we discuss general properties of function transformations, some of which, especially those pertaining to generalized functions, form the foundations of many of the results that will be given in the rest of this book.

4.2 Function Transformations

We begin by defining what we mean by a function transformation. Our definition is somewhat influenced by ideas from electric engineering; see Appendix A.

DEFINITION 4.1 *Let $\mathcal{A}$ be a set of functions f defined on a subset Ω of $\mathcal{R}^n(\mathcal{C}^n)$ and $\mathcal{B}$ be a non-empty set. A function transformation $\mathcal{L}$ is defined as a mapping from $\mathcal{A}$ into $\mathcal{B}$. The image of f under $\mathcal{L}$, which will be denoted by $\mathcal{L}[f]$, or $(\mathcal{L}f)$, is called the $\mathcal{L}$-transform of f. The set $\mathcal{A}$ is called the domain of $\mathcal{L}$ and the set $\{\mathcal{L}[f] : f \in \mathcal{A}\}$ is called the range. The set $\mathcal{B}$ is called the co-domain or the target space of $\mathcal{L}$.*

If $\mathcal{A}$ is a set of sequences of numbers (complex or real), then $\mathcal{L}$ is said to be a sequence transformation. The image of a sequence $\{a_n\}$ under $\mathcal{L}$ is denoted by $\mathcal{L}[\{a_n\}]$.

Sometimes it is convenient to denote an element of the domain of $\mathcal{L}$ by a lower case letter and the transform of this element by the same letter in capital, for example, $F = \mathcal{L}[f]$. Other notations to distinguish between functions and their $\mathcal{L}$-transforms are also used such as $\hat{f}$ or $\check{f}$. In most cases of interest, $\mathcal{B}$ will be a set of functions defined on a subset of $\mathcal{R}^m(\mathcal{C}^m)$ or a set of sequences of complex numbers.

DEFINITION 4.2 *The function transform $\mathcal{L}[f]$ of f will be called continuous or discrete according to whether it is a function depending on a continuous parameter or a sequence of numbers.*

The notion of the continuity of the transform $\mathcal{L}[f]$ should not be confused with that of the continuity of the transformation $\mathcal{L}$ itself, which as a mapping from $\mathcal{A}$ into $\mathcal{B}$ requires that both sets have given topologies. For example, if a notion of convergence is defined on both $\mathcal{A}$ and $\mathcal{B}$, we may declare $\mathcal{L}$ to be continuous if $\mathcal{L}[f_n] \to \mathcal{L}[f]$ in $\mathcal{B}$ whenever $f_n \to f$

in $\mathcal{A}$. Therefore, we shall refrain from classifying function transformations as continuous or discrete as in Definition 4.2 since this would lead to some awkward statements, such as "a discrete transformation is continuous" or worse "a continuous transformation is not continuous!" It should be mentioned that such a distinction between the terms "transform" and "transformation" is not emphasized in the literature; many authors use them synonymously.

DEFINITION 4.3 *A function transformation is called an integral transformation if there exists a pair of functions G and K such that*

$$\mathcal{L}[f](\mathbf{y}) = \int_{\Omega_1} G(\mathbf{x}, \mathbf{y}, f(\mathbf{x}), K(\mathbf{x}, \mathbf{y}))d\mathbf{x},$$

where Ω_1 is a measurable subset of $\Omega, \mathbf{x} \in \mathcal{R}^n, \mathbf{y} \in \Omega_2 \subseteq \mathcal{R}^m$ and the integral is assumed to exist for every $f \in \mathcal{A}$.

Examples of Continuous and Discrete Transforms:

1. **Continuous Transforms:**
If G and K are continuous functions, the integral transform $\mathcal{L}[f]$ of f is continuous since $\mathcal{L}[f]$ is a function defined on some subset Ω_2 of $\mathcal{R}^m$. In most of the classical linear integral transformations, e.g., Fourier, Laplace, etc., $m = n$; however, this is not always the case. For example, in the wavelet and the sliding-window Fourier transforms $m > n$ and in the Radon transform, the dimension of $\Omega_1 = n-1 < m$.

2. **Discrete Transforms:**

   (a) **Orthogonal Expansions:**
   Let $\{\phi_n(x)\}_{n=0}^{\infty}$ be an orthonormal basis of $L^2(I, W)$. Then the transform $\mathcal{L}[f] = \{\langle f, \phi_n \rangle\}_{n=0}^{\infty}$, where $f \in L^2(I, W)$, is discrete. The inverse transformation is given by $\mathcal{L}^{-1}\left(\{\langle f, \phi_n \rangle\}_{n=0}^{\infty}\right) = \sum_{n=0}^{\infty} \langle f, \phi_n \rangle \phi_n = f$.

   (b) **Sampling:**
   The sampling transform of a function f is defined by

   $$\mathcal{L}[f] = \{f(\mathbf{x}_n)\}_{n=0}^{\infty},$$

   where $\{\mathbf{x}_n\}_{n=0}^{\infty}$ is a sequence of points in the domain of definition of f. More generally, a sampling transformation can be defined by

   $$\mathcal{L}[f] = \{c_n(f)\}_{n=0}^{\infty},$$

where $f \in \mathcal{A}$ and the sequence $\{c_n(f)\}_{n=0}^{\infty}$ is obtained by evaluating f and (or) some continuous transforms thereof at a discrete set of points, provided that these points are in the domain of definition of f and its associated continuous transforms. For example, $c_n(f)$ can be the values of f and some of its derivatives at a discrete set of points.

Analogous to the continuous transform case, the discrete transform of a function f defined on a subset of $\mathcal{R}^n$ can be a sequence indexed by m parameters with m and n not necessarily equal. For example, in the case of the discrete wavelet and discrete Gabor transforms, we have a function transformation that assigns to a function of one variable a doubly-indexed sequence of complex numbers, i.e.,

$$\mathcal{L}[f] = \{c_{m,n}(f)\}_{m,n=-\infty}^{\infty}.$$

4.3 Relationship Between Continuous and Discrete Transforms

In some cases a continuous transform may give rise to a discrete transform. We shall give an example in which the three transforms given in the above examples are related.

Let there exist a function $K(\mathbf{x}, \mathbf{t})$ of $n+m$ variables, where $\mathbf{x} \in \Omega \subseteq \mathcal{R}^n, \mathbf{t} \in \Omega^* \subseteq \mathcal{R}^m$ such that as a function of $\mathbf{t}, K$ is continuous and as a function of $\mathbf{x}, K \in L^2(I, W)$ for every fixed $\mathbf{t}$, where $I \subseteq \Omega$ is an open, connected set. Assume that there exists a sequence of points $\{\mathbf{t}_k\}_{k=0}^{\infty}$ in Ω^* such that $\{K(\mathbf{x}, \mathbf{t}_k)\}_{k=0}^{\infty}$ is a complete orthogonal set in $L^2(I, W)$. Let $f \in L^2(I, W)$ and define the continuous $\mathcal{K}$ -transform of f as

$$F(\mathbf{t}) = \mathcal{K}_{con}[f](\mathbf{t}) = \int_I f(\mathbf{x}) K(\mathbf{x}, \mathbf{t}) W(\mathbf{x}) d\mathbf{x}.$$

The discrete $\mathcal{K}$-transform is given by

$$\mathcal{K}_{dis}[f] = \{\langle f, K_k \rangle\}_{k=0}^{\infty},$$

where

$$K_k(\mathbf{x}) = K(\mathbf{x}, \mathbf{t}_k),$$

and

$$\langle f, K_k \rangle = \int_I f(\mathbf{x}) K(\mathbf{x}, \mathbf{t}_k) W(\mathbf{x}) d\mathbf{x}.$$

Hence,

$$\mathcal{K}_{con}[f](\mathbf{t}_k) = \langle f, K_k \rangle \,.$$

If we denote the sampling transformation associated with the points $\{\mathbf{t}_k\}_{k=0}^{\infty}$ by $\mathcal{S}$, then

$$\mathcal{S}\mathcal{K}_{con} = \mathcal{K}_{dis}.$$

Moreover, it is easy to see that

$$\mathcal{K}_{con}[f](\mathbf{t}) = \sum_{k=0}^{\infty} \mathcal{K}_{con}(f)(\mathbf{t}_k) S_k(\mathbf{t}),$$

with

$$S_k(\mathbf{t}) = \langle K, K_k \rangle \Big/ _{\|K_k\|^2} \,.$$

Example A: Let $\mathbf{B}$ be the n-dimensional rectangle $\Pi_{i=1}^{n} \left[-\sigma_i, \sigma_i\right]$, where $\sigma_i = \pi$. Let $f \in L^2(\mathcal{R}^n)$ have support in $\mathbf{B}$, and consider its continuous Fourier transform

$$F(\mathbf{t}) = \frac{1}{\left(\sqrt{2\pi}\right)^n} \int_{\mathbf{B}} f(\mathbf{x}) e^{i(\mathbf{x}\cdot\mathbf{t})} d\mathbf{x}, \qquad \mathbf{t} \in \mathcal{R}^n.$$

The set of vectors of the form $\mathbf{t} = (t_1, \cdots, t_n)$ with t_i is an integer for $i = 1, \ldots, n$ is countable; therefore, it can be enumerated as $\{\mathbf{t}_k\}_{k\in\mathcal{Z}}$. It is easy to see that the kernel of the transform, $e^{i(\mathbf{x}\cdot\mathbf{t})}$, has the property that$\left\{e^{i(\mathbf{x}\cdot\mathbf{t}_k)}\right\}_{k=-\infty}^{\infty}$ is a complete orthogonal system in $L^2(\mathbf{B})$ and with some straightforward calculations we obtain

$$F(\mathbf{t}) = \sum_{k=-\infty}^{\infty} F(\mathbf{t}_k) S_k(\mathbf{t}),$$

where

$$S_k(\mathbf{t}) = \frac{\sin \pi(t_1 - k_1)}{\pi(t_1 - k_1)} \cdots \frac{\sin \pi(t_n - k_n)}{\pi(t_n - k_n)},$$

$\mathbf{t} = (t_1, \cdots, t_n)$, and k is related to $(k_1, \cdots, k_n)$ as $\mathbf{t}_k$ is related to k. This relation can be rewritten in the more convenient form

$$F(t_1, \cdots, t_n) = \sum_{k_1=-\infty}^{\infty} \cdots \sum_{k_n=-\infty}^{\infty} F(k_1, \cdots, k_n) \frac{\sin \pi(t_1 - k_1)}{\pi(t_1 - k_1)} \cdots \frac{\sin \pi(t_n - k_n)}{\pi(t_n - k_n)}.$$

This equation was first derived by E. Parzen in 1956 [186] and has recently been generalized by A. Zayed [258]. The right-hand side of the equation is known as the n-dimensional cardinal series. As a special case, we obtain, for

$n = 1$, the Whittaker-Shannon-Kotel'nikov sampling theorem [257], which states that if

$$F(t) = \frac{1}{\sqrt{2\pi}} \int_{-\pi}^{\pi} f(x)e^{ixt}dx, \qquad \text{for some } f \in L^2[-\pi, \pi],$$

then

$$F(t) = \sum_{k=-\infty}^{\infty} F(k)\frac{\sin \pi(t-k)}{\pi(t-k)}.$$

4.4 Linear Integral Transformations

Let us assume that the domain $\mathcal{A}$ of the integral transformation $\mathcal{L}$ is a vector space over the field of complex numbers $\mathcal{C}$ and that $\mathcal{L}$ is given by

$$\mathcal{L}[f](\mathbf{y}) = \int_{\Omega_1} f(\mathbf{x})K(\mathbf{x}, \mathbf{y})d\mathbf{x}, \qquad f \in \mathcal{A}.$$

From the linearity of the integral it will follow that $\mathcal{L}[\alpha f + \beta g] = \alpha\mathcal{L}[f] + \beta\mathcal{L}[g]$; for all $\alpha, \beta \in \mathcal{C}, f, g \in \mathcal{A}$, and hence $\mathcal{L}$ is a linear integral transformation. The function K is called the *kernel of the transformation.* If $\mathcal{L}$ is one-to-one, it follows from a result in Section 1.5 that $\mathcal{L}^{-1}$ exists and is also linear. But note that the inverse of an integral transform may not necessarily be an integral transform, especially in the nonlinear case.

One of the most useful properties of linear integral transformations is the convolution property. But unfortunately, it is also one of the most difficult to obtain. It is the property that allows us to express the inverse transformation of the product of two transforms in terms of the original (transformed) functions. That is to say, if $\mathcal{L}[f] = F, \mathcal{L}[g] = G$, then the convolution operation associated with $\mathcal{L}$, which is denoted by $*$, is defined as

$$(f * g) = \mathcal{L}^{-1}[FG],$$

or

$$\mathcal{L}[f * g] = FG.$$

The problem of finding an explicit representation for a convolution operation associated with a given transformation is a difficult one, yet a general procedure for finding a solution can be described as follows. Suppose that there exists an operator $T_{\mathbf{y}}$, which we will call the *generalized translation operator*, with the following two properties:

(i)

$$T_{\mathbf{y}}K(\mathbf{x},\mathbf{t}) = K(\mathbf{x},\mathbf{t})K(\mathbf{y},\mathbf{t}),$$

and

(ii)

$$\int \left(T_{\mathbf{y}}f(\mathbf{x})\right) g(\mathbf{y})d\mathbf{y} = \int f(\mathbf{y}) \left(T_{\mathbf{y}}g(\mathbf{x})\right) d\mathbf{y}.$$

Then, the convolution, $*$, can be formally defined by

$$(f * g)(\mathbf{y}) = \int \left(T_{\mathbf{y}}f(\mathbf{x})\right) g(\mathbf{x})d\mathbf{x}.$$

For

$$\left(\mathcal{L}[f * g]\right)(\mathbf{t}) = \int \left((f * g)(\mathbf{y})\right) K(\mathbf{y},\mathbf{t})d\mathbf{y} = \int K(\mathbf{y},\mathbf{t})d\mathbf{y} \int (T_{\mathbf{y}}f(\mathbf{x}))g(\mathbf{x})d\mathbf{x}$$

$$= \int g(\mathbf{x})d\mathbf{x} \int (T_{\mathbf{y}}f(\mathbf{x}))K(\mathbf{y},\mathbf{t})d\mathbf{y}$$

$$= \int g(\mathbf{x})d\mathbf{x} \int f(\mathbf{y})(T_{\mathbf{y}}K(\mathbf{x},\mathbf{t}))d\mathbf{y}$$

$$= \int g(\mathbf{x})d\mathbf{x} \int f(\mathbf{y})K(\mathbf{x},\mathbf{t})K(\mathbf{y},\mathbf{t})d\mathbf{y} = \left(\int g(\mathbf{x})K(\mathbf{x},\mathbf{t})d\mathbf{x}\right)\left(\int f(\mathbf{y})K(\mathbf{y},\mathbf{t})d\right.$$

$$= F(\mathbf{t})G(\mathbf{t}).$$

In many cases, the generalized translation operator can be given by an integral in the form

$$T_{\mathbf{y}}f(\mathbf{x}) = \int f(\mathbf{z})R(\mathbf{z},\mathbf{x},\mathbf{y})\, d\mathbf{z},.$$

For example, if $R(\mathbf{z},\mathbf{x},\mathbf{y})$ is a function symmetric in $\mathbf{z},\mathbf{y}$ and satisfies

$$\int K(\mathbf{z},\mathbf{t})R(\mathbf{z},\mathbf{x},\mathbf{y})d\mathbf{z} = K(\mathbf{x},\mathbf{t})K(\mathbf{y},\mathbf{t}),$$

then it is easy to see that the first property of $T_{\mathbf{y}}$ is automatically satisfied and the second now follows immediately since

$$\int \left(T_{\mathbf{y}}f(\mathbf{x})\right) g(\mathbf{y})\, d\mathbf{y} = \int g(\mathbf{y})\, d\mathbf{y} \int f(\mathbf{z})R(\mathbf{z},\mathbf{x},\mathbf{y})\, d\mathbf{z}$$

$$= \int f(\mathbf{z})\, d\mathbf{z} \int g(\mathbf{y})R(\mathbf{y},\mathbf{x},\mathbf{z})\, d\mathbf{y} = \int f(\mathbf{z}) \left(T_{\mathbf{z}}g(\mathbf{x})\right) d\mathbf{z}.$$

4.5 Linear Integral Transformations of Generalized Functions

4.5.1 General Methods

There is no universal agreement on how to extend a particular function transformation to generalized functions. Different definitions can be used depending on what kind of properties the transformation is required to have. But no matter what one's definition is, it should be general enough to admit the following property: if $f(\mathbf{x})$ is a suitably restricted conventional function for which the transform $\mathcal{L}[f]$ exists both in the conventional and the generalized function sense, the two should coincide.

Notwithstanding the diversity of the definitions and techniques used to extend linear integral transformations of the form

$$\mathcal{L}[f](\mathbf{t}) = \int_I f(\mathbf{x})K(\mathbf{x},\mathbf{t})d\mathbf{x} \tag{4.1}$$

to generalized functions, they can generally fit into one of the following methods.

I. The Embedding Method

If the kernel of the transformation is an infinitely differentiable function in $\mathbf{x}$, a testing-function space $V(\Omega)$ (see Section 3.3.1 for the definition) is concocted so that $K(\mathbf{x},\mathbf{t}) \in V(\Omega)$ for each fixed $\mathbf{t}$ in some region Ω^* in $\mathcal{R}^m$. The transform of any $f \in V^*(\Omega)$ is then defined by

$$\mathcal{L}[f](\mathbf{t}) = \langle f(\mathbf{x}), K(\mathbf{x},\mathbf{t})\rangle\,,$$

where $\langle f, K\rangle$ denotes the number that the linear functional f assigns to K as a function of $\mathbf{x}$. The definition of $\mathcal{L}$, of course, depends on how the space V is constructed. This method is used, for example, in defining the Laplace and Hankel transformations of generalized functions.

II. The Adjoint Method

In this method, two testing-function spaces $U(\Omega)$ and $V(\Omega^*)$ are sought so that

$$\mathcal{L}\left[U(\Omega)\right] = \{\mathcal{L}(\phi) : \phi \in U(\Omega)\} = V(\Omega^*).$$

Then the transform $\mathcal{L}[f]$ of any $f \in V^*(\Omega^*)$ is defined as an element of $U^*(\Omega)$ given by

$$\langle \mathcal{L}[f], \phi\rangle = \langle f, \mathcal{L}[\phi]\rangle\,, \text{ for all } \phi \in U(\Omega).$$

This method, which was first used by L. Schwartz to extend the Fourier transformation to distributions, is particularly useful when the kernel of the transformation is not necessarily an infinitely differentiable function in $\mathbf{x}$, as is the case with the Hilbert transformation.

III. The Sequence Method

This method is used when generalized functions are defined as limits of sequences of smooth functions as done in [148] and [117]; see also [32, 33]. In this case, if $\{f_n\}$ is a sequence of smooth functions converging to a generalized function f, and if $\lim_{n\to\infty} \mathcal{L}[f_n]$ exists in some sense, usually in the weak sense, then we define the transform of f as

$$\mathcal{L}[f] = \lim_{n\to\infty} \mathcal{L}[f_n].$$

The first method is much more popular, but the second is more elegant and admittedly more difficult. Since the first method will be used more often in the sequel, we shall give a more detailed procedure for its implementation. This will be done in one dimension, not only to simplify the notation, but also to concur with the applications in this book.

4.5.2 The Embedding Method

By a generalized integral transform we mean an integral transform of a generalized function. Neither the definition of a generalized integral transform, nor the class of generalized functions to which it applies is universal. Our definition of the generalized integral transform is by no means the most general, but it is general enough to cover the applications. The technique we shall use is similar to the one used by Zemanian for the Laplace transform [276, 278].

First, let us assume that the generalized function f belongs to at least one of the spaces $\mathcal{L}^*_{a,b}$ of Example 5, Section 3.2.3 with $a < b$. Then, for each such f, define

$$\lambda_f = \{\sigma : \text{there exist two real numbers } a_\sigma \text{ and } b_\sigma \text{ such that } a_\sigma < \sigma < b_\sigma$$
$$\text{and } f \in \mathcal{L}^*_{a_\sigma, b_\sigma}\}$$

It is readily seen that (a, b) is contained in λ_f. In fact, it is not hard to see that λ_f is an open set. Let $\sigma_1 = \inf \lambda_f$ (possibly $-\infty$) and $\sigma_2 = \sup \lambda_f$ (possibly ∞) and let Ω_f denote the open strip $\sigma_1 < Re\ t < \sigma_2$ if t is complex or the open interval (σ_1, σ_2) if t is real. Finally, let us assume that the kernel $K(x, t)$ of the integral transform (4.1) belongs to the space $\mathcal{L}_{a,b}$ for $t \in \Omega_f$. Now we can define the generalized integral transform of f as follows:

DEFINITION 4.4 *The generalized integral transform $\mathcal{L}[f]$ of f is a conventional function defined on Ω_f and is given by*

$$\mathcal{L}_g[f](t) = F_g(t) = \langle f(x), K(x,t) \rangle\,, \qquad t \in \Omega_f. \tag{4.2}$$

We call Ω_f the region of definition of $\mathcal{L}_g[f]$ and σ_1 and σ_2 the abscissas of definition.

The subscript g is used to emphasize the difference between (4.1) and (4.2). When f is a suitably restricted regular generalized function, (4.2) reduces to (4.1).

If, in addition to the previous assumptions, the seminorms $\{\gamma_k\}_{k=1}^{\infty}$ defined in Example 5, Section 3.2.3 have the property that

$$\gamma_k(\psi\phi) = \sum_{j=0}^{k} P_j(\psi)\gamma_j(\phi)\,,$$

where ψ is a C^∞-function on $I, \phi \in \mathcal{L}_{a,b}$ and $P_j(x)$ are bounded functions of x, then arguments similar to those used in [276, Ch. 3] show that f can be extended so that $f \in \mathcal{L}^*(\sigma_1, \sigma_2)$. Moreover, $f \in \mathcal{L}^*(w, z)$ if either $w > \sigma_1$ or $z < \sigma_2$. In this case, we require that $K(x,t) \in \mathcal{L}(\sigma_1, \sigma_2)$ for $t \in \Omega_f$ so that (4.2) remains valid. It should be borne in mind that the operator R as well as the seminorms $\{\gamma_k\}_{k=1}^{\infty}$ are tailored in such a way that the kernel function $K(x,t)$ is indeed in $\mathcal{L}_{a,b}$ or in $\mathcal{L}(\sigma_1, \sigma_2)$.

If $f(x)$ is a locally integrable function on I such that $f(x)/\eta(x,a,b)$ is absolutely integrable on I, then both the conventional and the generalized integral transforms exist and they are equal as conventional functions in Ω_f. For, by Item 7, Section 3.3.3, $f(x)$ defines a regular functional on $\mathcal{L}_{a,b}$ and hence,

$$\mathcal{L}_g[f](t) = \langle f(x), K(x,t) \rangle = \int_I f(x)K(x,t)dx = \mathcal{L}_c[f](t)$$

for $t \in \Omega_f$.

The next theorem gives us an estimate for the growth rate of the generalized integral transform near infinity.

THEOREM 4.1 (**Zayed**, [263])
Assume that

$$R_x K(x,t) = P(t)K(x,t)$$

where $P(t)$ is a polynomial in t and R_x means that the operator R acts on $K(x,t)$ as a function of x. Then,

1.
$$\mathcal{L}_g\left[(R^*)^k f\right](t) = P^k(t)\mathcal{L}_g[f](t)$$
for any non-negative integer k, where $\mathcal{R}^*$ is the adjoint operator.

2. *$|\mathcal{L}_g[f](t)| \leq G(|t|)$ for $\sigma_1 < a \leq Re\ t \leq b < \sigma_2$ where G is a polynomial that depends in general on a and b. In particular, if $\sigma_2 = \infty$, then $\mathcal{L}_g[f](t)$ is bounded by a polynomial at ∞.*

Proof: 1) First, observe that R^* is continuous on $\mathcal{L}^*(\sigma_1, \sigma_2)$ (see item 3, Section 3.4.1 and Example 5, Section 3.2.3). Hence,

$$\mathcal{L}_g\left[(R^*)^k f\right](t) = \left\langle (R^*)^k f(x), K(x,t) \right\rangle = \left\langle f(x), R^k K(x,t) \right\rangle$$
$$= P^k(t) \left\langle f(x), K(x,t) \right\rangle = P^k(t) \mathcal{L}_g[f](t).$$

2) Since f is a continuous linear functional on $\mathcal{L}_{a,b}$ for any a and b such that $\sigma_1 < a \leq b < \sigma_2$, then there exist a constant C and a non-negative integer m such that

$$|\mathcal{L}_g[f](t)| = |\langle f(x), K(x,t) \rangle| \leq C \max_{0 \leq k \leq m} \gamma_{k,a,b}(K(x,t))$$
$$\leq C \max_{0 \leq k \leq m} \sup_{x \in I} \left| \eta(x,a,b) R^k K(x,t) \right|$$
$$= C \max_{0 \leq k \leq m} \left| P^k(t) \right| \sup_{x \in I} |\eta(x,a,b) K(x,t)|$$
$$= C \max_{0 \leq k \leq m} |P(t)|^k \gamma_{0,a,b}(K(x,t)) \leq G(|s|).$$

∎

We close this section by giving some examples of integral transforms that are special cases of our formulation. We shall state the interval I, the functions η, the operator R and the space to which the kernel of the integral transform belongs. It should be pointed out that in some of these examples, the seminorms are not exactly the same as those used in [276].

Examples:

1. **The Two-Sided Laplace Transform:**
   The conventional two-sided Laplace transform of $f(x)$ is defined by

$$(\mathcal{L}f)(t) = \int_{-\infty}^{\infty} f(x) e^{-xt} dx.$$

   To extend it to generalized functions, we take $I = (-\infty, \infty), R = \frac{d}{dx}$, and

$$\eta(x,a,b) = \begin{cases} e^{ax}, & 0 \leq x < \infty \\ e^{bx}, & -\infty < x < 0 \end{cases}$$

   where a and b are real numbers. Then $K(x,t) = e^{-xt} \in \mathcal{L}_{a,b}$ for $a \leq Re\ t \leq b$ and the generalized two-sided Laplace transform of $f \in \mathcal{L}^*_{a,b}$ can be defined as

$$(\mathcal{L}_g f)(t) = \left\langle f(x), e^{-xt} \right\rangle.$$

2. **The Stieltjes Transform:**
The conventional Stieltjes transform of $f(x)$ is defined by

$$(\mathcal{L}f)(t) = \int_0^\infty \frac{f(x)}{x+t} dx, \qquad 0 < t < \infty.$$

The extension to generalized functions is done by taking $I = (0, \infty)$, $R = x\frac{d}{dx}$,

$$\eta(x, a, b) = \begin{cases} x^{a+1/2}, & 0 < x < 1 \\ x^{b+1/2}, & 1 \leq x < \infty \end{cases}$$

so that $(x+t)^{-1} \in \mathcal{L}_{a,b}$ for $a > 0$ and $b < 0$. For $f \in \mathcal{L}^*_{a,b}$ we define

$$(\mathcal{L}_g f)(t) = \left\langle f(x), \frac{1}{x+t} \right\rangle.$$

4.6 Discrete Transforms of Generalized Functions

Let I be an interval and R be the linear differential operator defined in Example 6, Section 3.1.3. Assume that there exists a sequence $\{\lambda_n\}_{n=0}^\infty$ of real eigenvalues of R and a sequence $\{\phi_n\}_{n=0}^\infty$ of infinitely differentiable eigenfunctions on I such that

$$R\phi_n = \lambda_n \phi_n,$$

with $\{\phi_n\}_{n=0}^\infty$ being a complete orthonormal family on $L^2(I)$. Let us define $\mathcal{A}$ as the set of all complex-valued C^∞-functions ϕ on I that satisfy the following two conditions:

1. $$\gamma_k(\phi) = \left[\int_I \left| \left(R^k \phi \right)(x) \right|^2 dx \right]^{\frac{1}{2}} < \infty,$$

   for $k = 0, 1, 2, \ldots$

2. $$\left\langle R^k \phi, \phi_n \right\rangle = \left\langle \phi, R^k \phi_n \right\rangle$$

   for each n and k.

It is not difficult to show that $\mathcal{A}$ is a linear subspace of $L^2(I)$ and that $\{\gamma_k\}_{k=0}^\infty$ is a separating family of seminorms; hence, $\mathcal{A}$ is a countably normed space. In fact, it is a Fréchet space. It can also be shown

[276, p. 253] that $\mathcal{A}$ is a subspace of $\mathcal{E}(I)$ and that the convergence in $\mathcal{A}$ implies uniform convergence on compact subsets of I. Therefore, $\mathcal{A}$ is a testing-function space. The eigenfunctions $\{\phi_m\}_{m=0}^{\infty}$ belong to $\mathcal{A}$ since by assumption they are in $C^{\infty}(I)$ and in addition they satisfy conditions (1) and (2) above. For,

$$\gamma_k(\phi_m) = |\lambda_m|^k \left[\int_I |\phi_m(x)|^2 \, dx \right]^{\frac{1}{2}} = |\lambda_m|^k < \infty,$$

for $k = 0, 1, 2, \ldots$. Furthermore, we have

$$\left\langle R^k \phi_m, \phi_n \right\rangle = \lambda_m^k \left\langle \phi_m, \phi_n \right\rangle = \lambda_m^k \delta_{m,n},$$

and because the eigenvalues are real we have

$$\left\langle \phi_m, R^k \phi_n \right\rangle = \lambda_n^k \left\langle \phi_m, \phi_n \right\rangle = \lambda_n^k \delta_{m,n},$$

therefore

$$\left\langle R^k \phi_m, \phi_n \right\rangle = \left\langle \phi_m, R^k \phi_n \right\rangle$$

for all m and n.

Let f be a generalized function in $\mathcal{A}^*$. Since $\{\phi_n\}_{n=0}^{\infty} \subseteq \mathcal{A}$, the numbers $\langle f, \phi_n \rangle$ are well defined, and hence we can define the transformation $\mathcal{L}$ on $\mathcal{A}^*$ by

$$\mathcal{L}[f] = \{\langle f, \phi_n \rangle\}_{n=0}^{\infty} .$$

To find an inversion formula for $\mathcal{L}$, we need the following result which is proved in [276, p. 254]:

If $\phi \in \mathcal{A}$, then

$$\phi = \sum_{n=0}^{\infty} \langle \phi, \phi_n \rangle \, \phi_n,$$

where the series converges in the sense of $\mathcal{A}$. Moreover, if $\{a_n\}_{n=0}^{\infty}$ is a sequence of complex numbers, then the series $\sum_{n=0}^{\infty} a_n \phi_n$ converges in $\mathcal{A}$ if and only if

$$\sum_{n=0}^{\infty} |\lambda_n|^{2k} \, |a_n|^2 < \infty,$$

for every non-negative integer k. As a consequence of this, one can easily prove the following result which provides an inversion formula for $\mathcal{L}$.

THEOREM 4.2

Let $f \in \mathcal{A}^*$, then

$$f = \sum_{n=0}^{\infty} \langle f, \phi_n \rangle \, \phi_n \, ,$$

where the series converges in $\mathcal{A}^*$. Moreover, if $\{b_n\}_{n=0}^{\infty}$ is a sequence of complex numbers, then the series $\sum_{n=0}^{\infty} b_n \phi_n$ converges in $\mathcal{A}^*$ if and only if there exists a non-negative integer q such that

$$\sum_{\lambda_n \neq 0} |\lambda_n|^{-2q} |b_n|^2 < \infty.$$

Furthermore, if we denote the sum of the series $\sum_{n=0}^{\infty} b_n \phi_n$ by f, then $b_n = \langle f, \phi_n \rangle$.

Finally, the transformation $\mathcal{L}$ is easily seen to be one-to-one and continuous, that is if $\mathcal{L}[f] = \mathcal{L}[g]$, then $f = g$ in $\mathcal{A}^*$ and if $\lim_{\nu \to \infty} f_\nu = f$ in $\mathcal{A}^*$, then $\lim_{\nu \to \infty} \mathcal{L}[f_\nu] = \lim_{\nu \to \infty} \{\langle f_\nu, \phi_n \rangle\}_{n=0}^{\infty} = \{\langle f, \phi_n \rangle\}_{n=0}^{\infty} = \mathcal{L}[f]$.

Other expansions of generalized functions in series of orthogonal and non-orthogonal functions can be found in [260, 262, 264, 265, 275].

5

The Laplace Transform

One of the oldest and most commonly used linear integral transformations is the Laplace transformation. It dates back to the work of the French mathematician, Pierre Simon Marquis de Laplace (1749-1827), who used it in his work on probability theory in the 1780s. His major work [136] *Théorie analytique des probabilités* of 1812 also contains some of his results on this integral transformation. Although the Laplace transformation was also cited in Fourier's famous paper [66] on heat conduction in 1811, it was the English electrical engineer Oliver Heaviside (1850-1925) who made it popular by using it to solve elementary ordinary differential equations and to develop the operational calculus [98, 99].

The importance of the Laplace transformation stems from the fact that it provides an alternative method for solving differential equations, initial and boundary-value problems by reducing them to algebraic equations. Unlike other standard techniques used in solving ordinary differential equations, the Laplace transformation technique enables one to solve a nonhomogeneous equation directly without having to solve the corresponding homogeneous one first. It can also be used to solve certain partial differential equations and some related initial and boundary-value problems.

Before we give the definition of the Laplace transformation, it should be noted that for this transform it is customary to use t as the variable of the function to be transformed and s as the variable of its Laplace transform.

5.1 Definitions

DEFINITION 5.1 *The Laplace transform of a function $f(t)$ is defined as*

$$\mathcal{L}\left[f(t)\right](s) = F(s) = (\mathcal{L}f)(s) = \int_0^\infty f(t)e^{-st}dt, \tag{5.1}$$

whenever the integral exists for at least one value of s. The transform variable, s, can be taken as a complex number. We shall say that f is Laplace transformable or the Laplace transformation is applicable to f if $(\mathcal{L}f)$ exists for at least one value of s. The integral on the right-hand side of (5.1) is called the Laplace integral of f.

5.2 Existence and Domain of Convergence

1. Functions of Exponential Type:

Sufficient conditions for the existence of the Laplace transform are:

(a) f is a locally integrable function, i.e.,

$$f \in L^1_{loc}[0,\infty),$$

and

(b) f is of (real) exponential type (cf. Section 1.3), i.e., for some constants $M, t_0 > 0$ and real γ, f satisfies

$$|f(t)| \leq Me^{\gamma t} \text{ for all } t \geq t_0. \tag{5.2}$$

For,

$$|F(s)| \leq \int_0^{t_0} |f(t)|\,dt + M\int_{t_0}^\infty e^{(\gamma-s)t}dt < \infty,$$

provided that $Re\, s > \gamma$. This also shows that the Laplace integral of f converges absolutely as well. In fact, we have the following theorem.

THEOREM 5.1
If f is a locally integrable function on $[0,\infty)$ of (real) exponential type γ, then the Laplace integral of f,

$$\int_0^\infty f(t)e^{-st}dt,$$

converges absolutely for $Re\, s > \gamma$ and uniformly for $Re\, s \geq \gamma_1 > \gamma$. Consequently by Theorem 1.1, *$F(s)$ is analytic in the half-plane $\Omega = \{s \in C : Re\, s > \gamma\}$.*

The actual domain of existence of the Laplace transform may be larger than the one given by the theorem. For example, the function $f(t) = \cos e^t$ is of real exponential type zero, but

$$F(s) = \int_0^\infty (\cos e^t)e^{-st}dt = \int_1^\infty x^{-(1+s)} \cos x \, dx$$

exists for $\mathrm{Re}\, s > -1$.

With a minor modification of the proof, one can show that if the Laplace transform, $F(s)$, of f exists for some $s_0 = \sigma_0 + i\tau_0$, then it also exists for any $s = \sigma + i\tau$ for which $\mathrm{Re}\, s > \mathrm{Re}\, s_0$.

2. Arbitrary Functions:
For functions that are not necessarily of exponential type, the following is true.

THEOREM 5.2
If $f(t)$ is a locally integrable function on $[0,\infty)$ such that

$$\int_0^\infty f(t)e^{-s_0 t}dt \tag{5.3}$$

converges for some complex number s_0, then the Laplace integral

$$\int_0^\infty f(t)e^{-st}dt \tag{5.4}$$

converges in the region $Re\, s > Re\, s_0$ and converges uniformly in the region $|\arg(s - s_0)| \leq \theta' < \frac{\pi}{2}$. Moreover, if (5.3) diverges, then so does (5.4) for $Re\, s < Re\, s_0$.

It is worth noting that if the Laplace integral converges for any s with $\mathrm{Re}\, s > \mathrm{Re}\, s_0$ and for $s = s_0$, it does not necessarily converge for every other

s having the same real part as s_0, i.e., $\operatorname{Re} s = \operatorname{Re} s_0$. For example, for the function

$$f(t) = \begin{cases} 0, & 0 \le t < 1 \\ 1/t, & 1 \le t \end{cases},$$

the Laplace integral converges for $s_0 = 0 + i\tau (\tau \neq 0)$ since

$$\int_0^\infty f(t)e^{-s_0 t}dt = \int_1^\infty \frac{e^{-i\tau t}}{t}dt = -\operatorname{Ci}(\tau) + i\operatorname{si}(\tau)$$

(cf. Section 2.3.3, item 5(i)), but it diverges for $s_0 = 0$.

3. Abscissa of Convergence.

For a given $f(t)$, consider the set E_f of all real numbers σ for which the integral (5.4) converges for some complex number $s = \sigma + i\tau$, and let $\sigma_c = \inf E_f$. The number σ_c is called the *abscissa of convergence* or the *index of convergence.* The line $\operatorname{Re} s = \sigma_c$ is called the *axis of convergence.* In view of Theorem 5.2, one of the following cases must hold:

(i) $-\infty < \sigma_c < \infty$, then in this case $F(s)$ exists and defines an analytic function in the half-plane $\operatorname{Re} s > \sigma_c$ and in addition, the integral (5.4) does not exist for any s with $\operatorname{Re} s < \sigma_c$;

(ii) $\sigma_c = -\infty$, in this case the integral (5.4) converges absolutely and uniformly for any value of s; hence, defines an analytic function in the entire complex plane, i.e., $F(s)$ is an entire function;

(iii) $\sigma_c = \infty$, here the integral diverges for any s and the Laplace transformation is not applicable to $f(t)$.

The three cases described above can actually arise: for $f(t) = e^{e^t}, \sigma_c = \infty$; for $f(t) = \chi_{[0,1]}(t), \sigma_c = -\infty$, and for $f(t) = 1, \sigma_c = 0$. For functions of real exponential type γ, we have $\sigma_c \le \gamma$. Not much can be said about convergence on the line $\operatorname{Re} s = \sigma_c$.

We can summarize the aforementioned results in the following theorem.

THEOREM 5.3

For every Laplace transformable function f, there exists a real number σ_c with $-\infty \le \sigma_c < \infty$ such that the Laplace integral (5.4) converges for any s with $\operatorname{Re} s > \sigma_c$ and diverges for $\operatorname{Re} s < \sigma_c$. Moreover, the Laplace transform $F(s)$ is an analytic function in the half-plane $\operatorname{Re} s > \sigma_c$.

5.3 Domain of Absolute and Uniform Convergence

If we strengthen the hypothesis of Theorem 5.2, we obtain a more general result.

THEOREM 5.4
If $f(t)$ is a locally integrable function on $[0,\infty)$ such that the Laplace integral (5.3) converges absolutely at $s_0 = \sigma_0 + i\tau_0$, then it converges absolutely and uniformly in the half-plane $Re\, s \geq \sigma_0$.

The proof follows immediately from the fact that for $\mathrm{Re}\, s \geq \sigma_0$,

$$\int_0^\infty \left|e^{-st} f(t)\right| dt \leq \int_0^\infty e^{-\sigma_0 t} \left|f(t)\right| dt < \infty.$$

Analogous to the abscissa of convergence, we can define the *abscissa of absolute convergence*, σ_a. Evidently, $\sigma_c \leq \sigma_a$. For example, if $f(t) = e^t \sin e^t$, then $\sigma_c = 0$ and $\sigma_a = 1$. For,

$$\int_0^\infty e^{-st} e^t \sin e^t dt = \int_1^\infty \frac{\sin x}{x^s} dx < \infty, \text{ for } 0 < \mathrm{Re}\, s,$$

thus $\sigma_c = 0$. But since

$$\int_0^\infty \left|e^{-st} e^t \sin e^t\right| dt \leq \int_0^\infty e^{-t(\sigma-1)} dt < \infty$$

if $\sigma > 1$, and

$$\int_0^\infty \left|e^{-st} e^t \sin e^t\right| dt = \int_1^\infty \frac{|\sin x|}{x} dx = \infty,$$

if $\sigma = 1$, it follows that $\sigma_a = 1$.

A more striking example is $f(t) = e^t e^{e^t} \sin e^{e^t}$, for which $\sigma_c = 0$ and $\sigma_a = \infty$. For,

$$\int_0^\infty e^{-st} e^t e^{e^t} \sin e^{e^t} dt = \int_e^\infty \frac{\sin x}{(\ln x)^s} dx,$$

which converges for $s > 0$ and diverges for $s = 0$; hence $\sigma_c = 0$. On the other hand, this integral does not converge absolutely for any s.

As for the uniform convergence of the Laplace integral, we have the following theorem.

THEOREM 5.5 [245, p. 50]
If the integral (5.4) converges uniformly on the line $Re\, s = \sigma_0$, then it converges uniformly in the half-plane $Re\, s \geq \sigma_0$.

The *abscissa of uniform convergence* σ_u of the Laplace integral of $f(t)$ is defined as follows: if for every $\epsilon > 0$, the integral (5.4) converges uniformly in the half-plane $\operatorname{Re} s \geq \sigma_u + \epsilon$ and fails to do so in the half-plane $\operatorname{Re} s \geq \sigma_u - \epsilon$, we shall say that σ_u is the abscissa of uniform convergence and the line $\operatorname{Re} s = \sigma_u$ is the *axis of uniform convergence.*

The relationship between σ_c, σ_a and σ_u can be summarized as follows: $\sigma_c \leq \sigma_u \leq \sigma_a$ (cf. Theorem 5.4). It should be mentioned that although $F(s)$ is analytic in $\operatorname{Re} s > \sigma_c$ and that the Laplace integral defining F diverges for $\operatorname{Re} s < \sigma_c$ (it may even diverge for $\operatorname{Re} s = \sigma_c$), F may be continued analytically to and beyond the axis of convergence $\operatorname{Re} s = \sigma_c$. Hence, every point of the line $\operatorname{Re} s = \sigma_c$ can be a regular point of the analytic function $F(s)$. To illustrate, consider the function $f(t) = e^t \sin e^t$ for which the axis of convergence is $\operatorname{Re} \sigma = 0$. By integration by parts, the Laplace transform

$$F(s) = \int_0^\infty e^{-st} e^t \sin e^t dt = \int_1^\infty \frac{\sin x}{x^s} dx, \qquad \operatorname{Re} s > 0$$

can be continued analytically to the region $\operatorname{Re} s > -1$,

$$F(s) = \cos 1 - s \int_1^\infty \frac{\cos x}{x^{s+1}} dx, \qquad \operatorname{Re} s > -1.$$

In fact, by successive integration by parts, one can show that F is an entire function.

5.4 Uniqueness

Let $f(t)$ and $g(t)$ be two Laplace transformable functions with

$$F(s) = (\mathcal{L}f)(s) = \int_0^\infty f(t)e^{-st}dt, \quad G(s) = (\mathcal{L}g)(s) = \int_0^\infty g(t)e^{-st}dt.$$

If $F(s) = G(s)$ in some common region of convergence, then $f(t) = g(t), 0 \leq t < \infty$. Furthermore, two functions which differ on a set of positive measure on the positive t-axis cannot have the same Laplace transform.

5.5 Elementary Properties

Let $\mathcal{L}[f] = F$ and $\mathcal{L}[g] = G$, and $H(t)$ be the Heaviside function. In the notation of Section 1.1, $H(t) = \chi_{[0,\infty)}(t)$.

1. **Linearity:**

$$\mathcal{L}[\alpha f + \beta g] = \alpha\mathcal{L}[f] + \beta\mathcal{L}[g] = \alpha F + \beta G, \text{for any constants } \alpha \text{ and } \beta$$

2. **Dilation (scaling):**

$$\mathcal{L}\left[(f(at))\right](s) = \frac{1}{a}F(s/a), a > 0, \tag{5.5}$$

or

$$\mathcal{L}\left[D_{1/a}f\right](s) = D_a F(s), \quad a > 0.$$

3. **Modulation:**

$$\mathcal{L}\left[e^{at}f(t)\right](s) = F(s-a) \tag{5.6}$$

4. **Translation:**

$$\mathcal{L}\left[f(t-a)H(t-a)\right](s) = e^{-as}F(s), \quad a > 0 \tag{5.7}$$

or

$$\mathcal{L}\left[T_a(fH)\right](s) = E_{ia}(F(s)).$$

Property (4) can be put in the form
4′.

$$\mathcal{L}\left[f(t)H(t-a)\right](s) = e^{-as}\mathcal{L}\left[f(t+a)\right](s)$$

or

$$\mathcal{L}\left[fT_aH\right](s) = e^{-as}\mathcal{L}\left[T_{-a}f\right](s).$$

Example A: If

$$g(t) = \begin{cases} 0, & 0 \le t \le a \\ (t-a)^\nu, \, a \le t, & \operatorname{Re}\nu > -1 \,, \end{cases}$$

then $g(t) = f(t-a)H(t-a)$ where $f(t) = t^\nu$ and thus $G(s) = e^{-as}\Gamma(\nu+1)/s^{\nu+1}\,, \operatorname{Re} s > 0$.

Example B: If $g(t) = \begin{cases} t, \; 0 \le t \le a \\ 0, \quad a < t \end{cases}$, we may write

$$g(t) = t\left[H(t) - H(t-a)\right] = tH(t) - (t-a)H(t-a) - aH(t-a);$$

thus by properties (1) and (4)

$$G(s) = \frac{1}{s^2} - \frac{1}{s^2}e^{-as} - \frac{a}{s}e^{-as}.$$

5. **Differentiation:** Let f be a differentiable function satisfying conditions (a) and (b) of Section 5.2 with the additional assumption that $\lim_{t\to 0^+} f(t) = f(0^+)$ exists. If f' is locally integrable on $[0, \infty)$, then the Laplace transform of f' exists and

$$\mathcal{L}\left[f'\right](s) = sF(s) - f(0). \tag{5.8}$$

Note that although f is assumed to be of exponential type, f' need not be, for example, $f(t) = \sin e^{t^2}$, is bounded, but $f'(t) = 2te^{t^2} \cos e^{t^2}$ is not of exponential type.

6. **Differentiation of Higher Orders:** Let f be an n differentiable function such that $f^{(k)}, k = 0, 1, \ldots, n-1$ satisfies conditions (a) and (b) of Section 5.2 with the additional assumption that $\lim_{t\to 0^+} f^{(k)}(t) = f^{(k)}(0^+)$ exists. If $f^{(n)}$ is locally integrable on $[0, \infty)$, then its Laplace transform exists and

$$\mathcal{L}\left[f^{(n)}\right](s) = s^n F(s) - s^{n-1} f(0) - s^{n-2} f'(0) - \cdots - f^{(n-1)}(0). \tag{5.9}$$

7. **Integration:** Let $g(t) = \int_0^t f(x)dx$ and assume that $F(s)$ and $G(s)$ exist. Then

$$\mathcal{L}\left[(g)\right](s) = \mathcal{L}\left[\int_0^t f(x)\,dx\right](s) = \frac{1}{s}\mathcal{L}\left[f\right](s) \tag{5.10}$$

or $G(s) = (1/s)F(s)$. By repeated applications of (5.10) we formally obtain

$$\mathcal{L}\left[f^{(-n)}\right](s) = \frac{1}{s^n} F(s)$$

where $f^{(-n)}$ is the nth antiderivative of f defined by $f^{(-n)}(t) = \int_0^t dt_n \int_0^{t_n} dt_{n-1} \ldots \int_0^{t_2} f(t_1)dt_1$.

We have seen that the Laplace transform is an analytic function in a half-plane; hence, it has derivatives of all orders at any point in that half-plane. The next property shows that we can evaluate these derivatives by direct differentiation.

8. **Differentiation of the Transform Function:** Let f be a locally integrable function and σ_c be the abscissa of convergence of its Laplace integral, then F is analytic in $\operatorname{Re} s > \sigma_c$ and

$$F^{(n)}(s) = (-1)^n \mathcal{L}\left[t^n f(t)\right](s), \qquad n = 0, 1, 2, \ldots, \operatorname{Re} s > \sigma_c.$$

By combining this with (5.10), we obtain

$$\mathcal{L}\left[\left(t\frac{d}{dt}\right)^n f(t)\right](s) = \left(-\frac{d}{ds}s\right)^n F(s), \tag{5.11}$$

where

$$\left(t\frac{d}{dt}\right)^n f(t) = \left(t\frac{d}{dt}\right)\left(t\frac{d}{dt}\right)^{n-1} f, \left(t\frac{d}{dt}\right)^0 = 1\ ,$$

and

$$\left(\frac{d}{ds}s\right)^n F(s) = \left(\frac{d}{ds}s\right)\left(\frac{d}{ds}s\right)^{n-1} F(s), \left(\frac{d}{ds}s\right) F(s) = \frac{d}{ds}(sF(s)).$$

Property (6) gives us the transform of the derivatives of f while (8) gives us the derivatives of the transform of f.

9. **Division by t:** If f satisfies conditions (a) and (b) of Section 5.2, together with the condition that $f(t)/t$ is a Laplace transformable function, then

$$\mathcal{L}\left[\frac{f(t)}{t}\right](s) = \int_s^\infty F(u)du.$$

Or more generally,

$$\mathcal{L}\left[\frac{f(t)}{t^n}\right](s) = \int_s^\infty \cdots \int_s^\infty F(s)(ds)^n, \tag{5.12}$$

nth repeated integral.

Example C: From $\mathcal{L}[\sin t](s) = 1/(s^2+1)$, it follows that

$$\mathcal{L}\left[\frac{\sin t}{t}\right](s) = \frac{\pi}{2} - \tan^{-1} s = \tan^{-1}\left(\frac{1}{s}\right).$$

It also follows from properties (7) and (9) that

$$\mathcal{L}\left[\int_0^t \frac{f(x)}{x}\,dx\right](s) = \frac{1}{s}\int_s^\infty F(u)du. \tag{5.13}$$

For example

$$\mathcal{L}[\mathrm{Si}(t)](s) = \mathcal{L}\left[\int_0^t \frac{\sin x}{x}\,dx\right](s) = \frac{1}{s}\tan^{-1}\frac{1}{s}.$$

Similarly, one can easily show that

$$\mathcal{L}\left[\int_0^\infty \frac{f(t)}{t}\,dt\right](s) = \frac{1}{s}\int_0^\infty F(s)ds,$$

which, when combined with (5.13), yields

$$\mathcal{L}\left[\int_t^\infty \frac{f(x)}{x}dx\right](s) = \frac{1}{s}\int_0^s F(u)du. \tag{5.14}$$

10. **Periodic Functions:** Let f be a locally integrable function that is periodic with period T. Then

$$\mathcal{L}[f](s) = \frac{1}{(1-e^{-Ts})}\int_0^T f(t)e^{-st}dt. \tag{5.15}$$

Example D: Let

$$f = \begin{cases} 1 \text{ if } & nT \leq t \leq (n+1/2)T \\ 0 \text{ if } & (n+1/2)T < t < (n+1)T \end{cases},$$

for $n = 0, 1, 2, \cdot$. Clearly f is periodic with period T. Thus,

$$F(s) = \frac{1}{(1-e^{-Ts})}\int_0^{T/2} e^{-st}dt = \frac{(1-e^{-sT/2})}{s(1-e^{-Ts})} = \frac{1}{s(1+e^{-Ts/2})}.$$

5.6 Analytic Properties and Asymptotic Behavior

If $f(t)$ is a real-valued locally integrable function on $[0,\infty)$ for which the Laplace integral converges at s_0, then $F(s)$ is analytic, real for real s, and

$$\lim_{|s|\to\infty} F(s) = 0$$

uniformly in the region

$$|\arg(s - s_0)| \leq \theta < \frac{\pi}{2}.$$

The behavior of F on the boundary of the region of convergence can be described as follows: If the Laplace integral of f has abscissa of convergence $\sigma_c < \infty$, then

$$F(\sigma + i\tau) = 0\left(|\tau|\right) \text{ as } |\tau| \to \infty$$

uniformly in $\sigma_c + \delta \leq \sigma$ for any $\delta > 0$.

But if we further assume that $f \in L^1_{loc}[0, \infty)$ is of real exponential type α, then F tends to zero uniformly on vertical lines as well, i.e.,

$$\lim_{|s| \to \infty} F(s) = 0$$

for

$$|\arg(s - s_0)| \leq \frac{\pi}{2}$$

where $\operatorname{Re} s_0 > \alpha$.

As for the location of the singularities of F, we have the following Theorem.

THEOREM 5.6 [245, p. 96]
If f is periodic with period T and of bounded variation on $[0, T]$, then $F(s)$ has no singularities except possibly for poles of order one at $s = \pm(2\pi i n/T), n = 0, 1, 2, \ldots$.

The fact that $F(s)$ is analytic and approaches zero as $|s| \to \infty$ may suggest that it is analytic near infinity. This turns out to be true if further assumptions are imposed on $f(t)$.

THEOREM 5.7 [245, p. 95]
The Laplace transform $F(s)$ is an entire function in $1/s$ if and only if $f(t)$ is an entire function of order 1 and minimal type, i.e., $f(t) = 0(e^{\epsilon|t|})$ as $|t| \to \infty$ for every $\epsilon > 0$ (cf. Section 1.3). *Moreover, if*

$$F(s) = \sum_{n=0}^{\infty} \frac{a_n}{s^{n+1}}, \qquad |s| > c,$$

then $f(t)$ is an entire function of exponential type α with $\alpha > c$ and

$$f(t) = \sum_{n=0}^{\infty} \frac{a_n t^n}{n!}.$$

This theorem also relates the Laplace transform to the Borel transform, see Chapter 7.

5.7 Characterization of the Laplace Transform

We have seen in the preceding sections that for $F(s)$ to be the Laplace transform of some locally integrable function $f(t)$ it is necessary that F be analytic in a half-plane of the form $\text{Re}\, s > \sigma_c$ for some $-\infty \leq \sigma_c < \infty$ and that $\lim_{|s|\to\infty} F(s) = 0$ on a non-vertical ray in that half-plane. Now we give sufficient conditions for F to be a Laplace transform of $f(t)$.

THEOREM 5.8 [245, p. 375]
A function $F(s)$ is the Laplace transform of some locally integrable function $f(t)$ if any of the following conditions is satisfied:

1. *F is analytic in some half-plane $\text{Re}\, s > \sigma_c$ with*

$$\lim_{|\tau|\to\infty} |\tau|\,|F(\sigma + i\tau)| = 0, \qquad \sigma > \sigma_c.$$

2. *F is a rational function of the form $F(s) = P(s)/Q(s)$ where $P(s)$ and $Q(s)$, are polynomials in s with degree $Q >$ degree P.*

3. *F is either of the form*

   (a) *$F(s) = G(s)H(s)$, where G and H are the Laplace transforms of some locally integrable functions g and h,*
   or

   (b) *$F(s) = G(s)e^{-sT}, T \geq 0$, where G is the Laplace transform of a locally integrable function g.*

5.8 Inversion Formula

5.8.1 Inversion by Integration

If $f(t)$ is a locally integrable function on $[0, \infty)$ such that

1. f is of bounded variation in a neighborhood of a point $t_0 \geq 0$ (a right-hand neighborhood if $t_0 = 0$); see Section 1.4,

2. the Laplace integral of f converges absolutely on the line $\text{Re}\, s = c$, then

$$\lim_{T\to\infty} \frac{1}{2\pi i} \int_{c-iT}^{c+iT} F(s)e^{st_0}ds = \begin{cases} 0, & \text{if } t_0 < 0, \\ f(0+)/2 & \text{if } t_0 = 0, \\ \left[f(t_0+) + f(t_0-)\right]/2 & \text{if } t_0 > 0, \end{cases}.$$

In particular, if f is differentiable on $(0, \infty)$ and satisfies conditions (1) and (2), then

$$\lim_{T \to \infty} \frac{1}{2\pi i} \int_{c-iT}^{c+iT} F(s) e^{st} ds = f(t), \quad 0 < t < \infty.$$

It is worth emphasizing that the value of the integral $\int_{c-i\infty}^{c+i\infty} F(s)e^{st}ds$ is taken as the Cauchy principle value (cf. Section 1.4) since in general this integral may be divergent. For example, if $f(t) = 1$, then $F(s) = 1/s$ and for $c = 1, t = 0$, the integral

$$\int_{1-i\infty}^{1+i\infty} \frac{1}{s} ds$$

diverges. The existence of the infinite integral in the inversion formula can be interpreted in a different sense, namely, in the summability sense.

Let f be a locally integrable function that is Laplace transformable, i.e., $\sigma_c < \infty$, then for any $c > \sigma_c$, the integral

$$\frac{1}{2\pi i} \int_{c-i\infty}^{c+i\infty} F(s) e^{st} ds$$

is $(C, 1)$ summable (cf. Section 1.11.2) to 0 if $t < 0$, to $f(0^+)/2$ if $t = 0$, and to $[f(t^+) + f(t^-)]/2$ if $t > 0$, provided that these expressions are well defined.

If f is a locally integrable function with $\sigma_c < \infty$, then for every t in the Lebesgue set of f, the integral

$$\frac{1}{2\pi i} \int_{c-i\infty}^{c+i\infty} F(s) e^{st} ds, \quad c > \sigma_c,$$

is $(C, 1)$ summable to $f(t)$.

5.8.2 Inversion by Partial Fractions

The inversion formula given above is quite general, but sometimes is impractical to use because it requires contour integration. Another method can be employed if F is a rational function, $P(s)/Q(s)$, in which the degree of the denominator Q is greater than that of the numerator P.

For instance, let F be represented in its most reduced form where P and Q have no common zeros, and let us assume that Q has only simple zeros at $a_1, \ldots, a_n$, then

$$f(t) = \mathcal{L}^{-1}[F(s)](t) = \mathcal{L}^{-1}[P(s)/Q(s)](t) = \sum_{k=1}^{n} \frac{P(a_k)}{Q'(a_k)} e^{a_k t}.$$

Example A: If $P(s) = s-5$ and $Q(s) = s^2+6s+13$, then $a_1 = -3+2i, a_2 = -3-2i$, and it follows that

$$f(t) = \mathcal{L}^{-1}\left(\frac{s-5}{s^2+6s+13}\right)(t) = \frac{(2i-8)}{4i} e^{(-3+2i)t} + \frac{(2i+8)}{4i} e^{(-3-2i)t}$$

$$= e^{-3t}(\cos 2t - 4\sin 2t).$$

More generally, let $F(s) = P(s)/Q(s)$ where P and Q are as before, but Q has zeros of order $k_1+1, \ldots, k_m+1$ at $a_1, \ldots, a_m$. Let us define $\phi_i(s)$ as

$$F(s) = \frac{\phi_i(s)}{(s-a_i)^{k_i+1}}.$$

Then, in view of the Mittag–Leffler expansion (cf. Section 1.3)

$$F(s) = \sum_{i=1}^{m} \sum_{j=0}^{k_i} \frac{\phi_i^{(k_i-j)}(a_i)}{(k_i-j)!} \frac{1}{(s-a_i)^{j+1}}.$$

Thanks to properties (3), (6) in Section 5.5, and the Leibniz rule for the nth derivative of the product of two functions, we can now invert $F(s)$ to obtain

$$f(t) = \sum_{i=1}^{m} \sum_{j=0}^{k_i} \frac{\phi_i^{(k_i-j)}(a_i)}{(k_i-j)!j!} t^j e^{a_i t}$$

$$= \sum_{i=1}^{m} \frac{1}{k_i!} \left\{ \frac{\partial^{k_i}}{\partial s^{k_i}} \phi_i(s) e^{ts} \right\}_{s=a_i}.$$

Example B: For $m = 2, k_1 = 1, k_2 = 2, a_1 = a, a_2 = b$, we have

$$F(s) = \frac{P(s)}{(s-a)^2(s-b)^3},$$

where $P(s)$ is a polynomial of degree less than 5. Then

$$f(t) = \left\{ \frac{\partial}{\partial s} \phi_1(s) e^{st} \right\}_{s=a} + \frac{1}{2} \left\{ \frac{\partial^2}{\partial s^2} \phi_2(s) e^{st} \right\}_{s=b}$$

$$= (\phi_1'(a) + t\phi_1(a)) e^{at} + \frac{1}{2} \left(\phi_2''(b) + 2t\phi_2'(b) + t^2\phi_2(b)\right) e^{bt},$$

where

$$\phi_1(s) = \frac{P(s)}{(s-b)^3} \quad \text{and} \quad \phi_2(s) = \frac{P(s)}{(s-a)^2}.$$

5.8.3 Inversion by Differentiation

A less known inversion formula for the Laplace transform is derived as an analogue of the inversion formula for Taylor series. First, let us observe that the Laplace transform may be viewed as a generalization of the Taylor series

$$F(z) = \sum_{n=0}^{\infty} a_n z^n.$$

For, if we replace n by t, z by e^{-s}, $F(e^{-s})$ by $F(s)$ and the summation $\sum$ by integration, we obtain

$$F(s) = \int_0^{\infty} a(t) e^{-st} dt.$$

We may combine the two in one formula as a Laplace-Stieltjes transform

$$F(s) = \int_0^{\infty} e^{-st} d\alpha(t),$$

where $\alpha(t)$ is of bounded variation on any finite interval $[0, R]$, $R > 0$. The familiar formula for the determination of the Taylor coefficients

$$a_n = \frac{F^{(n)}(0)}{n!},$$

can be regarded as an inversion formula for the Taylor series of F. Likewise, there is an inversion formula for the Laplace transform involving the derivatives of F. It is said [245, Ch. 7] that this formula was known to Stieltjes and Hermite, but it was first proved by E.L. Post [195] and later generalized to its present form by Widder [245, Ch. 7]. Since the inversion formula for Taylor series involves the derivatives of F at the origin and the transformation $z = e^{-s}$ maps a right-handed real neighborhood of the origin to a neighborhood of the point at infinity, we expect the inversion formula for the Laplace transform to involve the derivatives of F near infinity. Indeed, if

$$F(s) = \int_0^{\infty} e^{-st} f(t) dt,$$

converges for some $s > 0$, where f is a locally integrable function, then

$$f(t) = \lim_{k\to\infty} \frac{(-1)^k}{k!} F^{(k)}\left(\frac{k}{t}\right)\left(\frac{k}{t}\right)^{k+1}, \tag{5.16}$$

for all positive t in the Lebesgue set of f. In particular, if f is of bounded variation in a neighborhood of $t > 0$, then the limit is $(f(t^+) + f(t^-))/2$.

5.9 Convolution Theorems

5.9.1 Convolution in the t-plane

Let $f(t)$ and $g(t)$ be locally integrable functions on $[0, \infty)$ and assume that their Laplace integrals converge absolutely in some half-plane $\operatorname{Re} s > \alpha$. Then the convolution h of f and g, defined by

$$h(t) = (f * g)(t) = \int_0^t f(x)g(t-x)dx,$$

is a locally integrable function on $[0, \infty)$ that is continuous if either f or g is continuous.

THEOREM 5.9 **(The Convolution Theorem)**
Let $f(t)$ and $g(t)$ be locally integrable functions on $[0, \infty)$ and assume that their Laplace integrals converge absolutely in some half-plane $\operatorname{Re} s > \alpha$. Let h denote the convolution of f and g. Then

$$H(s) = (\mathcal{L}h)(s) = F(s)G(s), \tag{5.17}$$

where F, G and H denote the Laplace transforms of f, g and h, respectively.

Proof: From the definitions of h and H, we have

$$H(s) = \int_0^\infty h(t)e^{-st}dt = \int_0^\infty e^{-st}dt \int_0^t f(x)g(t-x)\,dx\,.$$

By interchanging the integration signs, we obtain

$$\begin{aligned} H(s) &= \int_0^\infty f(x)\,dx \int_x^\infty g(t-x)e^{-st}dt \\ &= \int_0^\infty f(x)\,dx \int_0^\infty g(y)e^{-s(y+x)}dy = F(s)G(s). \end{aligned}$$

∎

Note that if $\sigma_a(f)$ and $\sigma_a(g) < \infty$, then there exists a half-plane in which the Laplace integrals of f and g are absolutely convergent, namely, $\operatorname{Re} s > \max(\sigma_a(f), \sigma_a(g))$

5.9.2 Convolution in the s-plane

Let $f(t) \in L^1_{loc}[0,\infty)$ and g be piecewise continuous on $[0,\infty)$ such that $\sigma_a(f), \sigma_a(g) < \infty$. From Section 5.8.1,

$$g(t) = \lim_{T\to\infty} \frac{1}{2\pi i} \int_\gamma G(s)e^{st}ds,$$

where γ is the line segment joining $c - iT$ and $c + iT$, and $c > \sigma_a(g)$. If the integral $\int_{c-i\infty}^{c+i\infty} G(s)e^{st}ds$ converges absolutely, then by interchanging the order of integration one can prove that

$$\int_0^\infty f(t)g(t)e^{-st}dt = \frac{1}{2\pi i} \int_{c-i\infty}^{c+i\infty} G(z)F(s-z)dz, \quad \operatorname{Re} s > \sigma_a(f) + \sigma_a(g), \tag{5.18}$$

or

$$\mathcal{L}[fg](s) = (F \otimes G)(s),$$

in which the convolution operation $\otimes$ is defined by the right-hand side of (5.18).

Example A: If $g(t) = e^{-at}, G(z) = 1/(z+a)$, and we have

$$\mathcal{L}\left[e^{-at}f(t)\right](s) = \frac{1}{2\pi i} \int_{c-i\infty}^{c+i\infty} \frac{1}{z+a} F(s-z)dz, \quad \operatorname{Re} z > -a$$

$$= F(s+a),$$

which is in agreement with property 3 in Section 5.5. The last integral can be evaluated by using the calculus of residues.

5.10 Parseval's Relation

If $f(t) \in L^2[0,\infty)$, then its Laplace integral,

$$\int_0^\infty f(t)e^{-st}dt,$$

is absolutely convergent for $Re\, s > 0$. Moreover, we have the following theorem.

THEOREM 5.10 [245, p. 245]
If $f \in L^2[0,\infty)$, then

$$\lim_{\sigma\to 0^+} F(\sigma+i\tau) = F(i\tau),$$

$$\lim_{T\to\infty} \frac{1}{2\pi i} \int_{-iT}^{iT} F(s)e^{st}ds = \begin{cases} f(t) & \text{if } t \geq 0 \\ 0 & \text{if } t < 0 \end{cases},$$

and

$$\int_0^\infty e^{-2ct} |f(t)|^2 \, dt = \frac{1}{2\pi} \int_{-\infty}^{\infty} |F(c+i\tau)|^2 \, d\tau, \tag{5.19}$$

where $c > 0$ and the limits above are taken in the sense of L^2. The relation (5.19) is the Parseval equality for the Laplace transform.

5.11 Abelian and Tauberian Theorems

5.11.1 Abelian Theorems

If for some $\gamma > 0$

$$f(t) \sim \frac{At^{\gamma-1}}{\Gamma(\gamma)} \text{ as } t \to 0^+ \text{ or as } t \to \infty,$$

then

$$F(s) \sim \frac{A}{s^\gamma} \text{ as } s \to \infty \text{ or } s \to 0^+,$$

and if

$$f(t) = o(t^{\gamma-1}) \text{ as } t \to 0^+ \text{ or as } t \to \infty,$$

then

$$F(s) = o\left(\frac{1}{s^\gamma}\right) \text{ as } s \to \infty \text{ or as } s \to 0^+, \quad [245, \text{pp. } 180\text{-}185]$$

THEOREM 5.11 **(Watson's Lemma** [216, p. 188] **)**
Let $f \in L^1_{loc}[0,\infty)$ with $f(t) = 0\left(e^{\alpha t}\right)$ as $t \to \infty$ for some $\alpha > 0$. Assume that

$$f(t) \sim \sum_{n=1}^{N} \frac{a_n}{\Gamma(\gamma_n)} t^{\gamma_n - 1}, \text{ as } t \to 0^+,$$

where $0 < \gamma_1 < \gamma_2 < \cdots < \gamma_N$*, then*

$$F(s) \sim \sum_{n=1}^{N} \frac{a_n}{s^{\gamma_n}} \quad as \ |s| \to \infty \ \ in \ |\arg(s-\alpha)| < \frac{\pi}{2}.$$

5.11.2 Tauberian Theorem

THEOREM 5.12 [245, pp. 185-188]
If $f \in L^1_{loc}[0,\infty)$ *and its Laplace integral converges for* $s > 0$*, then the two conditions*

$$\lim_{s\to 0^+} F(s) = A \ \ and \ f(t) = o\left(\frac{1}{t}\right) \quad as \ t \to \infty,$$

imply that

$$F(0^+) = \int_0^\infty f(t)dt = A.$$

More generally, let $f \geq 0$ *almost everywhere and let its Laplace transform exist for* $s > 0$*. Then the condition*

$$F(s) \sim \frac{A}{s^\gamma} \quad as \ s \to \infty \ \ or \ as \ s \to 0^+,$$

implies

$$f(t) \sim \frac{At^{\gamma-1}}{\Gamma(\gamma)} \quad as \ t \to 0^+ \ \ or \ as \ t \to \infty.$$

5.12 Laplace Transform of Generalized Functions

5.12.1 Definition and Properties

Consider the space $\mathcal{L}_{a,b}$ of Example 6, Section 3.1.3 in which

$$\eta(t,a,b) = \begin{cases} e^{at}, & 0 \leq t < \infty \\ e^{bt}, & -\infty < t < 0 \end{cases},$$

and $R = D = \frac{d}{dt}$. Let $\mathcal{L}_a$ denote the set of all C^∞ -functions on $\mathcal{R}$ such that for each nonnegative integer k and real number T,

$$\gamma_{k,a,T}(\phi) = \sup_{T<t<\infty} \left|e^{at} D^k \phi(t)\right| < \infty.$$

The set $S = \{\gamma_{k,a,T}\}$ is a separating family of seminorms since $\gamma_{0,a,T}$ is a norm. It can be shown that the two families of seminorms $\{\gamma_{k,a,T}\}$ and $\{\gamma_{k,a,T_n}\}_{n=0}^{\infty}$ are equivalent, where $\{T_n\}$ is a sequence of numbers tending to $-\infty$ as $n \to \infty$. Hence, $(\mathcal{L}_a, S)$ is a countably normed space. In addition, it is a Fréchet and testing-function space; see Sections 3.1.1 and 3.3.1. Therefore, its dual $\mathcal{L}_a^*$ is a space of generalized functions.

Two important properties of the space $\mathcal{L}_a$ are

1. If $\phi \in \mathcal{L}_a$, then ϕ and all its derivatives are of order $O(e^{-at})$ as $t \to \infty$ but no restriction on their growth as $t \to -\infty$.

2. $e^{-st} \in \mathcal{L}_a$ if and only if $\operatorname{Re} s \geq a$.
   The latter property allows us to define the Laplace transform of $f \in \mathcal{L}_a^*$ as
$$F(s) = (\mathcal{L}f)(s) = \langle f(t), e^{-st} \rangle, \quad \operatorname{Re} s \geq a.$$

Yet, a more general definition can be obtained.

For any $b, \mathcal{L}_{a,b}$ is a subspace of $\mathcal{L}_a$ and the topology of $\mathcal{L}_{a,b}$ is stronger than the one it inherits from $\mathcal{L}_a$. For, if $\phi \in \mathcal{L}_{a,b}$, then

$$\gamma_{k,a,T}(\phi) = \sup_{T<t<\infty} \left| \frac{e^{at}}{\eta(t,a,b)} \eta(t,a,b) D^k \phi(t) \right| \leq C \gamma_{k,a,b}(\phi),$$

where $C = \sup_{T<t<\infty} |e^{at}/\eta(t,a,b)|$. Similarly, one can show that $\mathcal{L}_c$ is a subspace of $\mathcal{L}_a$ if $c > a$ and that the topology of $\mathcal{L}_c$ is stronger than the one induced on it by $\mathcal{L}_a$.

Let w be either a real number or $-\infty$ and choose a sequence $\{a_n\}_{n=1}^{\infty}$, such that $a_n > w$ for all n and $\lim_{n\to\infty} a_n = w$. Define $\mathcal{L}(w)$ as the countable-union space $\bigcup_{n=1}^{\infty} \mathcal{L}_{a_n}$ (cf. Section 3.2). It can be shown that the definition of $\mathcal{L}(w)$ is independent of the sequence $\{a_n\}_{n=1}^{\infty}$ and if $w < u$, then $\mathcal{L}(u)$ is a subspace of $\mathcal{L}(w)$. Moreover, $\mathcal{L}(w)$ is not only complete, but a testing-function space for which the space $\mathcal{L}(w,z)$ (Section 3.2.3, Example 5) is a subspace for every z.

The main advantage of working with the space $\mathcal{L}(w)$ rather than $\mathcal{L}_a$ is that the space $\mathcal{D}$ is dense in $\mathcal{L}(w)$ but it is not dense in $\mathcal{L}_a$ for any a. Thus; $\mathcal{L}^*(w)$ is a space of Schwartz distributions, i.e., $\mathcal{L}^*(w) \subset \mathcal{D}^*$

DEFINITION 5.2 *Let $f \in \mathcal{L}^*(w)$ for some w. The Laplace transform of f is a conventional function $F(s)$ defined by*

$$F(s) = \langle f(t), e^{-st} \rangle, \quad \operatorname{Re} s > w.$$

We shall call f a Laplace transformable generalized function.

The definition makes sense since $e^{-st} \in \mathcal{L}(w)$ for $\operatorname{Re} s > w$. Let

$$\sigma_f = \inf \{w : f \in \mathcal{L}^*(w)\} \quad \text{and} \quad \Omega_f = \{s : \operatorname{Re} s > \sigma_f\}.$$

It follows that the Laplace transform of f

$$F(s) = \langle f(t), e^{-st} \rangle, \quad \operatorname{Re} s > \sigma_f$$

is an analytic function in Ω_f.

This definition truly generalizes the definition of the Laplace transform for conventional functions since if f is a locally integrable function on some interval $[\sigma_f, \infty)$ with $f(t) = 0$ for $-\infty < t < \sigma_f$, and is of real exponential type α, then it generates a regular member of $\mathcal{L}_\alpha^*$ (cf. item 7, Section 3.3.3) which will also be denoted by f and is defined by

$$\langle f, \phi \rangle = \int_{\sigma_f}^{\infty} f(t)\phi(t)dt, \quad \phi \in \mathcal{L}_\alpha.$$

The Laplace transform of such a function is also given by

$$F(s) = \langle f(t), e^{-st} \rangle = \int_{\sigma_f}^{\infty} f(t)e^{-st}dt, \quad \operatorname{Re} s > \alpha.$$

This is sometimes called a *right-handed Laplace transform* of f. In particular, if $\sigma_f = 0$, we obtain the standard Laplace transform. In fact, the following is true.

THEOREM 5.13 [276, p. 86]
Every (right-sided) Laplace transformable generalized function $f(t)$ is concentrated on a semi-infinite interval of the form $\sigma_f \le t < \infty$, where σ_f is finite and depends on f.

Most of the elementary properties of the conventional Laplace transform hold for the Laplace transform of generalized functions.

Examples:

1. Since the delta function is concentrated on a single point, it is easy to show that it is Laplace transformable and that

$$\mathcal{L}[\delta](s) = 1, \qquad -\infty < \operatorname{Re} s < \infty,$$

$$\mathcal{L}[\delta^k](s) = s^k, \qquad -\infty < \operatorname{Re} s < \infty.$$

2. Since $f(t) = \chi_{(0,\infty)}(t) \ln t$ is a regular member of $\mathcal{L}_\alpha^*$,

$$\frac{df}{dt} = \chi_{(0,\infty)}(t)(1/t)$$

in the sense of generalized functions, and $\mathcal{L}(f)(s) = -(1/s)(\gamma + \ln s)$, $\operatorname{Re} s > 0$, it follows that

$$\mathcal{L}\left[\chi_{(0,\infty)}(t)/t\right] = -\gamma - \ln s, \quad \operatorname{Re} s > 0,$$

where γ is Euler's constant (cf. Equation (2.3)). Symbolically, we may write the last relation as

$$\int_0^\infty \frac{e^{-st}}{t} dt = -\gamma - \ln s, \qquad \operatorname{Re} s > 0.$$

5.12.2 Characterization and Asymptotic Behavior

For a function $F(s)$ to be the Laplace transform of a generalized function concentrated on $[T,\infty)$ it is necessary and sufficient that F be analytic in a half-plane $\operatorname{Re} s \geq a$ and that

$$|F(s)| \leq e^{(-\operatorname{Re} s)T} P(|s|), \quad \operatorname{Re} s \geq a,$$

for some polynomial P [278, p. 236].

5.12.3 Inversion Formula

THEOREM 5.14 [276, p. 72]
Let f be a Laplace transformable generalized function with $(\mathcal{L}f)(s) = F(s)$ for $s \in \Omega_f$. If $Q(s)$ is a polynomial with no zeros in the strip $\sigma_f < a \leq \operatorname{Re} s \leq b < \infty$ such that

$$\left|\frac{F(s)}{Q(s)}\right| \leq \frac{K}{|s|^2}, \quad a < \operatorname{Re} s < b,$$

for some constants a, b and K then in the sense of $\mathcal{L}^*(a,b)$

$$f(t) = Q(D_t)\left(\frac{1}{2\pi i}\int_{\sigma-i\infty}^{\sigma+i\infty} \frac{F(s)}{Q(s)} e^{st} ds\right), a < \sigma < b,$$

where $D_t = \frac{d}{dt}$. The integral converges in the conventional sense to a continuous function that generates a regular member of $L^*(a,b)$.

5.12.4 Convolution

The convolution of two generalized functions $f, g \in \mathcal{L}^*_{a,b} (a \leq b)$ denoted by $f * g$, is defined by

$$\langle f * g, \phi\rangle = \langle f(t), \langle g(\tau), \phi(t+\tau)\rangle\rangle, \quad \phi \in \mathcal{L}_{a,b}.$$

One can verify that $f * g$ is also a generalized function in $\mathcal{L}^*_{a,b}$. In fact, the same result is true if $\mathcal{L}^*_{a,b}$ is replaced by $\mathcal{L}^*(a,b)$. This definition of convolution entails the following result.

THEOREM 5.15 [276, p. 99]
If $f, g \in \mathcal{L}^*(a,b)$ with $(\mathcal{L}f)(s) = F(s)$ for $s \in \Omega_f$, $(\mathcal{L}g)(s) = G(s)$ for $s \in \Omega_g$ and $\Omega_f \cap \Omega_g$ is not empty, then $f * g$ exists in $\mathcal{L}^*(a,b)$ where (a,b) is the intersection of $\Omega_f \cap \Omega_g$ with the real axis. Moreover,

$$\mathcal{L}(f * g)(s) = F(s)G(s), \quad s \in \Omega_f \cap \Omega_g.$$

5.13 Applications

Among all linear integral transformations, the Laplace transformation stands out as the most versatile in its applications. This may be attributed to the fact that it can convert a system of differential, or integral, or integro-differential equations into algebraic equations. The conversion is achieved by using various properties listed in Section 5.5, in particular, properties 5–9. Because of the appearance of $y(0), \ldots, y^{(n)}(0)$ in the formulas, these values must be specified in order to obtain a unique solution. This entails that the Laplace transform is better suited for solving initial-value problems. Assuming that the system of algebraic equations can be solved uniquely and that the inverse Laplace transforms can be calculated explicitly, the original system of equations can be solved. Below is a list of prototype problems that can be solved by using the Laplace transform.

5.13.1 Differential Equations (Initial-Value Problems)

Consider the differential operator with constant coefficients

$$L = p_0 \frac{d^n}{dt^n} + \cdots + p_{n-1} \frac{d}{dt} + p_n,$$

where $p_i = \text{constant}$, $i = 0, \ldots, n$, and consider the initial-value problem

$$Ly = f$$

$$y^{(k)}(0) = c_k, \quad k = 0, 1, \ldots, n-1,$$

where f is a Laplace transformable function (cf. Section 1.7). If we denote the Laplace transform of the solution by Y, then by using (5.9) we can convert the above initial-value problem to an algebraic equation in $Y(s)$.

Upon solving the algebraic equation for Y and taking the inverse transform, we obtain the solution $y(t)$.

Example A:

$$y^{(3)} - y = f(t)$$

$$y(0) = y'(0) = y''(0) = 0,$$

where $f(t) = \chi_{[a,\infty)}(t)$. By taking the Laplace transform of the differential equation and incorporating the initial conditions, we obtain

$$(s^3 - 1)Y = \frac{e^{-as}}{s},$$

or

$$Y(s) = \frac{e^{-as}}{s(s^3-1)} = \frac{e^{-as}}{3}\left(\frac{-3}{s} + \frac{1}{s-1} + \frac{2s+1}{s^2+s+1}\right).$$

An application of the inverse Laplace transform gives us

$$y(t) = \left[-1 + \frac{1}{3}e^{(t-a)} + \frac{2}{3}e^{-\frac{1}{2}(t-a)}\cos\frac{\sqrt{3}}{2}(t-a)\right]\chi_{[a,\infty)}(t).$$

5.13.2 Integral Equations

1. *The Volterra integral equation of the first kind and convolution type* is an integral equation of the form

$$\int_0^t y(\tau)k(t-\tau)d\tau = f(t),$$

in which k and f are known functions and y is to be determined. By using (5.17), we obtain

$$Y(s) = \frac{F(s)}{K(s)}; \text{ hence } y(t) = \mathcal{L}^{-1}\left[\frac{F(s)}{K(s)}\right].$$

If $(1/K(s))$ has an inverse Laplace transform, say $g(t)$, then by (5.17) again, we have

$$y(t) = \int_0^t f(\tau)g(t-\tau)d\tau.$$

Example B: Consider the integral equation

$$\sin t = \int_0^t e^{t-\tau}y(\tau)d\tau.$$

By applying the Laplace transform, we obtain

$$Y(s) = \frac{s-1}{s^2+1} = \frac{s}{s^2+1} - \frac{1}{s^2+1};$$

hence

$$y(t) = \cos t - \sin t.$$

2. The *Volterra integral equation of the second kind and convolution type*

$$y(t) = f(t) + \int_0^t y(\tau)k(t-\tau)d\tau$$

can also be solved by using the Laplace transform. As in case (1), an application of the Laplace transform gives us

$$Y(s) = \frac{F(s)}{1-K(s)}.$$

If we denote the inverse Laplace transform of $(1-K(s))^{-1}$ (if it exists) by $g(t)$, we then obtain

$$y(t) = \int_0^t f(\tau)g(t-\tau)d\tau.$$

Example C: Consider

$$y(t) = \sin 2t + \int_0^t y(\tau)\sin 2(t-\tau)d\tau.$$

The Laplace transform, when applied to this equation, gives us

$$Y(s) = \frac{2}{s^2+4} + \frac{3}{s^2+4}Y(s),$$

or

$$Y(s) = \frac{2}{s^2+2}; \qquad \text{hence, } y(t) = \sqrt{2}\sin\sqrt{2}t.$$

5.13.3 Integrodifferential Equation

An integrodifferential equation of the form

$$a_n y^{(n)} + \cdots + a_1 y^{(1)} + a_0 y + \int_0^t y(\tau)k(t-\tau)\,d\tau = f(t),$$

with $y^{(k)}(0) = c_k$, for some constants c_k and $k = 0, 1, \cdots, n-1$, can be solved by using the Laplace transform to yield

$$y(t) = \mathcal{L}^{-1}\left[\frac{F(s) + P_{n-1}(s)}{Q_n(s) + K(s)}\right],$$

where P_{n-1} and Q_n are polynomials of degrees $n-1$ and n, respectively.

Example D: Consider

$$y' + y + \int_0^t y(\tau)e^{(t-\tau)}d\tau = 1, \quad y(0) = 0.$$

Applying the Laplace transform to this integrodifferential equation and using the initial condition yields

$$Y(s) = \frac{s-1}{s^3} = \frac{1}{s^2} - \frac{1}{s^3};$$

hence, by using the inverse Laplace transform, we obtain

$$y(t) = t - \frac{1}{2}t^2.$$

5.13.4 Systems of Linear Differential Equations

Suppose we have a system of n first-order linear differential equations

$$\begin{array}{l} a_{11}y_1' + b_{11}y_1 + \cdots\cdots\cdots + a_{1n}y_n' + b_{1n}y_n = f_1(t) \\ \vdots \qquad\qquad\qquad\qquad\qquad\qquad \vdots \\ \vdots \qquad\qquad\qquad\qquad\qquad\qquad \vdots \\ a_{n1}y_1' + b_{n1}y_1 + \cdots\cdots\cdots + a_{nn}y_n' + b_{nn}y_n = f_n(t), \end{array}$$

where $a_{ij}, b_{ij}, i, j = 1, \ldots, n$ are constants and $f_i, i = 1, \ldots, n$ are Laplace transformable functions. By applying the Laplace transform to this system we obtain

$$(s\mathbf{A} + \mathbf{B})\,\mathbf{Y}(s) = \mathbf{F}(s) + \mathbf{A}\mathbf{y}(0), \tag{5.20}$$

where

$$\mathbf{A} = \begin{pmatrix} a_{11} & \cdots & a_{1n} \\ \vdots & & \vdots \\ a_{n1} & \cdots & a_{nn} \end{pmatrix}, \mathbf{B} = \begin{pmatrix} b_{11} & \cdots & b_{1n} \\ \vdots & & \vdots \\ b_{n1} & \cdots & b_{nn} \end{pmatrix}, \mathbf{Y}(s) = \begin{pmatrix} Y_1(s) \\ \vdots \\ Y_n(s) \end{pmatrix}$$

$$\mathbf{F}(s) = \begin{pmatrix} F_1(s) \\ \vdots \\ F_n(s) \end{pmatrix} \quad , \quad \mathbf{y}(0) = \begin{pmatrix} y_1(0) \\ \vdots \\ y_n(0) \end{pmatrix}.$$

Assuming that the matrix $s\mathbf{A} + \mathbf{B}$ is invertible, we obtain

$$Y(s) = (\mathbf{sA} + \mathbf{B})^{-1} \left(\mathbf{F}(s) + \mathbf{Ay}(0)\right),$$

hence

$$y(t) = \mathcal{L}^{-1} \left[(s\mathbf{A} + \mathbf{B})^{-1} \left(\mathbf{F}(s) + \mathbf{Ay}(0)\right) \right].$$

Equation (5.20) can also be solved by Cramer's rule.

Example E: Consider the system

$$y_1' + y_1 + y_2' + y_2 = 3e^{2t},$$

$$y_1' - 2y_1 - 2y_2' + y_2 = 6e^{-t},$$

with $y_1(0) = 1$ and $y_2(0) = 2$.
It is easy to see that

$$\mathbf{A} = \begin{pmatrix} 1 & 1 \\ 1 & -2 \end{pmatrix}, \mathbf{B} = \begin{pmatrix} 1 & 1 \\ -2 & 1 \end{pmatrix}, \mathbf{F}(s) = \begin{pmatrix} \frac{3}{s-2} \\ \frac{6}{s+1} \end{pmatrix} \text{ and } \mathbf{y}(0) = \begin{pmatrix} 1 \\ 2 \end{pmatrix},$$

Hence,

$$\begin{pmatrix} s+1 & s+1 \\ s-2 & -2s+1 \end{pmatrix} \begin{pmatrix} Y_1(s) \\ Y_2(s) \end{pmatrix} = \begin{pmatrix} \frac{3(s-1)}{s-2} \\ \frac{3(1-s)}{s+1} \end{pmatrix},$$

which yields

$$\begin{pmatrix} Y_1(s) \\ Y_2(s) \end{pmatrix} = \mathbf{M}^{-1} \begin{pmatrix} \frac{3(s-1)}{s-2} \\ \frac{3(1-s)}{s+1} \end{pmatrix} = \begin{pmatrix} \frac{1}{s-2} \\ \frac{2}{1+s} \end{pmatrix},$$

where $\mathbf{M}^{-1}$ is the inverse of the matrix

$$\begin{pmatrix} s+1 & s+1 \\ s-2 & -2s+1 \end{pmatrix}.$$

Thus

$$\begin{pmatrix} y_1(t) \\ y_2(t) \end{pmatrix} = \begin{pmatrix} e^{2t} \\ 2e^{-t} \end{pmatrix}.$$

5.13.5 Partial Differential Equations

The Laplace transform can be employed to solve some initial-boundary value problems involving partial differential equations. Examples of such problems are:
the *heat conduction problem*:

$$\begin{aligned} u_{xx} &= a^{-2}u_t, && 0 < x < \infty, \quad t > 0, \\ u(0,t) &= f(t), && u(x,t) \to 0 \text{ as } x \to \infty, \\ u(x,0) &= 0, && 0 < x < \infty, \end{aligned}$$

and the *vibrating string problem*:

$$\begin{aligned} u_{xx} &= c^{-2}u_{tt}, && 0 < x < \infty,\ t > 0 \\ u(0,t) &= f(t), && u(x,t) \to 0 \text{ as } x \to \infty, \\ u(x,0) &= 0, && u_t(x,0) = 0, \quad 0 < x < \infty. \end{aligned}$$

In both problems, the Laplace transform is applied with respect to the variable t.

Example F: Applying the Laplace transform to the partial differential equation and the boundary conditions in the heat conduction problem yields

$$U_{xx} = \frac{s}{a^2}U, \quad 0 < x < \infty,$$

$$U(0,s) = F(s), \text{ and } \quad U(x,s) \to 0 \text{ as } x \to \infty,$$

where

$$U(x,s) = \mathcal{L}\left[u(x,t)\right](s) = \int_0^\infty u(x,t)e^{-st}dt.$$

Thus, in view of the boundary conditions, we have

$$U(x,s) = F(s)e^{-x\sqrt{s}/a},$$

which, when we apply the inverse Laplace transformation to it, yields

$$u(x,t) = \frac{x}{2a\sqrt{\pi}} \int_0^t \frac{f(\tau)}{(t-\tau)^{3/2}} \exp\left(-\frac{x^2}{4a^2(t-\tau)}\right) d\tau.$$

Here we have used the fact (cf. Appendix B) that

$$\mathcal{L}^{-1}\left[e^{-x\sqrt{s}/a}\right](t) = \frac{x}{2a\sqrt{\pi}t^{3/2}} \exp\left(-\frac{x^2}{4a^2t}\right).$$

5.13.6 Ordinary Differential Delay Equations

Some ordinary differential delay equations can be solved by using the Laplace transform. For example, when the Laplace transform is applied to the problem

$$y'(t) + ay(t-k) = 0,$$
$$y(t) = y_0 \text{ when } -k \leq t \leq 0,$$

where a and k are constants, we obtain

$$(sY(s) - y_0) + ae^{-ks}Y(s) + ay_0\left(\frac{1-e^{-ks}}{s}\right) = 0,$$

which in turn gives us

$$Y(s) = \frac{y_0}{s} - \frac{ay_0}{s(s+ae^{-ks})}.$$

To invert $Y(s)$, we expand it in a series form as

$$Y(s) = \frac{y_0}{s} - y_0\sum_{n=0}^{\infty}(-1)^n\frac{a^{n+1}}{s^{n+2}}e^{-nks},$$

then apply the inverse Laplace transform to yield

$$y(t) = y_0\sum_{n=0}^{N}\frac{(-1)^n}{n!}a^n(t-nk+k)^n,$$

where $N = \left[\frac{t}{k}\right] + 1$ and $[t]$ is the greatest integer less than or equal to t.

5.13.7 Fractional Differential Equations

The qth fractional derivative of $f(t)$ is defined as (cf. Chapter 20)

$$\frac{d^q f}{dt^q} = \frac{d^n}{dt^n}\left[\frac{1}{\Gamma(n-q)}\int_a^t \frac{f(\tau)}{(t-\tau)^{q-n+1}}d\tau\right],$$

for $n > q \geq 0$, and n is a non-negative integer. In view of the relation

$$\mathcal{L}\left[\frac{1}{\sqrt{t}}\right](s) = \frac{\Gamma\left(\frac{1}{2}\right)}{\sqrt{s}},$$

we can compute the Laplace transform of the one-half derivative of f as follows:

$$\mathcal{L}\left[\frac{d^{1/2}f}{dt^{1/2}}\right] = \mathcal{L}\left\{\frac{d}{dt}\frac{1}{\Gamma(1/2)}\int_a^t \frac{f(\tau)}{(t-\tau)^{1/2}}d\tau\right\} = \frac{1}{\Gamma(1/2)}\mathcal{L}(g'(t))$$

where $g(t) = \int_a^t f(\tau)(t-\tau)^{-1/2} d\tau$. But $(\mathcal{L}g)(s) = \Gamma\left(\frac{1}{2}\right) F(s)/\sqrt{s}$. Hence,

$$\mathcal{L}\left(\frac{d^{1/2}f}{dt^{1/2}}\right)(s) = \sqrt{s}F(s) - C,$$

where $c = g(0)/\Gamma(1/2)$.
Now consider the fractional differential equation

$$\frac{d^{1/2}f}{dt^{1/2}} + af = 0.$$

By applying the Laplace transform to this equation and using the above result, we obtain

$$\sqrt{s}F(s) - C + aF(s) = 0,$$

which yields

$$F(s) = \frac{C}{\sqrt{s} + a}.$$

Thus,

$$f(t) = C\left[\frac{1}{\sqrt{\pi t}} - ae^{a^2 t} \operatorname{Erfc}\left(a\sqrt{t}\right)\right],$$

See Section 2.3.3, item 5(f).

5.13.8 Convolution Equations

Let f, y and g be Schwartz distributions whose supports are bounded on the left, i.e., their supports are inside an interval of the form $[a, \infty)$ for some a (see Definition 3.13) and consider the convolution equation

$$f * y = g,$$

in which f and g are given and y is unknown. This equation can be solved for y by using the Laplace transform to yield

$$Y(s) = \frac{G(s)}{F(s)}, \quad \operatorname{Re} s > \max(\sigma_f, \sigma_g).$$

If $1/F(s)$ has an inverse Laplace transform, in $Re\, s > \sigma_0$, then

$$y(t) = \mathcal{L}^{-1}\left(\frac{G(s)}{F(s)}\right), \quad \operatorname{Re} s > \max\left(\sigma_f, \sigma_g, \sigma_0\right).$$

Example G: Let us solve the convolution equation

$$\left(\chi_{[0,\infty)}(t) \cos t\right) * y(t) = \left(e^{t^2} - 1\right) \chi_{[0,\infty)}.$$

By applying the Laplace transform, we obtain

$$\left(\frac{s}{s^2+1}\right) Y(s) = G(s),$$

where $G(s)$ is the Laplace transform of the right-hand side. Hence

$$Y(s) = \left(s + \frac{1}{s}\right) G(s).$$

But since

$$\mathcal{L}^{-1}\left[s + \frac{1}{s}\right](t) = \delta^{(1)}(t) + \chi_{[0,\infty)}(t),$$

it then follows from the convolution theorem that

$$y(t) = \left(\delta^{(1)}(t) + \chi_{[0,\infty)}\right) * \left(\left(e^{t^2} - 1\right) \chi_{[0,\infty)}\right)$$

$$= \chi_{[0,\infty)}(t) \left(2te^{t^2} + \int_0^t \left(e^{\tau^2} - 1\right) d\tau\right).$$

5.13.9 Differential and Integral Equations Involving Generalized Functions

Most of the applications discussed above can be extended with no difficulty to include generalized functions. For example, initial-value problems involving generalized functions can also be solved in a similar fashion by using the generalized Laplace transform.

Example H: Let us solve the initial-value problem

$$y^{(2)} - y = \delta^{(2)}$$

$$y(0) = y'(0) = 0.$$

The generalized Laplace transform yields

$$Y(s) = \frac{s^2}{s^2-1} = 1 + \frac{1}{2}\left(\frac{1}{s-1} - \frac{1}{s+1}\right),$$

and hence

$$y(t) = \delta(t) + \frac{1}{2}\left(e^t - e^{-t}\right).$$

5.13.10 Difference Equations

The Laplace transform can be used to solve difference equations even if they contain generalized functions. This is demonstrated in the following example.

Example I: Solve the difference equation

$$y(t+2) + 2y(t+1) + y(t) = \delta(t+1),$$

with

$$y(t) = 0, \quad 0 \leq t \leq 2.$$

By taking the Laplace transform of both sides of the difference equation and using (5.7), we have

$$Y(s) = \frac{e^s}{e^{2s} + 2e^s + 1} = \frac{e^s}{(e^s + 1)^2}$$

The series expansion of $Y(s)$ in powers of e^{-s} is

$$Y(s) = \sum_{n=1}^{\infty} (-1)^{n-1} n\, e^{-sn}.$$

A term by term application of the inverse Laplace transform gives us

$$y(t) = \sum_{n=1}^{\infty} (-1)^{n-1} n \delta(t-n).$$

6

The Two-Sided Laplace Transform

Among the various generalizations of the Laplace transform, the two-sided (bilateral) Laplace transform seems to be the most natural. It has been pointed out in Section 5.12.1 that in the so-called right-sided Laplace transform the lower limit of the Laplace integral, zero, may be replaced by an arbitrary finite number. In the two-sided Laplace integral, this arbitrary number is replaced by $-\infty$.

6.1 Definition

DEFINITION 6.1 *The two-sided (bilateral) Laplace transform of a function $f(t)$ is defined as*

$$F(s) = \int_{-\infty}^{\infty} f(t)^{-st} dt, \quad s = \sigma + i\tau. \tag{6.1}$$

If $f(t)$ vanishes on $(-\infty, 0)$, we obtain the Laplace transform. The integral in (6.1), which is called a two-sided Laplace integral, is said to exist if and only if the limits

$$\lim_{R \to \infty} \int_{0}^{R} f(t) e^{-st} dt,$$

and

$$\lim_{R \to \infty} \int_{-R}^{0} f(t) e^{-st} dt$$

both exist. In this case, the sum of these limits is taken as the value of $F(s)$.

Because

$$F(s) = \int_{-\infty}^{0} f(t)e^{-st}dt + \int_{0}^{\infty} f(t)e^{-st}dt = \int_{0}^{\infty} f(-t)e^{st}dt + \int_{0}^{\infty} f(t)e^{-st}dt$$

$$= F_1(-s) + F_2(s),$$

where $F_1(s) = \int_0^\infty f(-t)e^{-st}dt$, the study of the two-sided Laplace transform is reduced to that of the sum of two one-sided Laplace transforms. Consequently, we can deduce any properties we want about the two-sided transform from those of the one-sided transform. However, in some cases, the proofs are more complicated than might be expected.

The two-sided, Laplace transform generalizes the Laurent series

$$\sum_{n=-\infty}^{\infty} a_n z^n,$$

as much as the Laplace transform generalizes the Taylor series $\sum_{n=0}^{\infty} a_n z^n$. Since the domain of convergence of a Laurent series is the region between two concentric circles and the transformation $z = e^{-s}$ maps vertical lines into circles centered at the origin, it is natural to expect the domain of convergence of the two-sided Laplace integral to be a vertical strip. This is indeed the case.

6.2 Existence and Domain of Convergence

6.2.1 Domain of Convergence

If the integral

$$F(s) = \int_{-\infty}^{\infty} f(t)e^{-st}dt$$

converges for $s_1 = \sigma_1 + i\tau_1$ and $s_2 = \sigma_2 + i\tau_2 (\sigma_1 < \sigma_2)$, then it converges in the vertical strip $\sigma_1 < \sigma < \sigma_2$.

For example, if $F_2(s)$ exists in the half-plane $\sigma > \sigma_1$ and $F_1(s)$ exists in the half-plane $\sigma > \sigma_2$, then $F(s)$ will exist in the strip $\sigma_1 < \sigma < \sigma_2$. If the integral (6.1) converges in the strip $\sigma_c' < \sigma < \sigma_c''$ and diverges for $\sigma > \sigma_c''$ and for $\sigma < \sigma_c'$, then each of the lines $\sigma = \sigma_c'$ and $\sigma = \sigma_c''$ will be called an *axis of convergence* and each of the numbers σ_c' and σ_c'' will be called an *abscissa of convergence*. The strip $\sigma_c' < \sigma < \sigma_c''$ will be called the *strip of convergence.* The abscissa of uniform and absolute convergence can be defined similarly, see Section 5.3.

As is the case with Laplace transform (one-sided), it is easy to show that when the two-sided Laplace integral converges in a proper strip, i.e., $-\infty < \sigma_c'$ and $\sigma_c'' < \infty$, it converges uniformly in any closed bounded region inside the strip, provided that the region does not touch the boundary of the strip. Consequently, in view of Theorem 1.1, the two-sided Laplace transform $F(s)$ is an analytic function in the interior of the strip. Moreover,

$$F^{(k)}(s) = (-1)^k \int_{-\infty}^{\infty} t^k f(t) e^{-st} dt, \quad k = 0, 1, 2, \cdots.$$

The domain of convergence of (6.1) is one of the following:

1. A right-half plane if $-\infty < \sigma_c'$, $\sigma_c'' = \infty$,
2. A left-half plane if $\sigma_c' = -\infty, \sigma_c'' < \infty$,
3. The entire plane if $\sigma_c' = -\infty$ and $\sigma_c'' = \infty$,
4. A vertical strip if $-\infty < \sigma_c'$ and $\sigma_c'' < \infty$,
5. A vertical line or a part thereof if $-\infty < \sigma_c' = \sigma_c'' < \infty$,
6. The empty set if $\sigma_c' > \sigma_c''$.

Examples showing that all these cases can actually arise are easily constructed. We only give some important cases.

Example A: If

$$f(t) = \begin{cases} e^{\alpha t}, & t > 0 \\ e^{\beta t}, & t < 0 \end{cases},$$

then

$$F(s) = \frac{1}{s - \alpha} + \frac{1}{\beta - s}, \quad \alpha < \operatorname{Re} s < \beta,$$

provided that $\alpha < \beta$. If $\alpha > \beta$, the integral (6.1) converges nowhere and $F(s)$ does not exist.

Example B: If $f(t) = e^{-\alpha t^2}$, then

$$F(s) = \int_{-\infty}^{\infty} e^{-\alpha t^2} e^{-st} dt = 2 \int_{0}^{\infty} e^{-\alpha t^2} \cosh st dt = \sqrt{\frac{\pi}{\alpha}} e^{s^2/4\alpha},$$

where $-\infty < \operatorname{Re} s < \infty$.

Example C: For $f(t) = (1+t^2)^{-1}$, the two-sided Laplace integral converges absolutely on the whole line $\sigma = 0$ and nowhere else.

Example D: For $f(t) = 1/\sqrt{|t|}$, the two-sided Laplace integral diverges everywhere off the line $\sigma = 0$, yet on this line it converges everywhere except at the origin.

6.2.2 Sufficient Conditions for Existence

Let $f(t)$ be a locally integrable function on $(-\infty, \infty)$ such that

$$|f(t)| \leq M_1 e^{\alpha t} \text{ for all } t \geq t_1 > 0,$$

and

$$|f(t)| \leq M_2 e^{\beta t} \text{ for all } t \leq t_2 < 0,$$

for some $M_1, M_2, t_1, t_2, \alpha$ and β. If $\alpha < \beta$, then the two-sided Laplace integral converges for $\alpha < \operatorname{Re} s < \beta$. More precisely, if

$$\limsup_{t\to\infty} \frac{\ln|f(t)|}{t} = \alpha \neq 0$$

and

$$\limsup_{t\to-\infty} \frac{\ln|f(t)|}{t} = \beta \neq 0$$

with $\alpha < \beta$, then the two-sided Laplace integral (6.1) converges for $\alpha < \operatorname{Re} s < \beta$ and diverges for $\operatorname{Re} s < \alpha$ and $\operatorname{Re} s > \beta$, [245, p. 240].

6.2.3 Uniqueness

If f_1 and f_2 are locally integrable functions such that

$$\int_{-\infty}^{\infty} f_1(t)e^{-st}dt = \int_{-\infty}^{\infty} f_2(t)e^{-st}dt, \quad \alpha < \operatorname{Re} s < \beta,$$

then $f_1(t) = f_2(t)$ almost everywhere.

6.3 Summability

If $f(t)$ is a locally integrable function such that

$$F(s) = \int_{-\infty}^{\infty} f(t)e^{-st}dt$$

converges in the strip $\alpha < \operatorname{Re} s < \beta$, then the integral

$$\frac{1}{2\pi i} \int_{c-i\infty}^{c+i\infty} F(s) e^{st} ds, \quad \alpha < c < \beta$$

is $(C, 1)$ summable to $[f(t+) + f(t-)]/2$ for any t whenever $f(t+)$ and $f(t-)$ exist. In particular, it is summable to $f(t)$ at all points t where f is continuous.

6.4 Inversion Formula

If $f(t)$ is a locally integrable function such that

$$F(s) = \int_{-\infty}^{\infty} f(t) e^{-st} dt$$

converges absolutely on the line $\operatorname{Re} s = c$ and if f is of bounded variation in a neighborhood of t_0, then

$$\lim_{R\to\infty} \frac{1}{2\pi i} \int_{c-iR}^{c+iR} F(s) e^{st_0} ds = \frac{f(t_0^+) + f(t_0^-)}{2}.$$

6.5 Characterization and Representation

If $F(s)$ is analytic in the strip $\alpha < \operatorname{Re} s < \beta$ and satisfies

$$\int_{-\infty}^{\infty} |F(\sigma + i\tau)| \, d\tau < \infty, \quad \alpha < \sigma < \beta,$$

with

$$\lim_{|\tau|\to\infty} F(\sigma + i\tau) = 0$$

uniformly in every closed subinterval of (α, β), then it is the two-sided Laplace transform of a function $f(t)$, i.e.,

$$F(s) = \int_{-\infty}^{\infty} f(t) e^{-st} dt, \quad \alpha < \operatorname{Re} s < \beta,$$

where

$$f(t) = \frac{1}{2\pi i} \int_{\sigma - i\infty}^{\sigma + i\infty} F(s) e^{ts} ds, \quad \alpha < \operatorname{Re} s < \beta.$$

6.6 Convolution

The convolution operation associated with the two-sided Laplace transform is defined as

$$(f * g)(x) = \int_{-\infty}^{\infty} f(t) g(x - t) dt, \tag{6.2}$$

whenever it exists.

THEOREM 6.1 **(Convolution Theorem)**
Let f and g be two locally integrable functions for which

$$F(s) = \int_{-\infty}^{\infty} f(t) e^{-st} dt \quad \text{and} \quad G(s) = \int_{-\infty}^{\infty} g(t) e^{-st} dt$$

converge absolutely for a common value of s. Then,

$$F(s)G(s) = \int_{-\infty}^{\infty} h(t) e^{-st} dt, \tag{6.3}$$

where

$$h(x) = (f * g)(x).$$

Example A: Let

$$f(t) = \frac{1}{1 + e^{-t}},$$

and

$$g(t) = (f * f)(t) = \int_{-\infty}^{\infty} \frac{1}{1 + e^{-(t-u)}} \frac{1}{1 + e^{-u}} du = \int_{0}^{\infty} \frac{e^t}{(e^t + x)(1 + x)} dx = \frac{t}{1 - e^{-t}}.$$

Then,

$$F(s) = \int_{-\infty}^{\infty} \frac{e^{-st}}{1 + e^{-t}} dt = \int_{0}^{\infty} \frac{e^{-st}}{1 + e^{-t}} dt + \int_{0}^{\infty} \frac{e^{st}}{e^t (1 + e^{-t})} dt$$

$$= \int_0^\infty e^{-st} \sum_{n=0}^\infty (-1)^n e^{-nt} dt - \int_0^\infty e^{st} \sum_{n=1}^\infty (-1)^n e^{-nt} dt$$

$$= \sum_{n=-\infty}^{\infty} \frac{(-1)^n}{s+n} = \frac{\pi}{\sin \pi s}, \qquad 0 < \operatorname{Re} s < 1.$$

Thus, from (6.3), we have

$$G(s) = \int_{-\infty}^{\infty} \frac{e^{-st} t}{1-e^{-t}} dt = F^2(s) = \left(\frac{\pi}{\sin \pi s}\right)^2 = \sum_{n=-\infty}^{\infty} \frac{1}{(s+n)^2}, \qquad 0 < \operatorname{Re} s < 1.$$

By differentiating $F(s)$ and $G(s)$, we obtain

$$F^{(k)}(s) = \frac{d^k}{ds^k}\left(\frac{\pi}{\sin \pi s}\right) = (-1)^k k! \sum_{n=-\infty}^{\infty} \frac{(-1)^n}{(s+n)^{k+1}}, \qquad 0 < \operatorname{Re} s < 1,$$

and

$$G^{(k)}(s) = \frac{d^k}{ds^k}\left(\frac{\pi}{\sin \pi s}\right)^2 = (-1)^k (k+1)! \sum_{n=-\infty}^{\infty} \frac{1}{(s+n)^{k+2}}, \qquad 0 < \operatorname{Re} s < 1.$$

6.7 Parseval's Relation

If $f(t)$ is a locally integrable function such that

$$\int_0^\infty e^{-2\alpha t} |f(t)|^2 \, dt < \infty,$$

and

$$\int_{-\infty}^0 e^{-2\beta t} |f(t)|^2 \, dt < \infty,$$

with $\alpha < \beta$, then

$$\lim_{R\to\infty} \int_{-R}^{R} f(t) e^{-st} dt$$

exists for $\alpha \le \operatorname{Re} s \le \beta$ and defines an analytic function $F(s)$ for $\alpha < \operatorname{Re} s < \beta$, where the limit is taken in the sense of L^2. Moreover, we have

$$F(s) = \int_{-\infty}^{\infty} f(t) e^{-st} dt;$$

hence

$$e^{-ct}f(t) = \lim_{R\to\infty} \frac{e^{-ct}}{2\pi i} \int_{c-iR}^{c+iR} F(s)e^{st}ds, \quad a \le c \le \beta,$$

and

$$\int_{-\infty}^{\infty} e^{-2\sigma t} |f(t)|^2 \, dt = \frac{1}{2\pi} \int_{-\infty}^{\infty} |F(\sigma + i\tau)|^2 \, d\tau, \quad \alpha \le \sigma \le \beta.$$

6.8 Generalized Functions

6.8.1 Definition (the Embedding Method)

Consider the Fréchet space $\mathcal{L}_{a,b}$ of Example 6, Section 3.1.3, in which

$$\eta(t,a,b) = \begin{cases} e^{at}, & 0 \le t < \infty \\ e^{bt}, & -\infty < t < 0 \end{cases},$$

and $R = D = \frac{d}{dt}$. Let $\mathcal{L}(z,w)$ be the associated countable union space constructed as in Example 5, Section 3.2.3. We say that a generalized function f is a two-sided Laplace transformable generalized function if there exists a unique nonempty open interval (σ_1, σ_2) such that $f \in \mathcal{L}^*(\sigma_1, \sigma_2)$ and in addition $f \notin \mathcal{L}^*(z,w)$ if either $z < \sigma_1$, or $w > \sigma_2$. Consider the vertical open strip

$$\Omega_f = \{s : \sigma_1 < \operatorname{Re} s < \sigma_1\}.$$

DEFINITION 6.2 *Let f be a two-sided Laplace transformable generalized function. Then the two-sided Laplace transform of f is defined as*

$$F(s) = \langle f(t), e^{-st} \rangle, s \in \Omega_f. \tag{6.4}$$

The definition makes sense because for any fixed $s \in \Omega_f, e^{-st} \in \mathcal{L}(\sigma_1, \sigma_2)$. We call σ_1 and σ_2 the abscissas of definition.

If f is a locally integrable function such that $e^{-\sigma t}f(t)$ is bounded on $(-\infty, \infty)$ for each real σ with $\sigma_1 < \sigma < \sigma_2$, then its two-sided Laplace transform $\int_{-\infty}^{\infty} f(t)e^{-st}dt$ exists for at least $\sigma_1 < \operatorname{Re} s < \sigma_2$ and can be identified with (6.4).

6.8.2 Elementary Properties

1. **Differentiation:**

$$F^{(k)}(s) = (-1)^k \left\langle f(t), t^k e^{-st} \right\rangle, s \in \Omega_f.$$

The differentiation operator, D, is continuous on $\mathcal{L}_{a,b}$ since $\gamma_k(D\phi) = \gamma_{k+1}(\phi)$, hence by item 3(b) of Section 3.4.1, its adjoint, which is also a differentiation operator on $\mathcal{L}^*_{a,b}$ is continuous. Consequently, D is continuous on $\mathcal{L}(z,w)$ and $\mathcal{L}^*(z,w)$. This leads to the following formula

$$s^k F(s) = \left\langle f^{(k)}(t), e^{-st} \right\rangle.$$

For

$$\left\langle D^k f(t), e^{-st} \right\rangle = \left\langle f(t), (-D_t)^k e^{-st} \right\rangle = \left\langle f(t), s^k e^{-st} \right\rangle = s^k F(s).$$

2. **Multiplication by** t^k : It is easy to see that the transformation $\phi \to t^k \phi, k = 0, 1, \ldots,$ is continuous on $\mathcal{L}_{a,b}$. Hence, by item 3(a) of Section 3.4.1 and an argument similar to that used in 1, one can show that multiplication by t^k is continuous on $\mathcal{L}^*(z,w)$. Thus, if f is a two-sided Laplace transformable function, we have

$$\left\langle tf(t), e^{-st} \right\rangle = \left\langle f(t), te^{-st} \right\rangle, \quad s \in \Omega_f.$$

or

$$F^{(1)}(s) = -\left\langle tf(t), e^{-st} \right\rangle, \quad s \in \Omega_f.$$

More generally,

$$F^{(k)}(s) = (-1)^k \left\langle t^k f(t), e^{-st} \right\rangle, \quad s \in \Omega_f.$$

3. **Multiplication by** $e^{\gamma t}$: Using arguments similar to those used in 1 and 2 above, one can show that the transformation $f \to e^{\gamma t} f$ is continuous on $\mathcal{L}^*(z,w)$, and hence

$$\left\langle e^{\gamma t} f(t), e^{-st} \right\rangle = \left\langle f(t), e^{-(s-\gamma)t} \right\rangle = F(s-\gamma), \quad s - \gamma \in \Omega_f.$$

4. **Translation:** For a fixed real number $c, \eta(t,a,b)/\eta(t+c,a,b)$ is bounded on $(-\infty, \infty)$; hence

$$\eta(t,a,b) D_t^k \phi(t+c) = \frac{\eta(t,a,b)}{\eta(t+c,a,b)} \eta(t+c,a,b) D_{t+c}^k \phi(t+c).$$

This shows that the translation operator T_c is continuous on $\mathcal{L}_{a,b}$. Therefore, it is continuous on $\mathcal{L}^*(z,w)$ and so is its adjoint T_{-c}. It now follows that

$$\left\langle f(t+c), e^{-st} \right\rangle = e^{cs} \left\langle f(t), e^{-st} \right\rangle = e^{cs} F(s), \; s \in \Omega_f.$$

Most of the other elementary properties of the Laplace transform can be extended to the generalized Laplace transform.

Example A: The two-sided Laplace transform of the delta function is readily seen to be

$$\left\langle \delta(t), e^{-st} \right\rangle = 1, \quad -\infty < \operatorname{Re} s < \infty.$$

Example B: If $f(t) = \sum_{n=0}^{\infty} \delta(t-n)$, then

$$F(s) = \sum_{n=0}^{\infty} e^{-sn} = \frac{1}{1-e^{-s}} \quad , s \neq 0.$$

Example C: If $f(t) = \delta(t+a) + \delta(t-a)$, then $F(s) = 2\cosh as$, $-\infty < \operatorname{Re} s < \infty$.

6.8.3 Inversion Formula

Let $F(s)$ be the two-sided Laplace transform of a Laplace transformable generalized function $f(t)$ with $s \in \Omega_f$, and let R be real. Then in the sense of convergence in $\mathcal{L}^*(\sigma_1, \sigma_2)$

$$f(t) = \lim_{R\to\infty} \frac{1}{2\pi i} \int_{\sigma-iR}^{\sigma+iR} F(s) e^{st} ds,$$

where $\sigma_1 < \sigma < \sigma_2$, see [276, p. 68].

6.8.4 Uniqueness

Let $f(t)$ and $g(t)$ be Laplace transformable generalized functions and denote their respective two-sided Laplace transforms by $F(s)(s \in \Omega_f)$ and $G(s)(s \in \Omega_g)$. If $\Omega_f \cap \Omega_g$ is not empty and $F(s) = G(s)$ in $\Omega_f \cap \Omega_g$, then $f = g$ in $\mathcal{L}^*(z, w)$ where the real interval (z, w) is the intersection of $\Omega_f \cap \Omega_g$ with the real axis [276, p. 69].

6.8.5 Characterization and Representation

For a function $F(s)$ to be the two-sided Laplace transform of a generalized function $f \in \mathcal{L}^*(\sigma_1, \sigma_2)$, it is necessary and sufficient that $F(s)$ be analytic in the strip $\Omega_f = \{s : \sigma_1 < \operatorname{Re} s < \sigma_2\}$ and for each closed substrip $\Omega(a, b) = \{s : \sigma_1 < a \leq \operatorname{Re} s \leq b < \sigma_2\}$ there be a polynomial $P(s)$ that depends on a and b such that $|F(s)| \leq P(|s|)$ for all $s \in \Omega(a, b)$.

6.8.6 Convolution

DEFINITION 6.3 *The convolution of two generalized functions $f, g \in \mathcal{L}^*_{a,b}$ is defined by*

$$\langle (f * g)(t), \phi(t) \rangle = \langle f(t), \langle g(\tau), \phi(t+\tau) \rangle \rangle, \phi \in \mathcal{L}_{a,b}. \tag{6.5}$$

From property 4 of Section 6.8.2, $\phi(t+\tau) \in \mathcal{L}_{a,b}$ if $\phi(t) \in \mathcal{L}_{a,b}$,; hence $\psi(t) = \langle g(\tau), \phi(t+\tau) \rangle$ is a well-defined function. In fact, it can be shown [276, p. 75] that $\psi \in \mathcal{L}_{a,b}$ and in addition if $\{\phi_n\}_{n=1}^{\infty}$ converges in $\mathcal{L}_{a,b}$ to zero, then for any fixed $g \in \mathcal{L}^*_{a,b}$, the sequence $\psi_n(t) = \langle g(\tau), \phi_n(t+\tau) \rangle$ converges to zero in $\mathcal{L}_{a,b}$ as well. By combining these facts together, we immediately obtain that $f * g$ is an element of $\mathcal{L}^*_{a,b}$.

If f and g are locally integrable functions such that

$$f(t)/\eta(t,a,b) \text{ and } g(t)/\eta(t,a,b)$$

are absolutely integrable, then by Section 3.3.3 item 7 they generate regular members of $\mathcal{L}^*_{a,b}$. Then, for any $\phi \in \mathcal{L}_{a,b}$,

$$\begin{aligned} &\langle (f * g)(t), \phi(t) \rangle = \langle f(t), \langle g(\tau), \phi(t+\tau) \rangle \rangle = \\ &\int_{-\infty}^{\infty} f(t)\, dt \int_{-\infty}^{\infty} g(\tau)\phi(t+\tau)\, d\tau = \int_{-\infty}^{\infty} \phi(t)\, dt \int_{-\infty}^{\infty} f(\tau) g(t-\tau)\, d\tau \\ &= \langle h(t), \phi(t) \rangle, \end{aligned} \tag{6.6}$$

where

$$h(t) = \int_{-\infty}^{\infty} f(\tau) g(t-\tau) d\tau,$$

is the conventional convolution of f and g, as defined in (6.2). Interchanging the integration signs is possible by Fubini's theorem, since $f(t)g(\tau)\phi(t+\tau)$ is absolutely integrable on the $t-\tau$ plane, as can be seen from the relation

$$\begin{aligned} &|f(t)g(\tau)\phi(t+\tau)| \\ &= \left| \frac{f(t)}{\eta(t,a,b)} \right| \left| \frac{g(\tau)}{\eta(\tau,a,b)} \right| \left| \frac{\eta(t,a,b)\eta(\tau,a,b)}{\eta(t+\tau,a,b)} \right| |\eta(t+\tau,a,b)\phi(t+\tau)|. \end{aligned}$$

The first factor is integrable over the t-axis, the second over the τ-axis and the last two are bounded and continuous over the whole $t-\tau$ plane. Equation (6.6) shows that the conventional convolution of f and g generates a member of $\mathcal{L}^*_{a,b}$ that is equal to their convolution as generalized functions in the sense of (6.5). Definition 6.3 can be easily extended to the space $\mathcal{L}^*(z,w)$ so that if f and g are members of $\mathcal{L}^*(z,w)$, then so is $f * g$.

It is easy to verify that

$$\delta * f = f \quad \text{and } \delta^{(k)} * f = f^k.$$

The following result is the analogue of the convolution theorem (Theorem 6.1) for generalized functions.

THEOREM 6.2 [276, p. 78]
Let f, g be two-sided Laplace transformable generalized functions and denote their transforms by $F(s)(s \in \Omega_f)$ and $G(s)(s \in \Omega_g)$, respectively. If $\Omega_f \cap \Omega_g$ is nonempty, then $f * g$ exists in the sense of $\mathcal{L}^*(z,w)$, where the real open interval (z,w) is the intersection of $\Omega_f \cap \Omega_g$ with the real axis. Moreover, the two-sided Laplace transform of $f * g$ is $F(s)G(s), s \in \Omega_f \cap \Omega_g$, that is

$$\langle (f * g)(t), e^{-st} \rangle = F(s)G(s), \quad s \in \Omega_f \cap \Omega_g.$$

This result can be used to show that the convolution operation on $\mathcal{L}^*(z,w)$ is both commutative and associative. More precisely if f, g and h are two-sided Laplace transformable generalized functions such that

$$\Omega_f \cap \Omega_g \cap \Omega_h \cap R = (z,w),$$

then

$$f * g = g * f, \quad \text{and } (f * g) * h = f * (g * h),$$

in the sense of $\mathcal{L}^*(z,w)$. Moreover,

$$D(f * g) = (Df) * g = f * (Dg),$$

and

$$T_a(f * g) = (T_a f) * g = f * (T_a g).$$

6.8.7 The Adjoint Method

The two-sided Laplace transformation can be defined for Schwartz distributions using the adjoint method described in Section 4.5.1. Let $\phi \in \mathcal{D}$ and g be a locally integrable function such that $e^{-ct}g(t) \in L^1(\mathcal{R})$ for every c. Denote their two-sided Laplace transforms by $\Phi(s)$ and $G(s)$, respectively. In view of the convolution theorem and the inversion formula, we have

$$\frac{1}{2\pi i} \int_{c-i\infty}^{c+i\infty} G(s)\Phi(s)ds = \int_{-\infty}^{\infty} g(t)\phi(-t)dt. \tag{6.7}$$

This suggests defining the two-sided Laplace transform of $f \in \mathcal{D}^*$ as

$$\langle F, \Phi \rangle = 2\pi i \left\langle f, \breve{\phi} \right\rangle, \tag{6.8}$$

where $\phi \in \mathcal{D}$ and $\breve{\phi}(t) = \phi(-t)$. Thus, the two-sided Laplace transformation is the adjoint of the transformation $\phi \to 2\pi i\breve{\phi}$. When $f \in \mathcal{D}^*$ is such that $e^{-ct}f(t) \in L^1(\mathcal{R})$, the left-hand side of (6.8) should be interpreted as the integral on the left-hand side of (6.7).

6.9 Applications

Many of the applications of the Laplace transform have their counterparts for the two-sided Laplace transform. Whether the former or the latter should be used depends on whether the problem is given on a half-line or the whole line. To illustrate the idea, let us consider a similar example to the one given in Section 5.13.5, but here we shall assume that x runs over the whole real line:

Example A: Consider the initial boundary-value problem:

$$\begin{array}{ll} u_{xx} = \frac{1}{a^2}u_t & -\infty < x < \infty, t > 0 \\ u(x,0) = f(x) & -\infty < x < \infty \\ u(x,t) \to 0 \ , \ u_x(x,t) \to 0 \ \text{ as } |x| \to \infty. & \end{array}$$

Because $-\infty < x < \infty$, we invoke the two-sided Laplace transform to obtain

$$a^2 s^2 U(s,t) = \frac{\partial U(s,t)}{\partial t}$$

$$U(s,0) = F(s),$$

where

$$U(s,t) = \int_{-\infty}^{\infty} u(x,t)e^{-sx}dx.$$

Upon solving the last differential equation, we obtain

$$U(s,t) = F(s)e^{a^2s^2t}.$$

The convolution theorem and the formula (cf. [79, p.307])

$$\int_{-\infty}^{\infty} \exp\left(-p^2x^2 \pm qx\right)\, dx = \frac{\sqrt{\pi}}{p}\exp\left(\frac{q^2}{4p^2}\right) \quad p > 0 \tag{6.9}$$

can be used to show that

$$u(x,t) = \frac{1}{2a\sqrt{\pi t}} \int_{-\infty}^{\infty} f(x-y)e^{-y^2/(4a^2t)}dy.$$

Example B: The two-sided Laplace transform can be used to solve integral equations of the form

$$g(x) = f(x) - \int_{-\infty}^{\infty} k(x-y)f(y)dy$$

in which g and k are given functions and f is unknown. Consider the integral equation

$$f(x) = \int_{0}^{\infty} \sqrt{xy}J_v(xy)f(y)dy, \quad [234, \text{ p. } 311].$$

where J_v is the Bessel function of the first kind and order v. The substitution of $x = e^{-t}$ and $y = e^{\tau}$ transforms this equation into

$$f(e^{-t}) = \int_{-\infty}^{\infty} e^{-(t-\tau)/2}J_v(e^{\tau-t})f(e^{\tau})e^{\tau}d\tau. \tag{6.10}$$

Let us denote the two-sided Laplace transform of $f(e^{-t})$ by $F(s)$ and assume that it exists in $\alpha < \operatorname{Re} s < \beta$, where $\alpha < 0 < \beta$. Thus, the two-sided Laplace transform of $e^{\tau}f(e^{\tau})$ is easily seen to be $F(1-s)$. Equation (6.10) may be written in the form

$$f(e^{-t}) = (g * h)(t),$$

where

$$g(x) = e^{-x/2}J_v(e^{-x}) \text{ and } h(x) = e^{x}f(e^{x}).$$

The two-sided Laplace transform of $g(x)$ is

$$G(s) = \int_{-\infty}^{\infty} e^{-st}e^{-t/2}J_v(e^{-t})dt = \int_{0}^{\infty} y^{s-1/2}J_v(y)dy = \frac{2^{s-1/2}\Gamma\left(\frac{v+s}{2}+\frac{1}{4}\right)}{\Gamma\left(\frac{v-s}{2}+\frac{3}{4}\right)};$$

where $-\operatorname{Re} v - 1/2 < \operatorname{Re} s < 1$ $(\operatorname{Re} v > -3/2)$.

We now invoke the convolution theorem to obtain

$$F(s) = \frac{2^{s-1/2}\Gamma\left(\frac{v+s}{2}+\frac{1}{4}\right)}{\Gamma\left(\frac{v-s}{2}+\frac{3}{4}\right)}F(1-s),$$

which yields

$$E(s) = \frac{F\left(s+\frac{1}{2}\right)}{2^{s/2}\Gamma\left(\frac{v+s+1}{2}\right)} = \frac{2^{s/2}F\left(\frac{1}{2}-s\right)}{\Gamma\left(\frac{v-s+1}{2}\right)}. \tag{6.11}$$

The function $E(s)$ is clearly an even function. For $F(s)$ and $F(1-s)$ to be defined on a common strip, we require that $\beta > 1/2$. We also require that $\operatorname{Re} v > \max(\alpha - 3/2, -\beta - 1/2)$. Therefore, the solution of (6.10) is given in terms of an arbitrary even function in s in the form given by (6.11), i.e.,

$$f(x) = \frac{A}{\sqrt{x}} \int\limits_{c-i\infty}^{c+i\infty} \frac{2^{s/2}}{x^s} E(s)\Gamma\left(\frac{v+s+1}{2}\right) ds,$$

provided that the integral converges, where A is a constant.

7

The Borel Transform

The Borel transform, which can be viewed as a generalization of the Laplace transform, was first brought to light by George Polya in 1923 in one of his papers on entire functions [193]. But it was named after the French mathematician, Emile Borel, who was a professor at Paris, because he was the first to use it [21, 22] in his studies of divergent series which led to the Borel summability method. Borel suggested that if we had a function $f(z)$ defined by a power series

$$f(z) = \sum_{n=0}^{\infty} a_n z^n, |z| < r, \quad r > 0,$$

we might associate with it the function

$$F(s) = \sum_{n=0}^{\infty} \frac{a_n}{n!} s^n.$$

Then the function $\tilde{f}$ defined by the integral

$$\tilde{f}(z) = \int_0^{\infty} F(zt) e^{-t} dt$$

provides the analytic continuation of f across any arc of the circle of convergence where f is analytic. The integral defining $\tilde{f}$ is called the Borel integral and this method of analytic continuation is known as *Borel's method of continuation* [224, p. 164].

In the *Borel summability method*, a number α is assigned to the series $\sum_{n=0}^{\infty} a_n$ as its sum if

$$\alpha = \int_0^{\infty} e^{-t} \left(\sum_{n=0}^{\infty} a_n \frac{t^n}{n!} \right) dt$$

exists or more generally if

$$\alpha = \lim_{s \to 1+} \int_0^\infty e^{-st} \left(\sum_{n=0}^\infty a_n \frac{t^n}{n!} \right) dt = \lim_{s \to 1+} \sum_{n=0}^\infty \frac{a_n}{s^{n+1}}$$

exists.

Borel also gave another definition for the series $\sum_{n=0}^\infty a_n$ to be summable to α. It says that $\sum_{n=0}^\infty a_n$ is summable to α if

$$\lim_{x \to \infty} e^{-x} \sum_{n=0}^\infty \frac{s_n}{n!} x^n = \alpha,$$

where

$$s_n = \sum_{k=0}^n a_k.$$

The two definitions are not equivalent.

Although there was great interest in summability techniques during the first decades of this century, the Borel transform did not receive as much attention as the Laplace and the Fourier transforms, apparently, because of its limited scope of applications.

Borel himself became known not because of his work on the Borel transform, but for his work in the application of set theory to the theory of functions which culminated in the famous Heine-Borel theorem. Most of the known applications of the Borel transform are in the area of entire functions, where it is used to characterize classes of entire functions of exponential type and their behavior near infinity.

7.1 Definitions

DEFINITION 7.1 *The Borel transform of a function $f(z)$ is defined as*

$$F(s) = \int_\gamma e^{-sz} f(z) dz,$$

where the integration is along a ray $\gamma = \{z : z = re^{-i\theta_0}, r > 0\}$, emanating from the origin.

Caveat: by taking the integration path to be the positive x-axis, it would appear that the Borel transform is a generalization of the Laplace transform. As a matter of fact, it is not a genuine generalization. The reason is that the Borel integral is, clearly, path-dependent. Therefore, for the definition to make sense, it must be path-independent, which in turn requires that f satisfy more conditions than just being locally integrable, e.g., f be analytic in at least an angle $|\arg z| \leq \alpha$. Thus, we can say that the Borel transform is a generalization of the Laplace transform when f is analytic in a region$\{z : |\arg z| \leq \alpha, \alpha > 0\}$.

7.2 Existence and Domain of Convergence

If $f(z)$ is analytic and of exponential type c in the angle $|\arg z| \leq \alpha \leq \frac{\pi}{2}$, then its Borel transform

$$F(s) = \int_{\gamma} e^{-sz} f(z) dz$$

is analytic outside an unbounded convex set Ω.

There is an interesting relationship between the growth rate of f and the region Ω. If we denote the *indicator function* of f by $h_f(\theta)$ or $h(\theta)$ for short:

$$h(\theta) = \limsup_{r \to \infty} \frac{\ln \left| f(re^{-i\theta}) \right|}{r}, \qquad |\theta| \leq \alpha,$$

then $h(-\theta)$ is the supporting function of Ω. The *supporting function* $k(\theta)$ of a convex set Ω is defined by

$$k(\theta) = \sup_{z \in \Omega} (x \cos\theta + y \sin\theta) = \sup_{z \in \Omega} Re(ze^{-i\theta}).$$

It is clearly a 2π-periodic function. If Ω is bounded and closed, the function $k(\theta)$ has a geometric interpretation: if we draw a ray through the origin making an angle θ with the positive x-axis and project the set Ω on this line, then $k(\theta)$ is the distance from the origin to the furthest point of this projection in the positive direction. $k(\theta)$ is negative if the projection of Ω lies completely on the ray emanating from the origin and making an angle $\theta + \pi$ with the positive x-axis. Since Ω is closed, there is at least one point in Ω, say $x_0 + iy_0$, such that $k(\theta) = x_0 \cos\theta + y_0 \sin\theta$. This means that the line, $x \cos\theta + y \sin\theta - k(\theta) = 0$, contains at least one point of Ω and that Ω lies entirely on that side of the line where $x \cos\theta + y \sin\theta - k(\theta) \leq 0$. This line is called a *supporting line* of Ω and the points that a supporting line has in common with Ω are called the *supporting points* of the line. A supporting line of a set is a generalization of a tangent line.

A supporting line of a bounded convex set Ω has either a single point of support or a whole line segment of supporting points. Those boundary points of Ω that are not interior to any straight line segment in the boundary are called *extreme points* of Ω.

It follows easily from the definition that the supporting function of a single point $\{x_0 + iy_0\}$ is $k(\theta) = x_0 \cos\theta + y_0 \sin\theta$, while that of the segment $(-a, a)$ is $k(\theta) = a\left|\cos\theta\right|$, and that of the circle $|z| \leq R$ is $k(\theta) = R$.

Note that the supporting function of a convex set Ω is a 2π-periodic function, but the converse is not true, i.e., not every 2π-periodic function is a supporting function of a convex set. However, as indicated above, the indicator function, $h_f(\theta)$, of an analytic function f of exponential type in an angle is the supporting function of some unbounded convex set, called the *indicator diagram* of f. The indicator diagram of f will be denoted by I_f.

If f is an entire function of exponential type, we have the following result:

THEOREM 7.1
If $f(z)$ is an entire function of exponential type, its indicator function $h_f(\theta)$ is the supporting function of a bounded, *convex set, namely, the indicator diagram of f, I_f and the Borel transform of f is analytic outside the set which is obtained by reflecting I_f on the real axis.*

More generally, a 2π-periodic function $h(\theta)$ is the supporting function of some non-empty bounded closed convex set if and only if for $\theta_1 < \theta_2 < \theta_3$, $\theta_2 - \theta_1 < \pi, \theta_3 - \theta_2 < \pi$, we have

$$\begin{vmatrix} h(\theta_1) & \cos\theta_1 & \sin\theta_1 \\ h(\theta_2) & \cos\theta_2 & \sin\theta_2 \\ h(\theta_3) & \cos\theta_3 & \sin\theta_3 \end{vmatrix} \geq 0, \qquad [146, \text{ p. } 76]$$

7.3 Series Representation of the Borel Transform

When f is an entire function of exponential type, its Borel transform takes a simpler form. Let

$$f(z) = \sum_{n=0}^{\infty} a_n z^n,$$

be an entire function of exponential type α. Then its Borel transform $F(s)$ can be given by

$$F(s) = \sum_{n=0}^{\infty} \frac{n! a_n}{s^{n+1}}.$$

Indeed,

$$F(s) = \int_0^\infty f(z)e^{-sz}dz = \sum_{n=0}^\infty a_n \int_0^\infty z^n e^{-sz}dz = \sum_{n=0}^\infty \frac{n!}{s^{n+1}} a_n, \quad \operatorname{Re} s > \alpha.$$

Interchanging the summation and integration signs is justified by the Lebesgue dominated convergence theorem since

$$\left| \int_0^\infty f(z)e^{-sz}dz \right| \le \int_0^\infty \sum_{n=0}^\infty |a_n| \, z^n e^{-\sigma z}dz = \sum_{n=0}^\infty \frac{n! \, |a_n|}{\sigma^{n+1}} < \infty,$$

where $\operatorname{Re} s = \sigma$. The last series converges because

$$\sigma > \alpha = \limsup_{n\to\infty} \sqrt[n]{n! \, |a_n|};$$

see Section 1.3. In fact, we have the following stronger result:

THEOREM 7.2 [15, p. 73]
The function

$$f(z) = \sum_{n=0}^\infty a_n z^n,$$

is an entire function of exponential type if and only if

$$F(s) = \sum_{n=0}^\infty \frac{n! a_n}{s^{n+1}} \tag{7.1}$$

is convergent for some finite s. Moreover, f is of exponential type σ ($\sigma > 0$) if and only if the radius of convergence of the series (7.1) is σ. If the radius of convergence is zero, then either f is of order less than 1 or of order 1 and type zero.

7.4 Polya's Theorem

We have associated with every entire function $f(z)$ of exponential type a non-empty closed bounded convex set, I_f, called the indicator diagram of f, and for which the supporting function is the indicator function $h_f(\theta)$ of f. We now associate a similar convex set to the Borel transform of f. Let I_f^* be the smallest convex domain containing all the singularities of the Borel transform $F(s)$ of f. The set I_f^*, which is called the *conjugate diagram* of

f, can be obtained by taking the intersection of all closed bounded convex sets containing the singularities of F.

If f is of exponential type σ, F is analytic in the region $|s| > \sigma$. Therefore, the conjugate diagram I_f^* of f is inside the disc $|s| \leq \sigma$. Polya established the following interesting result.

THEOREM 7.3 [194] and [146, p. 85]
The conjugate diagram of an entire function of exponential type is the reflection of its indicator diagram on the real axis.

Note that if $h(\theta)$ is the supporting function of some nonempty closed bounded convex set Ω, i.e.,

$$h(\theta) = \sup_{(x,y)\in\Omega} (x\cos\theta + y\sin\theta)$$

then

$$h(-\theta) = \sup_{(x,y)\in\Omega} (x\cos\theta - y\sin\theta) = \sup_{(x,y)\in\Omega} (x\cos\theta + (-y)\sin\theta)$$

$$= \sup_{(x,y)\in\Omega^*} (x\cos\theta + y\sin\theta)$$

is the supporting function of the convex set Ω^* which is obtained by reflecting Ω on the real axis. Thus, in other words, Polya's theorem says that for an entire function of exponential type not only $h_f(\theta)$ is the supporting function of the indicator diagram, but also $h_f(-\theta)$ is the supporting function of the conjugate diagram (cf. Theorem 7.1).

7.5 Inversion Formula

Let f be an entire function of exponential type σ, and c be any contour containing the conjugate diagram of f. Then

$$f(z) = \frac{1}{2\pi i}\int_c F(s)e^{sz}ds.$$

To establish this formula, note that since F is analytic outside the conjugate diagram of f, which is contained in the disc $|s| \leq \sigma$, the integral is unchanged if we replace c by any contour containing the disc $|s| \leq \sigma$, in particular, if we replace c by a circle centered at the origin with radius $\gamma > \sigma$. On this circle the series (7.1) is uniformly and absolutely convergent, hence interchanging the summation and the integration signs is permitted. Thus,

$$\frac{1}{2\pi i}\int_c \sum_{n=0}^{\infty} \frac{n!}{s^{n+1}} a_n e^{sz} ds = \frac{1}{2\pi i}\int_c \left(\sum_{n=0}^{\infty} \frac{n!}{s^{n+1}} a_n\right)\left(\sum_{m=0}^{\infty} \frac{s^m z^m}{m!}\right) ds$$

$$= \frac{1}{2\pi i}\sum_{m,n=0}^{\infty} \frac{n! a_n}{m!} z^m \int_c s^{m-n-1} ds = \sum_{n=0}^{\infty} a_n z^n = f(z).$$

The following stronger result holds.

THEOREM 7.4 [225, p. 48]
A necessary and sufficient condition that $f(z)$ be of the form

$$f(z) = \frac{1}{2\pi i}\int_c F(s) e^{sz} ds,$$

where $F(s)$ is analytic for sufficiently large s and c is a closed contour surrounding the origin, is that f should be an entire function of exponential type.

The function F in Theorem 7.4 is not unique since if F_1 is an analytic function on and inside c, then $F_2 = F + F_1$ will also give the same representation for f. But if F is assumed to be analytic and zero at ∞, then it is unique.

Example A: If

$$f(z) = e^{\lambda z} = \sum_{n=0}^{\infty} \frac{\lambda^n z^n}{n!},$$

then

$$F(s) = \sum_{n=0}^{\infty} \frac{\lambda^n}{s^{n+1}} = \frac{1}{s-\lambda}.$$

More generally, if

$$f(z) = \sum_{k=1}^{n} a_k e^{\lambda_k z},$$

then

$$F(s) = \sum_{k=1}^{n} \frac{a_k}{s-\lambda_k}.$$

The conjugate diagram of f in this case is the smallest convex polygon containing the points λ_k.

Example B: It is possible to construct an entire function of exponential type having a given bounded convex set I as its indicator diagram. Let I^* be the reflection of I with respect to the real axis. Choose a countable set of points $\{\lambda_k\}_{k=1}^{\infty}$ dense in the boundary of I^* and define

$$F(s) = \sum_{k=1}^{\infty} \frac{a_k}{s - \lambda_k},$$

with

$$\sum_{k=1}^{\infty} |a_k| < \infty.$$

It can be easily shown that all the points $\lambda_k, k = 1, 2, \ldots,$ are singular points of $F(s)$ and hence I^* is the smallest convex set containing all the singularities of F [146, p. 87]. Evidently, F is the Borel transform of the entire function

$$f(z) = \sum_{k=1}^{\infty} a_k e^{\lambda_k z},$$

and consequently f is the desired entire function, i.e., $I_f = I$.

7.6 Analytic Properties of the Borel Transform

The Borel transform of an entire function f of exponential type is analytic outside the conjugate diagram of f. Example A may suggest that the extreme points of the conjugate diagram (the vertices of the polygon) are singular points of F. It turns out that this is true in general.

THEOREM 7.5 [15, p. 75]
The extreme points of the conjugate diagram of any entire function of exponential type are singular points of its Borel transform.

7.7 The Borel Transform of Generalized Functions

The Borel transform of generalized functions can be introduced in different ways. The one we are about to discuss is not the most general, but is one of the most elegant. In it the inversion formula for the Borel transform is taken as the basis for the definition of the Borel transform of generalized functions.

Consider Example 5 of Section 3.1.3 in which the open set G is taken as the whole complex plane $\mathcal{C}$, so that the space $\mathcal{H}(G)$ is the space of all entire functions $\mathcal{H}(C)$.

DEFINITION 7.2 *For $f \in \mathcal{H}^*(C)$, we define the Borel transform as*

$$F(s) = \langle f(z), e^{sz} \rangle, \quad s \in \mathcal{C}.$$

The mapping $f \to F(s)$ is a continuous homomorphism from the algebra of $\mathcal{H}^*(\mathcal{C})$ (under addition and convolution) to $\mathcal{C}$. Moreover, the Borel transformation $B : f \to F$ is injective since exponential functions are dense in $\mathcal{H}(\mathcal{C})$. If a measure μ represents $f \in \mathcal{H}^*(\mathcal{C})$, then

$$F(s) = \int_{\mathcal{C}} e^{sz} d\mu(z).$$

Since $\mathcal{H}(\mathcal{C})$ is a subspace of the space of all continuous functions on $\mathcal{C}$, i.e., $C(\mathcal{C})$, it follows from Riesz' representation theorem and the Hahn-Banach theorem that μ has compact support [208, p. 42, 111]. Hence, it easily follows from Theorem 1.1 that F is an entire function in s. Let supp $\mu \subset \{z : |z| \leq r\}$. Then F is an entire function of exponential type at most r. For,

$$|F(s)| \leq \|\mu\| e^{r|s|}, \quad s \in \mathcal{C},$$

where $\|\mu\|$ denotes the total variation of μ. If f is not represented by a measure, we use Theorem 3.2 to show that F is of exponential type.

Let us denote the vector space of all entire functions of exponential type by Exp($\mathcal{C}$). It is evident that Exp($\mathcal{C}$) is also an algebra under addition and function multiplication. Putting all these facts together, we have the following theorem.

THEOREM 7.6
The Borel transformation is an injective homomorphism from the algebra $\mathcal{H}^*(\mathcal{C})$ to Exp($\mathcal{C}$).

For a complex convex set K in $\mathcal{C}$, let us define the modified supporting function of K by

$$h_K(s) = \sup \{\mathrm{Re}\,(zs); z \in K\},$$

and for any entire function $F(s)$ set

$$\|F\|_{(K)} = \sup \left\{ |F(s)|\, e^{-h_K(s)} : s \in \mathcal{C} \right\}.$$

It is easy to show that

$$\operatorname{Exp}(\mathcal{C};K) = \left\{F \in \operatorname{Exp}(\mathcal{C}) : \|F\|_{(K)} < \infty\right\}$$

is a Banach space with respect to the norm $\|F\|_{(K)}$. For an open set Ω let Exp($\mathcal{C}$,Ω) be the countable union space of

$$\bigcup_{n=1}^{\infty} \operatorname{Exp}(\mathcal{C}, K_n),$$

where $K_1 \subset K_2 \subset \cdots$, and $\Omega = \bigcup_{n=1}^{\infty} K_n$.

We have the following theorem.

THEOREM 7.7 [170, p. 18]
Let K and Ω be as above. Then the Borel transformations

$$B : \mathcal{H}^*(K) \to \operatorname{Exp}(\mathcal{C}; K),$$

and

$$B : \mathcal{H}^*(\Omega) \to \operatorname{Exp}(\mathcal{C}, \Omega),$$

are continuous injective linear mappings.

7.8 Applications

Most of the known applications of the Borel transform are in the area of function theory. We shall give one such application, but first let us recall the following theorem of Wigert: the function

$$f(z) = \sum_{n=0}^{\infty} a_n z^n$$

is an entire function of $1/(1-z)$, i.e., it is analytic except at $z = 1$, if and only if $a_n = \phi(n)$, where ϕ is an entire function of minimal type, that is it satisfies

$$\phi(z) = 0\left(e^{\epsilon|z|}\right),$$

for every $\epsilon > 0$. As an application of the Borel transform we have an analogue of Wigert's theorem.

The function

$$F(s) = \int_0^{\infty} f(z)e^{-sz}dz$$

is an entire function of $1/s$ if and only if $f(z)$ is an entire function of minimal type.

8

The Stieltjes Transform

The first appearance of the Stieltjes transform was reported in Stieltjes' paper [219] on continued fractions. The Dutch mathematician T.J. Stieltjes made several notable contributions in mathematics; his ideas, together with those of Lebesgue have led to one of the well-known integrals of the twentieth century, the Lebesgue-Stieltjes integral. The Stieltjes transform arose also in connection with Stieltjes' moment problem for a semi-infinite interval.

8.1 Definitions

DEFINITION 8.1 *The Stieltjes transform of a function $f(t)$ is defined as*

$$S[f](s) = F(s) = \int_0^\infty \frac{f(t)}{s+t}dt, \tag{8.1}$$

whenever the integral exists for any complex number s. Similarly, the Stieltjes transform of a measure $d\alpha(t)$ is defined as

$$S[d\alpha](s) = F(s) = \int_0^\infty \frac{d\alpha(t)}{s+t}. \tag{8.2}$$

More generally, we have the following definition.

DEFINITION 8.2 *The Stieltjes transform of index ρ of a measure $d\alpha(t)$ is defined as*

$$S_\rho[d\alpha](s) = F_\rho(s) = \int_0^\infty \frac{d\alpha(t)}{(s+t)^\rho}, \qquad \rho > 0. \tag{8.3}$$

We adopt the convention that $F_1 = F$, and if $d\alpha(t) = f(t)dt$, $S_\rho(d\alpha) = S_\rho(f)$.

As a special case of (8.2) we obtain the series

$$F(s) = \sum_{n=0}^{\infty} \frac{(-1)^n}{s+n},$$

which shows that (8.2) may converge but not necessarily absolutely.

The Stieltjes transform is closely related to the Laplace transform. In fact, it arises as an iteration of the Laplace transform. For, if

$$F(s) = \int_0^\infty e^{-su} g(u)du\,,$$

where

$$g(u) = \int_0^\infty e^{-tu} f(t)dt,$$

then, we have

$$F(s) = \int_0^\infty e^{-su}du \int_0^\infty e^{-tu} f(t)dt = \int_0^\infty f(t)dt \int_0^\infty e^{-(s+t)u}du$$
$$= \int_0^\infty \frac{f(t)}{(s+t)}dt,$$

provided that interchanging the integration signs is permissible.

8.2 Existence and Domain of Convergence

8.2.1 Domain of Convergence

Hereafter we assume that $\alpha(t)$ is of bounded variation in any finite interval. If the integral (8.3) converges for a point $s = s_0$ not on the negative real axis, it converges for every such point. For,

$$f_R(s) = \int_0^R \frac{d\alpha(t)}{(s+t)^\rho} = \int_0^R \frac{(s_0+t)^\rho}{(s+t)^\rho} d\beta(t) = \beta(R)\frac{(s_0+R)^\rho}{(s+R)^\rho}$$
$$+\rho(s_0 - s)\int_0^R \beta(t)\frac{(s_0+t)^{\rho-1}}{(s+t)^{\rho+1}}dt,$$

where

$$\beta(t) = \int_0^t \frac{d\alpha(u)}{(s_0+u)^\rho}, \qquad 0 \le t < \infty.$$

But since $\lim_{t\to\infty}\beta(t) = f(s_0)$, it is easy to see that the integral

$$\int_0^\infty \beta(t)\frac{(s_0+t)^{\rho-1}}{(s+t)^{\rho+1}}dt$$

converges absolutely, and hence

$$f(s) = \lim_{R\to\infty} f_R(s) = f(s_0) + \rho(s_0 - s)\int_0^\infty \beta(t)\frac{(s_o+t)^{\rho-1}}{(s+t)^{\rho+1}}dt.$$

By modifying this argument, we obtain a stronger result that can be summarized in the following theorem.

THEOREM 8.1 [245, p. 326]
If the integral in (8.3) converges at some point $s = s_0$, not on the negative real axis, it converges uniformly on any compact region not containing a point of the negative real axis.

8.2.2 Uniqueness

Let f and g be two functions such that the integrals defining $F_\rho(s)$ and $G_\rho(s)$ are absolutely convergent. Then, $F_\rho(s) = G_\rho(s)$ implies that $f(t) = g(t)$ almost everywhere. Another important uniqueness result for the Stieltjes transform is the following:

THEOREM 8.2 [245, p. 336]
Let $\alpha(t)$ be of bounded variation in every finite interval $[0, R]$ $(R > 0)$ with

$$\alpha(0) = 0,$$

$$\alpha(t) = \frac{\alpha(t^+) + \alpha(t^-)}{2}.$$

If the integral

$$F(s) = \int_0^\infty \frac{d\alpha(t)}{s+t}$$

converges and

$$F(s_0 + nl) = 0,$$

for some $s_0, l > 0$, $n = 0, 1, 2, \ldots$, then $\alpha(t)$ is identically zero.

8.3 Analytic Properties

From Theorem 8.1 and Theorem 1.1, it follows that $F_\rho(s)$ is analytic in the complex plane cut along the negative real axis. Moreover,

$$F_\rho^{(k)}(s) = (-)^k (\rho)_k \int_0^\infty \frac{d\alpha(t)}{(s+t)^{\rho+k}}, \qquad k = 0, 1, 2, \ldots,$$

For the definition of $(\rho)_k$, see Section 2.3. In addition, if the integral in (8.3) converges for $s > 0$, then

$$\lim_{s\to\infty} F_\rho^{(k)}(s) = 0, \text{ [245, p. 185]} .$$

The points of the negative real axis need not be singular points of F_ρ, in which case F_ρ can be continued analytically across this axis. For example, if

$$\alpha(t) = \begin{cases} 1 \ , t > 0 \\ 0 \ , t = 0 \end{cases},$$

then $F_\rho(s) = 1/s^\rho$, which is analytic everywhere except at $s = 0$. The following theorem gives us a sufficient condition for the origin to be a singularity of the Stieltjes transform.

THEOREM 8.3 [245, p. 337]
If $\alpha(t)$ is a real-valued, non-decreasing function for which the point $t = 0$ is a point of increase and for which the integral

$$F_\rho(s) = \int_0^\infty \frac{d\alpha(t)}{(s+t)^\rho}$$

converges, then $s = 0$ is a singularity of F_ρ.

On the other hand, the negative real axis can be a natural boundary for F_ρ. That is, F_ρ cannot be continued analytically across any interval of that axis. For example, by choosing $\alpha(t)$ appropriately, we can obtain

$$F_\rho(s) = \sum_{n=0}^{\infty} \frac{\beta_n}{(s-a_n)^\rho},$$

where $\{a_n\}_{n=0}^\infty$ is a dense set on the negative real axis and $\{\beta_n\}_{n=0}^\infty$ is chosen so that

$$\sum_{n=0}^{\infty} |\beta_n| < \infty.$$

It is easy to see that F_ρ is defined for every s not on the negative real axis. Since F_ρ has a singularity at a_n for every n and the set $\{a_n\}_{n=0}^{\infty}$ is dense on the negative real axis, it is impossible to continue F_ρ analytically across that axis.

8.4 Elementary Properties

1. Translation:
If $f(t)$ is zero on the negative real axis, then

$$T_{-a}(S_\rho f)(s) = S_\rho[T_a f](s), \qquad a > 0$$

or

$$F_\rho(s+a) = S_\rho[f(t-a)](s), \qquad a > 0.$$

2. Dilation:

$$S_\rho[f(at)](s) = a^{\rho-1} S_\rho[f](as), \qquad a > 0$$

or

$$S_\rho[D_{\frac{1}{a}} f](s) = a^{\rho-1} D_{\frac{1}{a}}(F(s)).$$

3. Differentiation:

$$S_\rho[f^{(k)}](s) = (\rho)_k S_{\rho+k}[f](s) - \frac{f^{(k-1)}(0)}{s^\rho} - \frac{(\rho)_1 f^{(k-2)}(0)}{s^{\rho+1}}$$

$$- \cdots - \frac{(\rho)_{k-1} f(0)}{s^{\rho+k-1}}, \qquad k = 1, 2, \ldots.$$

4. Multiplication by t:

$$S_\rho[tf(t)](s) = S_{\rho-1}[f](s) - sS_\rho[f](s) = F_{\rho-1}(s) - sF_\rho(s).$$

5. Integration: Let

$$g(t) = \frac{1}{\Gamma(\mu)} \int_0^t f(u)(t-u)^{\mu-1} du, \qquad 0 < \operatorname{Re}\mu.$$

Then

$$S_\rho[g](s) = \frac{\Gamma(\rho-\mu)}{\Gamma(\rho)} S_{\rho-\mu}[f](s), \qquad \operatorname{Re}\rho > \operatorname{Re}\mu.$$

In particular, if $\mu = 1$, it follows that if

$$g(t) = \int_0^t f(u) du,$$

then

$$S_\rho\left[\int_0^t f(u)du\right](s) = \frac{1}{(\rho-1)} S_{(\rho-1)}[f](s).$$

6. Iteration: Let

$$F_\rho(s) = S_\rho[f](s) = \int_0^\infty \frac{f(t)}{(s+t)^\rho} dt, \qquad \rho > 0$$

and

$$G_\mu(z) = S_\mu[F_\rho](z) = \int_0^\infty \frac{F_\rho(s)}{(z+s)^\mu} ds, \qquad \mu > 0.$$

Then,

$$\begin{aligned} G_\mu(z) &= \int_0^\infty \frac{1}{(z+s)^\mu} ds \int_0^\infty \frac{f(t)}{(s+t)^\rho} dt \\ &= \int_0^\infty f(t)dt \int_0^\infty \frac{ds}{(z+s)^\mu (s+t)^\rho} \\ &= B(1, \mu+\rho-1) z^{-\mu} \int_0^\infty t^{1-\rho} f(t)\, {}_2F_1(\mu, 1; \mu+\rho; 1-\frac{t}{z}) dt. \end{aligned}$$

This is a generalization of the formula given in [245, p. 383].

Here we have made use of the formula

$$\int_0^\infty s^{\nu-1}(z+s)^{-\mu}(t+s)^{-\rho} ds = z^{-\mu} t^{\nu-\rho} B(\nu, \mu-\nu+\rho)\, {}_2F_1(\mu, \nu; \mu+\rho; 1-\frac{t}{z})$$

$$; \mathrm{Re}\,\nu > 0, \mathrm{Re}\,\mu > \mathrm{Re}\,(\nu-\rho).$$

cf. [79, p. 286, no. 3].

In the special case where $\rho = \mu = 1$, we obtain

$$G(z) = \int_0^\infty \frac{(\ln z/t) f(t)}{z-t} dt.$$

8.5 Relationships with Other Transforms

8.5.1 The Laplace Transform

We have seen in Section 8.1 that the Stieltjes transform of index 1 can be obtained as an iterated Laplace transform. Now we show that the Stieltjes transform of general index $\rho > 0$ can also be obtained by iteration of the Laplace transform. In fact,

$$\begin{aligned} S_{\rho+1}(f)(s) &= \frac{1}{\Gamma(\rho+1)} \mathcal{L}\left\{t^\rho \mathcal{L}(f)(t)\right\}(s) \\ &= \frac{1}{\Gamma(\rho+1)} \mathcal{L}\left\{t^\rho F(t)\right\}(s), \end{aligned} \tag{8.4}$$

where $\mathcal{L}(f) = F$ is the Laplace transform of f. For,

$$\mathcal{L}\{t^\rho F(t)\}(s) = \int_0^\infty t^\rho F(t) e^{-st} dt = \int_0^\infty t^\rho e^{-st} dt \int_0^\infty e^{-tx} f(x) dx$$

$$= \int_0^\infty f(x) dx \int_0^\infty t^\rho e^{-(s+x)t} dt = \Gamma(\rho+1) S_{\rho+1}(f)(s), \qquad (s > 0),$$

provided that interchanging the integration signs is justified. A sufficient condition for this to hold is that f be locally integrable on $[0, \alpha)$ and of order $O(t^a)$ as $t \longrightarrow \infty$ for some $a < \rho$.

The next theorem shows that every Stieltjes transform is in a sense a Laplace transform.

THEOREM 8.4

If the integral

$$F(s) = \int_0^\infty \frac{d\alpha(t)}{(s+t)^\rho}, \quad \rho > 0, \tag{8.5}$$

converges, then it can be written in the form

$$F(s) = \int_0^\infty x^{\rho-1} g(x) e^{-sx} dx, \qquad \operatorname{Re} s > 0, \tag{8.6}$$

where

$$g(x) = \frac{1}{\Gamma(\rho)} \int_0^\infty e^{-tx} d\alpha(t), \qquad t > 0. \tag{8.7}$$

Conversely, if the integral (8.7) converges for $t \geq 0$, and if $F(s)$ is defined by (8.6) for $\operatorname{Re} s > 0$, *then F can be represented as in (8.5) for all s in the complex plane cut along the negative real axis.*

Proof: From the relation

$$\frac{1}{(s+t)^\rho} = \frac{1}{\Gamma(\rho)} \int_0^\infty x^{\rho-1} e^{-(s+t)x} dx,$$

it follows that

$$F(s) = \int_0^\infty \frac{d\alpha(t)}{(s+t)^\rho} = \int_0^\infty d\alpha(t) \frac{1}{\Gamma(\rho)} \int_0^\infty x^{\rho-1} e^{-(s+t)x} dx$$
$$= \int_0^\infty x^{\rho-1} g(x) e^{-sx} dx,$$

where $g(x)$ is given by (8.7). Interchanging the integration signs is permissible because $\alpha(t) = o(t^\rho)$ as $t \to \infty$, see Equation (8.10) below. The proof of the converse statement is similar. ∎

It is worth emphasizing that the condition $t \geq 0$ in the converse statement is crucial. For, we may choose $d\alpha$ appropriately so that

$$g(t) = \sum_{n=0}^{\infty} (-1)^n (1+n) e^{-nt} = \frac{1}{(1+e^{-t})^2}, \qquad t > 0,$$

and hence

$$F(s) = \int_0^{\infty} \frac{e^{-st}}{(1+e^{-t})^2} dt.$$

The last integral converges for $\operatorname{Re} s > 0$, but (8.5) becomes

$$F(s) = \sum_{n=0}^{\infty} (-1)^n \frac{1+n}{s+n},$$

which diverges for all s.

8.5.2 The Convolution Transform

The convolution transform of a function $\phi(u)$ is defined by

$$H(x) = \int_{-\infty}^{\infty} \phi(u) G(x-u) du,$$

for some kernel function $G(u)$. Setting $G(x) = [2\cosh(x/2)]^{-\rho}$, we obtain

$$H(x) = \int_{-\infty}^{\infty} \frac{\phi(u)du}{\left(2\cosh\frac{(x-u)}{2}\right)^{\rho}} = \int_{-\infty}^{\infty} \frac{\phi(u)du}{\left[e^{(x-u)/2} + e^{-(x-u)/2}\right]^{\rho}}$$

$$= \int_{-\infty}^{\infty} \frac{\phi(u) e^{(x+u)\rho/2} du}{[e^x + e^u]^{\rho}}.$$

If we make the change of variables, $s = e^x$, $t = e^u$, the last equation becomes

$$F_\rho(s) = \int_0^{\infty} \frac{f(t)dt}{(s+t)^{\rho}},$$

where

$$f(t) = t^{\rho/2-1}\phi(\ln t) \quad \text{and} \quad F_\rho(s) = s^{-\rho/2} H(\ln s).$$

8.6 Inversion Formulas

8.6.1 Complex-Inversion Formula

An inversion formula for the Stieltjes transform may be obtained from (8.4). Symbolically, we have

$$f = \Gamma(\rho)\mathcal{L}^{-1}\left\{t^{-\rho+1}\mathcal{L}^{-1}(S_\rho(f))(t)\right\}, \qquad \rho > 0, \tag{8.8}$$

where $\mathcal{L}^{-1}(g)$ denotes the inverse Laplace transform of g. More explicitly, if $F_\rho(s) = S_\rho(f)(s)$, then

$$f(t) = \frac{\Gamma(\rho)}{(2\pi i)^2}\int_{c-i\infty}^{c+i\infty} z^{-\rho+1}e^{tz}dz \int_{d-i\infty}^{d+i\infty} e^{sz}F_\rho(s)ds, \qquad c, d > 0,$$

provided that the integrals converge absolutely and uniformly.

For $\rho = 1$, we have a more convenient inversion formula:

THEOREM 8.5 [245, p. 339]
If $f \in L^1_{loc}[0, \infty)$ is such that the integral

$$F(s) = \int_0^\infty \frac{f(t)}{s+t}dt$$

converges, then

$$\lim_{\delta\to o^+} \frac{F(-t - i\delta) - F(-t + i\delta)}{2\pi i} = \frac{f(t^+) + f(t^-)}{2},$$

for any positive t for which the right-hand side exists.

8.6.2 Real Inversion Formula

An inversion formula involving successive derivatives of the Stieltjes transform can be obtained in a similar fashion to that of the Laplace transform; see Section 5.8.3.

THEOREM 8.6 [245, p. 345]
Let f satisfy the same conditions as in Theorem 8.5. Then

$$\lim_{k\to\infty} \frac{(-t)^{k-1}}{k!(k-2)!}\frac{d^{2k-1}}{dt^{2k-1}}\left[t^k F(t)\right] = f(t), \qquad k = 1, 2, \ldots \tag{8.9}$$

at any Lebesgue point of f.

8.7 Asymptotic Behavior

If the integral (8.3) converges, then

$$\alpha(t) = o(t^{\rho}) \qquad as \qquad t \longrightarrow \infty, \tag{8.10}$$

but the converse is not true. For example, the function

$$\alpha(t) = \int_0^t \frac{du}{\ln(u+2)}, \qquad 0 \leq t < \infty,$$

is $o(t)$ as $t \longrightarrow \infty$, yet

$$\int_0^{\infty} \frac{d\alpha(t)}{(s+t)} = \int_0^{\infty} \frac{dt}{(s+t)\ln(t+2)}$$

diverges at $s = 2$ and hence everywhere. However, if

$$\alpha(t) = O(t^{\rho-\delta}) \quad , \delta > 0 ,$$

as $t \longrightarrow \infty$, then (8.3) converges.

If $F_\rho(s)$ is defined by the convergent integral (8.3) with $\alpha(0) = 0$, then

$$F_\rho^{(n)}(s) \sim \frac{(-1)^n(\rho)_n \alpha(0^+)}{s^{\rho+n}}$$

as $s \longrightarrow 0^+$, $n = 0, 1, 2, \ldots$, and

$$F_\rho^{(n)}(s) = o(s^{-n})$$

as $s \longrightarrow \infty$, where s is real and $n = 0, 1, 2, \ldots$.

8.8 Abelian and Tauberian Theorems

8.8.1 Abelian Theorem

THEOREM 8.7 [245, p. 183]
If the integral (8.3) converges for $s > 0$ and if $0 < \gamma \leq \rho$, then for any constant A,

$$\limsup_{s \longrightarrow 0^+} |s^{\gamma} F_\rho(s) - A| \leq \limsup_{t \longrightarrow 0^+} \left| \frac{\alpha(t)}{ct^{\rho-\gamma}} - A \right| ,$$

and

$$\limsup_{s\longrightarrow\infty}|s^{\gamma}F_{\rho}(s)-A|\leq\limsup_{t\longrightarrow\infty}\left|\frac{\alpha(t)}{ct^{\rho-\gamma}}-A\right|,$$

where

$$c=\frac{\Gamma(\rho)}{\Gamma(\gamma)\Gamma(\rho-\gamma+1)}.$$

As a corollary, we obtain that if $0<\gamma\leq\rho$, and

$$\alpha(t)\sim Act^{\rho-\gamma}\qquad\text{as}\qquad t\longrightarrow\infty\qquad\text{or}\qquad t\longrightarrow 0^{+},$$

then

$$F_{\rho}(s)\sim\frac{A}{s^{\gamma}}\qquad\text{as}\qquad s\longrightarrow\infty\qquad\text{or}\qquad s\longrightarrow 0^{+}.$$

8.8.2 Tauberian Theorem

THEOREM 8.8 [245, p. 199]
Let $f(t)\in L^{1}_{loc}(0,\infty)$ and $F(s)$ be given by the convergent integral (8.1). If there exist constants K and A such that

$$\frac{-K}{t}<f(t),\qquad 0<t<\infty,$$

and

$$F(s)\sim\frac{A}{s}\qquad as\qquad s\longrightarrow\infty,$$

then

$$\int_{0}^{\infty}f(t)dt=A.$$

8.9 Stieltjes Transform of Generalized Functions

8.9.1 Definition and Properties

Because of its connection with the Laplace and convolution transforms, the Stieltjes transform has received a fair amount of interest from researchers working on generalized functions. As a result, its extension to generalized functions has been achieved in a number of different ways. Zemanian [276, p. 244], for example, introduced it as a special case of the convolution transform of generalized functions, while Misra and Lavoine [165] introduced it as an iterated Laplace transform. Benedetto [12], on the other hand, used a more direct approach based on the embedding method.

Our approach is different from theirs and is based on the general procedure described in Section 4.5.2. Consider Example 6, Section 3.1.3 in which we shall take $I = (0, \infty)$, $R = t\frac{d}{dt}$, and $\eta(t, a, b)$ to be a positive continuous function satisfying

$$\eta(t, a, b) = \begin{cases} O(t^a) \text{ as } t \longrightarrow 0^+, \\ O(t^b) \ as\ t \longrightarrow \infty. \end{cases}$$

The space $\mathcal{L}_{a,b}$ is defined as the space of all infinitely differentiable functions ϕ on I that satisfy

$$\gamma_{k,a,b}(\phi) = \sup_{0<t<\infty} |\eta(t, a, b)R^k\phi(t)| < \infty.$$

It is easy to see that

$$(\mathcal{L}_{a,b}\ , \{\gamma_{k,a,b}\}_{k=0}^{\infty})$$

is a Fréchet space (see Section 3.1.3) and for each z in the plane cut along $(-\infty, 0]$, the function, $(z+t)^{-\rho}$, belongs to $\mathcal{L}_{a,b}$, provided that $a > 0$ and $b < \rho$.

Therefore, the following definition makes sense:

DEFINITION 8.3 *The Stieltjes transform, $S_\rho f$, of $f \in \mathcal{L}^*_{a,b}$ $(a > 0, b < \rho)$ is defined by*

$$(S_\rho f)(z) = F_\rho(z) =< f(t), \frac{1}{(z+t)^\rho} >, \qquad z \notin (-\infty, 0], \qquad \rho > 0. \tag{8.11}$$

It can be easily shown that F_ρ is an analytic function in the complex z-plane cut along $(-\infty, 0]$ and

$$\frac{d^k}{dz^k}(S_\rho f)(z) = F_\rho^{(k)}(z) = (-1)^k(\rho)_k F_{\rho+k}(z) = (-1)^k(\rho)_k(S_{\rho+k}f)(z).$$

Moreover,

$$|F_\rho(z)| \longrightarrow 0 \qquad \text{as} \qquad |s| \longrightarrow \infty$$

in the cut plane.

Example A:

$$S_\rho[\delta(t-c)](z) = \frac{1}{(z+c)^\rho}, \qquad c > 0,$$

and

$$\frac{d^k}{dz^k}\left(S_\rho\delta(t-c)\right)(z) = \frac{(-1)^k(\rho)_k}{(z+c)^{\rho+k}}.$$

More generally,

$$\frac{d^k}{dz^k}\left(S_\rho\delta^{(m)}(t-c)\right)(z)=\frac{(-1)^k(\rho)_{m+k}}{(z+c)^{\rho+m+k}},\quad k,m=0,1,2,\ldots.$$

If $f \in L^1_{loc}[0,\infty)$ is such that

$$\int_0^\infty \left|\frac{f(t)}{\eta(t,a,b)}\right| dt < \infty,$$

then f defines a regular element of $\mathcal{L}^*_{a,b}$. For any such element the conventional Stieltjes transform (8.3) coincides with the generalized transform (8.11); see Section 3.3.3 item 7.

8.9.2 Inversion Formula

There are a number of inversion formulas for the generalized Stieltjes transform; each depends on how the transform is defined, see [12, 165, 276]. The one we are about to give is based on (8.8) and is similar to the one given in [165].

Let $F_\rho(s)$ be the generalized Stieltjes transform of $f \in \mathcal{L}^*_{a,b}$ as given by (8.11). Let $|F_\rho(s)| \le c/|s|^2$ as $|s| \to \infty$ in the region $|\arg s| < \pi$. Then its inverse Laplace transform $g(t) = \mathcal{L}^{-1}(F)(t)$ is a continuous function in $(0,\infty)$, cf. Section 5.7. In fact, it is a regular member of the space $\mathcal{L}^*(0,\infty)$ defined in Section 5.12.3. If $t^{-\rho+1}g(t)$ is analytic in $\mathrm{Re}\, t > 0$ and of order $O(1/|t|^2)$ as $|t| \to \infty$, then

$$f(t) = \Gamma(\rho)\mathcal{L}^{-1}\{t^{-\rho+1}\mathcal{L}^{-1}(F_\rho)(t)\}.$$

8.10 Applications

In this section we consider two Stieltjes-type integral equations that are akin to the Stieltjes transform, the first of which is the Stjeltjes integral equation.

Example A: Consider the Stieltjes integral equation

$$g(x) = \lambda \int_0^\infty \frac{f(t)}{x+t} dt, \tag{8.12}$$

in which g is a known function, f is unknown and λ is a parameter.

This equation arose, as the Stieltjes transform did, in connection with the Stieltjes moment problem. It is the problem of determining for what sequences $\{\mu_n\}_{n=0}^\infty$ there exists a non-decreasing function $\alpha(t)$ such that

$$\mu_n = \int_0^\infty t^n d\alpha(t), \quad n = 0, 1, 2, \ldots.$$

The Stieltjes integral equation (8.12), which may be written in the form

$$g(x) = \lambda S(f)(x),$$

can be solved by first making the change of variables: $x = e^u$, $t = e^v$, $\psi(u) = e^{\frac{u}{2}} g(e^u)$, and $\phi(v) = e^{\frac{v}{2}} f(e^v)$ to obtain

$$\psi(u) = \frac{\lambda}{\sqrt{2\pi}} \int_{-\infty}^{\infty} h(u-v)\phi(v)dv,$$

where

$$h(z) = \sqrt{\frac{\pi}{2}}(\cosh \frac{z}{2})^{-1}. \tag{8.13}$$

We then invoke the convolution theorem for the Fourier transform (Theorem 12.14) to obtain

$$\widehat{\psi}(w) = \lambda \widehat{h}(w)\widehat{\phi}(w),$$

where

$$\hat{h}(w) = \pi(\cosh \pi w)^{-1}.$$

This leads to the formal solution

$$\begin{aligned}
\phi(x) = e^{\frac{x}{2}} f(e^x) &= \frac{1}{\pi\lambda\sqrt{2\pi}} \int_{-\infty}^{\infty} \widehat{\psi}(w) \cosh(\pi w)\, e^{-ixw} dw \\
&= \frac{1}{\lambda(2\pi)^{\frac{3}{2}}} \int_{-\infty}^{\infty} \widehat{\psi}(w) \left(e^{-i(x+i\pi)w} + e^{-i(x-i\pi)w}\right) dw \\
&= \frac{1}{\lambda(2\pi)}\{\psi(x+i\pi) + \psi(x-i\pi)\} \\
&= \frac{i}{\lambda(2\pi)}\{e^{\frac{x}{2}} g(e^x e^{i\pi}) - e^{\frac{x}{2}} g(e^x e^{-i\pi})\},
\end{aligned}$$

or

$$f(y) = \frac{i}{\lambda(2\pi)}\{g(ye^{i\pi}) - g(ye^{-i\pi})\}.$$

For this solution to make sense, g must satisfy certain analyticity conditions; see [225, p. 318] for details.

The second Stieltjes-type integral equation is given in the next example.

Example B: Consider the integral equation

$$f(x) = \lambda \int_0^\infty \frac{f(t)}{x+t} dt, \tag{8.14}$$

where f is unknown and λ is a positive parameter. We may write this equation in the form

$$f(x) = \lambda S(f)(x),$$

and as before, we can convert it into

$$\phi(u) = \frac{\lambda}{\sqrt{2\pi}} \int_{-\infty}^{\infty} h(u-v)\phi(v)dv, \tag{8.15}$$

where $h(z)$ is given by (8.13). It is easy to see that $\phi(v) = e^{av}$ is a solution of (8.15) provided that a is a solution of the equation

$$1 - \frac{\lambda}{\sqrt{2\pi}} \int_{-\infty}^{\infty} h(z)e^{-az}dz = 0,$$

or

$$1 - \lambda\widehat{h}(ia) = 1 - \lambda\pi \sec \pi a = 0, \text{ or } \cos \pi a = \lambda\pi.$$

Clearly, if a is a solution of this equation, then so is $-a$. Therefore, the solution of (8.15) may be given by

$$\phi(v) = Ae^{av} + Be^{-av} = e^{\frac{v}{2}} f(e^v),$$

or

$$f(t) = At^{a-\frac{1}{2}} + Bt^{-a-\frac{1}{2}},$$

with $-\frac{1}{2} < a < \frac{1}{2}$ if $0 < \lambda < 1/\pi$ and $\operatorname{Re} a = 0$ if $\lambda > 1/\pi$. If $\lambda = 1/\pi$, $f(t) = A/\sqrt{t}$ is clearly a solution, but the other solution, which is less obvious, is

$$f(t) = B\frac{\ln t}{\sqrt{t}}.$$

This can be verified directly as follows. Since $\phi(v) = e^{\frac{v}{2}} f(e^v)$, we immediately have $\phi(v) = v$. Therefore, the right-hand side of (8.15) becomes

$$\frac{1}{\pi\sqrt{2\pi}} \int_{-\infty}^{\infty} h(v)\phi(u-v)dv = \frac{1}{2\pi} \int_{-\infty}^{\infty} \frac{(u-v)}{\cosh(v/2)} dv$$

$$= \frac{1}{2\pi} \left\{ u \int_{-\infty}^{\infty} \frac{dv}{\cosh(v/2)} - \int_{-\infty}^{\infty} \frac{v}{\cosh(v/2)} dv \right\} = u.$$

The last equality follows from the fact that the second integral between the braces { } is zero (in the Cauchy sense) since the integrand is odd. A more rigorous derivation of the second solution can be found in [245, pp. 387-390].

9

The Lambert Transform

Some function transformations have been introduced as a generalization of sequence transformations. A typical example is the *Lambert transformation.* In his studies of the theory of numbers, J.H. Lambert [135], a Swiss-German mathematician, proved in 1771that

$$\sum_{n=1}^{\infty} \frac{nx^n}{1-x^n} = \sum_{n=1}^{\infty} \sigma(n)x^n,$$

where $\sigma(n)$ is the sum of the divisors of n, a formula that was conjectured earlier by Euler. Lambert also showed that

$$\sum_{n=1}^{\infty} \frac{x^n}{1-x^n} = x + 2x^2 + 2x^3 + 3x^4 + \cdots\cdots = \sum_{n=1}^{\infty} \tau(n)x^n, \ |x| < 1$$

where $\tau(n)$ is the number of the divisors of n, including 1 and n. Among many other contributions to number theory by Lambert was the first published proof that π is an irrational number (Euler had shown earlier that e is irrational).

These two series gave rise to the sequence transformation

$$T\left[\{a_n\}_{n=1}^{\infty}\right](x) = f(x) = \sum_{n=1}^{\infty} a_n \frac{x^n}{1-x^n}, \qquad |x| < 1, \tag{9.1}$$

whenever the series converges. The series on the right-hand side of (9.1) is called the *Lambert series.*

Almost one hundred and eighty years later, a Harvard professor, D.V. Widder, introduced the *Lambert integral* as a generalization of the Lambert series in much the same way as the Laplace integral is a generalization of the power series [246].

9.1 Definitions

DEFINITION 9.1 *The Lambert transform of a function $f(t)$ is defined as*

$$F(x) = \int_0^\infty \frac{f(t)}{e^{xt} - 1} dt. \tag{9.2}$$

Upon replacing n by t, x by e^{-x} and the summation sign by integration sign, it becomes evident that (9.2) is the continuous analogue of (9.1). The two equations can be combined together by using the Stieltjes integral

$$F(x) = \int_0^\infty \frac{d\alpha(t)}{e^{xt} - 1}, \tag{9.3}$$

which we shall call the *Lambert–Stieltjes transform.*

The Lambert transform is a special case of the convolution transform (see Chapter 17) as can be seen if we replace x by e^{-x}, t by $e^t, G(x)$ by $F(e^{-x})$ and $g(t)$ by $e^t f(e^t)$, so that (9.2) becomes

$$G(x) = \int_{-\infty}^\infty h(x - t)g(t)dt, \tag{9.4}$$

where

$$h(x) = \left(e^{e^{-x}} - 1\right)^{-1}.$$

It should be noted that the convolution transform has been studied thoroughly by Hirshman and Widder (see Chapter 17), who derived, among other things, inversion formulas for a large class of convolution transforms. Nevertheless, their results did not include kernels like $h(x)$ in (9.4). Thus, the inversion of the Lambert transform is not trivial and cannot be deduced directly from known results on the convolution transform.

Because of the singularity of the integrand at $t = 0$, it is convenient to replace $f(t)$ by $tf(t)$ in (9.2) and likewise to replace $d\alpha(t)$ by $td\alpha(t)$ in (9.3) so that the Lambert–Stieltjes transform of f is defined as

$$F(x) = \int_0^\infty \frac{td\alpha(t)}{e^{xt} - 1} = \frac{1}{x}\int_0^\infty k(xt)d\alpha(t), \tag{9.5}$$

where

$$k(t) = \frac{t}{e^t - 1}, \quad 0 < t < \infty; \; k(0) = 1.$$

We assume that $\alpha(t)$ is a function of bounded variations on every bounded interval $[0, R]$, and that $\alpha(0) = 1$. If $\alpha(t) = \int_0^t \phi(u)du$, where $\phi(u)$ is a Lebesgue integrable function, then

$$F(x) = \frac{1}{x}\int_0^\infty k(xt)\phi(t)dt, \tag{9.6}$$

which will be called the Lambert-Lebesgue transform to distinguish it from the Lambert–Stieltjes transform (9.5).

9.2 Existence

The Lambert–Stieltjes transform

$$F(x) = \int_0^\infty \frac{td\alpha(t)}{e^{xt} - 1}$$

converges for all $x \neq 0$ if

$$\psi(x) = \int_0^\infty te^{-xt}d\alpha(t)$$

converges at $x = 0$. Moreover, $F(x)$ and $\psi(x)$ converge or diverge for the same values of $x \neq 0$ if ψ diverges at $x = 0$.

The proof of the first statement is straightforward. Set

$$\beta(t) = \int_1^t ud\alpha(u)\,, \ \beta(1) = 0,$$

so that $\beta(\infty) = \psi(0)$, which is finite by hypothesis. Thus, for $x \neq 0$ we have

$$\int_1^R \frac{td\alpha(t)}{e^{xt} - 1} = \int_1^R \frac{d\beta(t)}{e^{xt} - 1} = \frac{\beta(R)}{e^{xR} - 1} + x\int_1^R \frac{e^{xt}\beta(t)}{(e^{xt} - 1)^2}dt, \ R > 1.$$

As $R \to \infty$, the first term on the right-hand side tends to 0 or $-\beta(\infty)$ according to $x > 0$ or $x < 0$. The integral on the right-hand side converges for $x \neq 0$ since the integrand is $0(e^{-|x|t})$ as $t \to \infty$. The proof of the second statement is similar, but a little bit more technical.

9.3 Relationships with Other Transforms

We have shown that the Lambert transform is related to the convolution transform. Here we shall show how it is related to the Laplace transform.

The two-sided Laplace transform of the kernel,

$$h(t) = \left(e^{e^{-t}} - 1\right)^{-1},$$

of the Lambert transform can be evaluated explicitly as

$$H(s) = \int_{-\infty}^{\infty} e^{-st}\left(e^{e^{-st}} - 1\right)^{-1} dt = \int_{0}^{\infty} \frac{t^{s-1}}{e^t - 1} dt = \zeta(s)\Gamma(s),$$

where $\zeta(s)$ is the Riemann zeta function; see Section 2.2.1 (b).

THEOREM 9.1
Consider the Lambert–Lebesgue transform

$$F(x) = \frac{1}{x}\int_{0}^{\infty} k(xt)\phi(t)dt,$$

in which k is as defined in (9.5), $\phi \in L^1(0,\infty)$ and

$$\lim_{t\to 0^+} t^{1-\delta}\phi(t) = 0, \text{ for some } \delta > 0.$$

Moreover, consider the corresponding Laplace transform

$$g(s) = \int_{0}^{\infty} e^{-st} t\phi(t)dt. \tag{9.7}$$

Then

$$F(x) = \sum_{n=1}^{\infty} g(nx), \quad x > 0. \tag{9.8}$$

Proof: Since

$$k(t) = \frac{t}{e^t - 1} = t\sum_{n=1}^{\infty} e^{-nt}, \quad 0 < t,$$

and

$$0 < k(t) \le 1, \quad 0 \le t < \infty, \tag{9.9}$$

we have

$$F(x) = \frac{1}{x}\int_0^\infty k(xt)\phi(t)dt = \sum_{n=1}^\infty \int_0^\infty e^{-nxt} t\phi(t)dt = \sum_{n=1}^\infty g(nx), \quad x > 0. \tag{9.10}$$

Interchanging the summation and integration signs is permissible in view of the Lebesgue dominated convergence theorem since

$$|F(x)| \le \frac{1}{x}\int_0^\infty |k(xt)|\,|\phi(t)|\,dt \le \frac{1}{x}\int_0^\infty |\phi(t)|\,dt < \infty$$

by (9.9). ∎

Moreover, we have the following result.

THEOREM 9.2

Every Lambert–Lebesgue transform is a Laplace transform. In fact, if $F(x)$ is given by (9.6), then

$$F(x) = \int_0^\infty f(t)e^{-xt}dt, \tag{9.11}$$

where

$$f(t) = t\sum_{n=1}^\infty \frac{1}{n^2}\phi\left(\frac{t}{n}\right), \tag{9.12}$$

and the series is absolutely convergent for $0 < t < \infty$.

Proof: As in (9.10)

$$F(x) = \sum_{n=1}^\infty \int_0^\infty e^{-nxt} t\phi(t)\,dt$$

$$= \sum_{n=1}^\infty \int_0^\infty e^{-ux}\frac{u}{n^2}\phi\left(\frac{u}{n}\right)du = \int_0^\infty f(u)e^{-ux}du\,,$$

where the interchange of the summation and integration signs is justified as before. ∎

The converse of Theorem 9.2 is also true.

THEOREM 9.3
Every Laplace transform of the form (9.7) is a Lambert-Lebesgue transform. More precisely, if g is given by (9.7), then

$$g(x) = \int_0^\infty \frac{tc(t)}{e^{xt}-1}dt\ ,\ 0 < x < \infty,$$

where

$$c(t) = \sum_{n=1}^\infty \frac{\mu(u)}{n^2}\phi\left(\frac{t}{n}\right),$$

and $\mu(n)$ is the Möbius function defined by (see Section 2.2.1 (b))

$$\mu(n) = \begin{cases} 1 \text{ if } n = 1 \\ 0 \text{ if } n \text{ is divisible by a square} \\ (-1)^k \text{ if } n \text{ is the product of } k \text{ distinct primes} \end{cases}.$$

The series is absolutely convergent for $0 < t < \infty$.

Proof: An important property of the Möbius function is this

$$\sum_{d/n} \mu(d) = 0, \quad n = 2, 3, \cdots$$

where the summation runs over all the divisors, d, of n, including 1 and n. Using this property, one can show that

$$\sum_{n=1}^\infty \sum_{k=1}^\infty \cdot\mu(n)g(knx) = g(x) \tag{9.13}$$

whenever the double series converges absolutely. From (9.7), we have

$$g(knx) = \int_0^\infty e^{-knxt}t\phi(t)dt, \quad 0 < x < \infty,$$

hence by (9.13)

$$g(x) = \sum_{k=1}^\infty \sum_{k=1}^\infty \mu(n)g(knx) = \sum_{k=1}^\infty \sum_{n=1}^\infty \mu(n) \int_0^\infty e^{-knxt}t\phi(t)dt$$

$$= \sum_{k=1}^\infty \sum_{n=1}^\infty \mu(n) \int_0^\infty e^{-kxt}\frac{t}{n^2}\phi\left(\frac{t}{n}\right)dt = \sum_{k=1}^\infty \int_0^\infty te^{-kxt}c(t)dt$$

$$= \int_0^\infty \frac{tc(t)}{e^{xt}-1} dt, \quad 0 < x.$$

Interchanging the integral and the summation signs is permissible as outlined before; for the details see [246]. ∎

9.4 Inversion Formulas

First, we start with the inversion formula for (9.8). This follows immediately from (9.13) to yield

$$g(x) = \sum_{n=1}^{\infty} \mu(n) F(nx), \quad 0 < x < \infty, \tag{9.14}$$

with the series being absolutely convergent.

Under the assumption that $d\alpha/dt = \phi(t)$ is continuous at $t_0 > 0$, and $F(x)$ is given by (9.6), we have

$$t_0\phi(t_0) = \sum_{n=1}^{\infty} \mu(n) \lim_{k\to\infty} \left\{ \frac{(-1)^k}{k!} \left(\frac{k}{t_0}\right)^{k+1} n^k F^{(k)}\left(\frac{nk}{t_0}\right) \right\}. \tag{9.15}$$

To this end, differentiate (9.8) k times to obtain

$$F^{(k)}(x) = \sum_{m=1}^{\infty} m^k g^{(k)}(mx)\,, \quad 0 < x < \infty,$$

hence

$$\sum_{n=1}^{\infty} \mu(n) n^k F^{(k)}(nx) = \sum_{n=1}^{\infty} \mu(n) \sum_{m=1}^{\infty} (nm)^k g^{(k)}(nmx) = g^{(k)}(x),$$

where the last equality follows from (9.13). The right-hand side of (9.15) is easily seen to be equal to

$$\lim_{k\to\infty} \frac{(-1)^k}{k!} \left(\frac{k}{t_0}\right)^{k+1} g^{(k)}\left(\frac{k}{t_0}\right),$$

which, in view of the inversion formula of the Laplace transform (5.16), is equal to $\phi(t_0)$ and this establishes (9.15). For a justification for interchanging the limit and summation signs, see [246].

Another inversion formula for the Lambert–Lebesgue transform

$$F(x) = \int_0^\infty \frac{t\phi(t)}{(e^{xt} - 1)} dt,$$

in which the integral is assumed to converge for all $x > 0$, was obtained by W.B. Pennington in 1960 [191]. It reads as follows: for all $x > 0$, and $\rho \geq 0$

$$\int_0^x (x - t)^\rho \phi(t) dt = \sum_{n=1}^\infty \mu(n) n^\rho \lim_{k \to \infty} \frac{(-1)^k}{k!} \left(\frac{kn}{x}\right)^{k+1} \times \left[\frac{d^k}{ds^k} \left(\frac{\Gamma(\rho + 1) F(s)}{s^{\rho+1}}\right)\right]_{s=kn/x}.$$

Differentiating this formula with respect to x for $\rho = 0$ gives us an expression for $x\phi(x)$ at all points of continuity of $\phi(x)$. It is worth noting that Pennington's formula is not a trivial generalization of Widder's formula (9.15), nor is Widder's formula included in Pennington's.

9.5 Generalized Functions

Extending the Lambert transform to generalized functions is straightforward since the Kernel of the transform $t/(e^{xt} - 1)$ is a bounded C^∞-function for any $x \neq 0$ and $t \geq 0$. The embedding method is applicable here, where one may take $R = D, \eta = 1$ in Example 5, Section 3.2.3.

9.6 Applications

The Lambert transform was discovered in connection with the theory of numbers and that is where most of the known applications are [245, p. 231]. But it also arises in some physical problems.

For example, if h denotes Planck's constant, c the speed of light, T the absolute temperature, k Botlzmann's constant, and $\rho(\nu)d\nu$ the energy of the radiation field per unit volume in the frequency range ν to $\nu + d\nu$, then Planck's law describing the spectral distribution of the radiation from a black body is

$$\rho(\nu) d\nu = \frac{8\pi h \nu^3 d\nu}{c^3 (\exp(h\nu/kt) - 1)} \qquad [215, \text{pp. } 49\text{-}59]$$

Therefore, the total energy radiated by a black body is proportional to the integral

$$\int_0^\infty \frac{\nu^3 d\nu}{\exp(h\nu/kT) - 1},$$

which is clearly related to the Lambert–Lebesgue transform of the function $f(\nu) = \nu^2$.

Another related problem is when the source consists of a mixture of black bodies at different temperatures. In this case the radiation from the source may be given by

$$c\nu^3 d\nu \int_0^\infty \frac{\Phi(T)dT}{\exp(h\nu/kT) - 1},$$

where c is a constant and $\Phi(T)$ represents the amount of radiation coming from black bodies at temperature T. If we have an observable radiation with frequency distribution given by $\Psi(\nu)d\nu$, then

$$\Psi(\nu) = c\nu^3 \int_0^\infty \frac{\Phi(T)dT}{\exp(h\nu/kT) - 1}.$$

If we put $h/kT = \mu$, $\Phi(T)dT = \mu\phi(\mu)d\mu$ and $\Psi(\nu)/\nu^3 = \psi(\nu)$, we obtain

$$\psi(\nu) = c \int_0^\infty \frac{\mu\phi(\mu)d\mu}{\exp(\mu\nu) - 1},$$

that is, ψ is the Lambert transform of ϕ.

10

The Mellin Transform

The Mellin transform pair, i.e., the Mellin transform and its inversion formula, appeared in Riemann's famous memoir [204] on prime numbers and it was later formulated more explicitly by Cahen [35]. But the first one to put it on a rigorous basis and point out some of its applications was H. Mellin [159], and that is why the transform bears his name. Because the Mellin transform can be obtained from the two-sided Laplace transform by a change of variables, many of its properties can be easily obtained from those of the two-sided Laplace transform. Therefore, we shall restrict our discussion to a few of its properties that will be convenient to have in the Mellin form.

10.1 Definitions

DEFINITION 10.1 *The Mellin transform of a function $f(t)$ is defined as*

$$\mathcal{M}(f)(s) = F(s) = \int_0^\infty t^{s-1} f(t)dt, \qquad (s \text{ complex}) \qquad (10.1)$$

whenever the integral exists for at least one value s_0 of s.

If $\alpha(t)$ is a function of bounded variation on the interval $[0, R]$ for every $R > 0$, the Mellin-Stieltjes transform of $f(t)$ is defined as

$$F(s) = \int_0^\infty t^s d\alpha(t).$$

In the bilateral Laplace–Stieltjes transform,

$$F(s) = \int_{-\infty}^{\infty} e^{-st} d\alpha(t),$$

if we make the change of variable $x = e^{-t}$, we obtain the Mellin-Stieltjes transform

$$F(s) = \int_{0}^{\infty} x^s d\beta(x), \text{ where } \beta(x) = -\alpha(-\ln x),$$

and the integral is taken as a Cauchy limit at the lower limit of integration. In particular, if $d\alpha(t) = f(t)dt$, then the bilateral-Laplace-Stieltjes transform becomes

$$F(s) = \int_{0}^{\infty} x^{s-1} g(x) dx, \text{ where } g(x) = f(-\ln x).$$

The Mellin transform arises naturally in the theory of Dirichlet series

$$\phi(s) = \sum_{n=1}^{\infty} \frac{a_n}{n^s}, \qquad Re\ s > 1.$$

For, formally we have

$$\phi(s) = \sum_{n=1}^{\infty} \frac{a_n}{\Gamma(s)} \int_{0}^{\infty} e^{-nx} x^{s-1} dx = \frac{1}{\Gamma(s)} \int_{0}^{\infty} x^{s-1} f(x) dx, \tag{10.2}$$

where

$$f(s) = \sum_{n=1}^{\infty} a_n e^{-nx}.$$

Conversely, in view of the formula

$$e^{-x} = \frac{1}{2\pi i} \int_{c-i\infty}^{c+i\infty} \Gamma(s) x^{-s} ds \qquad (c > 0),$$

we have

$$f(x) = \sum_{n=1}^{\infty} a_n e^{-nx} = \sum_{n=1}^{\infty} a_n \frac{1}{2\pi i} \int_{c-i\infty}^{c+i\infty} \Gamma(s)(nx)^{-s} ds$$

$$= \frac{1}{2\pi i} \int_{c-i\infty}^{c+i\infty} \phi(s)\Gamma(s)x^{-s}ds. \tag{10.3}$$

In particular, if $a_n = 1$, $n = 1, 2, \ldots$, then $\phi(x)$ becomes the Riemann zeta function $\zeta(x)$ (see Section 2.2.1.(b)) and $f(x) = (e^x - 1)^{-1}$. Thus, (10.2) and (10.3) become

$$\mathcal{M}\left(\frac{1}{e^x - 1}\right)(s) = \zeta(s)\Gamma(s),$$

and

$$\frac{1}{e^x - 1} = \frac{1}{2\pi i} \int_{c-i\infty}^{c+i\infty} \zeta(s)\Gamma(s)x^{-s}ds.$$

Upon replacing x by $2x$ in the last integral and using the identity

$$\frac{2}{e^{2x} - 1} = \frac{1}{e^x - 1} - \frac{1}{e^x + 1},$$

we have

$$\mathcal{M}\left(\frac{1}{e^x + 1}\right)(s) = \left(1 - 2^{1-s}\right)\Gamma(s)\zeta(s).$$

10.2 Existence and Domain of Convergence

10.2.1 Domain of Convergence

THEOREM 10.1
If the integral

$$\mathcal{M}(f)(s) = F(s) = \int_0^\infty t^{s-1} f(t)dt,$$

converges for two points $s_1 = \sigma_1 + i\tau_1$ and $s_2 = \sigma_2 + i\tau_2$ with $\sigma_1 < \sigma_2$, then it converges in the vertical strip $\sigma_1 < Re\ s < \sigma_2$.

If the integral converges uniformly and absolutely in the interior of the strip, then by Theorem 1.1 it defines an analytic function thereon.

In particular, if $f \in L^1_{loc}[0, \infty)$ is such that

$$f(t) = \begin{cases} 0(t^{-\alpha}) \text{ as } t \to 0 \\ 0(t^{-\beta}) \text{ as } t \to \infty, \end{cases}$$

then $\mathcal{M}(f)$ exists for $\alpha < Re\ s < \beta$.

Let us denote the domain of convergence of the Mellin transform of f by Ω_f.

10.2.2 Uniqueness

Let F and G denote the Mellin transforms of f and g, respectively. If $\Omega_f \cap \Omega_g$ is not empty and $F(s) = G(s)$ for $s \in \Omega_f \cap \Omega_g$, then $f = g$ almost everywhere on $[0, \infty)$.

10.3 Elementary Properties

1. **Linearity:** The Mellin transformation is a linear transformation.

2. **Scaling:**

$$\mathcal{M}[f(t/a)](s) = a^s F(s), \quad a > 0$$

and

$$\mathcal{M}\left[f(t^a)\right](s) = \frac{1}{a} F\left(\frac{s}{a}\right), \quad a > 0$$

3. **Translation:**

$$\mathcal{M}\left[t^a f(t)\right](s) = F(s+a), \quad s + a \in \Omega_f$$

and

$$\mathcal{M}\left[t^{-1} f(t^{-1})\right](s) = F\left(1-s\right), \quad 1 - s \in \Omega_f. \tag{10.4}$$

4. **Differentiation:** If $f \in L^1(0, \infty)$ is a differentiable function such that

$$\lim_{t \to 0^+} t^{s-1} f(t) = 0, \quad \alpha < Re\ s < \beta,$$

and

$$\lim_{t \to \infty} t^{s-1} f(t) = 0, \quad \alpha < Re\ s < \beta,$$

then

$$\mathcal{M}[f'(t)](s) = -(s-1)F(s-1), \quad s - 1 \in \Omega_f.$$

More generally, if $f^{(k)}, k = 0, \ldots, n-1$, satisfies the above conditions, then

$$\mathcal{M}\left[f^{(n)}(t)\right](s) = (-1)^n \frac{\Gamma(s)}{\Gamma(s-n)} F(s-n),$$

$$= (-1)^n (s-n)_n F(s-n), n = 1, 2, \ldots, s - n \in \Omega_f.$$

Similarly, we have

$$\mathcal{M}\left[(tD_t)^n f(t)\right](s) = (-1)^n s^n F(s), \quad s \in \Omega_f$$

and

$$\mathcal{M}\left[(D_t t)^n f(t)\right](s) = (-1)^n (s-1)^n F(s), s \in \Omega_f$$

and

$$\mathcal{M}\left[(\ln t)^n f(t)\right](s) = F^{(n)}(s),$$
$$\mathcal{M}\left[t^n D_t^n f(t)\right](s) = (-1)^n (s)_n F(s). \qquad (10.5)$$

where $D_t = \frac{d}{dt}$.

5. **Integration:**

$$\mathcal{M}\left[\int_0^x f(t)dt\right](s) = -\frac{1}{s}\mathcal{M}\left[f(t)\right](s+1),$$

and

$$\mathcal{M}\left[\int_x^\infty f(t)dt\right](s) = \frac{1}{s}\mathcal{M}\left[f(t)\right](s+1).$$

10.4 Inversion Formula

If the integral $F(s) = \int_0^\infty t^{s-1} f(t)dt$ converges absolutely on the line $Re\ s = c$, and if f is of bounded variation in a neighborhood of $t = x$ ($x > 0$), then

$$\lim_{R\to\infty} \frac{1}{2\pi i} \int_{c-iR}^{c+iR} F(s)x^{-s}ds = \frac{f(x^+) + f(x^-)}{2}.$$

In particular, if f is continuous at x, the left-hand side converges to $f(x)$.

Conversely, if $F(s)$ is absolutely integrable on the line $Re\ s = c$ and of bounded variation in a neighborhood of the point $c + it$ and if

$$f(x) = \frac{1}{2\pi i} \int_{c-i\infty}^{c+i\infty} F(s)x^{-s}ds,$$

then

$$\left[F(c+it^+) + F(c+it^-)\right]/2 = \lim_{R\to\infty} \int_{1/R}^{R} f(x)x^{c+it-1}dx.$$

10.5 Summability

Let

$$F(s) = \int_0^\infty x^{s-1} f(x) dx$$

be absolutely and uniformly convergent for $s = c + it$ with t in any finite interval.

Then, the integral

$$\frac{1}{2\pi i} \int_{c-i\infty}^{c+i\infty} F(s) x^{-s} ds$$

is $(C, 1)$ summable to $[f(x+0) + f(x-0)]/2$, i.e.,

$$\frac{1}{2\pi i} \lim_{R\to\infty} \int_{c-iR}^{c+iR} \left(1 - \frac{|t|}{R}\right) F(s) x^{-s} ds = \frac{f(x+0) + f(x-0)}{2},$$

whenever $f(x+0)$ and $f(x-0)$ have a meaning. In particular, the integral is summable to $f(x)$ for almost all x and to $f(x)$ wherever $f(x)$ is continuous.

Conversely, if the integral

$$f(t) = \frac{1}{2\pi i} \int_{c-i\infty}^{c+i\infty} F(s) t^{-s} ds$$

converges absolutely, then

$$\lim_{R\to\infty} \int_{1/R}^{R} \left(1 - \frac{|\ln x|}{\ln R}\right) f(x) x^{s-1} dx = F(s)$$

almost everywhere [225, p. 47].

10.6 Convolution

DEFINITION 10.2 *The convolution operation, *, associated with the Mellin transform is defined by*

$$(f * g)(x) = \int_0^\infty f\left(\frac{x}{u}\right) g(u) \frac{du}{u}, \tag{10.6}$$

whenever the integral exists.

THEOREM 10.2 (**Convolution Theorem**)
Let $x^k f(x)$ and $x^k g(x)$ be in $L^1(0, \infty)$, and let

$$h(x) = (f * g)(x).$$

Then, $x^k h(x) \in L^1(0, \infty)$ and

$$\mathcal{M}(h)(s) = F(s)G(s),$$

or

$$h(x) = \frac{1}{2\pi i} \int_{c-i\infty}^{c+i\infty} F(s)G(s)x^{-s} ds ,$$

where $F(s)$ and $G(s)$ are the Mellin transforms of f and g, respectively.

Another important relation is

$$\int_0^\infty t^k f(t) g(xt) dt = \frac{1}{2\pi i} \int_{c-i\infty}^{c+i\infty} G(s)F(k+1-s)x^{-s} ds, \tag{10.7}$$

or equivalently

$$\mathcal{M}\left[\int_0^\infty t^k f(t) g(xt) dt\right](s) = G(s)F(k+1-s), \tag{10.8}$$

provided that the integrals converge absolutely. The constant c in (10.7) is chosen so that the line of integration is in the intersection of the domain of analyticity of $G(s)$ and $F(k+1-s)$.

To prove (10.7), replace u by xu in (10.6) to obtain

$$(\tilde{f} * g)(x) = \int_0^\infty \tilde{f}\left(\frac{1}{u}\right) g(xu) \frac{du}{u} = \int_0^\infty u^k f(u) g(xu) du,$$

where

$$f(u) = u^{-k-1}\tilde{f}\left(\frac{1}{u}\right).$$

The result now follows from the convolution theorem since $\tilde{F}(s) = F(k+1-s)$ where $\tilde{F}(s)$ is the Mellin transform of $\tilde{f}(t)$.

As special cases of (10.7) and (10.8), we have for $k = 0$,

$$\mathcal{M}\left[\int_0^\infty f(t)g(xt)dt\right](s) = F(1-s)G(s), \tag{10.9}$$

and for $x = 1$, we obtain *a generalized Parseval's relation for the Mellin transform*

$$\int_0^\infty t^k f(t)g(t)dt = \frac{1}{2\pi i}\int_{c-i\infty}^{c+i\infty} G(s)F(k+1-s)ds. \tag{10.10}$$

For $k = 0, c = 1/2$, $f = g$, and real f, we obtain from (10.10) the Parseval relation

$$\int_0^\infty |f(t)|^2\, dt = \frac{1}{2\pi}\int_{-\infty}^{\infty}\left|F\left(\frac{1}{2}+it\right)\right|^2 dt,$$

since

$$F\left(\frac{1}{2}+it\right)F\left(\frac{1}{2}-it\right) = \left|F\left(\frac{1}{2}+it\right)\right|^2.$$

Example A: Since $\mathcal{M}[e^{-x}](s) = \Gamma(s)\,(\mathrm{Re}\, s > 0)$, it follows that $\mathcal{M}[x^a e^{-x}](s$ $\Gamma(s+a)$ and $\mathcal{M}[x^{b-1}e^{-x}](s) = \Gamma(s+b-1)$. Thus,

$$\frac{1}{2\pi i}\int_{k-i\infty}^{k+i\infty}\Gamma(a+s)\Gamma(b-s)ds = \int_0^\infty x^{a+b-1}e^{-2x}dx = 2^{-a-b}\Gamma(a+b), \quad (-a < k < b$$

Moreover,

$$\int_0^\infty |\Gamma(a+it)|^2\, dt = 2^{-2a}\pi\Gamma(2a), \quad (a > 0).$$

10.7 Mellin Transform of L^2-Functions

If $f(t)$ is such that $t^k f(t) \in L^2(0,1)$ and $t^m f(t) \in L^2(1,\infty)$ with $k < m$, then in the sense of $L^2(0,\infty)$

$$\lim_{R\to\infty} \int_{1/R}^{R} t^{s-1} f(t)dt$$

exists for $k \le \operatorname{Re} s \le m$ and defines a function $F(s)$ that is analytic in the strip $k < \operatorname{Re} s < m$ and for which we have

$$F(s) = \int_0^\infty t^{s-1} f(t)dt, \quad k < \operatorname{Re} s < m,$$

where the integral converges absolutely. Moreover, the limits

$$\lim_{\sigma\to k+} f(\sigma + i\tau) = f(k + i\tau),$$

$$\lim_{\sigma\to m^-} f(\sigma + i\tau) = f(m + i\tau),$$

exist in the sense of $L^2(0,\infty)$ with

$$\lim_{R\to\infty} \int_0^\infty \left| t^c f(t) - \frac{t^c}{2\pi i} \int_{c-iR}^{c+iR} F(s)t^{-s} ds \right|^2 \frac{dt}{t} = 0, \quad k \le c \le m,$$

and

$$\int_0^\infty t^{2c-1} |f(t)|^2 \, dt = \frac{1}{2\pi} \int_{-\infty}^\infty |F(c+i\tau)|^2 \, d\tau, \quad k \le c \le m.$$

In particular, if $k = 0$ and $m = 1$, we obtain Parseval's relation

$$\int_0^\infty |f(t)|^2 \, dt = \frac{1}{2\pi} \int_{-\infty}^\infty \left| F\left(\frac{1}{2} + i\tau\right)\right|^2 \, d\tau,$$

. In fact, (10.10) may also be regarded as a generalized Parseval's relation.

10.8 Mellin Transform of Generalized Functions

10.8.1 Definition

Consider the space $\mathcal{L}_{a,b}$ of Section 3.1.3, Example 6, in which $I = (0, \infty)$,

$$\eta(t, a, b) = \begin{cases} t^{-a} \text{ if } 0 < t \le 1 \\ t^{-b} \text{ if } 1 < t < \infty \end{cases},$$

and $R^k = t^{k+1} D^k$. The seminorms

$$\gamma_{k,a,b}(\phi) = \sup_{0<t<\infty} \left| \eta(t, a, b) t^{k+1} D^k \phi(t) \right|, \quad \phi \in \mathcal{L}_{a,b}$$

form a separating family of seminorms on $\mathcal{L}_{a,b}$, and $\mathcal{L}_{a,b}$ is a testing-function space with the following properties:

1. if $a \le c$ and $d \le b$, then $\mathcal{L}_{c,d} \subseteq \mathcal{L}_{a,b}$ and the topology of $\mathcal{L}_{c,d}$ is stronger than the topology it inherits from $\mathcal{L}_{a,b}$,

2. for any a and b, $\mathcal{L}_{a,b}$ is dense in $\mathcal{E}(I)$ and the topology of $\mathcal{L}_{a,b}$ is stronger than that induced on it by $\mathcal{E}(I)$,

3. if f is a locally integrable function on I such that $f(t)/\eta(t, a, b)$ is absolutely integrable on I, then f generates a regular member of $\mathcal{L}^*_{a,b}$ with

$$\langle f, \phi \rangle = \int_0^\infty f(t)\phi(t)dt, \quad \phi \in \mathcal{L}_{a,b}.$$

4. the kernel of the Mellin transformation, t^{s-1}, is in $\mathcal{L}_{a,b}$ for $a < \operatorname{Re} s < b$.

Let $f \in \mathcal{L}^*_{a,b}$ for some a and b and assume that $f \in \mathcal{L}^*_{c,d}$ for all $a \le c, d \le b$. Let σ_2 be the supremum of all such b's and σ_1 the infimum of all such a's and set

$$\Omega_f = \{ s : \sigma_1 < \operatorname{Re} s < \sigma_2 \}.$$

It is easy to see that f belongs to the corresponding space $\mathcal{L}(\sigma_1, \sigma_2)$ of Section 3.2.3, Example 5 and does not belong to $\mathcal{L}(z, w)$ if either $z < \sigma_1$ or $w > \sigma_2$. Any generalized function satisfying the above condition is said to be Mellin transformable.

DEFINITION 10.3 *For any Mellin transformable generalized function f we define its Mellin transform by*

$$\mathcal{M}(f)(s) = F(s) = \left\langle f(t), t^{s-1} \right\rangle, s \in \Omega_f. \tag{10.11}$$

The right-hand side has a meaning since $t^{s-1} \in \mathcal{L}(\sigma_1, \sigma_2)$ for any $s \in \Omega_f$. In particular, if f is a regular generalized function satisfying property (3) above, then definition (10.11) coincides with (10.1).

Moreover, $F(s)$ is analytic in Ω_f with

$$F^{(k)}(s) = \langle f(t), (\ln t)^k t^{s-1} \rangle, \quad s \in \Omega_f.$$

If $f \in \mathcal{E}^*(I)$, then $f \in \mathcal{L}(\sigma_1, \sigma_2)$ for any σ_1, σ_2 by property (2) of the space $\mathcal{L}_{a,b}$. Thus, Ω_f is the entire plane and consequently $\mathcal{M}(f)(s)$ is an entire function.

Example A: Although the δ-function is not in $\mathcal{L}^*_{a,b}$ for any a and b because members of $\mathcal{L}_{a,b}$ are not necessarily continuous at the origin, the translated δ-function, $\delta(t-a)$, for $a > 0$, is in $\mathcal{L}^*_{a,b}$ for any a. In fact,

$$\mathcal{M}(\delta(t-a))(s) = a^{s-1},$$

and hence

$$\mathcal{M}\left(\sum_{k=1}^{\infty} \delta(t-ka)\right)(s) = \sum_{k=1}^{\infty} (ak)^{s-1} = a^{s-1}\zeta(1-s), \quad Re\ s < 0.$$

10.8.2 Uniqueness

THEOREM 10.3 [276, p. 108]
Let f and g be two Mellin transformable functions. If $\mathcal{M}(f) = F$, $\mathcal{M}(g) = G$ and $\Omega_f \cap \Omega_g$ is not empty, then $F(s) = G(s)$ for $s \in \Omega_f \cap \Omega_g$ implies that $f = g$ in $\mathcal{L}^*(z, w)$ where (z, w) is the intersection of $\Omega_f \cap \Omega_g$ with the real axis.

10.8.3 Inversion Formula

Let f be a Mellin transformable generalized function and $\mathcal{M}(f)(s) = F(s)$ in $\Omega_f = \{s : \sigma_1 < Re\ s < \sigma_2\}$. Then in the sense of $\mathcal{D}^*(I)$ we have

$$f(t) = \lim_{R \to \infty} \frac{1}{2\pi i} \int_{c-iR}^{c+iR} F(s) t^{-s} ds,$$

for any c such that $\sigma_1 < c < \sigma_2$.

10.8.4 Characterization

THEOREM 10.4 [276, p. 108]
For a function $F(s)$ analytic in a vertical strip $\Omega_f = \{s : \sigma_1 < Re\ s < \sigma_2\}$ to be the Mellin transform of a generalized function f in $\mathcal{L}^*(\sigma_1, \sigma_2)$ accord-

ing to Definition 10.3, it is necessary and sufficient that F be bounded by a polynomial P in each closed substrip of Ω_f, *that is, there exists a polynomial* $P(s)$ *that depends on the closed substrip such that*

$$|F(s)| \leq P(|s|) \text{ for } \sigma_1 < a \leq Re\ s \leq b < \sigma_2.$$

Moreover, if $|F(s)| \leq K|s|^{-2}$ *in* $a < Re\ s < b$ *for some constant K and if*

$$\frac{1}{2\pi i}\int_{\sigma-i\infty}^{\sigma+i\infty} F(s)x^{-s}ds\,, \quad a < \sigma < b$$

exists, then it defines a continuous function $f(t)$ *on* $(0,\infty)$ *that generates a regular member of* $\mathcal{L}^*(a,b)$ *and for which* $\mathcal{M}(f)(s) = F(s)$ *in* $a < Re\ s < b$.

10.9 Applications

10.9.1 Differential Equations and Initial-Value Problems

The Laplace transformation, when applied to the differential equation,

$$b_n\frac{d^ny}{dt^n} + \cdots + b_1\frac{dy}{dt} + b_0y = \tilde{f}(t)$$

with constant coefficients, will transform it into an algebraic equation. On the other hand, the substitution, $t = -\ln x$, transforms the Laplace transform into a Mellin transform and the above differential equation into a Cauchy-Euler differential equation of the form

$$a_nx^n\frac{d^ny}{dx^n} + \cdots + a_1x\frac{dy}{dx} + a_0y = f(x).$$

Therefore, it is reasonable to expect the Mellin transform to transform a Cauchy-Euler differential equation into an algebraic one. This can be seen directly from (10.5). More precisely, consider Cauchy-Euler's equation

$$a_nx^n\frac{d^ny}{dx^n} + \cdots + a_1x\frac{dy}{dx} + a_0y = f(x)\,,$$

where f is a given Mellin transformable function. By taking the Mellin transform of both sides of this equation and employing (10.5), we obtain

$$P(s)Y(s) = F(s), \quad s \in \Omega_f$$

where Y, F are the Mellin transforms of y, f, respectively, and

$$P(s) = (-1)^na_n(s)_n + \cdots + (-1)a_1s + a_0,$$

is a polynomial in s of degree n. If $P(s)$ has no zeros in Ω_f, then

$$y(t) = \mathcal{M}^{-1}\left(\frac{F(s)}{P(s)}\right),$$

where $\mathcal{M}^{-1}$ is the inverse Mellin transformation. This method is valid even if f is a generalized function.

10.9.2 Partial Differential Equations

Consider the Dirichlet problem

$$r^2\frac{\partial^2 u}{\partial r^2} + r\frac{\partial u}{\partial r} + \frac{\partial^2 u}{\partial \theta^2} = 0, \quad 0 \le r < \infty, \quad -\alpha < \theta < \alpha, \tag{10.12}$$

$$\lim_{\theta\to\alpha^-} u(r,\theta) = f(r), \quad \lim_{\theta\to-\alpha^+} u(r,\theta) = f(r). \tag{10.13}$$

Let us denote the Mellin transform (with respect to r) of $u(r,\theta)$ by $U(s,\theta)$. By applying the Mellin transform to Laplace's equation (10.12), we obtain

$$U(s,\theta) = A(s)e^{is\theta} + B(s)e^{-is\theta},$$

and upon incorporating the initial conditions (10.13), we have

$$A(s) = \frac{F(s)}{2\cos(\alpha s)} = B(s).$$

Hence,

$$U(s,\theta) = \frac{F(s)\cos(\theta s)}{\cos(\alpha s)},$$

and consequently

$$u(r,\theta) = \frac{1}{2\pi i}\int_{\sigma-i\infty}^{\sigma+i\infty}\frac{F(s)\cos(\theta s)}{\cos(\alpha s)}r^{-s}ds, \quad s \in \Omega_f,$$

where σ is a real number in Ω_f that is not equal to $(2n+1)\pi/2\alpha$, $n = 0, \pm 1, \pm 2, \cdots$. Note that

$$\frac{\cos s\theta}{\cos s\alpha} = O\left(e^{-|\tau|(\alpha-\theta)}\right) \quad \text{as } |\tau| \to \infty$$

where $s = \sigma + i\tau$.

This technique, with some appropriate modifications, is valid even if F is a generalized function; see [276, p. 123] .

10.9.3 Integral Equations

Consider the integral equation

$$y(x) = f(x) + \int_0^\infty g(xt)y(t)dt, \tag{10.14}$$

in which f and g are given Mellin transformable functions. If we apply the Mellin transform to (10.14) and use (10.9), we obtain

$$Y(s) = F(s) + G(s)Y(1-s),$$

or

$$Y(1-s) = F(1-s) + G(1-s)Y(s).$$

By eliminating $Y(1-s)$ from these two equations, we obtain

$$Y(s) = \frac{F(1-s)G(s) + F(s)}{1 - G(s)G(1-s)},$$

which yields

$$y(t) = \frac{1}{2\pi i}\int_{c-i\infty}^{c+i\infty} \frac{F(1-s)G(s) + F(s)}{1 - G(s)G(1-s)} t^{-s} ds, \quad s \in \Omega_f \cap \Omega_g,$$

where c is a real number in $\Omega_f \cap \Omega_g$.

11

Finite Mellin-Type Transforms

In [177], D. Naylor established a procedure for generating integral transforms adapted to the removal of the polar differential operators that occur when the Laplace differential operator is expressed in either spherical or plane polar coordinates. In so doing, he introduced a set of eight integral transforms that are suited to solve initial and boundary-value problems involving Laplace's operator on regions bounded by the natural coordinate surfaces of cylindrical or spherical coordinate systems. Four of these transforms are suitable for bounded regions and the others for unbounded regions. Naylor called these transforms the *finite Mellin-type integral transforms* since they are extensions of the Mellin integral transform and their inversion formulas are similar to that of the Mellin transform.

11.1 Definitions

DEFINITION 11.1 *The finite Mellin-type transforms of $f(t)$ of the first, second, third, and fourth kinds are defined by*

$$F_1(s) = \mathcal{M}_1(f)(s) = \int_0^a \left(\frac{a^{2s}}{t^{s+1}} - t^{s-1}\right) f(t)dt,$$

$$F_2(s) = \mathcal{M}_2(f)(s) = \int_0^a \left(\frac{a^{2s}}{t^{s+1}} + t^{s-1}\right) f(t)dt,$$

$$F_3(s) = \mathcal{M}_3(f)(s) = \int_0^a \left(\frac{a^{2s+1}}{t^{s+1}} - t^{s}\right) f(t)dt,$$

and

$$F_4(s) = \mathcal{M}_4(f)(s) = \int_0^a \left(\frac{a^{2s+1}}{t^{s+1}} + t^s\right) f(t)dt,$$

respectively. The associated finite Mellin-type transforms of $f(t)$ of the first, second, third, and fourth kinds are defined similarly, except that the integrals are taken over the interval (a, ∞), e.g., the associated transform of the first kind is defined by

$$\tilde{F}_1(s) = \int_a^\infty \left(\frac{a^{2s}}{t^{s+1}} - t^{s-1}\right) f(t)dt.$$

11.2 Existence

The finite Mellin-type transforms exist at least in the strip $-\alpha < \text{Re } s \leq \alpha$ if f is a locally integrable function with $f(t) = 0(t^\alpha)$ as $t \to 0^+$. Similarly, the associated transforms exist at least in the strip $-\alpha < \text{Re } s \leq \alpha - 2$ provided that f is integrable on (a, ∞) and $f(t) = O(t^{-\alpha})$ as $t \to \infty$ for some $\alpha > 0$.

11.3 Elementary Properties

1. **Linearity:** All the eight kinds of the finite Mellin-type transforms are linear.

2. **Translation:**

$$F_1(-s) = -a^{-2s} F_1(s) \quad , \quad F_2(-s) = a^{-2s} F_2(s),$$

$$a^{2s-1} F_3(-s) = -F_3(s-1) \text{ and } a^{2s-1} F_4(-s) = F_4(s-1).$$

3. **Differentiation:**

$$\mathcal{M}_1\left[tf'(t)\right](s) = s\mathcal{M}_2[f](s),$$

provided that

$$\lim_{t \to 0+} \left(\frac{a^{2s}}{t^s} - t^s\right) f(t) = 0.$$

Similarly,

$$\mathcal{M}_2\left[tf'(t)\right] = 2a^s f(a) + s\mathcal{M}_1[f](s),$$

provided that

$$\lim_{t\to 0^+}\left(\frac{a^{2s}}{t^s}+t^s\right)f(t)=0.$$

From this it follows that if

$$L_1 = t\frac{d}{dt}t\frac{d}{dt} = t^2\frac{d}{dt} + t\frac{d}{dt},$$

then

$$\mathcal{M}_1[L_1 f](s) = s^2 F_1(s) + 2sa^s f(a), \tag{11.1}$$

provided that

$$\lim_{t\to 0^+}\left\{\left(\frac{a^{2s}}{t^s}-t^s\right)tf'(t)\right\}=0 \text{ and } \lim_{t\to 0^+}\left\{\left(\frac{a^{2s}}{t^s}+t^s\right)f(t)\right\}=0.$$

In addition,

$$\mathcal{M}_2\left[L_1 f\right](s) = s^2 F_2(s) + 2a^{s+1}f'(a) \tag{11.2}$$

provided that

$$\lim_{t\to 0^+}\left\{\left(\frac{a^{2s}}{t^s}+t^s\right)tf'(t)\right\}=0 \text{ and } \lim_{t\to 0^+}\left\{\left(\frac{a^{2s}}{t^s}-t^s\right)f(t)\right\}=0.$$

11.4 Relationship with the Mellin Transform

The finite Mellin-type transforms and their associated transforms are readily seen to be related to the Mellin transform. For example, if we denote the Mellin transform of $\tilde{f}(t)$ by $\tilde{F}(s)$, then

$$F_1(s) = \mathcal{M}_1(f)(s) = \int_0^a \left(\frac{a^{2s}}{t^{s+1}} - t^{s-1}\right) f(t)dt = a^{2s}\tilde{F}(-s) - \tilde{F}(s),$$

where

$$\tilde{f}(t) = \begin{cases} f(t) & \text{if } 0 \le t \le a \\ 0 & \text{if } a < t < \infty \end{cases}.$$

Similar results can be derived for the other types.

11.5 Convolution

Let $\tilde{F}_2(s)$ and $\tilde{G}_2(s)$ be the associated Mellin-type transforms of the second kind of $f(t)$ and $g(t)$, respectively, i.e.,

$$\tilde{F}_2(s) = \int_a^\infty \left(\frac{a^{2s}}{t^{s+1}} + t^{s-1} \right) f(t) dt,$$

$$\tilde{G}_2(s) = \int_a^\infty \left(\frac{a^{2s}}{t^{s+1}} + t^{s-1} \right) g(t) dt.$$

Then

$$\int_a^\infty \left(\frac{a^{2s}}{t^{s+1}} + t^{s-1} \right) f(t) g(t) dt = \frac{1}{2\pi i} \int_\gamma \tilde{F}_2(u) \tilde{G}_2(s-u)\, du\ ,$$

where $\tilde{F}_2(s)$ and $\tilde{G}_2(s)$ are assumed to be analytic in some vertical strip $|Re\, s| < c$ and γ is a vertical line chosen suitably in that strip.

11.6 Inversion Formulas

If F_1 is analytic in the strip $|Re\ s| \leq c$ and $\lim\limits_{|\tau| \to \infty} |F(\sigma + i\tau)| < \infty$ for $|\sigma| \leq d < c$, then

$$f(t) = \frac{1}{2\pi i} \int_\gamma F_1(s) t^{-s} ds, \quad 0 < t < a,$$

where γ is the line $|Re\ s| = d, \quad (0 < d < c)$.

Similarly,

$$f(t) = \frac{1}{2\pi i} \int_\gamma F_2(s) t^{-s} ds,$$

$$f(t) = \frac{1}{2\pi i} \int_\gamma F_3(s) t^{-s-1} ds,$$

and

$$f(t) = \frac{1}{2\pi i} \int_\gamma F_4(s) t^{-s-1} ds,$$

where γ has the same meaning as above; see [216, p. 465].

The inversion formulas for the associated transforms are the same as the aforementioned ones, except for some obvious modifications. For example,

$$f(t) = \frac{1}{2\pi i} \int_{\gamma} \tilde{F}_1(s) t^{-s} ds, \quad a < t < \infty.$$

11.7 Applications

The finite Mellin-type transforms can be used in problems involving the two-dimensional Laplacian operator in polar coordinates

$$\nabla = \frac{\partial^2}{\partial \rho^2} + \frac{1}{\rho}\frac{\partial}{\partial \rho} + \frac{1}{\rho^2}\frac{\partial^2}{\partial \phi^2},$$

or the axially symmetric three-dimensional Laplacian operator

$$\nabla = \frac{1}{r^2}\frac{\partial}{\partial r} r^2 \frac{\partial}{\partial r} + \frac{1}{\sin\theta}\frac{\partial}{\partial \theta}\sin\theta \frac{\partial}{\partial \theta}.$$

In the first case if we apply the finite Mellin-type transform of the first kind (taken with respect to ρ) to the equation

$$\rho^2 \nabla u = \rho^2 \frac{\partial^2 u}{\partial \rho^2} + \rho \frac{\partial u}{\partial \rho} + \frac{\partial^2 u}{\partial \phi^2} = 0\,, \tag{11.3}$$

we obtain, with the aid of (11.1),

$$\left(\frac{\partial^2}{\partial \phi^2} + s^2\right) U_1(s, \phi) + 2sa^s u(a, \phi) = 0\,, \tag{11.4}$$

where $0 \le \rho \le a$,

$$\mathcal{M}_1\left(u(\rho, \phi)\right)(s) = U_1(s, \phi),$$

and provided that

$$\lim_{\rho \to 0+} \left\{ \left(\frac{a^{2s}}{\rho^s} - \rho^s\right) \rho u_\rho(\rho, \phi) \right\} = 0,$$

and

$$\lim_{\rho \to 0+} \left\{ \left(\frac{a^{2s}}{\rho^s} + \rho^s\right) u(\rho, \phi) \right\} = 0.$$

Equation (11.4) can now be solved if $u(a, \phi)$ is given, say for $\alpha \le \phi \le \beta$. On the other hand, if $\partial u/\partial \rho$, instead of $u(a, \phi)$, is given at $\rho = a$, we then apply the finite Mellin-type transform of the second kind to (11.3) and use (11.2) to obtain

$$\left(\frac{\partial^2}{\partial\phi^2}+s^2\right)U_2(s,\phi)+2a^{s+1}\frac{\partial u(a,\phi)}{\partial\rho}=0,$$

provided that

$$\lim_{\rho\to 0+}\left\{\left(\frac{a^{2s}}{\rho^s}+\rho^s\right)\rho u_\rho(\rho,\phi)\right\}=0$$

and

$$\lim_{\rho\to 0+}\left\{\left(\frac{a^{2s}}{\rho^s}-\rho^s\right)u(\rho,\phi)\right\}=0.$$

For the axially symmetric three-dimensional Laplacian operator, we use the finite Mellin-type transform of the third or fourth kind depending on whether $u(a,\phi)$ or $u_r(a,\phi)$ is given. For details, see [216, p. 466]. The associated transforms are used to solve the same problems when $a\le\rho$.

Example A: Consider the boundary-value problem

$$\nabla u=0,\quad 0\le\rho\le a\,,-\alpha\le\phi\le\alpha\,,$$

with

$$u(a,\phi)=0\,,\quad u(\rho,\pm\alpha)=f(\rho).$$

From Equation (11.4), we have

$$\left(\frac{\partial^2}{\partial\phi^2}+s^2\right)U_1(s,\phi)=0,$$

which, upon solving, yields the solution

$$U_1(s,\phi)=A(s)\cos(s\phi)+B(s)\sin(s\phi).$$

If we denote the Mellin-type transform of the first kind of $f(\rho)$ by $f_1(s)$, we find that U_1 must satisfy

$$U_1(s,\pm\alpha)=f_1(s),$$

and hence

$$U_1(s,\phi)=f_1(s)\frac{\cos(s\phi)}{\cos(s\alpha)}.$$

Therefore, from the inversion formula of the Mellin-type transform of the first kind, we deduce that

$$u(\rho,\phi)=\frac{1}{2\pi i}\int_\gamma f_1(s)\frac{\cos(\phi s)}{\cos(\alpha s)}\rho^{-s}ds,$$

for some appropriate path γ.

12

The Fourier Transform

The origin of the *Fourier integral transformation* can be traced to Fourier's celebrated work on the *Analytical Theory of Heat, Théorie analytique de la chaleur,* which appeared in 1822 [66]. Joseph Fourier was a French physicist who taught at a local military school and later at the École Polytechnique. In 1798 he joined Napoleon in his invasion of Egypt and was appointed a secretary of the Institut d'Egypt which compiled the famous publication *Description de l'Egypte.* After he returned to France, he continued his scholarly pursuits and in 1807 he submitted a paper to the prestigious French Academy of Science on the problem of heat conduction. The paper was refereed by very prominent mathematicians such as Laplace, Lagrange and Legendre who rejected it for lack of rigor. Four years later, Fourier submitted a revised version of the paper, which was awarded a prize by the Academy. Nonetheless, the Academy still refused to publish it because some of Fourier's reasonings were described as unclear.

Fourier's major finding was to show that an "arbitrary" function defined on a finite interval could be expanded in a trigonometric series (series of sinusoidal functions) which now bears his name. In an attempt to extend his results to functions defined on the infinite interval $(0, \infty)$, Fourier introduced what is now known as the Fourier integral transform and gave an explicit formula for the Fourier cosine transform and its inversion formula, but his argument would hardly be called a proof today. Nevertheless, it is no exaggeration to say that his work had a profound impact on the development of mathematics.

It should be pointed out that the idea of the Fourier integral transformation was known to Cauchy and Laplace and it did appear in some of their early writings. A separate treatment of the Fourier integral transform also appeared in a paper by Poisson on water waves.

12.1 Definitions

DEFINITION 12.1 *The Fourier integral transform of a function $f(t)$ is defined by*

$$\mathcal{F}(f)(w) = \hat{f}(w) = F(w) = \frac{1}{\sqrt{2\pi}} \int_{-\infty}^{\infty} f(t)e^{itw}\, dt\,, \tag{12.1}$$

whenever the integral exists.

It should be made clear that there is no universal agreement on the definition of the Fourier integral transform; some authors take the kernel of the transformation to be e^{-itw}, so that the kernel of the inverse transformation is e^{itw}. In either case, if we define the Fourier transform as

$$\hat{f}(w) = a \int_{-\infty}^{\infty} f(t)e^{\pm itw}\, dt\,,$$

so that its inverse is

$$f(t) = b \int_{-\infty}^{\infty} \hat{f}(w)e^{\mp itw}\, dw\,,$$

for some constants a and b, then $ab = 1/2\pi$. Again there is no agreement on the choice of the constants; sometimes one of them is taken as 1 so that the other is $1/(2\pi)$. But for the sake of symmetry, we shall take $a = b = 1/\sqrt{2\pi}$. The functions f and $\hat{f}$ are called a *Fourier transform pair.*

Another definition that is more popular in the engineering literature is the one in which the kernel of the transformation is taken as $e^{2\pi itw}$ (or $e^{-2\pi itw}$) so that the kernel of the inverse transformation is $e^{-2\pi itw}$ (or $e^{2\pi itw}$). The main advantage of this definition is that the constants a and b disappear altogether and the Fourier transform pair becomes

$$\hat{f}(w) = \int_{-\infty}^{\infty} f(t)e^{\pm 2\pi itw}\, dt\,,$$

and

$$f(t) = \int_{-\infty}^{\infty} \hat{f}(w)e^{\mp 2\pi itw}\, dw\,.$$

Definition 12.1 implies certain restrictions on the behavior of $f(t)$ near $\pm\infty$; therefore, even if $F(w)$ does not exist, it is feasible that one of the integrals

$$F_+(w) = \frac{1}{\sqrt{2\pi}} \int_0^{\infty} f(t)e^{iwt} dt \quad , \quad F_-(w) = \frac{1}{\sqrt{2\pi}} \int_{-\infty}^{0} f(t)e^{iwt} dt \tag{12.2}$$

may exist. If both exist, then $F = F_+ + F_-$. Clearly, $F_\pm$ are related to the Laplace transform of f (see Chapter 5), for example, F_+ is the Laplace transform of the function $(1/\sqrt{2\pi})f(t)e^{itu}$, where $w = u + iv$. Also, F_+ can be regarded as the Fourier transform of the function$f(t)e^{-tv}H(t)$, where $H(t)$ is the Heaviside function. Similar assertions hold for F_-.

DEFINITION 12.2 *The Fourier cosine, $F_c(w)$, and sine, $F_s(w)$, transforms of $f(t)$ are defined as*

$$F_c(w) = \sqrt{\frac{2}{\pi}} \int_0^\infty f(t) \cos wt \, dt\,, \tag{12.3}$$

and

$$F_s(w) = \sqrt{\frac{2}{\pi}} \int_0^\infty f(t) \sin wt \, dt\,. \tag{12.4}$$

If f is even, i.e., $f(t) = f(-t)$, then $F(w) = F_c(w)$, while if f is odd, i.e.,$f(t) = -f(-t)$, then $F(w) = iF_s(w)$.

Two other integrals related to the Fourier integral transform are Fourier's repeated integral and the allied integral.

DEFINITION 12.3 *The Fourier cosine and sine coefficients of $f(t)$ are defined by*

$$a(w) = \frac{1}{\pi} \int_{-\infty}^\infty f(t) \cos wt \, dt\,,$$

and

$$b(w) = \frac{1}{\pi} \int_{-\infty}^\infty f(t) \sin wt \, dt\,.$$

Fourier's repeated integral, $S(f, t)$, of $f(t)$ is given by

$$S(f,t) = \int_0^\infty (a(w)\cos tw + b(w)\sin tw)\, dw = \frac{1}{\pi}\int_0^\infty dw \int_{-\infty}^\infty f(x)\cos w(t-x) dx\,. \tag{12.5}$$

The allied Fourier integral, $\tilde{S}(f,t)$, of f is defined by

$$\tilde{S}(f,t) = \int_0^\infty (b(w)\cos tw - a(w)\sin tw)\, dw = \frac{1}{\pi}\int_0^\infty dw \int_{-\infty}^\infty f(x)\sin w(x-t) dx\,. \tag{12.6}$$

Formally, we have

$$S(f,t) = \frac{1}{\pi}\int_{-\infty}^\infty f(x) dx \int_0^\infty \cos w(t-x) dw\,,$$

and

$$\tilde{S}(f,t) = \frac{1}{\pi}\int_{-\infty}^{\infty} f(x)dx \int_0^{\infty} \sin w(x-t)dw\,.$$

The Fourier cosine and sine coefficients are evidently related to the Fourier cosine and sine integral transforms. For example, if f is even,then $a(w) = \sqrt{2/\pi}F_c(w)$ and if f is odd, $b(w) = \sqrt{2/\pi}F_s(w)$.

Moreover, we have

$$\begin{aligned} S(f,t) &= \frac{1}{\pi}\int_0^{\infty} dw \int_{-\infty}^{\infty} f(x)\cos w(x-t)dx \\ &= \frac{1}{\pi}\lim_{\lambda\to\infty}\int_0^{\lambda} dw \int_{-\infty}^{\infty} f(x)\cos w(x-t)dx \\ &= \lim_{\lambda\to\infty}\frac{1}{\pi}\int_{-\infty}^{\infty} f(x)\frac{\sin\lambda(t-x)}{(t-x)}dx, \end{aligned} \tag{12.7}$$

and

$$\begin{aligned} \frac{1}{\sqrt{2\pi}}\int_{-\infty}^{\infty}\hat{f}(w)e^{-itw}dw &= \lim_{\lambda\to\infty}\frac{1}{\sqrt{2\pi}}\int_{-\lambda}^{\lambda}\hat{f}(w)e^{-itw}dw \\ &= \frac{1}{2\pi}\lim_{\lambda\to\infty}\int_{-\lambda}^{\lambda} e^{-itw}dw\int_{-\infty}^{\infty} f(x)e^{iwx}dx \\ &= \lim_{\lambda\to\infty}\frac{1}{\pi}\int_{-\infty}^{\infty} f(x)\frac{\sin\lambda(t-x)}{(t-x)}dx. \end{aligned} \tag{12.8}$$

Because of (12.7) and (12.8), the study of the convergence of Fourier's repeated integral and the integral

$$\int_{-\infty}^{\infty}\hat{f}(w)e^{-itw}dw$$

is reduced to the study of the existence of the limit

$$\lim_{\lambda\to\infty}\frac{1}{\pi}\int_{-\infty}^{\infty} f(x)\frac{\sin\lambda(t-x)}{(t-x)}dx\,,$$

provided that interchanging the integral signs in (12.7) and (12.8) is justified.

Similarly, because

$$\begin{aligned} \tilde{S}(f,t) &= \frac{1}{\pi}\int_0^{\infty} dw\int_{-\infty}^{\infty} f(x)\sin w(x-t)dx \\ &= \frac{1}{\pi}\lim_{\lambda\to\infty}\int_0^{\lambda} dw\int_{-\infty}^{\infty} f(x)\sin w(x-t)dx = \lim_{\lambda\to\infty}\frac{1}{\pi}\int_{-\infty}^{\infty} f(x)\frac{1-\cos\lambda(x-t)}{(x-t)}dx \\ &= \lim_{\lambda\to\infty}\int_{-\infty}^{\infty} f(x)\frac{2\sin^2\frac{\lambda}{2}(x-t)}{\pi(x-t)}dx\,, \end{aligned} \tag{12.9}$$

the study of the convergence of the allied integral is reduced to that of the existence of the limit

$$\lim_{\lambda\to\infty}\int_{-\infty}^{\infty} f(x)\frac{2\sin^2\frac{\lambda}{2}(x-t)}{\pi(x-t)}dx\,.$$

Likewise, the (C,1) sums of Fourier's repeated integral and the integral

$$\frac{1}{\sqrt{2\pi}}\int_{-\infty}^{\infty}\hat{f}(w)e^{-itw}dw,$$

are

$$\frac{1}{\lambda\pi}\int_0^{\lambda}dy\int_0^{y}dw\int_{-\infty}^{\infty}f(x)\cos w(t-x)dx=\frac{1}{\lambda\pi}\int_0^{\lambda}dy\int_{-\infty}^{\infty}f(x)\frac{\sin y(t-x)}{(t-x)}dx$$
$$=\frac{1}{\lambda}\int_{-\infty}^{\infty}f(x)\frac{2\sin^2\frac{\lambda}{2}(t-x)}{\pi(t-x)^2}dx\,, \tag{12.10}$$

and

$$\frac{1}{\lambda\sqrt{2\pi}}\int_0^{\lambda}dy\int_{-y}^{y}\hat{f}(w)e^{-itw}dw=\frac{1}{2\lambda\pi}\int_0^{\lambda}dy\int_{-y}^{y}e^{-itw}dw\int_{-\infty}^{\infty}f(x)e^{ixw}dx$$
$$=\frac{1}{\lambda\pi}\int_0^{\lambda}dy\int_{-\infty}^{\infty}f(x)\frac{\sin y(t-x)}{(t-x)}dx=\frac{1}{\lambda}\int_{-\infty}^{\infty}f(x)\frac{2\sin^2\frac{\lambda}{2}(t-x)}{\pi(t-x)^2}dx\,. \tag{12.11}$$

Therefore, the study of the (C,1) summability of Fourier's repeated integral and the integral

$$\frac{1}{\sqrt{2\pi}}\int_{-\infty}^{\infty}\hat{f}(w)e^{-itw}\,dw$$

is reduced to that of the limit

$$\lim_{\lambda\to\infty}\frac{1}{\lambda}\int_{-\infty}^{\infty}f(x)\frac{2\sin^2\frac{\lambda}{2}(t-x)}{\pi(t-x)^2}dx\,,$$

provided that interchanging the integral signs in (12.10) and (12.11) is permitted.

Similar statements can be made about the allied integral

$$\tilde{S}(f,t)=\frac{1}{\pi}\int_0^{\infty}\tilde{g}(w,t)\,dw\,,$$

where

$$\tilde{g}(w,t)=\int_{-\infty}^{\infty}f(x)\sin w(x-t)\,dx\,.$$

For example, the (C,1) sum of the allied integral is

$$\begin{aligned}
&\lim_{\lambda\to\infty} \frac{1}{\lambda\pi} \int_0^\lambda dy \int_0^y \tilde{g}(w,t)\,dw \\
&= \lim_{\lambda\to\infty} \frac{1}{\lambda\pi} \int_0^\lambda dy \int_0^y dw \int_{-\infty}^{\infty} f(x)\sin w(x-t)\,dx \\
&= \lim_{\lambda\to\infty} \frac{1}{\lambda\pi} \int_0^\lambda dy \int_{-\infty}^{\infty} f(x) \frac{1-\cos y(x-t)}{(x-t)} dx \\
&= \lim_{\lambda\to\infty} \frac{1}{\pi} \int_{-\infty}^{\infty} f(x) \left[\frac{1}{(x-t)} - \frac{\sin\lambda(x-t)}{\lambda(x-t)^2}\right] dx \\
&= \lim_{\lambda\to\infty} \frac{1}{\pi} \int_{-\infty}^{\infty} f(x+t) \left[\frac{1}{x} - \frac{\sin\lambda x}{\lambda x^2}\right] dx\,.
\end{aligned}$$

DEFINITION 12.4 *The Hilbert transform $\tilde{f}(x)$ of a function $f(t)$ defined on $(-\infty,\infty)$ is given by (see Chapter 14)*

$$\tilde{f}(x) = \frac{1}{\pi}\int_{-\infty}^{\infty} \frac{f(t)}{t-x} dt = \lim_{\epsilon\to 0^+} \frac{1}{\pi} \int_{|t-x|\geq\epsilon} \frac{f(t)}{t-x} dt\,,$$

whenever the integral exists. The integral is taken as a Cauchy principle value. The function $\tilde{f}$ is sometimes called the conjugate of f.

DEFINITION 12.5 *An integral transform of the form*

$$F(x) = \int_I f(t)k(xt)\,dt\,,$$

whose inverse is given by

$$f(t) = \int_I F(x)h(tx)\,dx\,,$$

for some appropriate functions $k\,,h$ and an infinite interval I, is called a Fourier-type integral transform. The pair of functions, f, F, is called a Fourier-type pair with unsymmetrical kernels. If $k = h$, it is called a Fourier-type pair with symmetrical kernels.

The Fourier transform is clearly of Fourier type. More examples of Fourier-type integral transforms and Fourier-type pairs can be found in Chapter 21.

DEFINITION 12.6 *Let $g(t)$ be a bounded real-valued function. The* *windowed-Fourier transform of $f(t)$ with respect to g is defined by*

$$F_g(w,t) = \frac{1}{\sqrt{2\pi}} \int_{-\infty}^{\infty} f(x)g(x-t)e^{iwx}\,dx\,.$$

The function g is called the window function.

In most cases of interest g either has compact support or is of rapid decay. Usually, it is taken to be even so that its Fourier transform is also a real-valued function. For more details, see Chapter 30.

The windowed-Fourier transform, which has been used extensively by electrical engineers, is an example of an integral transform that maps a function of one variable into a function of two variables. Admittedly, this results in redundancy in the information conveyed by F_W; nevertheless, this redundancy can be exploited to solve many problems with useful applications.

12.2 Existence and Uniqueness

12.2.1 Existence

For the Fourier integral transform to exist, it is sufficient that f be absolutely integrable on $(-\infty, \infty)$, i.e., $f \in L^1(\mathcal{R})$. In fact, more can be said in this case.

THEOREM 12.1
If $f \in L^1(\mathcal{R})$, then its Fourier transform $\hat{f}(w)$ is defined everywhere, uniformly continuous, and tends to zero as $w \to \pm\infty$.

The uniform continuity follows from the relation

$$\left|\hat{f}(w+h) - \hat{f}(w)\right| \le \frac{1}{\sqrt{2\pi}} \int_{-\infty}^{\infty} |f(t)| \left|e^{iht} - 1\right| dt\,,$$

and the tendency toward zero at $\pm\infty$ is a consequence of the Riemann-Lebesgue lemma [279, Vol. 1, p. 46].

THEOREM 12.2 (Riemann–Lebesgue Lemma)
Let $f \in L^1(I)$, where $I = (a,b)$ is finite or infinite and let λ be a real variable. Let $a \le a' < b' \le b$ and

$$\hat{f}_\lambda(w, a', b') = \int_{a'}^{b'} f(t)e^{i\lambda wt}dt\,.$$

Then

$$\lim_{\lambda \to \pm\infty} \hat{f}_\lambda(w, a', b') = 0\,,$$

and the convergence is uniform in a' and b'.

The following theorem extends the definition of the Fourier transform to the case where $f \in L^p(\mathcal{R})\,, 1 < p \leq 2$

THEOREM 12.3 [225, p. 96]
Let $f \in L^p(\mathcal{R})$, where $1 < p \leq 2$, and set

$$\hat{f}_\lambda(w) = \frac{1}{\sqrt{2\pi}} \int_{-\lambda}^{\lambda} f(t) e^{iwt} dt\,.$$

Then

$$\hat{f}(w) = \lim_{\lambda \to \infty} \hat{f}_\lambda(w)$$

exists in the sense of $L^q(\mathcal{R})$, where $\frac{1}{p} + \frac{1}{q} = 1$. The limit also exists almost everywhere for $1 \leq p < 2$. The limit function $\hat{f}$ is called the Fourier transform of f and it satisfies the inequality

$$\left(\frac{1}{\sqrt{2\pi}} \int_{-\infty}^{\infty} \left| \hat{f}(w) \right|^q dw \right)^{1/q} \leq \left(\frac{1}{\sqrt{2\pi}} \int_{-\infty}^{\infty} |f(t)|^p \, dt \right)^{1/p}.$$

Moreover,

$$\lim_{\lambda \to \infty} \frac{1}{\sqrt{2\pi}} \int_{-\lambda}^{\lambda} \hat{f}(w) e^{-itw} dw = f(t)\,,$$

in the sense of $L^p(\mathcal{R})$.

Functions in $L^p(\mathcal{R})$, with $p > 2$ do not in general possess Fourier transforms in the classical sense, yet they may have Fourier transforms in the sense of generalized functions; see Section 13.

12.2.2 Uniqueness

Let $f, g \in L^1(\mathcal{R})$. If $\hat{f}(w) = \hat{g}(w)$ for all w,then $f(t) = g(t)$ almost everywhere.

12.3 Analytic Properties

If f is analytic and satisfies certain growth conditions at $\pm\infty$, its Fourier transform can be continued analytically to the complex w-plane.

THEOREM 12.4 [225, p. 44]
Let $f(z)$ be analytic in the strip $-a < y < b$, where $a, b > 0$ and assume that in the interior of the strip it satisfies

$$f(z) = \begin{cases} O\left(e^{-(\lambda-\epsilon)x}\right) & \text{as } x \to \infty, \\ O\left(e^{(\mu-\epsilon)x}\right) & \text{as } x \to -\infty, \end{cases}$$

for any $\epsilon > 0$ and some $\lambda, \mu > 0$, where $z = x + iy$. Then the Fourier transform $\hat{f}$ of f exists, is analytic in the strip $-\lambda < v < \mu$ and satisfies

$$\hat{f}(w) = \begin{cases} O\left(e^{-(b-\epsilon)u}\right) & \text{as } u \to \infty, \\ O\left(e^{(a-\epsilon)u}\right) & \text{as } u \to -\infty, \end{cases}$$

where $w = u + iv$.

12.4 Elementary Properties

1. **Linearity**: The Fourier transformation is linear.

2. **Translation**:
$$(\widehat{T_a f(t)})(w) = e^{iaw}\hat{f}(w).$$

3. **Dilation**:
$$(\widehat{D_a f(t)})(w) = D_{1/a}\hat{f}(w), \quad a \neq 0.$$

4. **Translation and Dilation**:
$$(\widehat{f(at-b)})(w) = \frac{1}{a}e^{ibw/a}\hat{f}(\frac{w}{a}), \quad a > 0,$$
or
$$\left(\widehat{D_{1/a}T_b f(t)}\right)(w) = D_a\left(e^{ibw}\hat{f}(w)\right), \quad a > 0.$$

5. **Complex Conjugation**:
$$\left(\widehat{\bar{f}(t)}\right)(w) = \overline{\hat{f}(-w)}.$$

6. **Modulation**:
$$(\widehat{e^{iat}f(t)})(w) = \hat{f}(w+a)$$
and
$$(\widehat{e^{iat}f(bt)})(w) = \frac{1}{b}\hat{f}(\frac{w+a}{b}), \quad b > 0.$$

7. **Differentiation**: If $f^{(k)} \in L^1(\mathcal{R})\,, k = 0, 1, 2, \cdots, n$ and $\lim_{|t|\to\infty} f^{(k)}(t) = 0, k = 0, 1, 2, \cdots, n-1$, then

$$\left(\widehat{f^{(n)}(t)}\right)(w) = (-iw)^n \hat{f}(w).$$

8. **Integration**: Let $f \in L^1(\mathcal{R})$ and define

$$g(x) = \int_{-\infty}^{x} f(t)\,dt\,.$$

If $g \in L^1(\mathcal{R})$, then

$$\hat{g}(w) = \frac{-1}{iw}\hat{f}(w).$$

9. **Multiplication by Polynomials**: If $t^n f(t) \in L^1(\mathcal{R})$, then

$$(\widehat{t^k f(t)})(w) = \frac{1}{(i)^k}\hat{f}^{(k)}(w),$$

and hence

$$\left(\widehat{(\sum_{k=0}^{n} a_k t^k) f(t)}\right)(w) = \sum_{k=0}^{n} \frac{a_k}{(i)^k}\hat{f}^{(k)}(w).$$

Properties (7) and (8) are proved by integration by parts and in (8) one also uses the fact that $g(\infty) = 0$, which is a consequence of the hypothesis that f and g are in $L^1(\mathcal{R})$. The proof of the other properties is straightforward.

12.5 Summability

THEOREM 12.5 [225, p. 29]
Let $f(t)/\left(1 + |t|^2\right) \in L^1(\mathcal{R})$,then

$$\lim_{\lambda\to\infty} \frac{1}{\pi}\int_{-\infty}^{\infty} f(t)\frac{2\sin^2\frac{\lambda}{2}(x-t)}{\lambda(x-t)^2}dt$$

is equal to $[f(x+0) + f(x-0)]/2$ whenever the expression has a meaning; to $f(x)$ whenever $f(x)$ is continuous; and to $f(x)$ almost everywhere.

In view of Theorem 12.5 and (12.11) we have

THEOREM 12.6
Let f satisfy one of the following conditions:

A) $f \in L^1(\mathcal{R})$, *or*

B) $f(t)/\left(1+|t|^2\right) \in L^1(\mathcal{R})$ *and the integral*

$$\hat{f}(w) = \frac{1}{\sqrt{2\pi}} \int_{-\infty}^{\infty} f(t) e^{iwt} dt$$

converges uniformly over every finite interval of w.

Then the integral (the inverse Fourier transform)

$$\frac{1}{\sqrt{2\pi}} \int_{-\infty}^{\infty} \hat{f}(w) e^{-itw} dw \tag{12.12}$$

is (C,1) summable to $f(t)$ *almost everywhere. More precisely,*

$$\lim_{\lambda \to \infty} \frac{1}{\sqrt{2\pi}} \int_{-\lambda}^{\lambda} \left(1 - \frac{|w|}{\lambda}\right) \hat{f}(w) e^{-itw} dw$$

is equal to $[f(t+0) + f(t-0)]/2$ *whenever the expression has a meaning, to* $f(t)$ *whenever* $f(t)$ *is continuous, and to* $f(t)$ *almost everywhere.*

Because of (12.10) and Theorem 12.5, Theorem 12.6 remains valid if we replace (12.12) by Fourier's repeated integral

$$\frac{1}{\pi} \int_0^{\infty} dw \int_{-\infty}^{\infty} f(x) \cos w(t-x)\, dx.$$

The next theorem gives us an analogue of Theorem 12.6 for the allied integral.

THEOREM 12.7 [225, p. 147]
Let f *satisfy either one of the conditions of Theorem 12.6 and* $\tilde{f}$ *be its Hilbert transform. Then the integral*

$$\frac{1}{\pi} \int_0^{\infty} dw \int_{-\infty}^{\infty} f(x) \sin w(x-t) dx$$

is (C,1) summable to $\tilde{f}(t)$ *almost everwhere. More precisely*

$$\lim_{\lambda \to \infty} \frac{1}{\pi} \int_{-\infty}^{\infty} f(x) \left[\frac{1}{x-t} - \frac{\sin \lambda(x-t)}{\lambda(x-t)^2}\right] dx$$

is equal to $\left[\tilde{f}(t+0) + \tilde{f}(t-0)\right]/2$ *whenever the expression has a meaning; to* $\tilde{f}(t)$ *whenever* $\tilde{f}(t)$ *is continuous; and to* $\tilde{f}(t)$ *almost everywhere.*

For functions in L^2, we have the following:

THEOREM 12.8
If $f \in L^2(\mathcal{R})$, then both the integral

$$\frac{1}{\sqrt{2\pi}} \int_{-\infty}^{\infty} \hat{f}(w) e^{-itw} dw,$$

and Fourier's repeated integral are (C,1) summable to $f(t)$ almost everywhere.

12.6 Parseval's Relation

If F and G are the Fourier transforms of f and g, respectively, we formally have

$$\int_{-\infty}^{\infty} F(w)G(w)\, dw = \frac{1}{\sqrt{2\pi}} \int_{-\infty}^{\infty} G(w)dw \int_{-\infty}^{\infty} f(t)e^{iwt}dt$$
$$= \frac{1}{\sqrt{2\pi}} \int_{-\infty}^{\infty} f(t)dt \int_{-\infty}^{\infty} G(w)e^{iwt}dw = \int_{-\infty}^{\infty} f(t)g(-t)\, dt, \tag{12.13}$$

which is known as *Parseval's relation.* If we replace G by $\bar{G}$ so that $g(-t)$ is replaced by $\bar{g}(t)$,we obtain a more convenient form of Parseval's relation

$$\int_{-\infty}^{\infty} F(w)\bar{G}(w)dw = \int_{-\infty}^{\infty} f(t)\bar{g}(t)dt\,. \tag{12.14}$$

For $f = g$, we have

$$\int_{-\infty}^{\infty} |F(w)|^2\, dw = \int_{-\infty}^{\infty} \left|\hat{f}(w)\right|^2 dw = \int_{-\infty}^{\infty} |f(t)|^2\, dt\,. \tag{12.15}$$

The validity of (12.13) through (12.15) is easily established if f and g belong to $L^2(\mathcal{R})$ or if f and G belong to $L^1(\mathcal{R})$. If f and G belong to $L^p(\mathcal{R})\,, 1 < p < 2$, then by Theorem 12.3, F and g belong to $L^q(\mathcal{R})$, where $(1/p) + (1/q) = 1$. Hölder's inequality can now be used to show that both sides of (12.13) exist. In fact, in this case they not only exist, but coincide.

12.7 The Uncertainty Principle

THEOREM 12.9
Let T and W be two real numbers defined by

$$T^2 = \frac{1}{E}\int_{-\infty}^{\infty} t^2 |f(t)|^2 dt \quad , \quad W^2 = \frac{1}{E}\int_{-\infty}^{\infty} w^2 |\hat{f}(w)|^2 dw,$$

where

$$E = \int_{-\infty}^{\infty} |f(t)|^2 dt = \int_{-\infty}^{\infty} |\hat{f}(w)|^2 dw.$$

Assume that f is differentiable and $\lim_{|t|\to\infty} tf^2(t) = 0$, then $1/2 \leq TW$, or

$$\left(\frac{1}{2}\int_{-\infty}^{\infty} |f(t)|^2\, dt\right)^2 \leq \left(\int_{-\infty}^{\infty} t^2|f(t)|^2\, dt\right)\left(\int_{-\infty}^{\infty} w^2|\hat{f}(w)|^2\, dw\right). \quad (12.16)$$

This means that f and $\hat{f}$ cannot both be very small. Another related property of the Fourier transform is that if either one of the functions f or $\hat{f}$ vanishes outside some finite interval, then the other one must trail on to infinity. In other words, they cannot both vanish outside any finite interval.

Proof: To prove the uncertainty principle, we use integration by parts to deduce

$$\int_{-\infty}^{\infty} tf(t)\frac{df}{dt}\, dt = \frac{1}{2}\int_{-\infty}^{\infty} tdf^2 = \frac{1}{2}\left[tf^2(t)|_{-\infty}^{\infty} - \int_{-\infty}^{\infty} f^2(t)dt\right],$$

which, in view of the assumption on f, yields

$$\int_{-\infty}^{\infty} tf(t)\frac{df}{dt}\, dt = -\frac{1}{2}E.$$

Therefore, by applying the Cauchy–Schwarz inequality and using property 7 of Section 4, we obtain

$$\frac{1}{4}E^2 = \left|\int_{-\infty}^{\infty} tf(t)\frac{df}{dt}\, dt\right|^2 \leq \left(\int_{-\infty}^{\infty} t^2|f(t)|^2 dt\right)\left(\int_{-\infty}^{\infty} \left|\frac{df}{dt}\right|^2 dt\right) \leq (ETW)^2,$$

which immediately yields the result. ∎

12.8 Inversion Formula

Many of the theorems on the inversion of the Fourier transform are based on *Dini's condition*, which can be stated as follows:

If $f \in L^1(\mathcal{R})$, then a necessary and sufficient condition that

$$\lim_{\lambda\to\infty} \frac{1}{\pi} \int_{-\infty}^{\infty} f(t) \frac{\sin \lambda(x-t)}{(x-t)} dt = a$$

is that

$$\lim_{\lambda\to\infty} \int_0^{\delta} (f(x+y) + f(x-y) - 2a) \frac{\sin \lambda y}{y} dy = 0 , \tag{12.17}$$

for any fixed $\delta > 0$.

By the Riemann-Lebesgue Lemma, this condition is satisfied if

$$\int_0^{\delta} \left| \frac{f(x+y) + f(x-y) - 2a}{y} \right| dy < \infty , \tag{12.18}$$

for some $\delta > 0$. In particular, condition (12.18) holds for $a = f(x)$ if f is differentiable at x, and for $a = [f(x+0) + f(x-0)]/2$ if f is of bounded variation in a neighborhood of x. The restriction that $f \in L^1(\mathcal{R})$ can be weakened to $f(t)/(1+|t|) \in L^1(\mathcal{R})$ or $f(t)/t$ is of bounded variation and tends to zero as $|t| \to \infty$.

THEOREM 12.10 [280, p. 242]
Let f be a locally integrable function and of bounded variation in a neighborhood of x. In addition, if f satisfies either one of the following conditions:

A) $f(t)/(1+|t|) \in L^1(\mathcal{R})$, or

B) $f(t)/t$ is of bounded variation in (a,∞) and $(-\infty,-a)$ for some $a > 0$, and tends to zero as $|t| \to \infty$,

then

$$\lim_{\lambda\to\infty} \frac{1}{\pi} \int_{-\infty}^{\infty} f(t) \frac{\sin \lambda(x-t)}{(x-t)} dt$$

is equal to $[f(x+0) + f(x-0)]/2$ whenever the expression has meaning; to $f(x)$ whenever $f(x)$ is continuous; and to $f(x)$ almost everywhere. If f is continuous and of bounded variation in the interval (a,b), then the convergence is uniform in any interval interior to (a,b).

Condition (A) of the theorem is clearly satisfied if f is integrable. The inversion of the Fourier transform and the convergence of Fourier's repeated integral will follow if interchanging the integral signs in (12.7) and (12.8) is justified. The next theorem provides sufficient conditions under which the inversion of the Fourier transform is possible.

THEOREM 12.11
Let f be a locally integrable function and of bounded variation in a neighborhood of the point x. If f satisfies either one of the following conditions:

A) $f(t) \in L^1(\mathcal{R})$,

B) $f(t)/(1+|t|) \in L^1(\mathcal{R})$, *and the integral*

$$\int_{-\infty}^{\infty} f(t)e^{iwt}dt$$

converges uniformly on every finite interval of w,

then

$$\frac{1}{\sqrt{2\pi}}\int_{-\infty}^{\infty}\hat{f}(w)e^{-itw}dw = \lim_{\lambda\to\infty}\frac{1}{\sqrt{2\pi}}\int_{-\lambda}^{\lambda}\hat{f}(w)e^{-itw}dw$$

is equal to $[f(x+0)+f(x-0)]/2$ whenever the expression has meaning; to $f(x)$ whenever $f(x)$ is continuous; and to $f(x)$ almost everywhere. If f is continuous and of bounded variation in the interval (a,b), then the convergence is uniform in any interval interior to (a,b).

Because of (12.7), a similar result holds for Fourier's repeated integral. As for functions in $L^p(\mathcal{R})$, $1<p\le 2$, we have the following reciprocity relation.

THEOREM 12.12 [225, p. 69]
Let $f \in L^p(\mathcal{R})$, $1<p\le 2$. Then

$$f(t) = \frac{1}{\sqrt{2\pi}}\frac{d}{dt}\int_{-\infty}^{\infty}\hat{f}(w)\frac{e^{-itw}-1}{-iw}dw\,,$$

and

$$\hat{f}(w) = \frac{1}{\sqrt{2\pi}}\frac{d}{dw}\int_{-\infty}^{\infty}f(t)\frac{e^{iwt}-1}{it}dt\,,$$

almost everywhere.

The inversion formula for the functions $F_{\pm}$, given by (12.2) can be obtained from Theorem 12.11. It is readily seen that

$$\frac{1}{\sqrt{2\pi}}\int_{-\infty}^{\infty}F_+(u+iv)e^{-itu}du = \begin{cases} f(t)e^{-tv} & \text{if } t>0\\ 0 & \text{if } t<0\,,\end{cases}$$

or

$$\frac{1}{\sqrt{2\pi}}\int_{-\infty}^{\infty} F_+(u+iv)e^{-it(u+iv)}du = \begin{cases} f(t) & \text{if } t>0 \\ 0 & \text{if } t<0\,, \end{cases} \tag{12.19}$$

A similar formula holds for F_-, and together they yield

$$f(t) = \frac{1}{\sqrt{2\pi}}\int_{ia-\infty}^{ia+\infty} F_+(w)e^{-itw}dw + \frac{1}{\sqrt{2\pi}}\int_{ib-\infty}^{ib+\infty} F_-(w)e^{-itw}dw\,, \tag{12.20}$$

where a is a sufficiently large positive number and b is a negative number with sufficiently large absolute values.

The inversion formulas for the Fourier cosine and sine transforms also follow from Theorem 12.11. They are easily shown to be as follows: if

$$F_c(w) = \sqrt{\frac{2}{\pi}}\int_0^{\infty} f(t)\cos wt\,dt,$$

then

$$f(t) = \sqrt{\frac{2}{\pi}}\int_0^{\infty} F_c(w)\cos tw\,dw\,, \quad t \geq 0,$$

and if

$$F_s(w) = \sqrt{\frac{2}{\pi}}\int_0^{\infty} f(t)\sin wt\,dt,$$

then

$$f(t) = \sqrt{\frac{2}{\pi}}\int_0^{\infty} F_s(w)\sin tw\,dw\,, \quad t > 0.$$

12.9 Convolution

12.9.1 Convolution for the Fourier Transform

DEFINITION 12.7 *The convolution operation, $\star$, associated with the Fourier transform is defined as*

$$h(t) = (f\star g)(t) = \frac{1}{\sqrt{2\pi}}\int_{-\infty}^{\infty} f(x)g(t-x)dx,$$

where f and g are defined over the whole real line.

The next theorem provides important properties of the convolution.

THEOREM 12.13 [280, p. 252]
If $f \in L^s(\mathcal{R})$, $g \in L^t(\mathcal{R})$, where $s,t \geq 1$, $\frac{1}{s}+\frac{1}{t}>1$, then $h = f \star g$ exists almost everywhere as an absolutely convergent integral and belongs to $L^r(\mathcal{R})$, where $\frac{1}{r}=\frac{1}{s}+\frac{1}{t}-1$, with

$$\left(\frac{1}{\sqrt{2\pi}}\int_{-\infty}^{\infty}|h(x)|^r\,dx\right)^{1/r} \leq \left(\frac{1}{\sqrt{2\pi}}\int_{-\infty}^{\infty}|f(x)|^s\,dx\right)^{1/s}\left(\frac{1}{\sqrt{2\pi}}\int_{-\infty}^{\infty}|g(x)|^t\,dx\right)^{1/t}.$$

Clearly, if either one of the functions f or g is in $L^1(\mathcal{R})$ and the other is in $L^p(\mathcal{R})$, then their convolution is in $L^p(\mathcal{R})$.

Formally, we have

$$\begin{aligned}\hat{h}(w) = (\widehat{f\star g})(w) &= \frac{1}{\sqrt{2\pi}}\int_{-\infty}^{\infty}h(x)e^{iwx}dx\\ &= \frac{1}{2\pi}\int_{-\infty}^{\infty}e^{ixw}dx\left(\int_{-\infty}^{\infty}f(t)g(x-t)\,dt\right)\\ &= \frac{1}{2\pi}\int_{-\infty}^{\infty}f(t)\,dt\left(\int_{-\infty}^{\infty}g(x-t)e^{ixw}dx\right)\\ &= \frac{1}{2\pi}\left(\int_{-\infty}^{\infty}f(t)e^{iwt}dt\right)\left(\int_{-\infty}^{\infty}g(y)e^{iwy}dy\right) = \hat{f}(w)\hat{g}(w).\end{aligned}$$

Hence,

$$(\widehat{f\star g})(w) = \hat{f}(w)\hat{g}(w) \quad \text{or } \mathcal{F}^{-1}\left(\hat{f}\hat{g}\right)(x) = (f\star g)(x). \tag{12.21}$$

Similarly,

$$(\hat{f}\star\hat{g})(w) = (\widehat{fg})(w). \tag{12.22}$$

For,

$$\begin{aligned}(\hat{f}\star\hat{g})(w) &= \frac{1}{\sqrt{2\pi}}\int_{-\infty}^{\infty}\hat{f}(x)\hat{g}(w-x)dx = \frac{1}{2\pi}\int_{-\infty}^{\infty}\hat{g}(w-x)dx\int_{-\infty}^{\infty}f(t)e^{ixt}dt\\ &= \frac{1}{2\pi}\int_{-\infty}^{\infty}f(t)dt\int_{-\infty}^{\infty}\hat{g}(w-x)e^{itx}dx = \frac{1}{2\pi}\int_{-\infty}^{\infty}f(t)e^{iwt}dt\int_{-\infty}^{\infty}\hat{g}(x)e^{-itx}dx\\ &= \frac{1}{\sqrt{2\pi}}\int_{-\infty}^{\infty}f(t)g(t)e^{iwt}dt = (\widehat{fg})(w).\end{aligned}$$

The next theorem provides conditions under which this formal manipulation is valid [225, p. 106].

THEOREM 12.14 (**Convolution Theorem**)
Let $p,q \geq 1, (1/p)+(1/q) > 1$. Let p' and q' be the conjugate numbers of p and q, respectively, i.e., $(1/p)+(1/p') = 1, (1/q)+(1/q') = 1$. If

$f \in L^p(\mathcal{R}), \hat{f} \in L^{p'}(\mathcal{R})$ and $g \in L^q(\mathcal{R}), \hat{g} \in L^{q'}(\mathcal{R})$, then $f \star g \in L^r(\mathcal{R})$ and $\hat{f}\hat{g} \in L^{r'}(\mathcal{R})$ where $(1/r) + (1/r') = 1$ and

$$r = \frac{pq}{p+q-pq}.$$

If $1 < r \leq 2$, then $f \star g$ has Fourier transform and moreover

$$\mathcal{F}(f \star g)(w) = (\widehat{f \star g})(w) = \hat{f}(w)\hat{g}(w).$$

If $r > 2$ (hence, $1 < r' < 2$), then $\hat{f}\hat{g}$ has Fourier transform and in addition

$$\mathcal{F}^{-1}\left(\hat{f}\hat{g}\right)(t) = (f \star g)(t) \tag{12.23}$$

That $(f \star g) \in L^r(\mathcal{R})$ follows from Theorem 12.14 and that $\hat{f}\hat{g} \in L^{r'}(\mathcal{R})$ follows from

$$\int_{-\infty}^{\infty} \left|\hat{f}(w)\hat{g}(w)\right|^{r'} dw \leq \left(\int_{-\infty}^{\infty} \left|\hat{f}(w)\right|^{p'} dw\right)^{r'/p'} \left(\int_{-\infty}^{\infty} |\hat{g}(w)|^{q'} dw\right)^{r'/q'} \leq \infty,$$

which is a consequence of Hölder's inequality since $(r')/(p') + (r')/(q') = 1$.

Some important special cases in which (12.21) or (12.23) holds are:

1. $p = q = 1$, (then $r = 1$) ,
2. $f, \hat{g} \in L^p(\mathcal{R})$, and $\hat{f}, g \in L^{p'}(\mathcal{R})$,
3. $f \in L^p(\mathcal{R}), \hat{f} \in L^{p'}(\mathcal{R})$ and $g \in L^1(\mathcal{R})$

In the third case $f \star g \in L^p(\mathcal{R})$ and $\hat{f}\hat{g} \in L^{p'}(\mathcal{R})$.

12.9.2 Convolution for the Fourier Cosine and Sine Transforms

The following convolution formulas and Parseval's relations for the Fourier cosine and sine transforms can be easily verified:

$$\int_0^{\infty} F_c(w)G_c(w)\cos tw\, dw = \frac{1}{2}\int_0^{\infty} f(x)\left[g(t+x) + g(|t-x|)\right] dx,$$

which yields

$$\int_0^{\infty} F_c(w)G_c(w)dw = \int_0^{\infty} f(x)g(x)dx,$$

and

$$\int_0^\infty |F_c(w)|^2\, dw = \int_0^\infty |f(x)|^2\, dx\,.$$

Similarly,

$$\int_0^\infty F_s(w)G_s(w)\cos tw\, dw = \frac{1}{2}\int_0^\infty f(x)\,[g(t+x)+g(|t-x|)]\, dx,$$

which yields

$$\int_0^\infty F_s(w)G_s(w)dw = \int_0^\infty f(x)g(x)dx,$$

and

$$\int_0^\infty |F_s(w)|^2\, dw = \int_0^\infty |f(x)|^2\, dx.$$

12.10 Poisson Summation Formulas

12.10.1 Poisson Summation Formula for Uniform Sampling

THEOREM 12.15

The Poisson summation formula may be written in the form

$$\frac{1}{\sqrt{2\pi}}\sum_{k=-\infty}^{\infty} f(t+\frac{k\pi}{\sigma}) = \frac{\sigma}{\pi}\sum_{k=-\infty}^{\infty}\hat{f}(2k\sigma)e^{-2ikt\sigma}, \quad \sigma>0, \tag{12.24}$$

provided that the two series converge. A sufficient condition for the validity of (12.24) is that $f \in L^1(\mathcal{R})$ and $\hat{f} = O\left((1+|w|)^{-\alpha}\right)$ as $|w| \to \infty$ for some $\alpha > 1$.

Another version of the Poisson summation formula is

$$\sum_{k=-\infty}^{\infty}\hat{f}(w+k\sigma)\bar{\hat{g}}(w+k\sigma) = \frac{1}{\sigma}\sum_{k=-\infty}^{\infty}\langle f(.), g(.-2\pi k/\sigma)\rangle e^{2\pi ikw/\sigma}, \tag{12.25}$$

where

$$\langle f(.), g(.-2\pi k/\sigma)\rangle = \int_{-\infty}^{\infty} f(t)\bar{g}(t-\frac{2\pi k}{\sigma})\, dt\,.$$

Proof: To prove (12.24), consider the function $f_{per}(t)$ defined by

$$f_{per}(t) = \sum_{k=-\infty}^{\infty} f(t+2k\sigma),$$

which is clearly periodic with period 2σ. In addition, $f_{per} \in L^1(-\sigma, \sigma)$ since

$$\int_{-\sigma}^{\sigma} |f_{per}(t)|\, dt \leq \int_{-\sigma}^{\sigma} \sum_{k=-\infty}^{\infty} |f(t+2k\sigma)|\, dt = \sum_{k=-\infty}^{\infty} \int_{-\sigma}^{\sigma} |f(t+2k\sigma)|\, dt$$

$$= \sum_{k=-\infty}^{\infty} \int_{(2k-1)\sigma}^{(2k+1)\sigma} |f(t)|\, dt = \int_{-\infty}^{\infty} |f(t)|\, dt < \infty.$$

This equation also shows that the series defining f_{per} converges absolutely and uniformly. Therefore, from the Fourier series of f_{per}

$$f_{per}(t) = \sum_{k=-\infty}^{\infty} f(t+2k\sigma) = \sum_{n=-\infty}^{\infty} \hat{f}_{per}(n) e^{-in\pi t/\sigma},$$

where

$$\hat{f}_{per}(n) = \frac{1}{2\sigma} \int_{-\sigma}^{\sigma} f_{per}(t) e^{in\pi t/\sigma} dt,$$

we obtain

$$\hat{f}_{per}(n) = \frac{1}{2\sigma} \int_{-\sigma}^{\sigma} \left(\sum_{k=-\infty}^{\infty} f(t+2k\sigma) \right) e^{in\pi t/\sigma} dt$$

$$= \frac{1}{2\sigma} \sum_{k=-\infty}^{\infty} \int_{-\sigma}^{\sigma} f(t+2k\sigma) e^{in\pi t/\sigma} dt = \frac{1}{2\sigma} \sum_{k=-\infty}^{\infty} \int_{(2k-1)\sigma}^{(2k+1)\sigma} f(t) e^{in\pi t/\sigma} dt$$

$$= \frac{1}{2\sigma} \int_{-\infty}^{\infty} f(t) e^{in\pi t/\sigma} dt = \sqrt{\frac{\pi}{2\sigma^2}} \hat{f}(\frac{n\pi}{\sigma}).$$

Hence,

$$\sum_{k=-\infty}^{\infty} f(t+2k\sigma) = \sqrt{\frac{\pi}{2\sigma^2}} \sum_{n=-\infty}^{\infty} \hat{f}(\frac{n\pi}{\sigma}) e^{-in\pi t/\sigma},$$

which leads to (12.24) when we replace σ by $\pi/(2\sigma)$. The proof of (12.25) is similar. ∎

The Poisson summation formula is used to prove many useful identities, such as *Jacobi's identity* for the theta-function

$$\vartheta(z\,,\,\tau) = \sqrt{i/\tau} \exp\left(-i\pi z^2/\tau\right) \vartheta(z/\tau\,,\,-1/\tau)\,,$$

where

$$\vartheta(z\,,\,\tau) = \sum_{k=-\infty}^{\infty} \exp\left(i\pi k^2 \tau + 2i\pi k z\right)\,, \quad \operatorname{Im} z > 0\,.$$

To this end, note that the Fourier transform of the function $f(x) = \exp(-\alpha^2 x^2)$ is given by

$$\mathcal{F}\left[\exp(-\alpha^2 x^2)\right](w) = \frac{1}{\alpha\sqrt{2}} \exp\left(-w^2/(4\alpha^2)\right).$$

Therefore, by substituting f and its Fourier transform into formula (12.24), we obtain

$$\sum_{k=-\infty}^{\infty} \exp\left[-\frac{\alpha^2\pi^2}{\sigma^2}\left(k + \frac{\sigma t}{\pi}\right)^2\right] = \frac{\sigma}{\sqrt{\pi}\alpha} \sum_{k=-\infty}^{\infty} \exp\left[2ikt\sigma - \frac{k^2\sigma^2}{\alpha^2}\right].$$

Formally, upon setting

$$\sigma = \alpha\sqrt{\frac{i\pi}{\tau}} \quad \text{and} \quad t = \frac{\sqrt{\pi}z}{\alpha\sqrt{i\tau}},$$

and simplifying the result, we obtain Jacobi's identity.

12.10.2 Poisson Summation Formula for Non-Uniform Sampling

The Poisson summation formula can also be proved by using contour integration. Let $f(z)$ be analytic in the strip $\Omega = \{z = x + iy : |y| < y_0\}$ of the complex plane $z = x + iy$. By integrating the function

$$g(z) = \pi f(z)\cot(\pi z)$$

around the rectangle with corners at $\pm T \pm ib$ $(0 < b < y_0)$, then taking the limit as $T \to \infty$, we obtain, on one hand, the left-hand side of (12.24) (up to a constant) as the sum of the residues of $g(z)$. On the other hand, by replacing $i\cot(\pi z)$ in the integrand by $1 + 2\sum_{n=1}^{\infty} e^{2n\pi iz}$ when $y > 0$ and by $-\left(1 + 2\sum_{n=1}^{\infty} e^{2n\pi iz}\right)$ when $y < 0$ and interchanging the summation and integration signs, we obtain the right-hand side of (12.24) (up to a constant). The constants depend on those that appear in the definition of the Fourier transform.

Using this idea, the great mathematician S. Bochner was able to extend Poisson's summation formula to non-uniform sampling points as follows. Let $\phi(z)$ be a meromorphic function whose poles are all simple and located in the closed region $|y| \le \delta < y_0$. Let us denote the poles of ϕ by $\{a_n\}$ and the corresponding residues by $\{r_n\}$. Let us assume that ϕ is bounded on every vertical line segment $x = \pm T$ in the strip Ω and can be represented by an absolutely convergent two-sided Stieltjes–Laplace integral

$$\phi(z) = \sqrt{2\pi}\, i \int_{-\infty}^{\infty} e^{isz} d\mu_1(s), \quad \delta < y < y_0$$

and

$$\phi(z) = \sqrt{2\pi}\, i \int_{-\infty}^{\infty} e^{isz} d\mu_2(s) , \quad -y_0 < y < -\delta.$$

Set $\mu(s) = \mu_1(s) - \mu_2(s)$.

THEOREM 12.16 **(Bochner, [19])**
Let ϕ and μ be defined as above. If $f(z)$ is analytic in the strip Ω and tends to zero as $x \to \pm\infty$ and for each y in Ω the limit

$$e^{sy}\hat{f}(s) = \lim_{T\to\infty} \frac{1}{\sqrt{2\pi}} \int_{-T}^{T} e^{isx} f(x+iy)\, dx$$

exists boundedly in all s and uniformly in every finite interval, then

$$\sum r_n f(a_n) = \int_{-\infty}^{\infty} \hat{f}(s)\, d\mu(s) ,$$

where the sum extends over all the poles $\{a_n\}$. The integral converges absolutely. In particular, if μ is a pure jump function having jumps γ_n at the point α_n, then

$$\sum r_n f(a_n) = \sum \gamma_n \hat{f}(\alpha_n) ,$$

or more generally

$$\sum r_n f(a_n + t) = \sum \gamma_n \hat{f}(\alpha_n) e^{-i\alpha_n t} .$$

Note that although the second formula looks slightly more general than the first, it is, in fact, obtained from the first by observing that, for fixed t, the Fourier transform of $f(x+t)$ is $e^{-ist}\hat{f}(s)$.

12.11 The Fourier Transform of L^2-Functions

The Fourier transformation of L^2-functions has many interesting and symmetrical properties that warrant devoting a whole section to it. Some of these properties were mentioned in previous sections; however, we will restate them for emphasis.

The following three theorems are due to Paley and Wiener [182, pp. 1-13].

THEOREM 12.17
For $f \in L^2(\mathcal{R})$ the following two conditions are equivalent:

1. *f is the restriction to $\mathcal{R}$ of a function $F(z)$ analytic in the strip $|y| < \sigma$ that satisfies*

$$\int_{-\infty}^{\infty} |F(x+iy)|^2 \, dx \leq C, \quad |y| < \sigma,$$

2.

$$e^{\sigma|w|}\hat{f}(w) \in L^2(\mathcal{R}).$$

THEOREM 12.18
The following two conditions are equivalent for $f \in L^2(\mathcal{R})$:

1. *There exists a function $F(z)$ analytic in the upper half-plane $y > 0$ satisfying*

$$\int_{-\infty}^{\infty} |F(x+iy)|^2 \, dx \leq C, \quad y > 0,$$

and

$$\lim_{y \to 0+} \int_{-\infty}^{\infty} |F(x+iy) - f(x)|^2 \, dx = 0,$$

2. $\hat{f}(w) = 0$ *for* $w < 0$.

THEOREM 12.19 (**Paley–Wiener**)
A function $F(z)$ is an entire function of exponential type at most σ and belongs to $L^2(\mathcal{R})$ when restricted to the real line, i.e., $F \in B_\sigma^2$ if and only if it can be written in the form

$$F(x) = \int_{-\sigma}^{\sigma} f(t) e^{ixt} dt,$$

for some $f \in L^2(-\sigma, \sigma)$ with $f(t) = 0$ for $|t| > \sigma$.

For the definition of the space B_σ^2, see Section 1.3 . The mapping $f \to \hat{f}$ is an isomorphism from $L^2(\mathcal{R})$ onto itself and in view of (12.15) it is also an isometry. This result has interesting ramifications. For example:

1. If $n_1, n_2, \cdots$ is a sequence of numbers not necessarily integers such that the intervals $(n_k - \lambda, n_k + \lambda)$, $k = 1, 2, \cdots, \lambda > 0$ do not overlap, then the functions

$$\left\{ \frac{1}{\sqrt{\lambda\pi}} e^{in_k w} \frac{\sin \lambda w}{w} \right\}_{k=1}^{\infty},$$

form an orthonormal system over $(-\infty, \infty)$. For, if we denote the characteristic function of the interval $(n_k - \lambda, n_k + \lambda)$ by $\chi_k(t)$ and set

$$\psi_k(t) = \frac{1}{\sqrt{2\lambda}} \chi_k(t),$$

we obtain

$$\hat{\psi}_k(w) = \frac{1}{\sqrt{\lambda\pi}} e^{in_k w} \frac{\sin \lambda w}{w}.$$

Hence, in view of (12.14), we have

$$\delta_{k,m} = \langle \psi_k, \psi_m \rangle = \langle \hat{\psi}_k, \hat{\psi}_m \rangle = \frac{1}{\lambda\pi} \int_{-\infty}^{\infty} e^{i(n_k - n_m)w} \frac{\sin^2 \lambda w}{w^2} dw.$$

2. Similarly, the functions

$$\left\{ \frac{1}{\sqrt{\sigma\pi}} \frac{\sin \sigma(t - n\pi/\sigma)}{(t - n\pi/\sigma)} \right\}_{n=-\infty}^{\infty}$$

form an orthonormal basis for B_σ^2. This follows from the fact that the Fourier integral transform of the function

$$\psi_n(t) = \frac{1}{\sqrt{2\sigma}} e^{-in\pi t/\sigma}$$

is

$$\hat{\psi}_n(w) = \frac{1}{\sqrt{\sigma\pi}} \frac{\sin \sigma(w - n\pi/\sigma)}{(w - n\pi/\sigma)},$$

and that $\{\psi_n(t)\}_{n=-\infty}^{\infty}$ is an orthonormal basis of $L^2(-\sigma, \sigma)$.

3. More generally, if $\{t_n\}_{n=-\infty}^{\infty}$ is a sequence of real numbers such that $\{e^{it_n t}\}_{n=-\infty}^{\infty}$ is a basis (not necessarily orthonormal) of $L^2(-\sigma, \sigma)$, then

$$\left\{ \sqrt{\frac{2}{\pi}} \frac{\sin \sigma(t - t_n)}{(t - t_n)} \right\}_{n=-\infty}^{\infty}$$

is a basis of B_σ^2.

THEOREM 12.20 **(Levinson** [145, Ch. 4])
If $\{t_k\}_{-\infty}^{\infty}$ is a sequence of real numbers such that

$$D = \sup_k \left| t_k - \frac{k\pi}{\sigma} \right| < \frac{\pi}{4\sigma}, \tag{12.26}$$

then:

1. *The set* $\{e^{it_kx}\}_{-\infty}^{\infty}$ *is complete in* $L^2(-\sigma,\sigma)$ *in the sense that if*

$$\int_{-\sigma}^{\sigma} f(x)e^{-it_kx}dx = 0$$

   for all k*, then* f *is identically zero.*

2. *The set* $\{e^{it_kx}\}_{-\infty}^{\infty}$ *possesses a unique biorthonormal set* $\{h_k(x)\}_{-\infty}^{\infty}$*, i.e.,*

$$\int_{-\sigma}^{\sigma} h_m(x)e^{it_kx}dx = \delta_{k,m}.$$

3. *Moreover,*

$$\int_{-\sigma}^{\sigma} h_k(x)e^{itx}dx = \frac{G(t)}{(t-t_k)G'(t_k)},$$

   where

$$G(t) = (t-t_0)\prod_{n=1}^{\infty}\left(1-\frac{t}{t_n}\right)\left(1-\frac{t}{t_{-n}}\right) \tag{12.27}$$

   is an entire function of exponential type with $G(t)/(t-t_k) \in L^2(\mathcal{R})$ *for any* k.

4. *For any* $F \in L^2(-\sigma,\sigma)$ *we have*

$$F(x) = \sum_{k=-\infty}^{\infty} F_k h_k(x),$$

   in the mean, as well as uniformly on any closed subinterval of $(-\sigma,\sigma)$*, where*

$$F_k = \int_{-\sigma}^{\sigma} F(x)e^{it_kx}dx.$$

5. *The aforementioned results are not in general true if* $D \geq \pi/(4\sigma)$.

Some of these results were sharpened by Kadec; see [251, p. 29].

THEOREM 12.21 (Kadec)
If the sequence $\{t_k\}_{-\infty}^{\infty}$ *satisfies (12.26), then the set* $\{e^{it_kx}\}_{-\infty}^{\infty}$ *is a Riesz basis for* $L^2(-\sigma,\sigma)$ *and this result is not true if* $D \geq \pi/(4\sigma)$.

The constant $\pi/(4\sigma)$ is known as Kadec's constant. For a definition of a Riesz basis, see Definition 1.3. The following theorem, which is known as the Paley–Wiener-Levinson interpolation theorem for band-limited functions, can be obtained easily from Levinson's theorem (Theorem 12.20).

THEOREM 12.22 **(Paley–Wiener-Levinson)**
Let $\{t_k\}_{-\infty}^{\infty}$ be a sequence of real numbers satisfying (12.26) and let $G(t)$ be defined by (12.27). Then for any $f \in B_\sigma^2$, we have

$$f(t) = \sum_{k=-\infty}^{\infty} f(t_k) S_k^*(t),$$

where

$$S_k^*(t) = \frac{G(t)}{(t - t_k)G'(t_k)}.$$

The series converges to f uniformly on compact sets.

As a special case of this theorem we obtain, for $t_k = (k\pi)/\sigma = -t_{-k}$, $k = 0, 1, 2, \cdots$, the *Whittaker-Shannon-Kotel'nikov sampling theorem* [257].

THEOREM 12.23 **(Whittaker–Shannon–Kotel'nikov)**
If f is a band-limited signal, i.e., $f \in B_\sigma^2$, then it can be reconstructed from its sample values at the points $t_k = (k\pi)/\sigma$, $k = 0, \pm 1, \pm 2, \cdots$, via the formula

$$f(t) = \sum_{k=-\infty}^{\infty} f(t_k) \frac{\sin \sigma(t - t_k)}{\sigma(t - t_k)}, \tag{12.28}$$

with the series being absolutely and uniformly convergent on compact sets.

Proof: Since $f \in B_\sigma^2$, there exists a function $F \in L^2(-\sigma, \sigma)$ such that

$$f(t) = \frac{1}{\sqrt{2\pi}} \int_{-\sigma}^{\sigma} F(w) e^{-itw} dw. \tag{12.29}$$

First, we extend $F(w)$ and e^{itw} periodically to the entire line as functions with period 2σ. Second, we expand these extended functions in Fourier series to yield

$$F(w) = \sum_{n=-\infty}^{\infty} \hat{F}(n) e^{inw\pi/\sigma} \quad , |w| < \sigma,$$

where

$$\hat{F}(n) = \frac{1}{2\sigma} \int_{-\sigma}^{\sigma} F(w) e^{-inw\pi/\sigma} dw \, ,$$

and

$$e^{-itw} = \sum_{n=-\infty}^{\infty} \frac{\sin \sigma(t - t_n)}{\sigma(t - t_n)} e^{-inw\pi/\sigma} \quad , \quad |w| < \sigma.$$

By applying Parseval's equality to the right-hand side of (12.29), we obtain

$$f(t) = \frac{2\sigma}{\sqrt{2\pi}} \sum_{n=-\infty}^{\infty} \hat{F}(n) \frac{\sin \sigma(t - t_n)}{\sigma(t - t_n)}, \tag{12.30}$$

which, in view of the fact that

$$f(t_n) = \frac{2\sigma}{\sqrt{2\pi}} \hat{F}(n),$$

implies (12.28).

The convergence in (12.30) is pointwise; however, if we apply the Cauchy-Schwarz inequality to it, we can easily show that (12.28) converges absolutely and uniformly on compact subsets of the real line because $\sum_{n=-\infty}^{\infty} |\hat{F}(n)|^2 < \infty$. ∎

Equation (12.28) can be written in the form

$$f(t) = \sum_{k=-\infty}^{\infty} f(t_k)\phi(t - t_k),$$

where

$$\phi(t) = \text{Sinc}\left(\frac{\sigma t}{\pi}\right), \quad \text{and} \quad \text{Sinc}(x) = \frac{\sin(\pi x)}{\pi x}.$$

To the best of our knowledge, the term "Sinc function" was first used by Woodward in 1953 [249], and after that it became very popular in electrical engineering circles. The series on the right-hand side of (12.28) can also be written in the form

$$\sin \sigma t \sum_{k=-\infty}^{\infty} f(t_k) \frac{(-1)^k}{(\sigma t - k\pi)},$$

which is a special case of a more general type of series of the form

$$\sin \sigma t \sum_{k=-\infty}^{\infty} C_k \frac{(-1)^k}{(\sigma t - k\pi)},$$

known as *Cardinal series*.

Unlike the space $L^2(\mathcal{R})$, the space B_σ^2 is a reproducing-kernel Hilbert space with reproducing kernel

$$K(x, t) = \frac{\sin \sigma(x - t)}{\pi(x - t)}. \tag{12.31}$$

Hence if $f \in B_\sigma^2$, then

$$f(x) = \int_{-\infty}^{\infty} f(t) \frac{\sin \sigma(x - t)}{\pi(x - t)} dt.$$

To derive formula (12.31) using Equation (1.18), we note that

$$K(x,t) = \frac{1}{\sigma\pi} \sum_{k=-\infty}^{\infty} \frac{\sin\sigma(x-t_k)}{(x-t_k)} \frac{\sin\sigma(t-t_k)}{(t-t_k)},$$

where $t_k = (k\pi)/\sigma$. If we fix x, apply the Whittaker-Shannon-Kotel'nikov sampling theorem to the function $f_x(t) = [\sin\sigma(x-t)]/[\pi(x-t)]$ and compare the result with the last expression for $K(x,t)$, we obtain (12.31). But a more direct derivation of (12.31) goes as follows:

$$\begin{aligned} f(x) &= \frac{1}{\sqrt{2\pi}} \int_{-\sigma}^{\sigma} \hat{f}(w) e^{-ixw}\, dw \\ &= \frac{1}{\sqrt{2\pi}} \int_{-\sigma}^{\sigma} e^{-ixw} \left[\frac{1}{\sqrt{2\pi}} \int_{-\infty}^{\infty} f(t) e^{itw}\, dt \right] dw \\ &= \frac{1}{2\pi} \int_{-\infty}^{\infty} f(t) \left[\int_{-\sigma}^{\sigma} e^{iw(t-x)} dw \right] dt \\ &= \int_{-\infty}^{\infty} f(t) \frac{\sin\sigma(t-x)}{\pi(t-x)} dt\,. \end{aligned}$$

12.12 Analytic Representation of the Fourier Transform

Let $f \in L^p(\mathcal{R}), 1 \leq p < \infty$. We define the *analytic representation* $\tilde{f}$ of f by

$$\tilde{f}(z) = \frac{1}{2\pi i} \int_{-\infty}^{\infty} \frac{f(t)}{t-z} dt\,, \quad \mathrm{Im} z \neq 0.$$

This defines a pair of analytic functions, one in the upper and one in the lower half-planes. In general, they are not analytic continuations of each other. If $f \in L^2(\mathcal{R})$, then the analytic representation of its Fourier transform

$$\tilde{\hat{f}}(z) = \frac{1}{2\pi i} \int_{-\infty}^{\infty} \frac{\hat{f}(w)}{w-z} dw\,, \quad \mathrm{Im} z \neq 0,$$

is related to the functions $F_\pm$ introduced in Section 12.2. To this end, observe that

$$\mathcal{F}\left(e^{-itz} H(-t)\right)(w) = \frac{1}{\sqrt{2\pi}} \int_{-\infty}^{0} e^{-itz} e^{iwt} dt = \frac{1}{i\sqrt{2\pi}} \frac{1}{(w-z)},$$

for Im $z > 0$ and similarly

$$\mathcal{F}\left(e^{-itz} H(t)\right)(w) = \frac{1}{i\sqrt{2\pi}} \frac{-1}{(w-z)},$$

for $\operatorname{Im} z < 0$. Hence, in view of the convolution theorem, we have

$$\tilde{\hat{f}}(z) = \begin{cases} (2\pi)^{-1/2} \int_0^\infty f(t)e^{itz}dt\,, & \operatorname{Im} z > 0 \\ (2\pi)^{-1/2} \int_{-\infty}^0 f(t)e^{itz}dt\,, & \operatorname{Im} z < 0. \end{cases}$$

The result remains valid if the condition $f \in L^2(\mathcal{R})$ is replaced by $f \in L^1(\mathcal{R})$. Moreover,

$$\lim_{y\to 0} \left(\tilde{\hat{f}}(x+iy) - \tilde{\hat{f}}(x-iy)\right) = \frac{1}{\sqrt{2\pi}} \lim_{y\to 0} \int_{-\infty}^{\infty} f(t)e^{-|t|y}e^{ixt}dt = \hat{f}(x).$$

Likewise, for the analytic representation of f

$$\tilde{f}(z) = \frac{1}{2\pi i} \int_{-\infty}^{\infty} \frac{f(t)}{t-z}dt\,, \qquad z = x+iy,\ y \neq 0,$$

we can show that

$$\tilde{f}(z) = \begin{cases} (2\pi)^{-1/2} \int_0^\infty \hat{f}(w)e^{-izw}dw, & y \le 0 \\ (2\pi)^{-1/2} \int_{-\infty}^0 \hat{f}(w)e^{-izw}dw & y \ge 0, \end{cases}$$

and

$$\lim_{y\to 0} \left(\tilde{f}(x+iy) - \tilde{f}(x-iy)\right) = \frac{1}{\sqrt{2\pi}} \lim_{y\to 0} \int_{-\infty}^{\infty} \hat{f}(w)e^{-|w|y}e^{-ixw}dw = f(x).$$

12.13 Generalized Functions

Constant functions and polynomials do not belong to $L^p(\mathcal{R})$ for any $1 \le p < \infty$, and hence they do not have Fourier transforms in the ordinary sense. But such transforms are called for in many engineering and physics problems. This motivated many people in the early days of generalized functions to extend the definition of the Fourier transform to include functions of polynomial growth.

The Fourier transform of generalized functions may be defined by using either the embedding or the adjoint method. The domain of definition in the former case is smaller than that in the latter. The main advantage of using the embedding method, however, is its simplicity.

12.13.1 The Embedding Method

Since the kernel of the Fourier transform, e^{iwx}, is a C^∞-function, it is contained in the space $\mathcal{E}$ of Section 3.1.3, Example 2. As a result, we can define the Fourier transform of any generalized function $f \in \mathcal{E}^*$ as follows.

DEFINITION 12.8 *The Fourier transform of any $f \in \mathcal{E}^*$ is defined as*

$$\mathcal{F}(f)(w) = \frac{1}{\sqrt{2\pi}} \langle f(t), e^{iwt} \rangle .$$

Since $\mathcal{E}^*$ consists of generalized functions with compact support, the right-hand side may be interpreted as the value of

$$\frac{1}{\sqrt{2\pi}} \langle f(t), \chi(t) e^{iwt} \rangle ,$$

where χ is any testing-function in $\mathcal{D}$ that is equal to one over some neighborhood of the support of f. The definition can be shown to be independent of χ.

It is easy to see that $\hat{f}(w)$ is an infinitely differentiable function. The delta function and its derivatives of all orders have compact support; hence we can apply Definition 12.8 to obtain

$$\mathcal{F}(\delta)(w) = \frac{1}{\sqrt{2\pi}} \tag{12.32}$$

$$\mathcal{F}\left(\delta^{(k)}(t-a)\right)(w) = \frac{(-iw)^k}{\sqrt{2\pi}} e^{iaw} , k = 0, 1, 2, \cdots . \tag{12.33}$$

From these relations we obtain

$$\mathcal{F}\left(t^k e^{iat}\right)(w) = \sqrt{2\pi}(-i)^k \delta^{(k)}(w+a) \tag{12.34}$$

$$\mathcal{F}\left(\sum_{k=0}^{n} a_k t^k\right)(w) = \sqrt{2\pi} \sum_{k=0}^{n} (-i)^k a_k \delta^{(k)}(w). \tag{12.35}$$

Formally, we have

$$\delta(t-a) = \frac{1}{2\pi} \int_{-\infty}^{\infty} e^{i(t-a)w} dw ,$$

and

$$\delta^{(k)}(t-a) = \frac{1}{2\pi} \int_{-\infty}^{\infty} (iw)^k e^{i(t-a)w} dw , \quad k = 0, 1, \cdots .$$

The following theorem extends the analytic representation and the convolution theorem to generalized functions.

THEOREM 12.24 [278, p. 191]
Let f and g be generalized functions with compact supports, i.e., $f, g \in \mathcal{E}^*$, and let $\hat{f}, \hat{g}$ be the Fourier transforms of f and g as given by Definition 12.8. Then

$$\left(\widehat{f * g}\right)(w) = \hat{f}(w)\hat{g}(w) ,$$

and

$$\lim_{y\to 0}\langle \tilde{\hat{f}}(x+iy)-\tilde{\hat{f}}(x-iy),\phi(x)\rangle = \langle \hat{f}(x),\phi(x)\rangle\,,$$

for all $\phi \in \mathcal{E}$.

12.13.2 The Adjoint Method

Consider the space $\mathcal{S}$ of Section 3.1.3, Example 4. It is the Schwartz space of rapidly decreasing functions. One of its most important properties is given in the next theorem.

THEOREM 12.25
The Fourier transformation and its inverse are continuous linear transformations of S onto itself.

Sketch of the Proof: Let $\phi \in \mathcal{S}$. Because $t^k\phi(t) \in \mathcal{S}$, for all $k = 0, 1, 2, \cdots$, it follows that

$$\hat{\phi}^{(k)}(w) = \frac{1}{\sqrt{2\pi}}\int_{-\infty}^{\infty}(it)^k\phi(t)e^{iwt}dt\,, \tag{12.36}$$

exists, hence $\hat{\phi}$ is a bounded C^∞-function. Integrating the right-hand side of Equation (12.36) by parts m times gives us

$$(-iw)^m\hat{\phi}^{(k)}(w) = \mathcal{F}\left[\left((it)^k\phi(t)\right)^{(m)}\right](w). \tag{12.37}$$

But since $[(it)^k\phi(t)]^{(m)} \in \mathcal{S}$, we have

$$\left|w^m\hat{\phi}^{(k)}(w)\right| \le C_{k,m} \qquad \text{for all } w\,,$$

which implies that $\hat{\phi} \in \mathcal{S}$. If $\{\phi_n\}_{n=1}^{\infty}$ is a sequence converging to zero in $\mathcal{S}$ as $n \to \infty$, then $\left\{\hat{\phi}_n^{(k)}\right\}_{n=1}^{\infty}$ converges to zero uniformly as $n \to \infty$ for $k = 0, 1, 2, \cdots$. For, by (12.36)

$$\left|\hat{\phi}_n^{(k)}(w)\right| \le C \sup_{-\infty<t<\infty}\left|(1+t^2)t^k\phi_n(t)\right| \to 0$$

as $n \to \infty$. Upon observing that differentiation and multiplication by t^k are continuous operations on $\mathcal{S}$, it follows from (12.37) that if $\phi_n \to 0$ in $\mathcal{S}$, then $[(t)^k\phi_n(t)]^{(m)} \to 0$ in $\mathcal{S}$ for all $k, m = 0, 1, 2, \cdots$, and hence $w^m\hat{\phi}_n^{(k)}(w)$ converges to zero uniformly as $n \to \infty$, and this completes the proof. ∎

DEFINITION 12.9 *The Fourier transform of a tempered distribution* f*, i.e.,* $f \in \mathcal{S}^*$ *is defined by*

$$\langle \hat{f}, \phi \rangle = \langle f, \hat{\phi} \rangle \quad \phi \in \mathcal{S}. \tag{12.38}$$

By definition, the Fourier transform of a tempered distribution is also a tempered distribution. If f is a function with slow growth (polynomial growth), by Section 3.3.3 item 7 it defines a tempered distribution and its Fourier transform is given by the aforementioned definition. The following theorem can be easily established.

THEOREM 12.26

The Fourier transformation and its inverse are continuous linear transformations from $\mathcal{S}^*$ *onto itself.*

The Fourier transform of any Schwartz distribution can also be defined, but first let us recall the definition of the spaces $\mathcal{D}$ and $\mathcal{Z}$ of Section 3.2.3, Examples 1 and 3. Using a combination of the arguments used in the proof of Theorem 12.25 and the Paley–Wiener theorem, one can prove the following theorem.

THEOREM 12.27 [278, p. 201]

The Fourier transformation is a continuous, one-to-one linear transformation from $\mathcal{D}$ *onto* $\mathcal{Z}$*.*

This theorem enables us to define the Fourier transform of any Schwartz distribution.

DEFINITION 12.10 *The Fourier transform of a distribution* $f \in \mathcal{D}^*$ *is defined by*

$$\langle \hat{f}, \psi \rangle = \langle f, \hat{\psi} \rangle \quad \psi \in \mathcal{Z}.$$

Consequently, the Fourier transformation is a continuous linear transformation from $\mathcal{D}^*$ onto $\mathcal{Z}^*$, the *space of ultradistributions*, and the inverse Fourier transformation is also a continuous linear transformation from $\mathcal{Z}^*$ onto $\mathcal{D}^*$. It is also easy to see that $\mathcal{Z}$ is a proper, but dense, subspace of $\mathcal{S}$ and that the topology of $\mathcal{Z}$ is stronger than that of $\mathcal{S}$. Hence, we have the inclusions

$$\mathcal{Z} \subset \mathcal{S} \subset \mathcal{S}^* \subset \mathcal{Z}^*.$$

If f is a tempered distribution, its Fourier transform as an element of $\mathcal{S}^*$ coincides with its Fourier transform as an element of $\mathcal{Z}^*$. Because of the continuity of the Fourier transformation, it is true that if

$$\sum_n g_n = g \in \mathcal{D}^*$$

then

$$\sum_n \hat{g}_n = \hat{g} \in \mathcal{Z}^*.$$

In particular, since

$$e^{at} = \sum_{k=0}^{\infty} \frac{(at)^k}{k!}$$

we have

$$\mathcal{F}(e^{at}) = \sqrt{2\pi} \sum_{k=0}^{\infty} \frac{(-ia)^k}{k!} \delta^{(k)}(w) = \sqrt{2\pi}\delta(w - ia)\,.$$

As special cases, we have

$$\mathcal{F}(\cosh at)(w) = \sqrt{\frac{\pi}{2}}\left(\delta(w - ia) + \delta(w + ia)\right)\,,$$

$$\mathcal{F}(\sinh at)(w) = \sqrt{\frac{\pi}{2}}\left(\delta(w - ia) - \delta(w + ia)\right)\,,$$

$$\mathcal{F}(\cos at)(w) = \sqrt{\frac{\pi}{2}}\left(\delta(w + a) + \delta(w - a)\right)\,,$$

$$\mathcal{F}(\sin at)(w) = -i\sqrt{\frac{\pi}{2}}\left(\delta(w + a) - \delta(w - a)\right)\,.$$

The next theorem provides a characterization of the Fourier transforms of Schwartz distributions. It is a generalization of the Paley–Wiener theorem (Theorem 12.19).

THEOREM 12.28 **(Paley–Wiener–Schwartz,** [70, Vol. 2, p. 162])
A function $f(t)$ is an entire function of exponential type (at most) σ and does not grow on the real axis more rapidly than $|x|^p$ as $|x| \to \infty$ for some positive integer p if and only if it is the Fourier transform of a Schwartz distribution with support in $[-\sigma, \sigma]$.

The Convolution theorem for the Fourier transform can also be extended to Schwartz distributions.

THEOREM 12.29
If $f, g \in \mathcal{D}^*$, and f has compact support, then

$$\widehat{(f * g)} = \hat{f}\hat{g}.$$

The following theorem is particularly useful in solving homogeneous linear differential and integral equations.

THEOREM 12.30 [278, p. 210]
Let $p(w) = (w - \alpha_1)^{k_1} \cdots (w - \alpha_q)^{k_q}$, where α_i are complex numbers and k_i are positive integers. A necessary and sufficient condition for an ultradistribution f to satisfy the equation

$$p(w)f(w) = 0$$

is that

$$f(w) = \sum_{i=1}^{q} \sum_{j=0}^{k_i - 1} b_{i,j} \delta^{(j)}(w - \alpha_i),$$

where $b_{i,j}$ are arbitrary constants.

To illustrate the idea, let us consider the homogeneous linear differential equation

$$a_n y^{(n)} + \cdots + a_1 y^{(1)} + a_0 y = 0 \quad a_n \neq 0 \, , \; n \geq 1.$$

By taking the Fourier transform of both sides of this equation, we obtain $P(w)Y(w) = 0$, where $P(w) = \sum_{k=0}^{n} a_k(-iw)^k$, and Y is an ultradistribution. The polynomial P can be factored as

$$P(w) = (w - \alpha_1)^{k_1} \cdots (w - \alpha_q)^{k_q}.$$

Then, by Theorem 12.30, we have

$$Y(w) = \sum_{\mu=1}^{q} \sum_{j=0}^{k_\mu - 1} b_{\mu,j} \delta^{(j)}(w - \alpha_\mu),$$

which, after we apply the inverse Fourier transform, yields the solution

$$y(t) = \sum_{\mu=1}^{q} \sum_{j=0}^{k_\mu - 1} c_{\mu,j} (w)^j e^{-i\alpha_\mu t},$$

where $c_{\mu,j} = b_{\mu,j}(i)^j / \sqrt{2\pi}$.

12.14 Applications

The Fourier transform has numerous applications in physics, mathematical physics, and engineering. For applications in electrical engineering and optics, we refer the reader to [23, 183, 184, 185]. In this section, we concentrate on the mathematical applications.

12.14.1 Ordinary Differential Equations

A. Non-Homogeneous Equations

Consider the nth order non-homogeneous differential equation

$$a_n y^{(n)} + \cdots + a_1 y^{(1)} + a_0 y = h(t), \quad a_n \neq 0, n \geq 1,$$

with constant coefficients, where h is a Fourier transformable function or generalized function. By applying the Fourier transform to both sides of the above equation, the differential equation is converted into the algebraic equation

$$P(w)Y(w) = H(w),$$

where Y, H are the Fourier transforms of y, h, respectively, and

$$P(w) = \sum_{k=0}^{n} a_k (iw)^k.$$

Therefore, we formally have

$$y(t) = \mathcal{F}^{-1}\left(K(w)H(w)\right)(t) = \frac{1}{\sqrt{2\pi}} \int_{-\infty}^{\infty} k(x)h(t-x)dx,$$

where $K(w) = 1/P(w)$.

Example A: Consider the boundary-value problem

$$y'' - y = \chi_{(-1,1)}(t), \quad -\infty \leq t \leq \infty,$$

$$y(t), \quad y'(t) \to 0 \text{ as } |t| \to \infty.$$

When applied to the above problem, the Fourier transform method yields the algebraic equation

$$-(w^2+1)Y(w) = \sqrt{\frac{2}{\pi}} \frac{\sin w}{w},$$

hence

$$Y(w) = -\sqrt{\frac{2}{\pi}} \frac{\sin w}{w(1+w^2)},$$

But since

$$\mathcal{F}^{-1}\left(\frac{\sin w}{w}\right) = \sqrt{\frac{\pi}{2}} \chi_{(-1,1)}(t),$$

and

$$\mathcal{F}^{-1}\left(\frac{1}{1+w^2}\right) = \sqrt{\frac{\pi}{2}} e^{-|t|},$$

we obtain with the aid of the convolution theorem that

$$y(t) = \frac{1}{2}\int_{t-1}^{t+1} e^{-|x|}dx = \begin{cases} e^{-t}\sinh 1\,, & 1 < t < \infty, \\ 1 - e^{-1}\cosh t\,, & -1 \le t \le 1, \\ e^{t}\sinh 1\,, & -\infty < t < -1 \end{cases}$$

B. Homogeneous Equations

Consider the nth order homogeneous differential equation

$$a_n y^{(n)} + \cdots + a_1 y^{(1)} + a_0 y = 0\,, \quad a_n \neq 0\,, n \ge 1,$$

with constant coefficients. Repeating the procedure outlined above yields

$$P(w)Y(w) = 0.$$

Let $\gamma_1, \gamma_2, \cdots, \gamma_q$ denote the zeros of P, and assume without loss of generality that $a_n = 1$. Then,

$$P(w) = (w - \gamma_1)^{k_1} \cdots (w - \gamma_q)^{k_q},$$

where $k_1 + k_2 + \cdots + k_q = n$. Thus, by Theorem 12.30, we have

$$Y(w) = \sum_{\mu=1}^{q} \sum_{\nu=0}^{k_\mu - 1} b_{\mu,\nu}\delta^{(\nu)}(w - \gamma_\mu)\,,$$

which, after an application of the inverse Fourier transformation, yields

$$y(t) = \sum_{\mu=1}^{q} \sum_{\nu=0}^{k_\mu - 1} c_{\mu,\nu} t^{\nu} e^{-i\gamma_\mu t}\,,$$

where $c_{\mu,\nu} = (i)^{\nu} b_{\mu,\nu}/\sqrt{2\pi}$.

12.14.2 Integral Equations

Consider the integral equation of convolution type

$$f(t) = g(t) + \int_{-\infty}^{\infty} k(t-x)f(x)dx\,, \quad -\infty < t < \infty.$$

By taking the Fourier transform, the equation is converted into

$$F(w) = G(w) + \sqrt{2\pi}K(w)F(w)\,,$$

which yields

$$\begin{aligned} f(t) &= \mathcal{F}^{-1}\left(\frac{G(w)}{1 - \sqrt{2\pi}K(w)}\right) \\ &= \frac{1}{\sqrt{2\pi}}\int_{-\infty}^{\infty} \frac{G(w)}{1 - \sqrt{2\pi}K(w)} e^{-itw}dw. \end{aligned}$$

If we set

$$M(w) = \frac{K(w)}{1 - \sqrt{2\pi}K(w)},$$

then

$$\begin{aligned} f(t) - g(t) &= \mathcal{F}^{-1}\left[\frac{G(w)}{1 - \sqrt{2\pi}K(w)}\right](t) - \mathcal{F}^{-1}\left[G(w)\right](t) \\ &= \mathcal{F}^{-1}\left[\sqrt{2\pi}G(w)M(w)\right](t), \end{aligned}$$

which, in view of the convolution theorem, yields

$$f(t) = g(t) + \int_{-\infty}^{\infty} g(x)m(t-x)dx,$$

where m is the inverse Fourier transform of M.

12.14.3 Difference Equations

Some difference equations can also be solved by using the Fourier transformation. Take, for example, the difference equation

$$f(x+n) - f(x) = g(x), \quad n \text{ is a positive integer} \tag{12.39}$$

where g is a known function and f is unknown.

Let us assume that f and g are of real exponential type, i.e., they are of order $O(\exp c|x|)$ as $|x| \to \infty$ for some $c > 0$. Using the notation of $F_\pm$ introduced in (12.2), and the inversion formula (12.20), we obtain

$$\begin{aligned} f(x) = \frac{1}{\sqrt{2\pi}}&\left(\int_{ia-\infty}^{ia+\infty} F_+(w)e^{-ixw}dw\right. \\ &\left. + \int_{ib-\infty}^{ib+\infty} F_-(w)e^{-ixw}dw\right), \end{aligned}$$

where $a > 0$, $b < 0$. Upon substituting this into (12.39) and rearranging the resulting equation, we obtain

$$\begin{aligned} &\int_{ia-\infty}^{ia+\infty} e^{-ixw}\left[F_+(w)\left(e^{-inw} - 1\right) - G_+(w)\right]dw \\ &+ \int_{ib-\infty}^{ib+\infty} e^{-ixw}\left[F_-(w)\left(e^{-inw} - 1\right) - G_-(w)\right]dw = 0, \end{aligned}$$

where $G_\pm$ are related to g as $F_\pm$ are related to f. In view of Theorem 141 in [225, p. 255], there exists a function $\phi(w)$ analytic in the strip $b < v < a$, and tends to zero as $|u| \to \infty$ such that

$$F_+(w)\left(e^{-inw} - 1\right) - G_+(w) = \phi(w) = -F_-(w)\left(e^{-inw} - 1\right) + G_-(w),$$

where $w = u + iv$.

Hence

$$f(x) = \frac{1}{\sqrt{2\pi}} \left[\int_{ia-\infty}^{ia+\infty} e^{-ixw} \frac{\phi(w) + G_+(w)}{(e^{-inw} - 1)} dw \right.$$
$$\left. - \int_{ib-\infty}^{ib+\infty} e^{-ixw} \frac{\phi(w) - G_-(w)}{(e^{-inw} - 1)} dw \right]$$
$$= \frac{1}{\sqrt{2\pi}} \left[\int_{ia-\infty}^{ia+\infty} e^{-ixw} \frac{G_+(w)}{(e^{-inw} - 1)} dw \right.$$
$$\left. + \int_{ib-\infty}^{ib+\infty} e^{-ixw} \frac{G_-(w)}{(e^{-inw} - 1)} dw \right] + f^*(x),$$

where

$$f^*(x) = \frac{1}{\sqrt{2\pi}} \left\{ \int_{ia-\infty}^{ia+\infty} e^{-ixw} \frac{\phi(w)}{(e^{-inw} - 1)} dw - \int_{ib-\infty}^{ib+\infty} e^{-ixw} \frac{\phi(w)}{(e^{-inw} - 1)} dw \right\}.$$

12.14.4 Delayed Differential Equations

Consider the delayed differential equation

$$y^{(1)}(t) + ay(t+b) = g(t) \ , \quad a, b \text{ are constants,}$$

with the initial condition $y(0) = 0$, where g is a known function. Under the assumption that y and g are of exponential type c, we obtain as in the preceding example

$$y(t) = \frac{1}{\sqrt{2\pi}} \left[\int_{ia-\infty}^{ia+\infty} Y_+(w) e^{-itw} dw + \int_{ib-\infty}^{ib+\infty} Y_-(w) e^{-itw} dw \right].$$

Substituting this into the differential equation and simplifying the resulting equation yields

$$\int_{ia-\infty}^{ia+\infty} e^{-itw} \left[Y_+(w) \left(ae^{-ibw} - iw \right) - G_+(w) \right] dw$$
$$+ \int_{ib-\infty}^{ib+\infty} e^{-itw} \left[Y_-(w) \left(ae^{-ibw} - iw \right) - G_-(w) \right] dw = 0.$$

By Theorem 141 in [225, p. 255], there exists a function $\phi(w)$ analytic in the strip $-c \le v \le c$ that tends to zero as $|u| \to \infty$ such that

$$Y_+(w) \left(ae^{-ibw} - iw \right) - G_+(w) = \phi(w)$$
$$Y_-(w) \left(ae^{-ibw} - iw \right) - G_-(w) = -\phi(w).$$

Hence

$$\begin{aligned} y(t) &= \frac{1}{\sqrt{2\pi}}\left[\int_{ia-\infty}^{ia+\infty} e^{-itw}\frac{\phi(w)+G_+(w)}{(ae^{-ibw}-iw)}dw \right. \\ &\quad \left. - \int_{ib-\infty}^{ib+\infty} e^{-itw}\frac{\phi(w)-G_-(w)}{(ae^{-ibw}-iw)}dw\right] \\ &= \frac{1}{\sqrt{2\pi}}\left[\int_{ia-\infty}^{ia+\infty} e^{-itw}\frac{G_+(w)}{(ae^{-ibw}-iw)}dw \right. \\ &\quad \left. + \int_{ib-\infty}^{ib+\infty} e^{-itw}\frac{G_-(w)}{(ae^{-ibw}-iw)}dw\right] + f^*(t)\,, \end{aligned}$$

where

$$\begin{aligned} f^*(t) &= \frac{1}{\sqrt{2\pi}}\left[\int_{ia-\infty}^{ia+\infty} e^{-itw}\frac{\phi(w)}{(ae^{-ibw}-iw)}dw \right. \\ &\quad \left. - \int_{ib-\infty}^{ib+\infty} e^{-itw}\frac{\phi(w)}{(ae^{-ibw}-iw)}dw\right], \end{aligned}$$

and the constants a and b are chosen so that all the zeros of $ae^{-ibw} - iw$ are in the strip $-c \le v \le c$.

If we denote the zeros of the function $h(w) = ae^{-ibw} - iw$ in the strip $-c \le v \le c$ by w_k, then

$$f^*(t) = \sum_k c_k e^{-iw_k t}\,,$$

for some constants c_k.

12.14.5 Differential-Difference Equations

Consider the differential-difference equation

$$f'(x) = \frac{1}{2h}\left[f(x+h) - f(x-h)\right]\,, \tag{12.40}$$

where f is assumed to be of real exponential type c.

Since f is of exponential type, so is f'. By repeated application of (12.40), it can be shown that

$$f''(x) = \frac{1}{(2h)^2}\left[f(x+2h) + f(x-2h) - 2f(x)\right]\,,$$

and by induction it follows that f has derivatives of all orders and they are all of exponential type. It is easy to verify that

$$g(x) = f(x) - f(0) - xf'(0)$$

also satisfies (12.40) and that $g(0) = 0 = g'(0)$; hence, without loss of generality, we assume that $f(0) = 0 = f'(0)$.

Let us define $F_\pm$ as before. From the fact that $f^{(k)}$ is of exponential type, it follows that $F_\pm$ are of rapid decay at $\pm\infty$, and hence

$$f(x) = \frac{1}{\sqrt{2\pi}} \left\{ \int_{ia-\infty}^{ia+\infty} F_+(w) e^{-ixw} dw + \int_{ib-\infty}^{ib+\infty} F_-(w) e^{-ixw} dw \right\},$$

where $a > 0$, $b < 0$.

By substituting this into (12.40), we obtain after some calculations

$$\int_{ia-\infty}^{ia+\infty} F_+(w) \left(w - \frac{\sin hw}{h} \right) e^{-ixw} dw + \int_{ib-\infty}^{ib+\infty} F_-(w) \left(w - \frac{\sin hw}{h} \right) e^{-ixw} dw = 0 .$$

Once more by Theorem 141 in [225, p. 255] , there exists a function $\phi(w)$ analytic in the strip $b < v < a$, and tends to zero as $|u| \to \infty$ such that

$$F_+(w) \left(w - \frac{\sin hw}{h} \right) = \phi(w) ,$$

and

$$F_-(w) \left(w - \frac{\sin hw}{h} \right) = -\phi(w) .$$

Therefore, upon solving for $F_\pm$, we obtain

$$f(x) = \frac{h}{\sqrt{2\pi}} \left[\int_{ia-\infty}^{ia+\infty} \frac{\phi(w)}{hw - \sin hw} e^{-ixw} dw - \int_{ib-\infty}^{ib+\infty} \frac{\phi(w)}{hw - \sin hw} e^{-ixw} dw \right] .$$

Let us denote the function $hw - \sin hw$ by $g(w)$ Clearly, g has a zero of order 3 at the origin, which in view of Theorem 12.30 results in a quadratic polynomial in the inverse Fourier transform. If we denote the other zeros of g that are in the strip $|v| \le c$ by w_k , $k = \pm 1, \pm 2, \cdots$, where $w = u + iv$, then it can be verified that

$$f(x) = A + Bx + Cx^2 + \sum_{k=-\infty}^{\infty} C_k e^{-iw_k x}.$$

12.14.6 Convolution Equations

Consider a convolution equation of the form

$$y * f = g,$$

in which f, g are given functions or generalized functions and y is unknown. If f and g are Fourier transformable, we then invoke the Fourier transform method to obtain

$$Y(w) = \frac{G(w)}{F(w)},$$

which formally leads to the solution

$$y(t) = \mathcal{F}^{-1}\left(\frac{G(w)}{F(w)}\right),$$

provided that the inverse Fourier transform exists.

Example B: Consider the convolution equation

$$\left(\delta^{(1)} + \delta\right) * y = \delta.$$

Taking the Fourier transform of both sides of the equation yields

$$(-iw + +1)\, Y(w) = 1\,;$$

hence

$$y(t) = \mathcal{F}^{-1}\left(\frac{1}{1 - iw}\right) = e^{-t} H(t)\,.$$

12.14.7 Partial Differential Equations

A. Parabolic Equations (Heat Equation):

The problem of heat flow in an infinite medium with given initial temperature distribution can be described by the system

$$\begin{array}{ll} \frac{\partial^2 u}{\partial x^2} = a^{-2}\frac{\partial u}{\partial t}\,, & -\infty < x < \infty \quad , t \geq 0\,, \\ u(x,t)\ ,\ u_x(x,t) \to 0 & \text{as } |x| \to \infty \\ u(x,0) = f(x)\ , & -\infty < x < \infty\,. \end{array}$$

Taking the Fourier transform with respect to x and using property (7) of Section 12.4, we find that

$$w^2 U + a^{-2}\frac{\partial U}{\partial t} = 0\,, \tag{12.41}$$

with $U(w,0) = F(w)$, where

$$U(w,t) = \frac{1}{\sqrt{2\pi}}\int_{-\infty}^{\infty} u(x,t) e^{iwx} dx\,.$$

Therefore, by solving the differential equation (12.41) with the given initial condition, we obtain

$$U(w,t) = F(w)e^{-a^2w^2t},$$

and hence

$$u(x,t) = \frac{1}{\sqrt{2\pi}}\int_{-\infty}^{\infty} F(w)e^{-a^2w^2t}e^{-ixw}\,dw.$$

Upon observing that

$$\mathcal{F}^{-1}\left(e^{-a^2w^2t}\right)(x) = \frac{1}{a\sqrt{2t}}e^{-x^2/(4a^2t)}$$

we employ the convolution theorem to obtain

$$u(x,t) = \frac{1}{2a\sqrt{t\pi}}\int_{-\infty}^{\infty} f(y)e^{-(x-y)^2/(4a^2t)}\,dy.$$

B. Elliptic Equations (Laplace's Equation):

Consider the problem of finding the steady-state temperature distribution in a large rectangular plate whose surfaces are insulated and the temperature along one of its edges is prescribed by f and along the other sides the temperature tends to zero. This problem can be described by the system

$$\begin{aligned}
&\tfrac{\partial^2 u}{\partial x^2} + \tfrac{\partial^2 u}{\partial y^2} = 0 \quad -\infty < x < \infty \quad , y \geq 0\,,\\
&u(x,y) \to 0 \qquad \text{as } x^2 + y^2 \to \infty\\
&u(x,0) = f(x)\ ,\ -\infty < x < \infty\,.
\end{aligned}$$

By taking the Fourier transform with respect to x, the problem is converted into

$$-w^2U + U_{yy} = 0\,,$$

with $U(w,0) = F(w)\,,\quad U(w,y) \to 0$ as $y \to \infty$.

The solution of this problem is easily seen to be

$$U(w,y) = F(w)e^{-|w|y}\,,$$

which in turn yields

$$u(x,y) = \frac{y}{\pi}\int_{-\infty}^{\infty}\frac{f(t)}{[(x-t)^2+y^2]}dt\,,\quad y > 0$$

since

$$\mathcal{F}^{-1}\left(e^{-|w|y}\right)(x) = \sqrt{2/\pi}\left(\frac{y}{x^2+y^2}\right).$$

C. Hyperbolic Equations (Vibrating String):

$$\begin{aligned}
&\frac{\partial^2 u}{\partial x^2} = \frac{\partial^2 u}{\partial t^2}\,, \qquad -\infty < x < \infty \quad , t \geq 0\,,\\
&u(x,0) = f(x)\,,\ -\infty < x < \infty\\
&u_t(x,0) = g(x) \quad -\infty < x < \infty\,.
\end{aligned}$$

This problem describes the motion of an infinite string with a given initial displacement f and initial velocity g, and u being the displacement at distance x along the string at time t.

We take the Fourier transform with respect to x and convert the problem into

$$U_{tt} = -w^2 U\,,$$

with

$$U(w,0) = F(w) \quad , \quad U_t(w,0) = G(w).$$

The solution of this problem is easily seen to be

$$U(w,t) = F(w)\cos wt + \frac{G(w)}{w}\sin wt\,.$$

If we apply the inverse Fourier transformation, we find

$$u(x,t) = \frac{1}{\sqrt{2\pi}}\int_{-\infty}^{\infty} F(w)\cos(tw)e^{-ixw}dw + \frac{1}{\sqrt{2\pi}}\int_{-\infty}^{\infty} \frac{G(w)}{w}\sin(tw)e^{-ixw}dw.$$

The first term can be reduced to

$$\frac{1}{2\sqrt{2\pi}}\int_{-\infty}^{\infty} F(w)\left(e^{-iw(x-t)} + e^{-iw(x+t)}\right)dw = \frac{f(x+t)+f(x-t)}{2}\,,$$

while the second can be reduced to

$$\frac{1}{2i\sqrt{2\pi}}\int_{-\infty}^{\infty} \frac{G(w)}{w}\left(e^{-iw(x-t)} - e^{-iw(x+t)}\right)dw = \frac{1}{2}\int_{x-t}^{x+t} g(\xi)\,d\xi\,.$$

Therefore,

$$u(x,t) = \frac{f(x+t)+f(x-t)}{2} + \frac{1}{2}\int_{x-t}^{x+t} g(\xi)\,d\xi\,.$$

13

The Hartley Transform

Closely related to the Fourier transform is the Hartley transform, which was introduced by Ralph V. Hartley in 1942 in a paper that was published in the Proceedings of the Institute of Radio Engineers [93]. Hartley began his professional career as a researcher at the Engineering Department of Western Electric and was in charge of the early development of transatlantic radio telephone tests in 1915. Many practical and theoretical works bear his name, such as Hartley's oscillating circuit (known as Hartley's oscillator) and Hartley's law in information theory which states that the quantity of information transmitted by a telephone line is proportional to the product of the bandwidth and the duration of the transmission. His work in information theory laid the basis for Claude E. Shannon's work which revolutionized information theory [257]. Hartley is considered a founding member of the famous Bell Telephone Laboratories.

In his 1942 paper, Hartley tried to overcome what he thought to be a drawback of the Fourier transformation, namely, the representation of a real-valued function $f(t)$ by a complex-valued function, $\hat{f}(w)$. To dispense with that complex-valued representation, he introduced an integral transformation that mapped a real-valued function into another real-valued function using the sum of the cosine and sine functions as a kernel.

His formulation went unnoticed until 1984 when Ronald N. Bracewell, a Professor of Electrical Engineering at Stanford University, introduced an algorithm, known as the fast Hartley transform (FHT), to compute the Hartley transform in a fast and efficient way. The fast Hartley transform (FHT) computes the discrete Hartley transform as efficiently as the fast Fourier transform (FFT) computes the discrete Fourier transform (DFT) [24]. The introduction of the fast Hartley transform has made the Hartley transform much more accessible for practical implementation. Therefore, it is fair to say that Hartley gets the credit for introducing the transform and Bracewell for popularizing it.

13.1 Definitions

DEFINITION 13.1 *The Hartley transform of a function $f(x)$ is defined by*

$$H[f](\omega) = \frac{1}{\sqrt{2\pi}} \int_{-\infty}^{\infty} f(x)\, Cas\,(\omega x)\, dx\,,$$

where

$$Cas\, z = \cos\, z + \sin\, z\,.$$

For the Hartley transform to exist, it is sufficient that $f \in L^1(\mathcal{R})$. The function $\mathrm{Cas}\, z$ has similar properties to those of the sine and cosine functions. For example, one can easily verify that

$$\mathrm{Cas}(A+B) = \cos A\, \mathrm{Cas}\, B + \sin A\, \mathrm{Cas}'\, B\,, \tag{13.1}$$

$$\mathrm{Cas}(A-B) = \cos\, A\, \mathrm{Cas}'\, B + \sin A\, \mathrm{Cas}\, B\,, \tag{13.2}$$

where

$$\mathrm{Cas}'\, z = \frac{d}{dz} \mathrm{Cas}\, z = \mathrm{Cas}(-z)\,. \tag{13.3}$$

Moreover,

$$\mathrm{Cas}\, z = \frac{1}{2}\left[(1-i)e^{ix} + (1+i)e^{-ix}\right]\,,$$

and that $\left\{(1/\sqrt{2\pi})\mathrm{Cas}\, nx\right\}_{n=0}^{\infty}$ is an orthonormal set on $[-\pi, \pi]$. It is evident that

$$H[f](\omega) = \frac{1}{\sqrt{\pi}} \int_{-\infty}^{\infty} f(x) \cos\,(\omega x - \frac{\pi}{4})\, dx\,.$$

Example A: Let $f(t)$ be the characteristic function of the interval $[0, 1]$. Then it is easy to see that the Fourier and the Hartley transforms of f are given, respectively, by

$$F(\omega) = \frac{1}{\sqrt{2\pi}} \frac{\sin \omega - i(\cos \omega - 1)}{\omega}$$

and

$$H[f](\omega) = \frac{1}{\sqrt{2\pi}} \frac{\sin \omega - (\cos \omega - 1)}{\omega}.$$

13.2 Elementary Properties

1) **Linearity:** The Hartley transformation is a linear transformation.

2) **Translation:**

$$H[f(x-a)](\omega) = \cos(a\omega)\, H[f](\omega) + \sin(a\omega)\, H[f](-\omega)\,.$$

3) **Dilation:**

$$H[f(x/a)](\omega) = |a| H[f(x)](a\omega)\,,$$

in particular

$$H[f(-x)](\omega) = H[f(x)](-\omega)\,.$$

4) **Transform of the Derivative:**

$$H[f'(x)](\omega) = H[xf(-x)](\omega)\,.$$

5) **Derivative of the Transform:**

$$\frac{d}{d\omega} H[f(x)](\omega) = -H[xf(-x)](\omega)\,.$$

6) **Parity:** The Hartley transform of an even function is even and that of an odd function is odd.

7) **The Value at the Origin:**

$$H(0) = \frac{1}{\sqrt{2\pi}} \int_{-\infty}^{\infty} f(x)\, dx \quad \text{and} \quad f(0) = \frac{1}{\sqrt{2\pi}} \int_{-\infty}^{\infty} H(\omega)\, d\omega\,.$$

Properties (1), (3) and (6) are trivial, (2) follows from (13.1), (4) and (5) from (13.2) and (13.3). The first part of (7) is trivial and the second is a consequence of the inversion formula given below.

13.3 Relationship with Other Transforms

13.3.1 The Fourier Transform

Let F, F_c and F_s denote the Fourier, Fourier cosine and sine transforms of f, respectively, and let $H_e[f]$ and $H_0[f]$ denote the even and odd parts of the Hartley transform of f. That is

$$H_e[f](\omega) = \frac{H[f](\omega) + H[f](-\omega)}{2} = \frac{1}{\sqrt{2\pi}} \int_{-\infty}^{\infty} f(x) \cos \omega x dx, \quad (13.4)$$

$$H_O[f](\omega) = \frac{H[f](\omega) - H[f](-\omega)}{2} = \frac{1}{\sqrt{2\pi}} \int_{-\infty}^{\infty} f(x) \sin \omega x dx. \quad (13.5)$$

Hence

$$F(\omega) = H_e[f](\omega) + iH_O[f](\omega) , \quad (13.6)$$

$$\operatorname{Re} F(\omega) = H_e[f](\omega) , \quad \operatorname{Im} F(\omega) = H_O[f](\omega) ,$$

and

$$H(\omega) = \operatorname{Re} F(\omega) + \operatorname{Im} F(\omega) . \quad (13.7)$$

Moreover, if f is even, then

$$H[f](\omega) = H_e[f](\omega) = F_c(\omega) = F(\omega) ,$$

and if f is odd, then

$$H[f](\omega) = H_O[f](\omega) = F_s(\omega) = -iF(\omega) ,$$
$$F(\omega) = \frac{1}{\sqrt{2}} \left(H[f](\omega) e^{i\pi/4} + H[f](-\omega) e^{-i\pi/4} \right) .$$

13.3.2 The Hardy Transform

The Hardy transform of a function f is defined by (cf. Chapter 22)

$$\tilde{H}_\nu[f](\omega) = \int_0^\infty [\cos a\pi J_\nu(\omega x) + \sin a\pi Y_\nu(\omega x)]\, x f(x)\, dx ,$$

where J_ν and Y_ν are the Bessel functions of order ν of the first and second kind, respectively. For $\nu = \frac{1}{2}$ and $a = \frac{1}{4}$, it reduces to

$$\tilde{H}_{\frac{1}{2}}[f](\omega) = \frac{1}{\sqrt{\pi\omega}} \int_0^\infty \sqrt{x}\, f(x) \operatorname{Cas} \omega x\, dx .$$

Therefore, if f vanishes on $(-\infty, 0]$, then

$$H\left[\sqrt{x}\, f(x)\right](\omega) = \sqrt{\pi\omega}\, \tilde{H}_{\frac{1}{2}}[f(x)](\omega) .$$

13.4 Parseval's Relation

From (13.6) and Parseval's relation for the Fourier transform, it follows that

$$\begin{aligned}\int_{-\infty}^{\infty} f(x)\overline{g}(x)\,dx &= \int_{-\infty}^{\infty} F(\omega)\overline{G}(\omega)\,d\omega \\ &= \int_{-\infty}^{\infty} \left(H_e[f](\omega)H_e[g](\omega) + H_0[f](\omega)H_0[g](\omega)\right)\,d\omega\ ,\end{aligned}$$

in particular,

$$\int_{-\infty}^{\infty} |f(x)|^2 dx = \int_{-\infty}^{\infty} |H_e[f](\omega)|^2\,d\omega + \int_{-\infty}^{\infty} |H_0[f](\omega)|^2\,d\omega\ .$$

13.5 Inversion Formula

Using (13.4)–(13.6) and the inversion formula of the Fourier transform, we obtain

$$\begin{aligned}f(x) &= \frac{1}{\sqrt{2\pi}}\int_{-\infty}^{\infty} F(\omega)e^{-ix\omega}d\omega \\ &= \frac{1}{\sqrt{2\pi}}\int_{-\infty}^{\infty} \left(H_e[f](\omega) + iH_0[f](\omega)\right)(\cos x\omega - i\sin x\omega)\,d\omega \\ &= \frac{1}{\sqrt{2\pi}}\int_{-\infty}^{\infty} \left(H_e[f](\omega)\cos x\omega + H_0[f](\omega)\sin x\omega\right)d\omega \\ &= \frac{1}{2\sqrt{2\pi}}\int_{-\infty}^{\infty} H[f](\omega)(\cos x\omega + \sin x\omega) + H[f](-\omega)(\cos x\omega - \sin x\omega)\,d\omega\ .\end{aligned}$$

Changing ω to $-\omega$ in the second group of terms inside the integral yields the inversion formula of the Hartley transform

$$f(x) = \frac{1}{\sqrt{2\pi}}\int_{-\infty}^{\infty} H[f](\omega)\mathrm{Cas}(x\omega)d\omega\ .$$

13.6 Convolution

Let us define the convolution operations $*$ and $\star$ by

$$(f * g)(x) = \frac{1}{\sqrt{2\pi}} \int_{-\infty}^{\infty} f(t)g(x-t)dt \ ,$$

and

$$(f \star g)(x) = \frac{1}{\sqrt{2\pi}} \int_{-\infty}^{\infty} f(t)g(x+t)dt \ .$$

Hence,

$$\begin{aligned} H[f * g](\omega) &= \frac{1}{\sqrt{2\pi}} \int_{-\infty}^{\infty} \mathrm{Cas}(\omega x) \left(\frac{1}{\sqrt{2\pi}} \int_{-\infty}^{\infty} f(t)g(x-t)dt \right) dx \\ &= \frac{1}{2\pi} \int_{-\infty}^{\infty} f(t)dt \left(\int_{-\infty}^{\infty} g(z)\mathrm{Cas}\omega(t+z)dz \right) . \end{aligned}$$

If we employ (13.1), (13.4) and (13.5), we find

$$\begin{aligned} &H[f * g](\omega) = \\ &\frac{1}{\sqrt{2\pi}} \left\{ H[f](\omega) \int_{-\infty}^{\infty} g(z)\cos(\omega z)\, dz + H[f](-\omega) \int_{-\infty}^{\infty} g(z)\sin(\omega z)\, dz \right\} \\ &= H[f](\omega)H_e[g](\omega) + H[f](-\omega)H_0[g](\omega) \\ &= \frac{1}{2} \left\{ H[f](\omega)\left(H[g](\omega) + H[g](-\omega)\right) + H[f](-\omega)\left(H[g](\omega) - H[g](-\omega)\right) \right\} . \end{aligned}$$

Similarly, one can verify that

$$H[f \star g](\omega) = H[f](\omega)H_0[g](\omega) + H[f](-\omega)H_e[g](\omega) \ ,$$

in particular

$$H[f \star f](\omega) = \frac{1}{2} \left(H^2[f](\omega) + H^2[f](-\omega) \right) .$$

If either f or g is even, then

$$H[f * g](\omega) = H[f](\omega)H[g](\omega) \ , \quad H[f \star g](\omega) = H[f](-\omega)H[g](\omega)$$

and if either f or g is odd, then

$$H[f * g](\omega) = H[f](-\omega)H[g](\omega) \ , \quad H[f \star g](\omega) = H[f](\omega)H[g](\omega) \ .$$

13.7 Generalized Functions and Applications

13.7.1 Generalized Functions

The Hartley transform of generalized functions can be easily obtained as a special case of the Fourier transform of generalized functions by using (13.7). For generalized functions with support in $[0, \infty)$, it can be obtained as a special case of the Hardy transform of generalized functions; see Chapter 22.

13.7.2 Applications

The Hartley transform has many interesting applications in electrical engineering and optics, of which a good example is the optical Hartley transformer [24, Ch. 9], [25]. Bracewell designed an optical system that presents the Hartley transform of a given object and showed how a Hartley-based hologram of a real object could be constructed. Because of the close connection between the Fourier and the Hartley transforms, Bracewell purports that for every application of the Fourier transform there is a corresponding one of the Hartley transform [24, p. 105].

14

The Hilbert Transform

Closely related to the Fourier, Stieltjes and convolution transforms is the *Hilbert transform*. It is named after one of the greatest mathematicians of the twentieth century, David Hilbert (1862-1943), who, in his studies of integral equations, was the first to observe what is now known as the Hilbert transform pair [100]; see also [119, 120]. Although it was named after Hilbert, the Hilbert transform and its properties were developed mainly by E. C. Titchmarch [227, 228, 229] and G. H. Hardy [91]. It was Hardy who named it after Hilbert.

14.1 Motivation

Let $c_n = a_n - ib_n$ and consider the function

$$F(z) = \sum_{n=0}^{\infty} c_n e^{in\theta} r^n , \quad z = re^{i\theta} ;$$

which we will assume to be analytic in the unit disc. By separating the real and imaginary parts of F, we immediately obtain

$$F(r, \theta) = U(r, \theta) + i\tilde{U}(r, \theta) , \quad |r| < 1$$

where

$$U(r, \theta) = \sum_{n=0}^{\infty} (a_n \cos n\theta + b_n \sin n\theta) r^n ,$$

and

$$\tilde{U}(r, \theta) = \sum_{n=0}^{\infty} (a_n \sin n\theta - b_n \cos n\theta) r^n ,$$

and $c_n = a_n - ib_n$. Being the real and imaginary parts of an analytic function, U and $\tilde{U}$ are harmonic in the unit disc. They are called *harmonic conjugates*. We have formally,

$$\lim_{r\to 1^-} U(r,\theta) = f(\theta) = \sum_{n=0}^{\infty} a_n \cos n\theta + b_n \sin n\theta \,,$$

and

$$\lim_{r\to 1^-} \tilde{U}(r,\theta) = -\tilde{f}(\theta) = -\sum_{n=0}^{\infty} b_n \cos n\theta - a_n \sin n\theta \,.$$

Since they are the limits of conjugate functions, f and $\tilde{f}$ are also called conjugate functions. The transformation $\mathcal{H}$ that maps f into $\tilde{f}$, i.e., $\mathcal{H}f = \tilde{f}$ is called the *Hilbert transformation.*

Analogously, let f be an integrable function and define $a(t)$ and $b(t)$ by

$$a(t) = \frac{1}{\pi}\int_{-\infty}^{\infty} f(x)\cos tx\, dx \;, \quad b(t) = \frac{1}{\pi}\int_{-\infty}^{\infty} f(x)\sin tx\, dx. \tag{14.1}$$

Consider the function $F(z)$ defined by the Fourier integral

$$F(z) = \int_0^{\infty} (a(t) - ib(t))e^{izt} dt = U(z) + i\tilde{U}(z) \,, \tag{14.2}$$

where $z = x+iy$. It is easily seen that F is analytic in the upper-half plane, $y > 0$. The real and imaginary parts of F are

$$U(z) = \int_0^{\infty} (a(t)\cos xt + b(t)\sin xt)e^{-yt} dt \;,$$

and

$$\tilde{U}(z) = \int_0^{\infty} (a(t)\sin xt - b(t)\cos xt)e^{-yt} dt \;.$$

Formally, we have

$$\lim_{y\to 0} U(z) = f(x) = \int_0^{\infty} (a(t)\cos xt + b(t)\sin xt) dt \;, \tag{14.3}$$

and

$$\lim_{y\to 0} \tilde{U}(z) = -\tilde{f}(x) = \int_0^{\infty} (a(t)\sin xt - b(t)\cos xt) dt \;, \tag{14.4}$$

In view of (14.1), $\tilde{f}$ can be written in the form

$$\begin{aligned}
\tilde{f}(x) &= \frac{1}{\pi}\int_0^{\infty} dt \int_{-\infty}^{\infty} f(u)\sin t(u-x) du \\
&= \lim_{\lambda\to\infty} \frac{1}{\pi}\int_0^{\lambda} dt \int_{-\infty}^{\infty} f(u)\sin t(u-x) du \\
&= \lim_{\lambda\to\infty} \frac{1}{\pi}\int_{-\infty}^{\infty} f(u)\frac{1-\cos\lambda(u-x)}{u-x} du \\
&= \lim_{\lambda\to\infty} \frac{1}{\pi}\int_{-\infty}^{\infty} f(u+x)\frac{1-\cos\lambda u}{u} du \;.
\end{aligned}$$

If $f(u)/u$ is integrable over $(-\infty, \infty)$, it follows from the Riemann–Lebesgue Lemma (Theorem 12.2) that

$$\lim_{\lambda\to\infty}\int_{-\infty}^{\infty} f(u+x)\frac{\cos\lambda u}{u}du = \lim_{\lambda\to\infty}\int_{0}^{\infty}\frac{f(x+u)-f(x-u)}{u}\cos\lambda u\, du = 0 .$$

Therefore,

$$\tilde{f}(x) = \frac{1}{\pi}\int_{-\infty}^{\infty}\frac{f(u+x)}{u}du = \frac{1}{\pi}\int_{0}^{\infty}\frac{f(x+u)-f(x-u)}{u}du$$
$$= \frac{1}{\pi}\int_{-\infty}^{\infty}\frac{f(u)}{u-x}du ,$$

which is the Hilbert transform of f. More precisely, we have the following definition.

14.2 Definition

DEFINITION 14.1 *The Hilbert transform of f is defined as*

$$(\mathcal{H}f)(x) = \tilde{f}(x) = \frac{1}{\pi}\int_{-\infty}^{\infty}\frac{f(t)}{t-x}dt ,$$

where the integral is a Cauchy principle value.

As shown above, the Hilbert transform can also be written in the form

$$(\mathcal{H}f)(x) = \frac{1}{\pi}\int_{0}^{\infty}\frac{f(x+t)-f(x-t)}{t}dt = \frac{1}{\pi}\int_{-\infty}^{\infty}\frac{f(x+t)}{t}dt .$$

14.3 Existence

If $f \in L^1(\mathcal{R})$, then its Hilbert transform $(\mathcal{H}f)(x)$ exists for almost all x. But if $f \in L^p(\mathcal{R})\, , p > 1$, we have the following stronger result.

THEOREM 14.1 [225, p. 132]
Let $f \in L^p(\mathcal{R})$, $p > 1$. Then $(\mathcal{H}f)(x)$ exists for almost all x and defines a function that also belongs to $L^p(\mathcal{R})$ with

$$\int_{-\infty}^{\infty}|(\mathcal{H}f)(x)|^p\, dx \le C_p\int_{-\infty}^{\infty}|f(x)|^p\, dx .$$

In the special case when $p = 2$, we have

$$\int_{-\infty}^{\infty} |(\mathcal{H}f)(x)|^2\, dx = \int_{-\infty}^{\infty} |f(x)|^2\, dx \,.$$

The theorem is not valid if $p = 1$ because although it is true that $(\mathcal{H}f)(x)$ is defined almost everywhere, it is not necessarily in $L^1(\mathcal{R})$. The function $f(t) = (t \log^2 t)^{-1} H(t)$ provides a counter-example.

14.4 Elementary Properties

1. **Linearity:** The Hilbert transformation is linear.

2. **Translation:** The Hilbert transformation commutes with the translation operator
$$T_a[\mathcal{H}f] = \mathcal{H}[T_a f] \,.$$

3. **Dilation:** The Hilbert transformation also commutes with the dilation operator
$$D_a[\mathcal{H}f] = \mathcal{H}[D_a f] \quad a > 0 \,,$$
but
$$D_a[\mathcal{H}f] = -\mathcal{H}[D_a f] \quad \text{for} \quad a < 0 \,.$$

4. **Multiplication by t:**
$$\mathcal{H}[tf(t)](x) = x(\mathcal{H}f)(x) + \frac{1}{\pi}\int_{-\infty}^{\infty} f(t)dt \,.$$

5. **Differentiation:**
$$\mathcal{H}[f'] = (\mathcal{H}f)' \,,$$
provided that $f(t) = 0(t)$ as $|t| \to \infty$.

6. **Orthogonality:** The Hilbert transform of $f \in L^2(\mathcal{R})$ is orthogonal to f in the sense
$$\int_{-\infty}^{\infty} f(x)(\mathcal{H}f)(x)dx = 0 \,.$$

7. **Parity:** The Hilbert transform of an even function is odd and that of an odd function is even.

14.5 Relationships with Other Transforms

14.5.1 The Fourier Transform

The relationship between the Fourier and Hilbert transforms has been pointed out in Section 14.1. From (14.2)–(14.4), we obtain

$$\lim_{y\to 0} F(z) = F(x) = \int_0^\infty (a(t) - ib(t))e^{ixt}dt = f(x) - i(\mathcal{H}f)(x) \ ,$$

where $a(t)$ and $b(t)$ are given by (14.1).

Let g be a real-valued integrable function and consider its Fourier transform

$$\hat{g}(x) = \frac{1}{\sqrt{2\pi}} \int_{-\infty}^\infty g(t)e^{ixt}dt \ .$$

If we denote the real and imaginary parts of $\hat{g}$ by f and $\tilde{f}$, respectively, then

$$f(x) = \frac{1}{\sqrt{2\pi}} \int_{-\infty}^\infty g(t)\cos xtdt \ ,$$

and

$$\tilde{f}(x) = \frac{1}{\sqrt{2\pi}} \int_{-\infty}^\infty g(t)\sin xtdt \ .$$

Let us split g into its even and odd parts, g_e and g_o, respectively:

$$g_e(t) = \frac{g(t)+g(-t)}{2} \quad \text{and} \quad g_o(t) = \frac{g(t)-g(-t)}{2} \ ;$$

hence

$$f(x) = \sqrt{\frac{2}{\pi}} \int_0^\infty g_e(t)\cos xt\,dt \ , \ \tilde{f}(x) = \sqrt{\frac{2}{\pi}} \int_0^\infty g_o(t)\sin xt\,dt \ , \qquad (14.5)$$

or

$$f(x) = \frac{1}{\sqrt{2\pi}} \int_{-\infty}^\infty g_e(t)e^{ixt}dt \quad \text{and} \quad \tilde{f}(x) = \frac{-i}{\sqrt{2\pi}} \int_{-\infty}^\infty g_o(t)e^{ixt}dt \ .$$

This shows that if the Fourier transform of the even part of a real-valued function g represents a function $f(x)$, then the Fourier transform of the odd part of g represents the Hilbert transform of f (up to a multiplication by $-i$).

THEOREM 14.2

Let $f \in L^1(\mathcal{R})$ and assume that $\mathcal{H}f$ is also in $L^1(\mathcal{R})$. Then

$$\mathcal{F}(\mathcal{H}f)(w) = -i\ \text{sgn}(w)\mathcal{F}(f)(w) \qquad (14.6)$$

where $\mathcal{F}$ denotes the Fourier transformation. Similarly, if $f \in L^2(\mathcal{R})$, then by Theorem 14.1, $(\mathcal{H}f) \in L^2(\mathcal{R})$ and (14.6) remain valid.

Proof: From the definition of the Fourier transform we have

$$\begin{aligned}
\mathcal{F}(\mathcal{H}f)(w) &= \frac{1}{\sqrt{2\pi}}\frac{1}{\pi}\int_{-\infty}^{\infty} e^{ixw}\left(\int_{-\infty}^{\infty}\frac{f(t)}{t-x}dt\right)dx \\
&= \frac{1}{\pi}\int_{-\infty}^{\infty} f(t)\left(\frac{1}{\sqrt{2\pi}}\int_{-\infty}^{\infty}\frac{e^{ixw}}{t-x}dx\right)dt \\
&= \frac{1}{\pi}\int_{-\infty}^{\infty} f(t)e^{iwt}dt\left(\frac{1}{\sqrt{2\pi}}\int_{-\infty}^{\infty}\frac{e^{-iwy}}{y}dy\right) .
\end{aligned}$$

The result now follows immediately in view of the relation

$$\mathcal{F}^{-1}\left(\frac{1}{t}\right)(w) = \mathcal{F}\left(\frac{1}{t}\right)(-w) = -i\sqrt{\frac{\pi}{2}}\,\mathrm{sgn}(w).$$

∎

Equation (14.6) also shows that the Hilbert transform of a band-limited function is also band-limited.

14.5.2 The Stieltjes Transform

Let us denote the Stieltjes transform of f by $\mathcal{S}(f)$. Then

$$(\mathcal{H}f)(x) = \frac{1}{2\pi}\left\{\mathcal{S}(f)(xe^{i\pi}) + \mathcal{S}(f)(xe^{-i\pi})\right\} - \frac{1}{\pi}\mathcal{S}(f(-t))(x)\,, \quad x > 0$$

and

$$\begin{aligned}
(\mathcal{H}f)(x) &= \frac{1}{\pi}\,\mathcal{S}(f)(-x) \\
&\quad -\frac{1}{2\pi}\left\{\mathcal{S}(f(-t))(|x|e^{i\pi}) + \mathcal{S}(f(-t))(|x|e^{-i\pi})\right\}\,, \quad x < 0\,.
\end{aligned}$$

The first relation follows from the observation that

$$\begin{aligned}
(\mathcal{H}f)(x) &= \frac{1}{\pi}\int_{-\infty}^{\infty}\frac{f(t)}{t-x}dt = \frac{1}{\pi}\left\{\int_0^{\infty}\frac{f(t)}{t-x}dt - \int_0^{\infty}\frac{f(-t)}{t+x}dt\right\} \\
&= \frac{1}{2\pi}\left\{\int_0^{\infty}\frac{f(t)}{t+xe^{i\pi}}dt + \int_0^{\infty}\frac{f(t)}{t+xe^{-i\pi}}dt\right\} - \frac{1}{\pi}\int_0^{\infty}\frac{f(-t)}{t+x}dt\,,
\end{aligned}$$

and the second follows similarly.

14.5.3 The Boas Transform

If we denote the Boas transform of f by Bf (cf. Chapter 15), then

$$(Bf)(x) = (\mathcal{H}f)(x) - ((\mathcal{H}f) * g)(x) , \tag{14.7}$$

where $*$ denotes the convolution operation associated with the Fourier transform and

$$g(x) = \sqrt{\frac{2}{\pi}} \left(\frac{1 - \cos x}{x^2} \right) , \quad -\infty < x < \infty .$$

In particular, (14.7) holds if $f \in L^2(\mathcal{R})$. In Equation (14.7) it is tacitly assumed that all the terms are well defined.

14.5.4 The Convolution Transform

Although the Hilbert transform can be formally written as a convolution transform, namely,

$$(\mathcal{H}f)(x) = (f * g)(x) ,$$

where

$$g(t) = \sqrt{\frac{2}{\pi}} \left(\frac{-1}{t} \right) ,$$

it is not classified as such since unlike other convolution transforms the kernel g has a non-integrable singularity. Nonetheless, we have the following relations.

THEOREM 14.3
If f and $g \in L^1(\mathcal{R})$ are such that their Hilbert transforms are also in $L^1(\mathcal{R})$, then

$$\mathcal{H}(f * g) = \mathcal{H}f * g = f * \mathcal{H}g , \tag{14.8}$$

and

$$(f * g) = -(\mathcal{H}f * \mathcal{H}g) . \tag{14.9}$$

Proof: To prove (14.8), consider

$$\begin{aligned}
\mathcal{H}(f * g)(x) &= \frac{1}{\pi\sqrt{2\pi}} \int_{-\infty}^{\infty} \frac{dt}{t - x} \left(\int_{-\infty}^{\infty} f(y) g(t - y) dy \right) \\
&= \frac{1}{\pi\sqrt{2\pi}} \int_{-\infty}^{\infty} f(y) dy \left(\int_{-\infty}^{\infty} \frac{g(z)dz}{z - (x - y)} \right) \\
&= \frac{1}{\sqrt{2\pi}} \int_{-\infty}^{\infty} f(y) (\mathcal{H}g)(x - y) dy \\
&= (f * \mathcal{H}g)(x) .
\end{aligned}$$

Similarly, one can show that $\mathcal{H}(f*g) = \mathcal{H}f*g$. The proof of (14.9) follows from (14.8) by replacing g by $\mathcal{H}g$ and noting that $\mathcal{H}^2g = -g$; see Section 6 below. ∎

14.5.5 The Cauchy Transform

From Equations (14.1), it follows that

$$\hat{f}(t) = \sqrt{\frac{\pi}{2}}(a(t) + ib(t)) \quad \text{and} \quad \hat{f}(-t) = \sqrt{\frac{\pi}{2}}(a(t) - ib(t)) \, .$$

Following the notation of Section 14.1, we have

$$F(z) = \sqrt{\frac{2}{\pi}} \int_0^\infty \hat{f}(-t)e^{izt}dt \, .$$

But in view of Parseval's relation for the Fourier transform (see Section 12.6)

$$\int_{-\infty}^\infty \hat{f}(-t)\hat{h}(t)dt = \int_{-\infty}^\infty f(t)h(t)dt \, ,$$

and the fact that the inverse Fourier transform of $e^{izt}\chi_{[0,\infty)}(t)$ is $(i/\sqrt{2\pi})(z-w)^{-1}$, we obtain that F is the *Cauchy transform* of f, i.e.,

$$F(z) = \frac{1}{i\pi}\int_{-\infty}^\infty \frac{f(t)}{t-z}dt \, , \quad Imz > 0 \, .$$

By separating the real and imaginary parts of F, we have

$$U(z) = \frac{y}{\pi}\int_{-\infty}^\infty \frac{f(t)}{(t-x)^2 + y^2}dt \, ,$$

and

$$\tilde{U}(z) = \frac{-1}{\pi}\int_{-\infty}^\infty \frac{(t-x)f(t)}{(t-x)^2 + y^2}dt \, .$$

If $f \in L^1(0,1)$ and $f(x)/x \in L^1(1,\infty)$, then

$$\lim_{y\to 0} \tilde{U}(x,y) = -(\mathcal{H}f)(x) \, .$$

14.6 Inversion Formula

Let g and $\tilde{f}$ be defined as in Section 14.5.1. Since

$$g_e(t) = g_o(t)\text{sgn}(t) \, , \quad g_o(t) = g_e(t)\text{sgn}(t) \, ,$$

and the inverse Fourier transform of sgn t is given by

$$\mathcal{F}^{-1}(\operatorname{sgn}(t))(x) = \sqrt{\frac{2}{\pi}}\,\frac{1}{ix} ,$$

it follows from the convolution theorem for the Fourier transform (Theorem 12.14) that

$$\begin{aligned}\tilde{f}(x) &= \frac{-i}{\sqrt{2\pi}} \int_{-\infty}^{\infty} g_e(t)\ \operatorname{sgn}(t) e^{ixt} dt = -i\sqrt{\frac{2}{\pi}} \left(f(t) * \frac{1}{it} \right)(x) \\ &= \frac{1}{\pi} \int_{-\infty}^{\infty} \frac{f(t)}{t-x} dt ,\end{aligned}$$

and

$$\begin{aligned}f(x) &= \frac{1}{\sqrt{2\pi}} \int_{-\infty}^{\infty} g_0(t)\ \operatorname{sgn}(t) e^{ixt} dt = \sqrt{\frac{2}{\pi}} \left(i\tilde{f}(t) * \frac{1}{it} \right)(x) \\ &= \frac{1}{\pi} \int_{-\infty}^{\infty} \frac{\tilde{f}(t)}{x-t} dt = \frac{-1}{\pi} \int_{-\infty}^{\infty} \frac{\tilde{f}(t)}{t-x} dt .\end{aligned}$$

This relation forms a basis for the following theorem.

THEOREM 14.4 [225, p. 132]
Let $f \in L^p(\mathcal{R})$, where $p \geq 1$, then

$$\tilde{f}(x) = (\mathcal{H}f)(x) = \frac{1}{\pi} \int_{-\infty}^{\infty} \frac{f(t)}{t-x} dt , \tag{14.10}$$

is defined for almost all x and the inversion formula

$$f(x) = -\frac{1}{\pi} \int_{-\infty}^{\infty} \frac{(\mathcal{H}f)(t)}{t-x} dt \tag{14.11}$$

also holds almost everywhere. Moreover,

$$\mathcal{H}^2 f = \mathcal{H}(\mathcal{H}f) = -f .$$

Equations (14.10) and (14.11) form the Hilbert transform pair which was first observed by Hilbert and to which we have alluded in the introduction of this chapter. Let us recall that if f is even, then $\mathcal{H}f$ is odd. For,

$$(\mathcal{H}f)(-x) = \frac{1}{\pi} \int_{-\infty}^{\infty} \frac{f(t)}{t+x} dt = -(\mathcal{H}f)(x) .$$

An alternate form of the Hilbert transform pair can be obtained if f and $\mathcal{H}f$ are given by (14.5), or if f is even. In either case we have

$$(\mathcal{H}f)(x) = \frac{2x}{\pi} \int_0^\infty \frac{f(t)}{t^2 - x^2} dt .$$

Similarly, since

$$f(x) = \frac{-1}{\pi} \int_{-\infty}^\infty \frac{(\mathcal{H}f)(t)}{t - x} dt = f(-x) ,$$

we have

$$f(x) = \frac{-2}{\pi} \int_0^\infty \frac{t(\mathcal{H}f)(t)}{t^2 - x^2} dt .$$

Another version of the inversion formula is given in the next theorem.

THEOREM 14.5 [225, p. 122]
Let $f \in L^2(\mathcal{R})$. Then the formula

$$(\mathcal{H}f)(x) = -\frac{1}{\pi} \frac{d}{dx} \int_{-\infty}^\infty f(t) \ln \left| 1 - \frac{x}{t} \right| dt$$

defines almost everywhere a function $(\mathcal{H}f)(x)$ belonging to $L^2(\mathcal{R})$. Moreover,

$$f(x) = \frac{1}{\pi} \frac{d}{dx} \int_{-\infty}^\infty (\mathcal{H}f)(t) \ln \left| 1 - \frac{x}{t} \right| dt$$

also holds almost everywhere, with

$$\int_{-\infty}^\infty |f(x)|^2 \, dx = \int_{-\infty}^\infty |(\mathcal{H}f)(x)|^2 \, dx .$$

If the differentiation operator can be taken under the integral signs, we will obtain the Hilbert transform pair (14.10) and (14.11).

14.7 Generalized Functions

Since the kernel of the Hilbert transformation is not even a continuous function, extending the Hilbert transform to generalized functions by using the embedding method is futile, and hence one is led to use the adjoint method described in Section 4.5.1. A number of different structures for the adjoint spaces have been introduced. The one presented here is due to Gelfand and Shilov [70, Vol. 2, p. 151].

Let X be the set of all functions $\psi(s)$ defined on $\mathcal{R}$ such that $\psi(s)$ and its derivative $\psi'(s)$ satisfy the following conditions:

1) $\psi(s)$ and $\psi'(s)$ are continuous everywhere on $\mathcal{R}$ except for a possible discontinuity of the first kind at $s = 0$,

2) the functions $s^k\psi(s)$ and $s^k\psi'(s)$ are absolutely continuous and integrable on $\mathcal{R}$ for $k = 0, 1, 2, \dots$.

Provided with the countable family of norms

$$\|\psi\|_k = \int_{-\infty}^{\infty} |s^k\psi(s)|ds + \int_{-\infty}^{\infty} |s^k\psi'(s)|ds + \max_{s\in\mathcal{R}} |\psi(s)| + \max_{s\in\mathcal{R}} |\psi'(s)| \, ,$$

the space $(X, \| - \|_k)$ is a Fréchet space.

It is easy to see that the operator $A : X \to X$ defined by

$$A(\psi(s)) = \operatorname{sgn}(s)\psi(s) \, , \tag{14.12}$$

is bounded on X. Let $\tilde{X}$ be the subspace of X consisting of all functions $\psi(s)$ such that ψ and ψ' are continuous everywhere. Because $s^k\psi(s)$ and $s^k\psi'(s)$ are absolutely integrable and continuous for every $k = 0, 1, 2, \dots$, the Fourier transform, $\phi(x) = \hat{\psi}(x)$, of any $\psi \in \tilde{X}$ is infinitely differentiable and of order $O(1/|x|)$ as $|x| \to \infty$. We claim that the same is true for any $\psi \in X$. To this end, let $\psi \in X$ and set

$$\psi_0(s) = \begin{cases} \psi(s) - (c + c_1 s)e^{-s}, & \text{if} \quad s > 0, \\ \psi(s), & \text{if} \quad s < 0 \, . \end{cases}$$

By choosing c and c_1 suitably, for example,

$$c = \psi(0^+) - \psi(0^-) \, ,$$

and

$$c_1 = \psi'(0^+) - \psi'(0^-) + \psi(0^+) - \psi(0^-) \, ,$$

we can remove the discontinuities of ψ and ψ' at $s = 0$ and hence obtain a function ψ_0 belonging to $\tilde{X}$. If we denote by ψ_1 the function

$$\psi_1(s) = \begin{cases} (c + c_1 s)e^{-s} \, , & \text{if} \quad s > 0 \\ 0, & \text{if} \quad s < 0 \, , \end{cases}$$

we then have $\psi(s) = \psi_0(s) + \psi_1(s)$. But since the Fourier transform of ψ_1

$$\begin{aligned} \hat{\psi}_1(x) &= \frac{1}{\sqrt{2\pi}} \left\{ c \int_0^{\infty} e^{-s} e^{isx} ds + c_1 \int_0^{\infty} s e^{-s} e^{isx} ds \right\} \\ &= \frac{1}{\sqrt{2\pi}} \left\{ \frac{c}{-1 + ix} - \frac{c_1}{(-1 + ix)^2} \right\} \end{aligned}$$

is also infinitely differentiable on the real line and of order $O(1/|x|)$ as $|x| \to \infty$, it follows that the same is true for the Fourier transform of ψ. Let

$$Y = \hat{X} = \{\phi : \phi(x) = \hat{\psi}(x) \quad \text{where} \quad \psi \in X\} .$$

A sequence $\{\phi_n\}_{n=1}^{\infty}$ in Y is said to converge to zero if and only if $\{\psi_n\}_{n=1}^{\infty}$ converges to zero in X, where $\phi_n = \hat{\psi}_n$. Members of the spaces X and Y are easily seen to be Hilbert transformable functions. If we denote the Fourier and Hilbert transformations by $\mathcal{F}$ and $\mathcal{H}$, then in view of (14.6) and (14.12), we have for any $\phi \in Y$

$$\mathcal{F}(\mathcal{H}(\phi))(s) = -i \operatorname{sgn}(s)\hat{\phi}(s) = -i \operatorname{sgn}(s)\psi(-s) \in X .$$

Thus, $\mathcal{H}(\phi) \in Y$. This implies that the Hilbert transformation is a bounded transformation on Y. Moreover, $\mathcal{H}(\mathcal{H}(\phi)) = -\phi$ for any $\phi \in Y$. The Hilbert transformation of a class of generalized functions can now be defined.

DEFINITION 14.2 *The Hilbert transform $\mathcal{H}(f)$ of $f \in Y^*$ is defined as an element of Y^* given by*

$$\langle \mathcal{H}(f), \phi \rangle = \langle f, \mathcal{H}(\phi) \rangle ,$$

for all $\phi \in Y$.

If f is a locally integrable function with $f(x) = O(|x|^{-\epsilon})$ as $|x| \to \infty$, for some $\epsilon > 0$, then it defines a regular member of Y^* via

$$\langle f, \phi \rangle = \int_{-\infty}^{\infty} f(x)\phi(x)dx .$$

Evidently, $\mathcal{H}(\mathcal{H}(f)) = -f$ since

$$\begin{aligned} \langle \mathcal{H}(\mathcal{H}(f)), \phi \rangle &= \langle \mathcal{H}(f), \mathcal{H}(\phi) \rangle = \langle f, \mathcal{H}(\mathcal{H}(\phi)) \rangle \\ &= -\langle f, \phi \rangle . \end{aligned}$$

14.8 Applications

14.8.1 Signal Analysis

If $f(t)$ represents a signal in the time domain, its Fourier transform $F(w)$ represents its amplitude spectrum; see Appendix A. Since the Hilbert transform of $f(t)$ is obtainable from $f(t)$ by convolution with $(-1)/(\pi t)$, applying the Hilbert transformation to f is equivalent to passing it through a filter. If we rewrite Equation (14.6) in the form

$$\mathcal{F}(\mathcal{H}f)(w) = \begin{cases} e^{-i\pi/2}\mathcal{F}(f)(w) & \text{if } w > 0 \\ e^{i\pi/2}\mathcal{F}(f)(w) & \text{if } w < 0\,, \end{cases}$$

it becomes evident that the Hilbert transformation is equivalent to a filtering process in which the amplitudes of the spectral components are left unchanged, but their phases are altered by $\pi/2$; the phase is decreased by $\pi/2$ if w is positive and increased by $\pi/2$ if w is negative. Thus, the application of two Hilbert transformations in succession decreases the phase by π for positive w and increases it by π for negative w which results in reproducing the negative of the original signal.

Had the kernel of the Hilbert transformation been chosen as $[i\pi(t-x)]^{-1}$ instead of $[\pi(t-x)]^{-1}$, the transformation and its inversion formula would have been identical and consequently we would have had $\mathcal{H}(\mathcal{H}f) = f$. This also would have implied that in the equivalent filtration process the phase is decreased by π if w is positive and increased by π if w is negative, and hence two successive applications of the Hilbert transformation would have reproduced the original signal. The shortfall of this choice of the kernel is that the Hilbert transform of a real-valued function would have been complex-valued.

In some signal analysis applications, one often needs to determine a real-valued signal $f(t)$ when only the real or imaginary part of its frequency spectrum (its Fourier transform) is known. We shall now show that this can be done if $f(t)$ is causal, i.e., $f(t) = 0$ if $t < 0$. For, in this case we have

$$F(w) = \frac{1}{\sqrt{2\pi}} \int_0^\infty f(t)e^{itw}\,dt = R(w) + iX(w)\,,$$

where

$$R(w) = \frac{1}{\sqrt{2\pi}} \int_0^\infty f(t)\cos tw\,dt\,, \quad X(w) = \frac{1}{\sqrt{2\pi}} \int_0^\infty f(t)\sin tw\,dt\,.$$

In view of the fact that R is even and X is odd, we have

$$f(t) = \sqrt{\frac{2}{\pi}} \int_0^\infty (R(w)\cos tw + X(w)\sin tw)\,dw\,.$$

If f is even, then

$$f(t) = \sqrt{\frac{2}{\pi}} \int_0^\infty R(w)\cos tw\,dw\,,$$

while if f is odd

$$f(t) = \sqrt{\frac{2}{\pi}} \int_0^\infty X(w)\sin tw\,dw\,,$$

and in the general case $f(t) = f_e(t) + f_o(t)$. But since f is causal

$$f(t) = 2f_e(t) = 2f_o(t) , \quad t > 0 ,$$

and hence

$$f(t) = 2\sqrt{\frac{2}{\pi}} \int_0^\infty R(w) \cos tw \, dw = 2\sqrt{\frac{2}{\pi}} \int_0^\infty X(w) \sin tw \, dw .$$

14.8.2 Analytic Signals

Let $f(t)$ be a real-valued signal. In signal analysis and optics, when the independent variable t is time, a complex-valued signal $F(t)$ of the form

$$F(t) = f(t) + i(\mathcal{H}f)(t)$$

is called an *analytic signal* or the *analytic part of the signal* f. The imaginary part of F is called the quadrature function of $f(t)$ and the electric system whose input is $f(t)$ and output is $(\mathcal{H}f)(t)$ is known as the *quadrature filter.* The *quadrature function* of $\cos(at)$ is $-\sin(at)$ and that of $\sin(at)$ is $\cos(at)$. An analytic signal contains no negative-frequency components; in fact it is obtainable by suppressing the negative frequencies of f. For, if we denote the Fourier transform of f by $\hat{f}$, then

$$\hat{f}(w) = R(w) + iX(w) = \frac{1}{\sqrt{2\pi}} \int_{-\infty}^\infty f(t) e^{itw} dt ,$$

where

$$R(w) = \frac{1}{\sqrt{2\pi}} \int_{-\infty}^\infty f(t) \cos tw \, dt , \quad X(w) = \frac{1}{\sqrt{2\pi}} \int_{-\infty}^\infty f(t) \sin tw \, dt ,$$

and hence

$$f(t) = \sqrt{\frac{2}{\pi}} \int_0^\infty (R(w) \cos tw + X(w) \sin tw) \, dt .$$

As before (Section 14.1) it is easy to see that the Hilbert transform of f can be written as

$$(\mathcal{H}f)(t) = \sqrt{\frac{2}{\pi}} \int_0^\infty (X(w) \cos wt - R(w) \sin wt) dw ;$$

therefore the analytic signal F can be written as

$$F(t) = f(t) + i(\mathcal{H}f)(t) = \sqrt{\frac{2}{\pi}} \int_0^\infty \hat{f}(w) e^{-itw} dw .$$

Examples of how analytic signals can be used in signal analysis applications and modulation can be found in [23, p. 269] and [184, pp. 251-255].

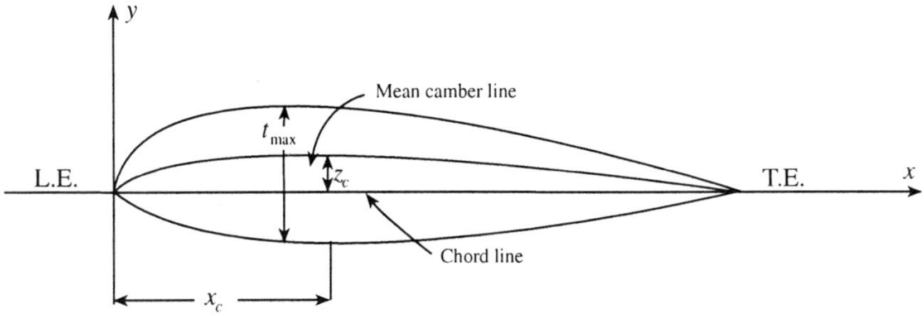

FIGURE 14.1
Cross section of an airfoil

14.8.3 Airfoil Design

Another important application of the Hilbert transform is in the area of airfoil design [133, 178, 232, 233]. An *airfoil* is a body designed to produce a desired lifting force when it moves in the surrounding air. A prototype of an airfoil is an airplane wing. Consider an airfoil extended along the z-axis with its cross section in the $x - y$ plane; see Figure 14.1. Let the *main chord of the airfoil*, which is the line joining the leading edge (L.E.) and the trailing edge (T.E.) of the foil, be along the x-axis. The *mean camber line* is a curve that lies halfway between the upper and lower surfaces of the airfoil and intersects the main chord at the leading and trailing edges. Let the deviation of the camber line from the main chord be denoted by z and its maximum value be denoted by z_c. The maximum camber z_c measures the *convexity of the airfoil.*

Any one family of airfoils has the same general shape for the thickness envelope and the mean camber line, but the individual members of the family are partly distinguished by the amount of their maximum cambers z_c and their locations relative to the airfoil, which is usually measured in terms of the distance x_c from the camber to the leading edge. The individual members of the family are also partly distinguished by the value of the ratio $(t_{\max}/c)$ of the maximum thickness $t_{\max}$ to the length of the chord c.

An airfoil is considered to be thin if $t_{\max}/c$ and z_c are small. The overall lifting characteristics of a thin airfoil can be approximated by replacing the airfoil by its mean camber line and the kinematic problem for the airfoil is solved by finding an air flow pattern that has one streamline coincident with the mean camber line. This is where the Hilbert transform comes into play.

In order for the resultant of the uniform stream and the velocity induced by the thin airfoil to be parallel to the surface, the normal components of the uniform stream $(V_\infty)_n$ and induced velocity $(V_i)_n$ must add to zero. If we denote the x-coordinates of the leading and trailing edges by x_L and x_T, respectively, then the normal component of the induced velocity at a point $P = (x_0, y_0)$ on the surface of the airfoil is given by

$$(V_i)_n = \frac{1}{2\pi}\int_{x_L}^{x_T} \frac{f(x)dx}{x - x_0} ,$$

for some function f which depends on the curl of the velocity vector. Clearly, $(V_i)_n$ is the Hilbert transform of a function with support in the interval $[x_L, x_T]$. The normal component of the uniform stream is easily shown to be

$$(V_\infty)_n = V_\infty \sin\left(\alpha - \tan^{-1}\left(\frac{dz}{dx}\right)\right) ,$$

where α is the angle that the uniform stream makes with the x-axis and $\frac{dz}{dx}$ is the slope of the tangent line to the mean camber line at P.

Therefore, for the sum of the normal components to be zero, we must have

$$\frac{1}{2\pi}\int_{x_L}^{x_T} \frac{f(x)dx}{x_0 - x} = V_\infty \sin\left(\alpha - \tan^{-1}\left(\frac{dz}{dx}\right)\right)\Big|_{x=x_0} \tag{14.13}$$

together with the boundary condition $f(x_T) = 0$, which is known as *the Kutta condition.*

The central problem of thin airfoil theory is to find a function f that satisfies the above equation. If α and dz/dx are small, Equation (14.13) becomes

$$\frac{1}{2\pi}\int_{x_L}^{x_T} \frac{f(x)dx}{x_0 - x} = V_\infty \left(\alpha - \frac{dz}{dx}\right)\Big|_{x=x_0} , \tag{14.14}$$

which, for a symmetrical airfoil, reduces to

$$\frac{1}{2\pi}\int_{x_L}^{x_T} \frac{f(x)dx}{x_0 - x} = \alpha V_\infty , \tag{14.15}$$

since z is constant.

The last integral equation can be solved explicitly. To this end, let us assume without loss of generality that $x_L = 0$ and set $c = x_T - x_L =$ the length of the cord. Then by letting

$$x = c(1 - \cos\theta)/2 \quad \text{and} \quad x_0 = c(1 - \cos\theta_0)/2 ,$$

we transform Equation (14.15) to

$$\frac{1}{2\pi}\int_0^\pi \frac{f(\theta)\sin\theta d\theta}{\cos\theta-\cos\theta_0} = \alpha V_\infty ,$$

with $f(\pi)=0$.

With the help of the relation

$$\int_0^\pi \frac{\cos n\theta}{\cos\theta-\cos\theta_0} d\theta = \pi \frac{\sin n\theta_0}{\sin\theta_0} , \tag{14.16}$$

it is easy to show that

$$f(\theta) = 2\alpha V_\infty \frac{1+\cos\theta}{\sin\theta} . \tag{14.17}$$

The lift per unit span is given by

$$\begin{aligned} L &= \int_0^c \rho V_\infty f(\theta) dx = \rho V_\infty \int_0^\pi 2\alpha V_\infty \left(\frac{1+\cos\theta}{\sin\theta}\right)\left(\frac{c}{2}\right)\sin\theta d\theta \\ &= \alpha c \rho V_\infty^2 \pi , \end{aligned}$$

where ρ is the air medium density.

The general solution of (14.14) may be represented as a sum of two parts, the first of which has the form of $f(\theta)$ for the symmetric airfoil given by Equation (14.17), and the second is represented as a Fourier sine series. Thus,

$$f(\theta) = 2V_\infty \left[A_0\left(\frac{1+\cos\theta}{\sin\theta}\right) + \sum_{n=1}^\infty A_n \sin n\theta\right] ,$$

which upon its substitution in (14.14) yields

$$\begin{aligned} \frac{1}{\pi}\left\{A_0 \int_0^\pi \frac{1+\cos\theta}{\cos\theta-\cos\theta_0} d\theta + \sum_{n=1}^\infty A_n \int_0^\pi \frac{\sin n\theta \sin\theta}{\cos\theta-\cos\theta_0} d\theta\right\} \\ = \left(\alpha - \frac{dz}{dx}\right)|_{x=x_0} , \end{aligned}$$

where $A_n, n = 0, 1, 2, \ldots$ are constants to be determined.

By using the identity

$$\sin n\theta \sin\theta = \frac{1}{2}\left[\cos(n-1)\theta - \cos(n+1)\theta\right] ,$$

and the relation (14.16), we finally obtain

$$\frac{dz}{dx}|_{x=x_0} = (\alpha - A_0) + \sum_{n=1}^\infty A_n \cos n\theta_0 .$$

Since for a given mean camber line $\frac{dz}{dx}$ is known, we can calculate the coefficients, A_n, as follows:

$$A_0 = \alpha - \frac{1}{\pi} \int_0^{\pi} \left(\frac{dz}{dx} \right) d\theta \ ,$$

and

$$A_n = \frac{2}{\pi} \int_0^{\pi} \left(\frac{dz}{dx} \right) \cos n\theta d\theta \ .$$

15

The Boas Transform

The Boas transform is closely related to the Fourier and Hilbert transforms. It was first presented in 1936 at a meeting of the American Mathematical Society by R.P. Boas, Jr., who later introduced it in a paper [18] that appeared in the same year. Though not very well known and only a handful of papers have been written about it, the Boas transform, as it turns out, has been indirectly used in some electrical engineering problems. In the application section of this chapter, we shall discuss the relevance of the Boas transform to the theory of filters in electrical engineering.

15.1 Motivation

The Paley-Wiener space B_σ^2 introduced in Section 1.3 consists of square integrable functions whose Fourier transforms vanish outside the interval $[-\sigma, \sigma]$. Paley and Wiener studied these functions in great details. The Boas transform was born when Boas attempted to study the properties of square integrable functions whose Fourier transforms vanish on $[-\sigma, \sigma]$. Let us denote the class of all such functions by $B^2_{\mathcal{R}-[-\sigma,\sigma]}$.

Although a function f whose Fourier transform vanishes on $[-\sigma, \sigma]$ can be written as a difference of two functions, one of which has Fourier transform vanishing outside $[-\sigma, \sigma]$, Boas has found out that some properties of f did not seem to follow trivially from those of functions in the Paley-Wiener space. As a result, he decided to study the class $B^2_{\mathcal{R}-[-\sigma,\sigma]}$ directly without much reference to the Paley-Wiener class B_σ^2. His approach has led to very interesting discoveries.

15.2 Definitions

DEFINITION 15.1 *The Boas transform $F(x)$ of a function $f(t) \in L^2(\mathcal{R})$ is defined by*

$$(Bf)(x) = F(x) = \frac{1}{\pi}\int_0^\infty \frac{f(x+t)-f(x-t)}{t^2}\sin t\; dt, \tag{15.1}$$

for any x for which the integral exists.

If

$$f(x) = \int_{-\sigma}^{\sigma} e^{ixt} d\alpha(t), \tag{15.2}$$

where $\alpha(t)$ is a complex-valued function of bounded variation with

$$\alpha(t) = \frac{1}{2}\left[\alpha(t+0)+\alpha(t-0)\right], \; -\sigma < t < \sigma,$$

then

$$(Bf)(x) = \frac{2i}{\pi}\int_0^\infty \frac{\sin t}{t^2}dt \int_{-\sigma}^{\sigma} e^{ixu}\sin tu\; d\alpha(u) = \frac{2i}{\pi}\int_{-\sigma}^{\sigma} e^{ixu} d\alpha(u)\int_0^\infty \frac{\sin t \sin tu}{t^2}dt.$$

In addition, if α is constant on $-1 < u < 0$ and $0 < u < 1$, then

$$(Bf)(x) = i\int_{-\sigma}^{\sigma} e^{ixu}\mathrm{sgn}(u)\; d\alpha(u), \; (\sigma > 1)$$

where $\mathrm{sgn}(u) = u/\left|u\right|, u \neq 0$.

Formula (15.1) can be put in a different form

$$\begin{aligned}(Bf)(x) &= \frac{1}{\pi}\left\{\int_0^\infty \frac{f(x+t)}{t^2}\sin t dt - \int_0^\infty \frac{f(x-t)}{t^2}\sin t dt\right\} \\ &= \frac{1}{\pi}\int_{-\infty}^{\infty} f(x+t)\frac{\sin t}{t^2}dt. \end{aligned} \tag{15.3}$$

The above integrals are taken in the Cauchy principle value sense. A more general definition of the Boas transform is given by

DEFINITION 15.2 *The generalized Boas transform of a function $f(t)$ is defined by*

$$(Bf)(x) = F(x) = \frac{1}{\pi}\int_0^\infty \frac{f(x+t)-f(x-t)}{t} H(t)dt, \tag{15.4}$$

for any x for which the integral exists, where $H(t)$ is a function satisfying the following conditions:

1. *there exists a function $h(s)$ of bounded variation on $-1 \le s \le 1$, such that*

$$H(t) = \int_{-1}^{1} \cos st \; dh(s),$$

2. *$H(t)$ is either of bounded variation on (T, ∞) for some T, or $h(s)$ is the integral of a function $k(s)$ of bounded variation on $(-1, 1)$.*

If $H(t)$ is even, then Bf may be written as

$$(Bf)(x) = \frac{1}{\pi}\int_{-\infty}^{\infty} f(x+t)\frac{H(t)}{t}dt.$$

15.3 Existence

Let

$$f(x) = \int_{-\sigma}^{\sigma} e^{ixt} d\alpha(t), \quad \sigma > 1,$$

and

$$f_1(x) = \int_{-1}^{1} e^{ixt} d\alpha(t)\,, \quad f_2(x) = \left(\int_{-\sigma}^{-1} + \int_{1}^{\sigma}\right) e^{ixt} d\alpha(t),$$

so that $f = f_1 + f_2$, where $\alpha(t)$ is defined as in (15.2). The definition of f_1 and f_2 can be easily modified if $0 < \sigma \le 1$. If $f \in L^2(\mathcal{R})$, then the Boas transform (15.4) exists for all x and

$$(Bf)(x) = H(0)(\mathcal{H}f_2)(x) + i\int_{-1}^{1} e^{ixt}\left[h(t) - h(-t)\right] d\alpha(t), \tag{15.5}$$

where $(\mathcal{H}f_2)$ is the Hilbert transform of f_2. The Hilbert transform of a function g is defined as

$$(\mathcal{H}g)(x) = \frac{1}{\pi}\int_{-\infty}^{\infty} \frac{g(t)}{t-x}\,dt = \frac{1}{\pi}\int_{0}^{\infty} \frac{g(x+t)-g(x-t)}{t}\,dt,$$

and is known to exist for almost all x if $g \in L^2(\mathcal{R})$, see Chapter 14. In particular, if $H(0) = 1$ and $\alpha(t)$ is constant on $-1 < t < 1$, then $f = f_2$ and

$$(Bf)(x) = \mathcal{H}(f)(x),$$

while if $H(0) = 0$,

$$(Bf)(x) = i\int_{-1}^{1} e^{ixt}\,[h(t) - h(-t)]\,d\alpha(t).$$

15.4 The Fundamental Relation for the Boas Transform

THEOREM 15.1
The fundamental relation for the Boas transform is

$$(Bf)(x) = (\mathcal{H}f)(x) - ((\mathcal{H}f) * g)(x), \tag{15.6}$$

where

$$g(x) = \sqrt{\frac{2}{\pi}}\left(\frac{1-\cos x}{x^2}\right), \qquad -\infty < x < \infty. \tag{15.7}$$

Proof: It is easy to show that the Fourier transform of $g(x)$ is

$$G(w) = \begin{cases} 1 - |w|, & \text{if } |w| \le 1 \\ 0, & \text{if } |w| > 1 \end{cases}.$$

Let $f \in L^2(\mathcal{R})$ and set

$$h(x) = (\mathcal{H}f)(x) - (Bf)(x). \tag{15.8}$$

Then

$$h(x) = \frac{1}{\pi}\int_{0}^{\infty} [f(x+t) - f(x-t)]\left(\frac{t-\sin t}{t^2}\right)dt = \frac{1}{\pi}\int_{-\infty}^{\infty} f(x+t)\left(\frac{t-\sin t}{t^2}\right)dt.$$

Since $f(t)$ and $(t - \sin t)/t^2$ are in $L^2(\mathcal{R})$, $h(x)$ is defined for almost all x. In view of the fact that $(\mathcal{H}f)(x)$ exists almost everywhere, it follows that the same must be true for $(Bf)(x)$. Moreover,

$$\begin{aligned} h(x) &= -\frac{1}{\pi}\int_{-\infty}^{\infty} f(t)\left[\frac{(x-t)-\sin(x-t)}{(x-t)^2}\right]dt \\ &= -(f*\phi)(x), \end{aligned}$$

where

$$\phi(x) = \sqrt{\frac{2}{\pi}}\left(\frac{x - \sin x}{x^2}\right),$$

and $*$ stands for the convolution operation associated with the Fourier transform, see Section 12.9. Noting that the Fourier transform $\hat{\phi}$ of ϕ is given by

$$\hat{\phi}(w) = \begin{cases} i(1-w), & 0 \le w \le 1 \\ -i(1+w), & -1 \le w \le 0 \\ 0, & \text{otherwise} \end{cases},$$

we can write

$$\hat{\phi}(w) = i\,\mathrm{sgn}(w)G(w).$$

Thus, by the convolution theorem for the Fourier transform (Theorem 12.14), we obtain

$$\hat{h}(w) = -i\hat{f}(w)\mathrm{sgn}(w)G(w), \qquad (15.9)$$

which in view of the relation (cf. Equation 14.6)

$$(\mathcal{H}f\hat{)}(w) = -i\,\mathrm{sgn}(w)\hat{f}(w), \qquad (15.10)$$

gives us

$$\hat{h}(w) = G(w)(\mathcal{H}f\hat{)}(w);$$

hence

$$h(x) = ((\mathcal{H}f) * g)(x).$$

Finally, by combining this and (15.8) we arrive at the fundamental relation for the Boas transform

$$(Bf)(x) = (\mathcal{H}f)(x) - ((\mathcal{H}f) * g)(x).$$

∎

15.5 Elementary Properties

1. The Boas transformation is a linear transformation that preserves translation. In fact, if $g(x) = [Bf(t)](x)$, then

$$T_a g(x) = g(x-a) = B[f(t-a)](x) = B[T_a f](x).$$

2. The transformation $U : L^2(\mathcal{R}) \to L^2(\mathcal{R})$, defined by $U(f) = Bf$, is a non-surjective bounded linear transformation on $L^2(\mathcal{R})$. In fact, from (15.6), Theorems 12.14 and 14.1, it follows that

$$\begin{aligned}\|Bf\|_2 &\le \|\mathcal{H}f\|_2 + \|(\mathcal{H}f * g)\|_2 \le \|\mathcal{H}f\|_2 + \|\mathcal{H}f\|_2 \|g\|_1 \\ &\le \|f\|_2 + \|f\|_2 \|g\|_1 ,\end{aligned}$$

hence

$$\|Bf\|_2 \le K \|f\|_2 ,$$

where $K = 1 + \|g\|_1$.

That U is not surjective will be seen below.

3. If $f \in L^2$, then

$$\int_{-\infty}^{\infty} (Bf)(x)dx = 0,$$

where the integral is taken as the Cauchy principle value. For the proof see [97].

15.6 Relationships with Other Transforms

1. **The Fourier Transform:** The Fourier transform of the Boas transform of f, $(B\hat{f})(w)$, can be derived with the aid of (15.6) and (15.10)

$$(B\hat{f})(w) = (\mathcal{H}\hat{f})(w)\,(1 - G(w)) = -i\ \text{sgn}(w)\ \hat{f}(w)\,(1 - G(w)) . \tag{15.11}$$

2. **The Hilbert Transform:** Let $g(x)$ be as defined by (15.7) and set $g_1(x) = g(x),\ \ g_n(x) = (g_{n-1} * g)(x), n = 2, 3, \ldots$. Then if $f \in L^2(\mathcal{R})$,

$$(\mathcal{H}f)(x) = (Bf)(x) + \sum_{n=1}^{\infty}((Bf) * g_n)(x),$$

where the series converges in the L^2-norm. This can be established from the relation (15.11) by writing

$$(H\hat{f})(w) = \frac{(B\hat{f})(w)}{1 - G(w)} = (B\hat{f})(w) \sum_{n=0}^{\infty} G^n(w),$$

and then taking the inverse Fourier transform; for the details of the proof see [76].

If $\hat{f} = 0$ almost everywhere on $(-1, 1)$, then from (15.9) $\hat{h} = 0$ and by (15.8) we have

$$B[f] = \mathcal{H}[f].$$

By replacing f by $\mathcal{H}f$, we obtain $B(\mathcal{H}f) = -f$, since $\mathcal{H}(\mathcal{H}f) = -f$; see Theorem 14.4.

3. **Convolution Transform:** If $f \in L^2$, then

$$B(Bf) = -f + 2f * g - f * g * g. \tag{15.12}$$

For, if we replace f by Bf in (15.11), we obtain

$$[B\widehat{(Bf)}](w) = -i \operatorname{sgn}(w)\,(1 - G(w))(B\hat{f})(w)$$
$$= -(1 - G(w))^2 \hat{f}(w) = -\hat{f}(w)\left(1 - 2G(w) + G^2(w)\right).$$

Upon taking the inverse Fourier transform of this equation, we obtain (15.12).

15.7 Parseval's Relation

Let f and g be in $L^2(\mathcal{R})$. Then

$$\int_{-\infty}^{\infty} f(x)(Bg)(x)dx = -\int_{-\infty}^{\infty} (Bf)(x)g(x)dx.$$

In particular, if $f = g$, we have

$$\int_{-\infty}^{\infty} f(x)(Bf)(x)dx = 0,$$

and if $g = Bf$, we have

$$\int_{-\infty}^{\infty} f(x)\,(B(Bf))\,(x)dx = -\int_{-\infty}^{\infty} \left[(Bf)\,(x)\right]^2 dx.$$

15.8 Inversion Formulas

1. The first inversion formula follows from (15.11) and is rather symbolic:

$$f(x) = \mathcal{F}^{-1}\left\{\frac{i\widehat{(Bf)}(w)\mathrm{sgn}w}{(1-G(w))}\right\}, \tag{15.13}$$

where $\mathcal{F}^{-1}$ denotes the inverse Fourier transformation.

2. The second formula is more explicit and requires more work to establish [97]: If $f \in L^2(\mathcal{R})$, then it can be recovered from its Boas transform by means of the formula

$$f(x) = \frac{1}{\pi}\int_0^\infty \{(Bf)(x-t) - (Bf)(x+t)\}\,\psi(t)dt,$$

or

$$f(x) = -\frac{1}{\pi}\int_{-\infty}^\infty (Bf)(x+t)\psi(t)dt,$$

where

$$\psi(t) = \frac{\cos t}{t} + \int_0^1 \frac{\sin tu}{u}du.$$

15.9 Characterization of the Boas Transform

For a function $g \in L^2(\mathcal{R})$ to be the Boas transform of a function $f \in L^2(\mathcal{R})$ i.e., $g = Bf$, it is necessary and sufficient that $\hat{g}(w)/w \in L^2(-1,1)$.
Necessity follows from (15.11). For, by (15.11) we have

$$\hat{f}(w) = \frac{i\hat{g}(w)}{(1-G(w))\mathrm{sgn}(w)} = \frac{i\hat{g}(w)}{w} \text{ if } |w| \le 1,$$

but since $\hat{f} \in L^2(\mathcal{R})$, $\hat{g}(w)/w$ must be in $L^2(-1,1)$. Conversely, if $g \in L^2(\mathcal{R})$ is such that $\hat{g}(w)/w \in L^2(-1,1)$, then the function

$$F(w) = \frac{i\hat{g}(w)\mathrm{sgn}(w)}{(1-G(w))}.$$

is in $L^2(\mathcal{R})$, and hence so is its inverse Fourier transform $f(x)$. Formula (15.13) and the uniqueness theorem for the Fourier transform imply that $g = Bf$.

From this characterization, it follows that the Boas transformation is not a surjective (onto) transformation on $L^2(\mathcal{R})$ as we pointed out in Section 15.5 item 2.

15.10 Boas' Theorems

We have pointed out in Section 15.1 that the Boas transform was a by-product of Boas' efforts to characterize square integrable functions whose Fourier transforms vanish on intervals of the form $[-\sigma, \sigma], (\sigma > 0)$. Below is Boas' major finding.

THEOREM 15.2
If $f \in L^2(\mathcal{R})$ and $\hat{f}$ is its Fourier transform, then a necessary and sufficient condition that $\hat{f}$ vanishes almost everywhere on $(-1, 1)$ is that

$$(B(Bf))(x) = -f(x).$$

Note that if g is given by (15.7), then $G = 0$ outside $(-1, 1)$ and $\neq 0$ in $(-1, 1)$. Therefore, in view of the relation, $\widehat{(f * g)}(w) = \hat{f}(w)G(w)$, it follows that $(f * g)(x) = 0$ almost everywhere if and only if $\hat{f} = 0$ almost everywhere in $(-1, 1)$. Thus, from (15.12) we have $B(Bf) = -f$. Conversely, if $B(Bf) = -f$, we have by (15.12) once more that

$$2f * g - f * g * g = 0,$$

or

$$(2 - G(w))G(w)\hat{f}(w) = 0.$$

But neither $(2 - G(w))$ nor $G(w)$ vanishes on $(-1, 1)$; thus $\hat{f}(w) = 0$ thereon.

THEOREM 15.3
If $\hat{f}$ vanishes a.e. on $(-1, 1)$, then the Boas and Hilbert transforms of f are equal, i.e., $Bf = \mathcal{H}f$.

It should be noted that Boas actually proved these theorems for the generalized Boas transform (15.4).

15.11 Applications

One of the applications of the Boas transform is in the area of electrical engineering. Recall that the system transfer function of a high pass filter is given by

$$H(w) = \begin{cases} Ae^{itow}, & \text{for } |w| \geq w_2 > 0 \\ 0, & \text{otherwise;} \end{cases}$$

see Appendix A. Without loss of generality we may take $w_2 = 1$.

Therefore, passing a finite energy signal f through a high pass filter produces an output signal g satisfying

$$\hat{g}(w) = H(w)\hat{f}(w).$$

But since H vanishes on $(-1, 1)$, then so does $\hat{g}$.

Boas' first theorem gives a characterization for the output of a high pass filter: a signal g is the output of a high pass filter if and only if $B(Bg) = -g$. The second theorem, on the other hand, tells us that if g is an output of high pass filter, then $Bg = Hg$.

16

The Mittag–Leffler Transform

The Mittag–Leffler transform is an integral transform whose kernel is related to the Mittag–Leffler function

$$E_\alpha(z) = \sum_{k=0}^{\infty} \frac{z^k}{\Gamma(\alpha k + 1)}, \qquad \alpha > 0,$$

which was introduced by Mittag–Leffler in 1903 [169] and was further investigated by Mittag–Leffler and Pollard in [168] and [192]. It furnishes an example of an entire function of order $\rho = 1/\alpha$. In fact, given any arbitrary but finite numbers $0 < \rho, \sigma < \infty$, we can construct an entire function of order ρ and type σ, namely,

$$f(z) = \sum_{k=0}^{\infty} \frac{(\sigma z)^k}{\Gamma\left(k/\rho + 1\right)}.$$

In a sense, this function is the simplest entire function having the prescribed order and type. Another function having similar properties to those of the Mittag–Leffler function is the function

$$E_\alpha(z;\beta) = \sum_{k=0}^{\infty} \frac{z^k}{\Gamma(\alpha k + \beta)}, \quad \alpha > 0, -\infty < \beta < \infty$$

which reduces to $E_\alpha(z)$ when $\beta = 1$. *The Mittag–Leffler transformation*, which has been developed primarily by M. Djrbashian and his Soviet colleagues [53], is an integral transformation whose kernel is $E_\alpha(z;\beta)$ for some special values of α, β and z.

Recall from Section 2.2 that

$$E_2(z;1) = \cosh\sqrt{z}, \quad E_2(z;2) = \frac{\sinh\sqrt{z}}{\sqrt{z}},$$

$$E_1(z;1) = e^z, \quad E_1(z;2) = \frac{e^z - 1}{z},$$

$$\int_0^z E_\alpha(\lambda t^\alpha;\beta)t^{\beta-1}dt = z^\beta E_\alpha(\lambda z^\alpha;\beta+1), \quad \beta>0,$$

and

$$\frac{1}{\Gamma(\mu)}\int_0^z (z-t)^{\mu-1}E_\alpha(\lambda t^\alpha;\beta)t^{\beta-1}dt = z^{\beta+\mu-1}E_\alpha(\lambda z^\alpha;\beta+\mu), \quad \beta,\mu>0.$$

For $0<\alpha\leq 2$, $-\infty<\beta<1+\alpha$, $\pi\alpha/2<\delta<\min(\pi,\alpha\pi)$,

$$E_\alpha(z;\beta) = \frac{1}{\alpha}z^{(1-\beta)/\alpha}e^{z^{1/\alpha}} - \sum_{k=1}^{n}\frac{z^{-k}}{\Gamma(\beta-\alpha k)} + 0\left(|z|^{-n-1}\right) \tag{16.1}$$

as $z\to\infty$ in $\Delta_\gamma = \{z : |\arg z|\leq\gamma\}$.

16.1 Definitions

DEFINITION 16.1 *The Mittag–Leffler transforms of a function $f(x)$ are defined as*

$$F^+(y) = \sqrt{\frac{\alpha}{2\pi}}\int_0^\infty E_\alpha\left(e^{i\pi\alpha/2}y^\alpha x^\alpha;\beta\right)x^{\beta-1}f(x)dx,$$

and

$$F^-(y) = \sqrt{\frac{\alpha}{2\pi}}\int_0^\infty E_\alpha\left(e^{-i\pi\alpha/2}y^\alpha x^\alpha;\beta\right)x^{\beta-1}f(x)dx.$$

As special cases, we obtain for $\alpha=2,\beta=1$ a multiple of the Fourier cosine transform

$$F^\pm(y) = \frac{1}{\sqrt{\pi}}\int_0^\infty f(x)\cos(xy)dx,$$

and for $\alpha=1,\beta=1$,

$$F^+(y) = \frac{1}{\sqrt{2\pi}}\int_0^\infty e^{ixy}f(x)dx, \text{ and } F^-(y) = \frac{1}{\sqrt{2\pi}}\int_0^\infty e^{-ixy}f(x)dx.$$

Since the Mittag–Leffler function is a generalization of the exponential function, the Mittag–Leffler transform is usually called a Fourier (or Laplace)-type transform with Mittag–Leffler kernel. We denote by $L^2_\beta(0,\infty)$ the class of all complex-valued functions $g(x)$ for which $g(x)x^{\beta-1} \in L^2(0,\infty)$. A sequence $\{g_n\}_{n=1}^\infty$ converges to g in $L^2_\beta(0,\infty)$ if

$$\lim_{n\to\infty} \int_0^\infty |g(x) - g_n(x)|^2 \, x^{2(\beta-1)} dx = 0,$$

and we shall say that $\lim\limits_{n\to\infty} g_n(x) = g(x)$ in the sense of $L^2_\beta(0,\infty)$. The class $L^2_1(0,\infty)$ is the same as $L^2(0,\infty)$.

16.2 Existence

For $0 < \alpha \le 2, -\infty < \beta < 1+\alpha$, it follows, in view of (16.1), that $F^\pm(y)$ exists if $f \in L^1(0,\infty)$. Moreover, if $f \in L^2(0,\infty)$, then $F^\pm(y)$ exists in the sense of $L^2_\beta(0,\infty)$, that is

$$\lim_{R\to\infty} \sqrt{\frac{\alpha}{2\pi}} \int_0^R E_\alpha \left(e^{\pm i\pi\alpha/2} y^\alpha x^\alpha; \beta\right) x^{\beta-1} f(x) dx$$

exists in $L^2_\beta(0,\infty)$.

16.3 Inversion Formula

For $F \in L^2(\mathcal{R})$ $, 0 < \alpha \le 2$, and $1/2 < \beta < 1/2 + \alpha$, we have

$$\sqrt{\frac{\alpha}{2\pi}} \left\{ e^{-i\pi(1-\beta)/2} \int_0^\infty e^{-ixy} F^+(y) y^{\beta-1} dy + e^{i\pi(1-\beta)/2} \int_0^\infty e^{ixy} F^-(y) y^{\beta-1} dy \right\} = \tilde{f}(x)$$

in the sense of $L^2(-\infty,\infty)$, where

$$\tilde{f}(x) = \begin{cases} f(x) & \text{if } x \in (0,\infty) \\ -\frac{\alpha}{2\pi}\frac{d}{dx} \int_0^\infty \psi_{\alpha,\beta}\left(-\frac{y}{x}\right) f(y) dy & \text{if } x \in (-\infty, 0), \end{cases}$$

and

$$\psi_{\alpha,\beta}(x) = 2 \int_x^\infty \frac{\sin \pi\beta + [\sin \pi (\alpha - \beta)]\, t^\alpha}{1 - 2t^\alpha \cos(\pi/\alpha) + t^{2\alpha}} t^{\beta-2} dt, \quad x \in (0, \infty).$$

The functions $F^{\pm}(y) \in L^2_\beta(0,\infty)$ and $\tilde{f} \in L^2(-\infty,\infty)$ are representable almost everywhere in $(0,\infty)$ and $(-\infty,\infty)$ by

$$F^{\pm}(y) = \sqrt{\frac{\alpha}{2\pi}} y^{1-\beta} \frac{d}{dy} \left\{ y^\beta \int_0^\infty E_\alpha \left(e^{\pm i\pi\alpha/2} x^\alpha y^\alpha; \beta + 1 \right) x^{\beta-1} f(x) dx \right\},$$

and

$$\tilde{f}(x) = \sqrt{\frac{\alpha}{2\pi}} \left\{ e^{-i\pi(1-\beta)/2} \frac{d}{dx} \int_0^\infty \frac{e^{-ixy} - 1}{-iy} F^+(y) y^{\beta-1} dy \right.$$

$$\left. + e^{i\pi(1-\beta)/2} \frac{d}{dx} \int_0^\infty \frac{e^{ixy} - 1}{iy} F^-(y) y^{\beta-1} dy \right\},$$

with

$$\int_{-\infty}^\infty \left|\tilde{f}(x)\right|^2 dx = \alpha \left\{ \int_0^\infty \left|F^+(y)\right|^2 y^{2(\beta-1)} dy + \int_0^\infty \left|F^-(y)\right|^2 y^{2(\beta-1)} dy \right\}.$$

16.4 Transforms of L^2-Functions

It was shown in the preceding sections that for a given $f \in L^2(0,\infty)$ there exist two Mittag–Leffler transforms $F^{\pm} \in L^2_\beta(0,\infty)$. We now show that some kind of a converse to that also holds. That is for a given $g \in L^2_\beta(0,\infty)$ there exist two functions $G^{\pm} \in L^2(0,\infty)$ that are related to g as $F^{\pm}$ related to f.

THEOREM 16.1
Let $0 < \alpha \le 2, 1/2 < \beta < \alpha + \frac{1}{2}$ and $g \in L^2_\beta(0,\infty)$. Then the function

$$G^{\pm}(x) = \sqrt{\frac{\alpha}{2\pi}} \int_0^\infty e^{\pm ixy} g(y) y^{\beta-1} dy \tag{16.2}$$

exists in the sense of $L^2(0,\infty)$ *and*

$$g(y) = \sqrt{\frac{\alpha}{2\pi}} \left\{ e^{-i\pi(1-\beta)/2} \int_0^\infty E_\alpha \left(e^{i\pi\alpha/2} y^\alpha x^\alpha; \beta \right) x^{\beta-1} G^-(x)\, dx \right.$$

$$\left. + e^{i\pi(1-\beta)/2} \int_0^\infty E_\alpha \left(e^{-i\pi\alpha/2} y^\alpha x^\alpha; \beta \right) x^{\beta-1} G^+(x)\, dx \right\} \tag{16.3}$$

in the sense of $L^2_\beta(0,\infty)$. *Moreover, the following Parseval-type equality holds*

$$\int_0^\infty |g(y)|^2 y^{2(\beta-1)} dy = \frac{1}{\alpha} \left\{ \int_0^\infty \left| G^+(x) \right|^2 dx + \int_0^\infty \left| G^-(x) \right|^2 dx \right\}.$$

For pointwise convergence, we have

$$G^\pm(x) = \sqrt{\frac{\alpha}{2\pi}} \frac{d}{dx} \int_0^\infty \frac{e^{\pm ixy} - 1}{\pm iy} g(y) y^{\beta-1} dy,$$

and

$$g(y) = \sqrt{\frac{\alpha}{2\pi}} y^{1-\beta} \left\{ e^{-i\pi(1-\beta)/2} \frac{d}{dy} \left[y^\beta \int_0^\infty E_\alpha \left(e^{i\pi\alpha/2} y^\alpha x^\alpha; \beta+1 \right) x^{\beta-1} G^-(x) dx \right] \right.$$

$$\left. + e^{i\pi(1-\beta)/2} \frac{d}{dy} \left[y^\beta \int_0^\infty E_\alpha \left(e^{-i\pi\alpha/2} y^\alpha x^\alpha; \beta+1 \right) x^{\beta-1} G^+(x) dx \right] \right\}$$

almost everywhere in $(0,\infty)$.

As a special case of (16.3), we obtain for $\alpha = 2$,

$$g(y) = \sqrt{\frac{2}{\pi}} \int_0^\infty E_2(-y^2x^2; \beta) x^{\beta-1} G(x) dx, \tag{16.4}$$

where

$$G(x) = \frac{1}{\sqrt{2}} \left(e^{-i\pi(1-\beta)/2} G^-(x) + e^{i\pi(1-\beta)/2} G^+(x) \right), \tag{16.5}$$

which, with the aid of (16.2), can be represented in the form

$$G(x) = \sqrt{\frac{2}{\pi}} \int_0^\infty \cos\left[\frac{\pi(1-\beta)}{2} + xy \right] g(y) y^{\beta-1} dy.$$

In addition, if $\beta = 1$, we have the familiar relations

$$g(y) = \sqrt{\frac{2}{\pi}} \int_0^\infty \cos(yx) G(x) dx,$$

and

$$G(x) = \sqrt{\frac{2}{\pi}} \int_0^\infty \cos(xy) g(y) dy.$$

Similarly, if $\beta = 2$, we have

$$\tilde{g}(y) = \sqrt{\frac{2}{\pi}} \int_0^\infty \sin(yx) G(x) dx,$$

and

$$G(x) = \sqrt{\frac{2}{\pi}} \int_0^\infty \sin(xy) \tilde{g}(y) dy,$$

where $\tilde{g}(y) = yg(y) \in L^2(0, \infty)$.

16.5 Applications

As is well known, the Laplace and the one-sided Fourier transforms play a vital role in harmonic analysis and the theory of H^p spaces [54]. It was also pointed out in Section 12.11, that the Paley-Wiener theorem is considered a milestone in signal analysis inasmuch as it gives a useful characterization of the class of band-limited signals.

As a generalization of the Laplace and the one-sided Fourier transforms, the Mittag–Leffler transform provides us with a tool for generalizing many of the classical results in harmonic analysis and the H^p theory. Several of these generalizations have been derived by M. Djrbashian and some of his colleagues from the former Soviet Union, and can be found in [53].

As an example, we shall give the following generalization of the Paley-Wiener theorem. But first let us introduce the following function space.

DEFINITION 16.2 *For* $1 < p < \infty, -1 < w < p - 1, 0 < \alpha, \sigma < \infty$, *we define* $B^{p,w}_{\alpha,\sigma}(\mathcal{R}^+)$ *as the set of all entire functions* $f(z)$ *of order* $\leq \alpha$ *and type* $\leq \sigma$ *that satisfy*

$$\int_0^\infty |f(x)|^p \, x^w dx < \infty.$$

For any $f \in B^{p,w}_{\alpha,\sigma}$, we set

$$\|f\|_{p,w} = \left\{ \int_0^\infty |f(x)|^p \, x^w dx \right\}^{1/p} .$$

The space $B^{p,w}_{\alpha,\sigma}(\mathcal{R})$ is defined similarly, except that for $f \in B^{p,w}_{\alpha,\sigma}(\mathcal{R})$ we have

$$\int_{-\infty}^{\infty} |f(x)|^p \, x^w dx < \infty.$$

THEOREM 16.2 [53, p. 23]
A function $f(z)$ belongs to $B^{2,w}_{1/2,\sigma}(\mathcal{R}^+)$, with $-1 < w < 1$, if and only if it can be represented in the form

$$f(z) = \int_0^\sigma E_2(-zt^2; \beta) t^{\beta-1} \phi(t) dt, \quad z \in \mathcal{C},$$

where $\beta = w + 3/2$ and $\phi \in L^2(0, \sigma)$. Moreover, the function ϕ is unique and can be recovered by means of the inversion formula

$$\phi_\sigma(t) = \frac{2}{\pi} \frac{d}{dt} \int_0^\infty \frac{\cos(tx - \pi\beta/2) - \cos(\pi\beta/2)}{x} f(x^2) x^{\beta-1} dx ,$$

where

$$\phi_\sigma(t) = \begin{cases} \phi(t) & \text{if } t \in (0, \sigma) \\ 0 & \text{if } t \in (\sigma, \infty). \end{cases}$$

In particular, if $w = \pm 1/2$, we have the following assertions:

1) the set $B^{2,-1/2}_{1/2,\sigma}(\mathcal{R}^+)$ coincides with the set of functions $f(z)$ of the form

$$f(z) = \int_0^\sigma \cos(\sqrt{z}t) \phi(t) dt ,$$

with $\phi \in L^2(0, \sigma)$. In addition,

$$\frac{2}{\pi} \frac{d}{dt} \int_0^\infty \frac{\sin(tx)}{x} f(x^2) dx = \phi(t) \chi_{(0,\sigma)}(t) ,$$

almost everywhere,

2) the set $B^{2,1/2}_{1/2,\sigma}(\mathcal{R}^+)$ coincides with the set of functions $f(z)$ of the form

$$f(z) = \int_0^\sigma \frac{\sin(\sqrt{z}t)}{\sqrt{z}} \phi(t) dt ,$$

with $\phi \in L^2(0,\sigma)$. In addition,

$$\frac{2}{\pi}\frac{d}{dt}\int_0^\infty \frac{1-\cos(tx)}{x} f(x^2)x dx = \phi(t)\chi_{(0,\sigma)}(t),$$

almost everywhere.

The following theorem is a generalization of the Paley-Wiener theorem which deals with square integrable functions over the whole real line.

THEOREM 16.3 [53, p. 25]
The set of all functions $f(z) \in B^{2,w}_{1,\sigma}(\mathcal{R})$, where $-1 < w < 1$, coincides with the set of functions $f(z)$ of the form

$$f(z) = \int_{-\sigma}^{\sigma} E_1(izt;\beta)\,|t|^{\beta-1}\,\phi(t)dt, \quad z \in C,$$

where $\beta = 1 + w/2$ and $\phi \in L^2(-\sigma,\sigma)$. Moreover,

$$\frac{1}{2\pi}\frac{d}{dt}\int_{-\infty}^{\infty} \frac{e^{-itx}-1}{-ix} \exp\left[i\frac{\pi}{2}(\beta-1)\mathrm{sign}(xt)\right] f(x)|x|^{\beta-1}dx = \phi(t)\chi_{(-\sigma,\sigma)}(t),$$

almost everywhere.

As a special case of this theorem we obtain, for $w = 0$, hence $\beta = 1$, the well-known Paley-Wiener theorem since $E_1(izt;1) = e^{izt}$, and $B^{2,0}_{1,\sigma}(\mathcal{R})$ is the Paley-Wiener space of entire functions. The Mittag–Leffler transformation may also be used to solve some initial and boundary-value problems associated with fractional order differential equations; for more details, see [53].

17

The Convolution Transform

The theory of convolution transformations was pioneered by I. Hirschman and D. Widder in the late 1940s in a series of papers [104–108] which were later compiled in a book form in the mid-1950s [103]. They showed, among other things, that for a certain class of kernels, the convolution transform encompasses a variety of other known integral transforms, such as the Laplace and the Stieltjes transforms. Chief among their discoveries of various properties of the convolution transform is the inversion formula.

To justify some definitions to come, Hirschman and Widder offered the following heuristic approach to the convolution transform, which is based on *operational calculus.* Operational calculus is a technique in which an operational symbol, such as the differentiation operator D, is treated as if it were a number throughout some algebraic manipulation. After the manipulation process is completed the final result is then interpreted in terms of the operational symbol. To illustrate, let $D^k = \frac{d^k}{dx^k}$, and define

$$e^{aD} = \sum_{k=0}^{\infty} \frac{a^k D^k}{k!} ,$$

so that

$$e^{aD} f(x) = \sum_{k=0}^{\infty} \frac{a^k f^{(k)}(x)}{k!} = f(x+a) .$$

Guided by this, we define

$$e^{aD} f(x) = f(x+a) \tag{17.1}$$

as long as the right-hand side has meaning and regardless of whether f is differentiable or not. Now we use this idea to solve the problem

$$(1-D)F(x) = f(x) ,$$

$$F(x) = o(e^x) \quad \text{as} \quad x \to \infty ,$$

where f is a given continuous and bounded function. Formally, we have

$$F(x) = (1-D)^{-1} f(x) .$$

For this solution to make sense, it is evident that an interpretation of the operator $(1-D)^{-1}$ is needed. To this end, observe that

$$(1-s)^{-1} = \frac{1}{1-s} = \int_{-\infty}^{0} e^t e^{-st} dt , \quad s < 1 .$$

This suggests interpreting $(1-D)^{-1}$ as $\int_{-\infty}^{0} e^t e^{-tD} dt$, which in turn, together with (17.1), leads to the formal solution

$$F(x) = \int_{-\infty}^{0} e^t e^{-tD} f(x) dt = \int_{-\infty}^{0} e^t f(x-t) dt = e^x \int_{x}^{\infty} e^{-y} f(y) dy .$$

Of course this solution can be verified by using more rigorous methods.

Let us rewrite the last equation in the form

$$F(x) = \int_{-\infty}^{\infty} G(t) f(x-t) dt ,$$

where $G(t) = e^t \chi_{(-\infty,0)}(t)$ and then reverse the process. That is, suppose that F is given by the above integral in which f is now unknown, can we solve the above convolution integral equation for f?

Clearly, $f(x) = (1-D)F(x)$, is a solution, but here it is crucial to note that the two-sided (bilateral) Laplace transform of the kernel G is

$$(\mathcal{L}G)(s) = \int_{-\infty}^{\infty} G(t) e^{-st} dt = \frac{1}{1-s} ,$$

and that

$$f(x) = (1-D)F(x) = [(\mathcal{L}G)(D)]^{-1} F(x) .$$

More generally, let

$$F(x) = \int_{-\infty}^{\infty} G(x-t) f(t) dt ,$$

for some appropriate functions, F and G and let

$$\int_{-\infty}^{\infty} G(t) e^{-st} dt = \frac{1}{E(s)}$$

be the two-sided Laplace transform of G. Then by the operational calculus developed above,

$$\frac{1}{E(D)} f(x) = \int_{-\infty}^{\infty} G(t) e^{-tD} f(x) dt = \int_{-\infty}^{\infty} G(t) f(x-t) dt = F(x) ,$$

or

$$f(x) = E(D) F(x) .$$

This symbolic manipulation of the operator D, which can be justified and proved rigorously, is the basis for the inversion formula of the convolution transform as developed by Hirschman and Widder.

17.1 Definitions

DEFINITION 17.1 *The convolution transform of a function f is defined as*

$$F(x) = \int_{-\infty}^{\infty} G(x-t)f(t)dt\ , \tag{17.2}$$

for a suitably restricted function, G, called the kernel function.

In the theory of the convolution transform as developed by Hirschman and Widder, the kernel function is assumed to be in the form

$$G(t) = \frac{1}{2\pi i}\int_{-i\infty}^{i\infty} \frac{1}{E(s)} e^{st} ds\ , \tag{17.3}$$

where

$$E(s) = e^{-cs^2+bs}\prod_{k=1}^{\infty}\left(1-\frac{s}{a_k}\right)e^{s/a_k}\ , \tag{17.4}$$

$a_k(k=1,2,...)$, b and c are real numbers with $c \geq 0$, and

$$\sum_{k=1}^{\infty}\frac{1}{a_k^2} < \infty\ .$$

The last condition on the $a_k's$ ensures that the infinite product in (17.4) is convergent and hence defines an entire function. The function G is the two-sided Laplace transform of $1/E(s)$, i.e.,

$$\frac{1}{E(s)} = \int_{-\infty}^{\infty} G(t)e^{-st}dt\ . \tag{17.5}$$

DEFINITION 17.2 *The kernel function G is said to belong to the class of finite kernels if E is a polynomial in s with real zeros so that*

$$E(s) = \left(1-\frac{s}{a_1}\right)\cdots\left(1-\frac{s}{a_n}\right)\ .$$

If E is not a polynomial, G is said to belong to

i) class I if there are positive and negative a_k's,

ii) class II if there are only positive a_k's and $\sum_k 1/a_k = \infty$,

iii) class III if there are only positive a_k's and $\sum_k 1/a_k < \infty$.

If $c = 0$, then $G(t)$ or $G(-t)$ belongs to one of these classes. Upon observing that

$$\frac{d}{ds} \ln E(s)\,|_{s=0} = b \quad \text{and} \quad \frac{d^2}{ds^2} \ln E(s)|_{s=0} = -\left(2c + \sum_{k=1}^{\infty} \frac{1}{a_k^2}\right),$$

the following relations are easily established from (17.4) and (17.5):

1) $\int_{-\infty}^{\infty} G(t)dt = 1$,

2) $\int_{-\infty}^{\infty} G(t)(t-b)dt = 0$,

3) $\int_{-\infty}^{\infty} G(t)(t-b)^2 dt = 2c + \sum_{k=1}^{\infty}(1/a_k^2)$.

Moreover, if G belongs to the class of finite kernels, then

$$G(t) = (g_1 * g_2 * \cdots * g_n)(t) ,$$

where

$$g_k(t) = |a_k| g(a_k t) , \quad a_k \neq 0 , \quad k = 1, ..., n ,$$

and

$$g(t) = \begin{cases} e^t \text{ if } -\infty < t < 0 \\ \frac{1}{2} \text{ if } \quad t = 0 \\ 0 \text{ if } \quad t > 0. \end{cases}$$

This is an easy consequence of the fact that the two-sided Laplace transform of g_k is

$$\int_{-\infty}^{\infty} g_k(t)^{-st} dt = \frac{1}{(1 - s/a_k)}$$

for $\operatorname{Re} s < a_k$ if $a_k > 0$ and for $\operatorname{Re} s > a_k$ if $a_k < 0$. Therefore,

$$\int_{-\infty}^{\infty} e^{-st} G(t)dt = \frac{1}{E(s)} \quad \text{for} \quad \alpha_1 < \operatorname{Re} s < \alpha_2$$

where α_1 is the largest negative a_k (or $-\infty$ if all the a_k are positive), and α_2 is the smallest positive a_k (or ∞ if all the a_k are negative). From the inversion formula for the two-sided Laplace transform

$$G(t) = \frac{1}{2\pi i} \int_{-i\infty}^{i\infty} \frac{e^{ts}}{E(s)} ds ,$$

and the fact that $E(iy) = O(y^n)$ as $|y| \to \infty$, it follows that the kth derivative of G,

$$G^{(k)}(t) = \frac{1}{2\pi i} \int_{-i\infty}^{i\infty} \frac{s^k e^{ts}}{E(s)} ds ,$$

exists for $k = 1, 2, ..., n-2$.

Some of the above properties of G can be extended to the non-finite kernel case, but first we need the following definition.

DEFINITION 17.3 *An entire function $E(s)$ is said to belong to the class E if and only if it is in the form (17.4) with*

$$\sum_{k=1}^{\infty} \frac{1}{a_k^2} < \infty .$$

To include polynomials in the class E, we shall allow all the a_k's starting from a certain term to be infinite.

A useful characterization of the class E is this [103, p.42]: a function $E(s)$ belongs to the class E if and only if there exists a sequence of polynomials $\{P_n(s)\}_{n=1}^{\infty}$, each having real zeros only such that

$$\lim_{n\to\infty} P_n(s) = E(s)$$

uniformly on compact subsets of the complex s-plane.

If $E(s) \in E$ and $E(s)$ is not the product of a polynomial by an exponential function, then

1) for any positive numbers p and R

$$\frac{1}{E(\sigma + i\tau)} = O\left(\frac{1}{|\tau|^p}\right) \quad \text{as} \quad |\tau| \to \infty$$

uniformly in the strip $|\sigma| \le R$,

2) the integral

$$\frac{1}{2\pi i}\int_{-i\infty}^{i\infty} \frac{e^{st}}{E(s)} ds \, , \quad -\infty < t < \infty$$

converges and defines a C^{∞}-function $G(t)$ and in addition

$$\frac{1}{E(s)} = \int_{-\infty}^{\infty} e^{-st} G(t) dt$$

where the integral converges in the largest strip which contains the origin and is free of zeros of $E(s)$,

3) moreover,

$$\int_{-\infty}^{\infty} tG(t)\, dt = b \, ,$$

and

$$\int_{-\infty}^{\infty} (t-b)^2 G(t)\, dt = 2c + \sum_{k=1}^{\infty} \frac{1}{a_k^2}$$

4) the function $-\log G(t)$ is convex.

17.2 Existence

Because G is integrable, it follows immediately that if f is bounded and continuous, then its convolution transform exists everywhere. More generally, we have the following theorem

THEOREM 17.1 [103, p. 123]
Let $\alpha(t)$ be of bounded variation in every interval and

$$\int_{-\infty}^{\infty} G(x_0 - t)d\alpha(t)$$

converges. Then
1) if G is of class I,

$$\int_{-\infty}^{\infty} G(x - t)d\alpha(t)$$

converges uniformly in any interval,
2) if G is of class II, then

$$\int_{-\infty}^{\infty} G(x - t)d\alpha(t)$$

converges uniformly in any interval bounded on the left by x_0.
3) if G is of class III, $\alpha(t)$ is of bounded variation in every interval of the form (t_1, t_2) where $T < t_1$ for some T and $x_0 > T + b + \sum a_k^{-1}$, then

$$\int_{-\infty}^{\infty} G(x - t)d\alpha(t)$$

converges uniformly in any closed interval lying to the right of

$$T + b + \sum_{n=0}^{\infty} a_k^{-1} .$$

The existence of the convolution transform is similar to that of the Stieltjes or the Laplace transforms depending on whether the kernel function is of class I or II. If it is of class I, then the existence of the convolution transform at only one point implies its existence at every other point. If it is of class II, then there exists a constant γ_c, which may be $\pm\infty$, such that the convolution integral converges for every $x > \gamma_c$ and diverges for every $x < \gamma_c$, and for $x = \gamma_c$ it may converge or diverge. The number, γ_c, is called the abscissa of convergence.

17.3 Elementary Properties

Let **C** denote the convolution transformation.

1. **Linearity:** The convolution transformation is linear.

2. **Translation:**

$$\mathbf{C}\left[f(t+a)\right](x) = \mathbf{C}\left[f(t)\right](x+a).$$

3. **Differentiation**

$$\mathbf{C}\left[f^{(n)}\right](x) = \left(\mathbf{C}[f]\right)^{(n)}(x).$$

17.4 Relationships with Other Transforms

The class of convolution transforms encompasses many of the known integral transforms.

17.4.1 The Laplace Transform

In the Laplace transform of ϕ

$$\Phi(y) = \int_0^\infty e^{-yu}\phi(u)du$$

if we make the change of variables, $y = e^x$ and $u = e^{-t}$, we obtain

$$\Phi(e^x) = \int_{-\infty}^\infty \exp(-e^{x-t})e^{-t}\phi(e^{-t})dt \,,$$

which can be put in the convolution form

$$F(x) = \int_{-\infty}^\infty G(x-t)f(t)dt \,,$$

with

$$F(x) = e^x\Phi(e^x) \,, \quad f(t) = \phi(e^{-t}) \,, \quad \text{and}$$
$$G(t) = e^{-e^t}e^t \,.$$

Moreover,

$$\frac{1}{E(s)} = \int_{-\infty}^\infty e^{-st}G(t)dt = \int_{-\infty}^\infty e^{-st}e^{-e^t}e^t dt = \int_0^\infty u^{-s}e^{-u}du = \Gamma(1-s) \,,$$

for $\operatorname{Re} s < 1$.

17.4.2 The Stieltjes Transform

The Stieltjes transform

$$\Phi(y) = \int_0^\infty \frac{\phi(u)}{y+u} du$$

can be transformed into the convolution transform by first setting $y = e^x, u = e^t$, to obtain

$$\Phi(e^x) = \int_{-\infty}^\infty \frac{\phi(e^t)e^t}{e^x + e^t} dt \ ,$$

then setting

$$F(x) = e^{x/2}\Phi(e^x) \quad \text{and} \quad f(t) = e^{t/2}\phi(e^t)$$

to yield

$$F(x) = \int_{-\infty}^\infty G(x-t)f(t)dt \ ,$$

where

$$G(t) = \frac{1}{2}\text{sech}\frac{t}{2} \ .$$

Moreover,

$$\frac{1}{E(s)} = \int_{-\infty}^\infty \frac{e^{-st}}{e^{\frac{t}{2}} + e^{-\frac{t}{2}}} dt = \frac{\pi}{\cos \pi s} \ , \quad -\frac{1}{2} < \text{Re } s < \frac{1}{2}$$

17.4.3 The Generalized Stieltjes Transform

The generalized Stieltjes transform

$$\Phi(y) = \int_0^\infty \frac{\phi(u)}{(y+u)^\nu} du \ , \quad \nu > 0$$

becomes

$$\Phi(e^x) = \int_{-\infty}^\infty \frac{\phi(e^t)e^t dt}{[e^x + e^t]^\nu}$$

if we set $y = e^x$ and $u = e^t$. This in turn can be transformed into

$$F(x) = \int_{-\infty}^\infty G(x-t)f(t)dt \ ,$$

where $F(x) = e^{\nu x/2}\Phi(e^x), f(t) = e^{(1-\nu/2)t}\phi(e^t)$ and $G(t) = 2^{-\nu}\text{sech}^\nu \frac{t}{2}$. Moreover,

$$\begin{aligned}\frac{1}{E(s)} &= \int_{-\infty}^\infty \frac{e^{-st}}{(e^{\frac{t}{2}} + e^{-\frac{t}{2}})^\nu} dt = \int_0^\infty \frac{u^{(\frac{\nu}{2}-1-s)}}{(1+u)^\nu} du \\ &= \frac{\Gamma(\frac{\nu}{2}-s)\Gamma(\frac{\nu}{2}+s)}{\Gamma(\nu)}.\end{aligned}$$

17.4.4 The K (Meijer) Transform

The K (Meijer) transform of a function ϕ is defined by (cf. Chapter 23)

$$\Phi(y) = \int_0^\infty \sqrt{yu}\, K_\nu(yu)\phi(u)du$$

can be transformed into

$$F(x) = \int_{-\infty}^\infty G(x-t)f(t)dt\ ,$$

upon setting $y = e^x, u = e^{-t}, F(x) = e^{x/2}\Phi(e^x), f(t) = e^{-t/2}\phi(e^{-t})$ and $G(t) = e^t K_\nu(e^t)$, where K_ν is the modified Bessel function of the third kind. In addition, we have

$$\begin{aligned}\frac{1}{E(s)} &= \int_{-\infty}^\infty e^{-st}e^t K_\nu(e^t)dt = \int_0^\infty z^{-s}K_\nu(z)dz \\ &= \frac{1}{2^{1+s}}\Gamma\left(\frac{1-s-\nu}{2}\right)\Gamma\left(\frac{1-s+\nu}{2}\right) .\end{aligned}$$

17.4.5 Fourier-Type Integral Transforms

Formally, any Fourier-type integral transform (see Definition 12.5)

$$\Phi(y) = \int_0^\infty k(yu)\phi(u)du$$

can be transformed into a convolution transform by setting $y = e^x, u = e^{-t}$ to obtain

$$F(x) = \int_{-\infty}^\infty G(x-t)f(t)dt\ ,$$

where $F(x) = e^x\phi(e^x), f(t) = \phi(e^{-t})$ and $G(t) = e^t k(e^t)$. Moreover,

$$\frac{1}{E(s)} = \int_{-\infty}^\infty e^{-st}e^t k(e^t)dt = \int_0^\infty v^{-s}k(v)dv\ .$$

17.5 Inversion Formula

The following theorem provides an inversion formula for the convolution transform.

THEOREM 17.2 [103, p. 57]
Let $E(s) \in E$ be of the form

$$E(s) = e^{bs} \prod_{k=1}^{\infty} \left(1 - \frac{s}{a_k}\right) e^{s/a_k} , \tag{17.6}$$

and

$$G(t) = \frac{1}{2\pi i} \int_{-i\infty}^{i\infty} \frac{e^{st}}{E(s)} ds , \quad -\infty < t < \infty . \tag{17.7}$$

If

$$F(x) = \int_{-\infty}^{\infty} G(x-t) f(t) dt ,$$

for some continuous, bounded function f, then

$$f(x) = E(D)F(x) = \lim_{n\to\infty} P_n(D)F(x) , \tag{17.8}$$

where

$$P_n(s) = e^{(b-\epsilon_n)s} \prod_{k=1}^{n} \left(1 - \frac{s}{a_k}\right) e^{s/a_k} , \quad \epsilon_n = o(1) \quad as \quad n \to \infty , \tag{17.9}$$

and $D = \frac{d}{dx}$.
In particular, if $E(s) \in E$ is of the form

$$E(s) = \prod_{k=1}^{\infty} \left(1 - \frac{s}{a_k}\right) , \quad with \quad \sum_{n=0}^{\infty} \frac{1}{|a_k|} < \infty ,$$

then

$$f(x) = E(D)F(x) = \lim_{n\to\infty} P_n(D)F(x) ,$$

where

$$P_n(s) = \prod_{k=1}^{n} \left(1 - \frac{s}{a_k}\right) .$$

To show that the second part of the theorem is a special case of the first, set $b = -\sum_{n=0}^{\infty} 1/a_k$ and $\epsilon_n = \sum_{k=1}^{n} (1/a_k) + b$. Then

$$E(s) = \prod_{k=1}^{\infty} \left(1 - \frac{s}{a_k}\right) = e^{bs} \prod_{k=1}^{\infty} \left(1 - \frac{s}{a_k}\right) e^{s/a_k} ,$$

and

$$P_n(s) = \prod_{k=1}^{n} \left(1 - \frac{s}{a_k}\right) = e^{(b-\epsilon_n)s} \prod_{k=1}^{n} \left(1 - \frac{s}{a_k}\right) e^{s/a_k} .$$

Example A: For the Laplace transform,

$$\Phi(y) = \int_0^\infty e^{-yu}\phi(u)du$$

we have $F(x) = e^x\Phi(e^x)\,, f(x) = \phi(e^{-x})$ and $E(s) = 1/\Gamma(1-s)$ (cf. Section 17.4.1). To write E as an infinite product, we appeal to formula (2.6) to deduce that

$$E(s) = \frac{1}{\Gamma(1-s)} = e^{-\gamma s}\prod_{k=1}^{\infty}\left(1-\frac{s}{k}\right)e^{s/k}\,,$$

which is in the form (17.6) with $b = -\gamma$.

Clearly, by Equation (2.3), $\epsilon_n = \sum_1^n 1/a_k - \ln n - \gamma$ goes to zero as $n \to \infty$. Hence, if we substitute this ϵ_n in (17.9), we obtain

$$P_n(D) = e^{D\ln n}\prod_{k=1}^{n}\left(1-\frac{D}{k}\right).$$

By induction on n, one can verify that

$$\prod_{k=1}^{n}\left(1-\frac{D}{k}\right)[e^x\Phi(e^x)] = \frac{(-1)^n}{n!}e^{(n+1)x}\Phi^{(n)}(e^x)\,.$$

Therefore, in view of (17.1) and (17.8), we have the following inversion formula for the Laplace transform

$$\lim_{n\to\infty}\frac{(-1)^n}{n!}e^{(n+1)(x+\ln n)}\Phi^{(n)}(e^{x+\ln n}) = \phi(e^{-x})\,,$$

or alternatively

$$\lim_{n\to\infty}\frac{(-1)^n}{n!}\left(\frac{n}{y}\right)^{n+1}\Phi^{(n)}\left(\frac{n}{y}\right) = \phi(y)\,;$$

cf. Equation (5.16).

Example B: For the Stieltjes transform (cf. Section 17.4.2)

$$\Phi(y) = \int_0^\infty \frac{\phi(u)}{y+u}du\,,$$

we have $F(x) = e^{x/2}\Phi(e^x)\,, f(x) = e^{x/2}\phi(x)$ and

$$\begin{aligned}E(s) &= \frac{\cos\pi s}{\pi} = \frac{1}{\pi}\prod_{k=0}^{\infty}\left(1-\frac{s^2}{(k+\frac{1}{2})^2}\right)\\ &= \frac{1}{\pi}\prod_{k=0}^{\infty}\left(1-\frac{s}{(k+\frac{1}{2})}\right)\left(1+\frac{s}{(k+\frac{1}{2})}\right).\end{aligned}$$

To evaluate $E(D)e^{x/2}\Phi(e^x)$, let

$$P_n(s) = \frac{1}{\pi}\prod_{k=0}^{n}\left(1 - \frac{s^2}{(k+\frac{1}{2})^2}\right) = \frac{1}{\pi}Q_n(s)R_n(s) ,$$

where

$$Q_n(s) = \prod_{k=0}^{n}\left(1 + \frac{s}{(k+\frac{1}{2})}\right) \quad \text{and} \quad R_n(s) = \prod_{k=0}^{n}\left(1 - \frac{s}{(k+\frac{1}{2})}\right) .$$

By induction, one can verify that for any suitable function η

$$R_{n-1}(D)\left[e^{\frac{x}{2}}\eta(e^x)\right] = (-1)^n c_n e^{(n+\frac{1}{2})x}\eta^{(n)}(e^x)$$

and

$$Q_{n-1}(D)\left[e^{-(n-\frac{1}{2})x}\eta(e^x)\right] = c_n e^{\frac{x}{2}}\eta^{(n)}(e^x) ,$$

where $c_n = 4^n n!/(2n)!$. Thus,

$$R_{n-1}(D)\left[e^{\frac{x}{2}}\Phi(e^x)\right] = (-1)^n c_n e^{(n+\frac{1}{2})x}\Phi^{(n)}(e^x) ,$$

and if we define $H(e^x)$ by

$$e^{-(n-\frac{1}{2})x}H(e^x) = e^{(n+\frac{1}{2})x}\Phi^{(n)}(e^x) ,$$

it follows that

$$\begin{aligned} P_{n-1}(D)\left[e^{\frac{x}{2}}\Phi(e^x)\right] &= \frac{(-1)^n}{\pi}c_n Q_{n-1}(D)\left[e^{(n+\frac{1}{2})x}\Phi^{(n)}(e^x)\right] \\ &= \frac{(-1)^n}{\pi}c_n Q_{n-1}(D)\left[e^{-(n-\frac{1}{2})x}H(e^x)\right] \\ &= \frac{(-1)^n}{\pi}c_n^2 e^{x/2}H^{(n)}(e^x) . \end{aligned}$$

After eliminating H from this equation, we obtain

$$P_{n-1}(D)\left[e^{\frac{x}{2}}\Phi(e^x)\right] = \frac{(-1)^n}{\pi}c_n^2 e^{\frac{x}{2}}\left[e^{2nx}\Phi^{(n)}(e^x)\right]^{(n)} .$$

The inversion formula for the Stieltjes transform now takes the form

$$\lim_{n\to\infty}\frac{(-1)^n}{\pi}c_n^2\left[e^{2nx}\Phi^{(n)}(e^x)\right]^{(n)} = \phi(e^x) ,$$

which, in view of Stirling's formula (see Section 2.1.2), can be converted into

$$\lim_{n\to\infty}\frac{(-1)^n}{2\pi}\left(\frac{e}{n}\right)^{2n}\left[y^{2n}\Phi^{(n)}(y)\right]^{(n)} = \phi(y) .$$

17.6 Generalized Functions

To extend the convolution transform to generalized functions we use the embedding method as introduced by Zemanian [276, Ch. 8]. First, we assume that to every permissible convolution kernel G there exists a sequence of operators $\{P_n(D)\}_{n=0}^{\infty}$ of the form

$$P_n(D) = e^{d_n D} Q_n(D) , \quad n = 1, 2, ..., \tag{17.10}$$

where $D = \frac{d}{dt}$, d_n is a real constant, and Q_n is a polynomial of degree n with real coefficients. We adopt the notation that $P_0(D)$ is the identity operator. Define a sequence of functions $\{G_n\}_{n=0}^{\infty}$ by

$$G_n(t) = P_n(D)G(t) , \quad \text{and} \quad G_0(t) = P_0(D)G(t) = G(t) .$$

Furthermore, we assume that the convolution kernel G satisfies the following conditions:

1) G is a C^∞-function on the real line.

2) $G_n(t) \geq 0$, $n = 0, 1, 2, ...$

3) $\int_{-\infty}^{\infty} G_n(t)dt = 1$, $n = 0, 1, 2, ...$

4) $\lim_{n\to\infty} G_n(t) = 0$, $0 < |t| < \infty$.

5) For every $\delta > 0$,
$$\lim_{n\to\infty} \int_{|t|>\delta} G_n(t)dt = 0.$$

6) Given any real numbers γ and $\delta > 0$, there exists a positive integer N that depends on γ and δ such that $e^{-\gamma t}G_n(t)$ is a monotone increasing function on $-\infty < -t < -\delta$ and a monotone decreasing function on $\delta < t < \infty$.

7) G possesses the following asymptotic properties:
   There exist $\beta < 0$ and $\alpha > 0$ such that for each $k = 0, 1, 2, ...$

   i) $D^k G(t) = O(e^{\alpha t})$ as $t \to -\infty$,
   ii) $D^k G(t) = O(e^{\beta t})$ as $t \to \infty$.

Let α_1 denote the infimum of all β for which (ii) holds and α_2 denote the supremum of all α for which (i) holds. Thus, α_1 is either a negative number or $-\infty$ and α_2 is either a positive number or ∞.

It can be shown that there exists a convolution kernel G satisfying all the above 7 conditions [276, p. 233]. Conditions (2)–(5) ensure that $\{G_n\}_{n=0}^{\infty}$ is a δ-sequence.

Let $E(s), G(t)$ and $P_n(s)$ be as given by Theorem 17.2. Define

$$E_n(s) = e^{\epsilon_n s} \prod_{n+1}^{\infty} \left(1 - \frac{s}{a_k}\right) e^{s/a_k} = \frac{E(s)}{P_n(s)} ,$$

where $\epsilon_0 = b$, and

$$G_n(t) = \frac{1}{2\pi i} \int_{-i\infty}^{\infty} \frac{e^{st}}{E_n(s)} ds .$$

Clearly, $P_n(D)$ has the same form given by (17.10), where $d_n = b - \epsilon_n + \sum_{k=1}^{n} \frac{1}{a_k}$, and

$$Q_n(D) = \prod_{k=1}^{n} \left(1 - \frac{D}{a_k}\right) .$$

Moreover, since $P_n(D)e^{st} = e^{st}P_n(s)$, we obtain from (17.7) that

$$P_n(D)[G(t)] = \frac{1}{2\pi i} \int_{-i\infty}^{\infty} \frac{e^{st}}{E_n(s)} ds = G_n(t) .$$

Let G be a given kernel satisfying conditions (1)–(7), and let α_1, α_2 be as defined in condition (7). Consider the space $\mathcal{L}_{a,b}$ defined in Section 3.1.3, Example 6 in which $R = D = \frac{d}{dt}$, and

$$\eta(t, a, b) = \begin{cases} e^{at} & , \quad 0 \le t < \infty \\ e^{bt} & , \quad -\infty < t < 0 . \end{cases}$$

For any fixed x and any $c < \alpha_2, d > \alpha_1$, the kernel $G(x - t)$ as a function of t belongs to $\mathcal{L}_{c,d}$. For, if we choose α and β such that $c < \alpha < \alpha_2$, and $\alpha_1 < \beta < d$, then

$$\eta(t, c, d) D^k[G(x - t)] = \begin{cases} O(e^{(c-\alpha)t}) = 0(1) & \text{as} \quad t \to \infty \\ O(e^{(d-\beta)t}) = 0(1) & \text{as} \quad t \to -\infty , \end{cases}$$

which proves our claim.

Now we define the convolution transform with kernel G for any member of $f \in \mathcal{L}^*_{c,d}$ where c and d are chosen as before.

DEFINITION 17.4 *Let G be defined as above and $f \in \mathcal{L}^*_{c,d}$. Then the convolution transform with of f with respect to the kernel G is defined as*

$$F(x) = \langle f(t), G(x - t) \rangle .$$

Since G is infinitely differentiable it can be easily shown that so is F, and

$$F^{(k)}(x) = \langle f(t), G^{(k)}(x-t)\rangle \ , \quad k = 0, 1, 2, \ldots \ .$$

The inversion formula (17.8) can be extended to generalized functions as follows.

THEOREM 17.3
Let $G, P_n(D), \alpha_1$ and α_2 be defined as above. If $f \in \mathcal{L}^*_{c,d}$ and

$$F(x) = \langle f(t), G(x-t)\rangle \ ,$$

then $\lim_{n\to\infty} P_n(D)[F] = f$ in the sense of D^*, i.e.,

$$\lim_{n\to\infty} \langle P_n(D)F, \psi\rangle = \langle f, \psi\rangle$$

for every $\psi \in \mathcal{D}$.

17.7 Applications

Because of its great generality, the convolution transform does not have many direct applications; however, many of its special cases, such as the Laplace, Stieltjes and Weierstrass transforms, have numerous direct applications. We refer the reader to the chapters on those transforms for examples of applications.

18

The Weierstrass Transform

The *Weierstrass transform*, also known as the *Gauss transform* [206, 207], the *Gauss-Weierstrass transform* [101] and the *Hille transform* [77], is intimately related to the solution of the heat equation for one-dimensional flow. It is also a special case of the convolution transform, yet it is considered a singular case of the convolution-transform theory developed by Hirschmann and Widder [103] which was discussed in Chapter 17. Because the Weierstrass transform is a singular case of their convolution-transform theory, Hirschmann and Widder dedicated a whole chapter in their book to its investigation.

18.1 Definitions

DEFINITION 18.1 *The Weierstrass transform $F(x)$ of a function $f(y)$ is defined as*

$$F(x) = \frac{1}{\sqrt{4\pi}} \int_{-\infty}^{\infty} f(y) e^{-(x-y)^2/4} dy \, , \tag{18.1}$$

whenever the integral converges. The Weierstrass-Stieltjes transform of a function $f(y)$ with respect to a measure $d\alpha$ is defined as

$$F(x) = \frac{1}{\sqrt{4\pi}} \int_{-\infty}^{\infty} f(y) e^{-(x-y)^2/4} d\alpha(y) \, .$$

Sometimes it is convenient to write the Weierstrass transform in the form

$$F(y) = \frac{1}{\sqrt{4\pi}} \int_{-\infty}^{\infty} f(y) k(x-y, 1) dy \, ,$$

where

$$k(x,t) = \frac{1}{\sqrt{4\pi t}} e^{-x^2/4t} \, , \quad 0 < t < \infty \, , \quad -\infty < x < \infty \, .$$

The function

$$k(x, ct) = \frac{1}{\sqrt{4\pi ct}} e^{-x^2/4ct}$$

is the fundamental solution (or the Green's function) of the heat equation

$$c\frac{\partial^2 u}{\partial x^2} = \frac{\partial u}{\partial t}\,, \quad 0 < t < \infty\,, \quad -\infty < x < \infty\,.$$

Without loss of generality, we shall take $c = 1$. From the fundamental solution of the heat equation, we obtain the associated function

$$k(x, y; t) = \frac{1}{\sqrt{4\pi t}} e^{-(x-y)^2/4t}\,.$$

The associated function is easily seen to satisfy the following properties:

1) $k(x, y; t) = k(y, x; t)$,

2) as a function of x and t or as a function of y and t, it is a solution of the heat equation, that is

$$\frac{\partial^2 k}{\partial x^2} = \frac{\partial u}{\partial t} \quad \text{and} \quad \frac{\partial^2 k}{\partial y^2} = \frac{\partial u}{\partial t}\,,$$

3) $k(x, y; 0) = \delta(x - y)$,

4) it satisfies the semi-group property

$$\int_{-\infty}^{\infty} k(x, z; t_1)k(z, y; t_2)dz = k(x, y; t_1 + t_2)\,.$$

The associated function provides a solution to the initial-value problem

$$\frac{\partial^2 u}{\partial x^2} = \frac{\partial u}{\partial t}\,, \quad 0 < t < \infty\,, \quad -\infty < x < \infty\,,$$

and

$$u(x, 0) = f(x)\,.$$

In fact, from properties (2) and (3), it follows that

$$F(x, t) = \int_{-\infty}^{\infty} f(y)k(x, y; t)dy$$

is one such solution. The function $F(x, t)$ is called the *Poisson transform* of f. The Weierstrass transform is equal to the Poisson transform at $t = 1$.

Now we offer a new definition of the Poisson and the Weierstrass transformations. The Poisson transformation is an integral transformation whose kernel is the associated function of the heat equation and the Weierstrass transformation is an integral transformation whose kernel is the associated function of the heat equation evaluated at $t = 1$. In fact, the choice $t = 1$ is insignificant. For, by fixing t and replacing x by $\sqrt{t}x$ and y by $\sqrt{t}y$, the Poisson transform can be converted into the Weierstrass transform

$$G(x) = \frac{1}{\sqrt{4\pi}} \int_{-\infty}^{\infty} g(y)k(x,y;1)dy ,$$

where $G(x) = F(\sqrt{t}x, t)$ and $g(y) = f(\sqrt{t}y)$.

Following this definition of the Poisson and Weierstrass transforms, F. Cholewinski and D. T. Haimo [37], D. T. Haimo [83–87] and A. Zayed and D. T. Haimo [268, 269] introduced other generalizations of the Weierstrass transformation.

Given below are three different generalizations of the heat equation, together with their fundamental solutions and associated functions:

1) **The Hankel heat equation**

$$\frac{\partial^2 u}{\partial x^2} + \frac{\partial \nu}{x}\frac{\partial u}{\partial x} = \frac{\partial u}{\partial t}, \quad \nu > 0 , \quad t > 0 , \quad 0 \le x < \infty ,$$

with the fundamental solution

$$g_\nu(x,t) = \int_0^\infty e^{-ty^2} \mathcal{J}_{\nu-\frac{1}{2}}(xy) d\mu_{\nu-\frac{1}{2}}(y) = \frac{1}{(2t)^{\nu+\frac{1}{2}}} e^{-x^2/4t} ,$$

where

$$\mathcal{J}_{\nu-\frac{1}{2}}(z) = 2^{\nu-\frac{1}{2}}\Gamma(\nu+\frac{1}{2}) z^{-\nu+\frac{1}{2}} J_{\nu-\frac{1}{2}}(z) , \quad d\mu_{\nu-\frac{1}{2}}(y) = \frac{y^{2\nu}}{2^{\nu-\frac{1}{2}}\Gamma(\nu+\frac{1}{2})} dy ,$$

and $J_\nu(z)$ is the Bessel function of the first kind and order ν.

The associated function is

$$g_\nu(x,y;t) = \int_0^\infty e^{-tz^2} \mathcal{J}_{\nu-\frac{1}{2}}(xz) \mathcal{J}_{\nu-\frac{1}{2}}(yz) d\mu_{\nu-\frac{1}{2}}(z) = \frac{1}{(2t)^{\nu+\frac{1}{2}}} e^{-(x^2+y^2)/4t} \mathcal{I}_{\nu-\frac{1}{2}}\left(\frac{xy}{2t}\right) ,$$

where

$$\mathcal{I}_{\nu-\frac{1}{2}}(z) = 2^{\nu-\frac{1}{2}}\Gamma(\nu+\frac{1}{2}) z^{-\nu+\frac{1}{2}} I_{\nu-\frac{1}{2}}(z) ,$$

and $I_\alpha(z)$ is the modified Bessel function of the first kind.

2) **The Laguerre heat equation**

$$x\frac{\partial^2 u}{\partial x^2} + (\alpha+1-x)\frac{\partial u}{\partial x} = \frac{\partial u}{\partial t}, \quad \alpha > -1 , \quad t > 0 , \quad x > 0 ,$$

with the fundamental solution

$$g_\alpha(x,t) = \sum_{n=0}^{\infty} e^{-nt} L_n^\alpha(x) ,$$

where L_n^α is the Laguerre polynomial of degree n, and the associated function

$$\begin{aligned} g_\alpha(x,y;t) &= \sum_{n=0}^{\infty} e^{-nt} L_n^\alpha(x) L_n^\alpha(y) [\Gamma(\alpha+1)\rho_\alpha(n)] \\ &= \left(\frac{e^t}{e^t-1}\right)^{\alpha+1} \exp\left(-\frac{x+y}{e^t-1}\right) \mathcal{I}_\alpha\left(\frac{2\sqrt{xye^t}}{e^t-1}\right) , \end{aligned}$$

where $\rho_\alpha(n) = n!/\Gamma(n+\alpha+1)$.

3) **The Sturm-Liouville heat equation**

$$-\frac{1}{r(x)}\frac{\partial}{\partial x}\left(p(x)\frac{\partial u}{\partial x}\right) + l(x)u = \frac{\partial u}{\partial t} ,$$

for some suitable functions, r, p, and l. This equation can be transformed into the more convenient form

$$\frac{\partial^2 u}{\partial x^2} - q(x)u = \frac{\partial u}{\partial t} ,$$

for some function q; see Section 1.8. The fundamental solution of this last equation is

$$g(x,t) = \int_{-\infty}^{\infty} e^{-\lambda t} \phi(x,\lambda) d\rho(\lambda) ,$$

where ϕ is a solution of the Sturm-Liouville differential equation

$$y'' - q(x)y = -\lambda y , \quad 0 < x < \infty$$

satisfying

$$\phi(0,\lambda) = \sin\alpha , \quad \phi'(0,\lambda) = -\cos\alpha$$

and $d\rho(\lambda)$ is the measure associated with the Sturm-Liouville problem, i.e., if

$$\hat{f}(\lambda) = \int_0^\infty f(x)\phi(x,\lambda)dx ,$$

for some $f \in L^2(0,\infty)$, then

$$\int_0^\infty |f(x)|^2 dx = \int_{-\infty}^{\infty} |\hat{f}(\lambda)|^2 d\rho(\lambda) .$$

The associated function $g(x, y; t)$ is given by

$$g(x, y; t) = \int_{-\infty}^{\infty} e^{-\lambda t} \phi(x, \lambda) \phi(y, \lambda) d\rho(\lambda) .$$

Now we introduce the following generalizations of the Weierstrass transformation.

DEFINITION 18.2 *The Poisson-Hankel, Poisson-Laguerre and Poisson-Sturm-Liouville integral transforms of f are defined, respectively, by*

$$F_\nu(x, t) = \int_0^\infty f(y) g_\nu(x, y; t) d\mu_{\nu - \frac{1}{2}}(y) ,$$

$$F_\alpha(x, t) = \int_0^\infty f(y) g_\alpha(x, y; t) d\sigma_\alpha(y) ,$$

and

$$F(x, t) = \int_0^\infty f(y) g(x, y; t) dy ,$$

where

$$d\sigma_\alpha(y) = \frac{y^\alpha e^{-y}}{\Gamma(\alpha + 1)} dy ,$$

g_ν, g_α and g are as defined in the above examples of generalized heat equations. The Weierstrass-Hankel, Weierstrass-Laguerre and Weierstrass-Sturm-Liouville transforms are defined, respectively, as

$$\tilde{F}_\nu(x) = F_\nu(x, 1) , \quad \tilde{F}_\alpha(x) = F_\alpha(x, 1) \quad \textit{and} \quad \tilde{F}(x) = F(x, 1) .$$

The Poisson-Laguerre and Weierstrass-Laguerre transforms are sometimes called the *dual Poisson-Laguerre* and *dual Weierstrass-Laguerre* transforms to distinguish them from similar transforms with the same names, but in which the integrals are replaced by series. Also the Weierstrass-Hankel transform is sometimes called the *Weierstrass-Hankel convolution transform.*

As shown before, the Poisson transform can be reduced to the Weierstrass transform by a change of variables. The same is true for the Poisson-Hankel and the Poisson-Laguerre transforms. For example, in the Poisson-Hankel transform of f

$$F_\nu(x, t) = \frac{1}{(2t)^{\nu + \frac{1}{2}} 2^{\nu - \frac{1}{2}} \Gamma(\nu + \frac{1}{2})} \int_0^\infty f(y) \exp[-(x^2 + y^2)/4t] \mathcal{I}_{\nu - \frac{1}{2}} \left(\frac{xy}{2t} \right) y^{2\nu} dy$$

if we replace x by $\sqrt{t}\, x$ and y by $\sqrt{t}\, y$, we find that $F_\nu(\sqrt{t} x, t)$ is the Weierstrass-Hankel transform of $f(\sqrt{t}\, y)$.

18.2 Existence

If f is a locally integrable function, such that $f(x) = 0(e^{-\alpha x^2})$ as $|x| \to \infty$, for $0 < \alpha < 1/4$, then its Weierstrass transform exists for almost all x. If the integral in (18.1) converges for $a < x < b$, then $F(x)$ can be extended as an analytic function $F(x+iy)$ in the strip $a < x < b$ with $F(x+iy) = 0(|y|e^{y^2/4})$ as $|y| \to \infty$ uniformly in every subinterval of (a, b).

If $f \in L^2(0,\infty)$, then its Weierstrass-Sturm-Liouville transform exists almost everywhere and also belongs to $L^2(0,\infty)$.

18.3 Elementary Properties

Let us denote by W_c the Weierstrass transformation

$$W_c[f](x) = \frac{1}{\sqrt{4\pi c}} \int_{-\infty}^{\infty} f(y) e^{-(x-y)^2/4c} dy\,,$$

with $W[f] = W_1[f]$.

1. **Linearity:** The Weierstrass transformation W_c is linear.

2. **Translation:**

$$W_c\left[f(y-a)\right](x) = W_c\left[f(y)\right](x-a).$$

3. **Dilation:**

$$W_c\left[f(ay)\right](x) = W_{a^2c}\left[f(y)\right](ax)\,, \quad a > 0.$$

18.4 Relationships with Other Transforms

18.4.1 The Two-Sided Laplace Transform

If we replace x by $-2x$ in (18.1), we obtain

$$e^{x^2} F(-2x) = \frac{1}{\sqrt{4\pi}} \int_{-\infty}^{\infty} f(y) e^{-\frac{1}{4}y^2} e^{-xy} dy\,,$$

which can be transformed into the two-sided Laplace transform

$$G(x) = \int_{-\infty}^{\infty} g(y)e^{-xy}dy ,$$

where $G(x) = e^{x^2}F(-2x)$, and $g(y) = \frac{1}{\sqrt{4\pi}}f(y)e^{-y^2/4}$.

This representation of the Weierstrass transform enables us to deduce many of its properties from the theory of the Laplace transformation. For example, if x is complex, say $x = \sigma + i\tau$, then the region of convergence of the Weierstrass transform is a vertical strip as mentioned above.

An inversion formula for the Weierstrass transform based on the inversion formula for the Laplace transform can, thus, be obtained.

18.4.2 The Convolution Transform

Clearly, the Weierstrass transform is a convolution transform. Even more, it is a special case of the Hirschman-Widder convolution transform (cf. Chapter 17). Recall that the latter is an integral transform of the form

$$F(x) = \int_{-\infty}^{\infty} G(x-y)f(y)dy ,$$

where

$$G(x) = \frac{1}{2\pi i}\int_{-i\infty}^{i\infty} \frac{e^{xs}}{E(s)}ds ,$$

and

$$E(s) = e^{-cs^2+bs}\prod_{k=1}^{\infty}\left(1-\frac{s}{a_k}\right)e^{s/a_k} ,$$

with $c \geq 0$, b, a_k $(k = 1, 2, \ldots)$ are real and $\sum_{k=1}^{\infty} 1/|a_k|^2 < \infty$.

The special case in which $E(s) = e^{-cs^2}$, yields the Weierstrass transform since

$$G_c(x) = \frac{1}{2\pi i}\int_{-i\infty}^{i\infty} e^{xs}e^{cs^2}ds = \frac{1}{2\pi}\int_{-\infty}^{\infty} e^{-cs^2}e^{ixs}ds = \frac{1}{\sqrt{4\pi c}}e^{-x^2/4c} . \tag{18.2}$$

When $c = 1$, we write $G_1(x) = G(x)$.

18.5 Inversion Formulas

Because of its affinity to the Laplace and the convolution transforms, the Weierstrass transform has two inversion formulas, one is based on the inversion formula for the Laplace transform and the other on the inversion formula for the convolution transform. We start with the former.

THEOREM 18.1
Let F be the Weierstrass transform of f given by

$$F(x) = \int_{-\infty}^{\infty} G(x-y)f(y)dy \ ,$$

where G is given by (18.2). If

$$\int_{-\infty}^{\infty} e^{-x^2/4}|f(x)|dx < \infty \ ,$$

and f is of bounded variation in a neighborhood of a point x, then

$$\lim_{R\to\infty} \frac{1}{\sqrt{4\pi}} \int_{-R}^{R} e^{(x-iy)^2/4} F(iy)dy = \frac{f(x+)+f(x-)}{2} \ .$$

To derive the second inversion formula, we note that in view of (18.2) and the inversion formula for the Laplace transform

$$e^{cs^2} = \frac{1}{\sqrt{4\pi}} \int_{-\infty}^{\infty} e^{-sy} e^{-y^2/4c} dy \ . \tag{18.3}$$

Replacing s by $D = \frac{d}{dx}$ and interpreting $e^{aD}f(x)$ as $f(x+a)$ (see Equation 17.1), we have formally

$$\begin{aligned} e^{cD^2} f(x) &= \frac{1}{\sqrt{4\pi}} \int_{-\infty}^{\infty} e^{-yD} f(x) e^{-y^2/4c} dy = \frac{1}{\sqrt{4\pi}} \int_{-\infty}^{\infty} f(x-y) e^{-y^2/4c} dy \\ &= \frac{1}{\sqrt{4\pi}} \int_{-\infty}^{\infty} f(y) e^{-(x-y)^2/4c} dy = F(x) \ , \end{aligned}$$

which suggests that

$$f(x) = e^{-cD^2} F(x) \ .$$

Because (18.3) is not valid for $c < 0$, we use (18.2) instead to give a meaning to the operator e^{-cD^2}. We obtain from (18.2)

$$e^{-y^2/4c} = \frac{\sqrt{c}}{\sqrt{\pi}\, i} \int_{a-i\infty}^{a+i\infty} e^{ys} e^{cs^2} ds \ ,$$

which, when we replace c by $1/4t$, yields

$$e^{-ty^2} = \frac{1}{2\sqrt{\pi t}\, i} \int_{a-i\infty}^{a+i\infty} e^{ys} e^{s^2/4t} ds \ .$$

Thus, we may define e^{-cD^2} as

$$e^{-tD^2}F(x) = \frac{1}{2\sqrt{\pi t}\, i}\int_{a-i\infty}^{a+i\infty} F(x+s)e^{s^2/4t}ds$$
$$= \frac{1}{2\pi i}\int_{d-i\infty}^{d+i\infty} F(u)k(u-x,t)du\ ,$$

where $k(u,t) = \sqrt{\pi/t}\, e^{u^2/4t}$, for some real numbers a and d.

Thus,

$$\lim_{t\to 1^-} e^{-tD^2}F(x) = \lim_{t\to 1^-}\frac{1}{2\pi i}\int_{d-i\infty}^{d+i\infty} F(u)k(u-x,t)du = f(x)\ .$$

Now we state the second inversion formula for the Weierstrass transform.

THEOREM 18.2

Let

$$F(x) = \int_{-\infty}^{\infty} G(x-y,1)f(y)dy$$

be the Weierstrass transform of f, where G is given by (18.2). If f is bounded and continuous, then

$$\lim_{t\to 1^-} e^{-tD^2}F(x) = e^{-D^2}F(x) = f(x)\ , \quad -\infty < x < \infty\ .$$

18.6 Generalized Functions

We can extend the Weierstrass transformation to generalized functions in much the same way as we did with the Laplace transformation. Because the extension of the Laplace transformation was discussed in detail in Chapter 5 , we shall describe briefly the procedure for the Weierstrass transformation. Consider the space $\mathcal{L}_{a,b}$ of Section 3.1.3, Example 6, in which $R = D = \frac{d}{d\tau}$, and

$$\eta(\tau,a,b) = \begin{cases} e^{(-2a\tau+\tau^2)/4}\ , & -\infty, \tau < 0 \\ e^{(-2b\tau+\tau^2)/4}\ , & 0 \le \tau < \infty\ . \end{cases}$$

Hence, $\phi \in \mathcal{L}_{a,b}$ if and only if ϕ is infinitely differentiable and satisfies

$$\gamma_k(\phi) = \sup_{\tau\in\mathcal{R}} \left|\eta(\tau,a,b)D^k\phi(\tau)\right| < \infty\ , \quad k = 0,1,2,\dots\ .$$

It is routine to show that $(\mathcal{L}_{a,b}\, , \{\gamma_k\}_{k=0}^{\infty})$ is a testing-function space, and that the function

$$k(x,t) = \frac{e^{-x^2/4t}}{\sqrt{4\pi t}}$$

belongs to $\mathcal{L}_{a,b}$ for any $0 < t < 1$. Moreover, if $a \leq c$ and $d \leq b$, then $\mathcal{L}_{c,d}$ is a subspace of $\mathcal{L}_{a,b}$ and that the topology of $\mathcal{L}_{c,d}$ is stronger than that induced on it by $\mathcal{L}_{a,b}$. From the space $\mathcal{L}_{a,b}$ we construct the space $\mathcal{L}(w,z)$ of Section 3.2.3, Example 5 as the countable-union space $\bigcup_{k=1}^{\infty} \mathcal{L}_{a_k,b_k}$, where $\{a_k\}_{k=1}^{\infty}$ and $\{b_k\}_1^{\infty}$ are monotone sequences of real numbers such that $a_k \to w^+$ and $b_k \to z^-$.

It is easy to see that for a fixed complex number, s, the function $k(s-\tau,1)$ belongs to $\mathcal{L}(w,z)$ if and only if $w < Re\, s < z$ and $k(s-\tau,t)$ belongs to $\mathcal{L}(w,z)$ for any s, provided that $0 < t < 1$. Moreover, the space $\mathcal{D}$ is a dense subspace of $\mathcal{L}(w,z)$ and $\mathcal{L}(w,z)$ is a dense subspace of $\mathcal{E}$.

Now we are able to extend the Weierstrass transform to generalized functions.

DEFINITION 18.3 *We define the Weierstrass transform of a generalized function $f \in L^*(w,z)$ as*

$$F(s) = \langle f(\tau), k(s-\tau,1)\rangle\ , \quad w < Re\, s < z\ .$$

If f is a locally integrable function such that $f(\tau)/\eta(\tau,c,d)$ is absolutely integrable, then it defines a regular member of $\mathcal{L}_{c,d}$ for every $w < c$ and $d < z$, and its conventional and generalized Weierstrass transforms coincide.

The next theorem provides an inversion formula for the generalized Weierstrass transform.

THEOREM 18.3 [276, p. 212]
Let $F(s)$ be the Weierstrass transform of some $f \in \mathcal{L}^*(w,z)$, for $w < Re\, s < z$. Then,

$$f(\tau) = \lim_{r\to\infty} \frac{1}{i\sqrt{4\pi}} \int_{\sigma-ir}^{\sigma+ir} F(s) e^{(s-\tau)^2/4} ds\ ,$$

in the sense of $\mathcal{D}^*$, where $w < \sigma < z$.

To extend the Poisson or Weierstrass-Sturm-Liouville transform to generalized functions, let us denote the Sturm-Liouville differential operator by R_x

$$R_x = -\frac{d^2}{dx^2} + q(x)\ .$$

We assume that q is a C^∞-function on $(0,\infty)$. From the definition of the associated function of the Sturm-Liouville heat equation

$$g(x,y;t) = \int_{-\infty}^{\infty} e^{-\lambda t}\phi(x,\lambda)\phi(y,\lambda)d\rho(\lambda)\ , \quad t > 0\ ,$$

where the support of $d\rho$ is assumed to be bounded from below, it follows that

$$R_x^n g(x,y;t) = \int_{-\infty}^{\infty} e^{-\lambda t}\lambda^n \phi(x,\lambda)\phi(y,\lambda)d\rho(\lambda) .$$

Since $\rho'(\lambda) = 0(\sqrt{\lambda})$, $\phi(x,\lambda) = 0(1)$, as $\lambda \to \infty$, and $\frac{\partial^k \phi(x,\lambda)}{\partial x^k} = 0(\lambda^p)$ as $\lambda \to \infty$, for some $p > 0, k = 0, 1, 2, \cdots$, it follows that g is infinitely differentiable in both x and y.

Let us consider the space $\mathcal{A} = \mathcal{L}_{a,b}$ in which $\eta(x,a,b) = 1$ and R is the Sturm-Liouville differential operator. Hence, $\psi \in \mathcal{A}$ if and only if ψ is a C^∞-function on $I = (0, \infty)$ satisfying

$$\gamma_k(\psi) = \sup_{x \in I} |R_x^k \psi(x)| < \infty .$$

It is easy to see that $(\mathcal{A}, \{\gamma_k\}_{k=0}^{\infty})$ is a testing-function space and that $g(x,y;t) \in \mathcal{A}$ as a function of x or y. Hence, we have the following definition.

DEFINITION 18.4 *The Poisson-Sturm-Liouville transform of a generalized function $f \in \mathcal{A}^*$ is defined as*

$$F(x,t) = \langle f(y), g(x,y;t)\rangle .$$

We close this section by pointing out that an inversion formula for the Poisson-Sturm-Liouville transform can be obtained by a simple modification of the argument given in Theorem 4.1 in [268].

18.7 Applications

Let $u(x,t)$ denote the temperature at time t in an infinitely long rod lying along the x-axis $(-\infty < x < \infty)$. Under the assumption that the rod is perfectly insulated and is made of homogeneous material in a uniform shape, the differential equation governing the temperature distribution is the homogeneous heat equation for one-dimensional flow

$$c\frac{\partial^2 u}{\partial x^2} = \frac{\partial u}{\partial t} , \quad t \geq 0 , \quad -\infty < x < \infty ,$$

where c is a positive constant depending on the properties of the rod material. The Cauchy-problem for the heat equation is to solve for $u(x,t)$ under the initial condition $u(x,0) = f(x)$.

We have seen in Section 18.1 that the Poisson (Weierstrass) transform of the initial data $f(x)$ provides a solution for the Cauchy problem. Similarly, the Poisson-Hankel, Poisson-Laguerre and Poisson-Sturm-Liouvile transforms provide solutions for the Cauchy problem of the Hankel, Laguerre and Sturm-Liouville heat equations.

19

The Abel Transform

The Abel transform is intimately related to the Abel integral equation. They are named after the Norwegian mathematician Niels Henrick Abel (1802-1829) who is regarded as the father of the subject of integral equations [1]. Although he died very young at the age of twenty-six, Abel accomplished many great mathematical achievements in his short life. His name is perpetuated also in the subject of Abelian groups.

Among the many topics in which Abel was interested were elliptic integrals and a particular type of Volterra integral equations that had several important applications. The most important of these applications is the tautochrone problem, which is the problem of determining a curve passing through the origin for which the time required for a particle to slide down the curve is independent of the starting point. The only forces acting on the particle are the gravity and the reaction of the curve on which the particle is constrained to move. The solution of this problem requires solving the equation

$$T = \frac{1}{\sqrt{2g}} \int_0^y \frac{f(x)}{\sqrt{y-x}} dx,$$

for the unknown function f, where T is a fixed constant and g is the acceleration of gravity. This equation is called Abel's integral equation. A more general form of Abel's integral equation is

$$g(y) = \int_0^y (y-x)^{-\alpha} f(x) dx, \quad 0 < \alpha < 1,$$

where g is a known function.

19.1 Definitions

DEFINITION 19.1 *The Abel transforms of the first and second kinds,* $F_{\alpha,1}(x)$, $F_{\alpha,2}(x)$ *or* $\tilde{A}_{\alpha,1}[f](x)$, $\tilde{A}_{\alpha,2}[f](x)$ *of* $f(r)$ *are defined as*

$$\tilde{A}_{\alpha,1}[f](x) = F_{\alpha,1}(x) = 2\int_x^\infty (r^2 - x^2)^{-\alpha} f(r) r dr, \quad 0 < \alpha < 1, \tag{19.1}$$

and

$$\tilde{A}_{\alpha,2}[f](x) = F_{\alpha,2}(x) = 2\int_0^x (r^2 - x^2)^{-\alpha} f(r) r dr, \quad 0 < \alpha < 1.$$

The choice of the symbols x and r is suggested by many applications in which r stands for a radius and x for an abscissa. Putting

$$\tilde{k}(x,r) = \begin{cases} 2r(r^2 - x^2)^{-\alpha}, & r > x \\ 0, & r \leq x \end{cases}$$

we find

$$\tilde{A}_{\alpha,1}[f](x) = \int_0^\infty \tilde{k}(x,r) f(r) dr.$$

Upon making the change of variable, $\rho = r^2, y = x^2$ and setting $g(\rho) = f(r), F_{\alpha,1}(x) = G_{\alpha,1}(y)$ we obtain

$$A_{\alpha,1}[g](y) = G_{\alpha,1}(y) = \int_y^\infty (\rho - y)^{-\alpha} g(\rho) d\rho, \tag{19.2}$$

or

$$A_{\alpha,1}[g](y) = G_{\alpha,1}(y) = \int_0^\infty k(y,\rho) g(\rho) d\rho,$$

where

$$k(y,\rho) = \begin{cases} (\rho - y)^{-\alpha}, & \rho > y \\ 0, & \rho \leq y \end{cases}.$$

Similar expressions can be easily obtained for the Abel transform of the second kind. Because of the close similarities between the two transforms, we shall restrict ourselves mainly to the Abel transform of the first kind and leave to the reader the derivation of the analogous results for the second kind.

To simplify the notation, we shall denote the transform of the first kind by $\tilde{A}_\alpha[f] = F_\alpha$. We shall also call (19.2) the modified Abel transform of the first kind and denote it by $A_\alpha[f] = G_\alpha$. The modified Abel transform of the second kind is defined similarly. It is worth noting that some authors (cf. [23]) define the Abel transform by (19.1), with $\alpha = 1/2$, and call (19.2), when $\alpha = 1/2$, the modified Abel transform.

19.2 Existence

If $g(\rho)$ is a bounded, integrable function, then the modified Abel transform of the first kind, $G_\alpha(y)$, is well defined. For, if we choose $\delta > 0$, then

$$|G_\alpha(y)| \le \int_y^{y+\delta} \left|\frac{g(\rho)}{(\rho-y)^\alpha}\right| d\rho + \int_{y+\delta}^{\infty} \left|\frac{g(\rho)}{(\rho-y)^\alpha}\right| d\rho$$

$$\le \frac{\delta^{1-\alpha}}{(1-\alpha)} \sup_{y\le\rho\le y+\delta} |g(\rho)| + \frac{1}{\delta^\alpha}\int_{y+\delta}^{\infty} |g(\rho)|\, d\rho < \infty.$$

The existence of the Abel transform of the first kind can be established similarly.

Example A: Let $g(\rho)$ be the characteristic function of the interval $[0, b]$. Then

$$A_\alpha[g](y) = G_\alpha(y) = \int_y^b (\rho - y)^{-\alpha} d\rho = \frac{1}{(1-\alpha)}(b-y)^{1-\alpha}.$$

If we revert to the variables r and x, we obtain

$$F_\alpha(x) = \tilde{A}_\alpha[f](x) = \frac{1}{(1-\alpha)}(a^2 - x^2)^{1-\alpha}, \quad b = a^2$$

in which f can be viewed as the characteristic function of the disc $|z| \le a$. In particular, for $\alpha = 1/2, F_{1/2}(x) = 2\sqrt{a^2 - x^2}$ describes a semi-ellipse.

19.3 Elementary Properties

1. **Linearity:** The Abel and modified Abel transformations are linear

transformations.

2. **Translation and Dilation:** The modified Abel transformation satisfies the following properties:

$$T_a(A_\alpha f) = A_\alpha[T_a f], \quad \text{for all } a,$$

and

$$D_{1/a}(A_\alpha f) = a^{1-\alpha} A_\alpha[D_{1/a} f], \quad a > 0$$

where T_a and D_a are the translation and dilation operators defined in Section 1.1. These relations may be written explicitly as

$$A_\alpha[f](x-a) = A_\alpha[f(t-a)](x),$$

$$A_\alpha[f](ax) = a^{1-\alpha} A_\alpha[f(at)](x), \; a > 0.$$

3. **Differentiation:** If f is differentiable and of order $o(y^{1-\alpha})$ as $y \to \infty$, then

$$A_\alpha[f](y) = -\frac{1}{(1-\alpha)}(A_{\alpha-1}[f'])(y).$$

19.4 Relationships with Other Transforms

If we denote the Abel, Fourier and Hankel transformations by $\tilde{A}_\alpha, \mathcal{F}$, and $\mathcal{H}_\alpha$, respectively, then

$$\mathcal{F}(\tilde{A}_\alpha[f])(w) = \Gamma(1-\alpha)\left(\frac{w}{2}\right)^{\alpha-1} \mathcal{H}_{1/2-\alpha}\left[y^{1-\alpha} f(y)\right](w).$$

For, by (19.1) and (2.17), we have

$$\begin{aligned}
\mathcal{F}(\tilde{A}_\alpha[f])(w) &= \frac{1}{\sqrt{2\pi}} \int_{-\infty}^{\infty} e^{ixw} dx \; 2\int_x^{\infty} (r^2 - x^2)^{-\alpha} f(r) r \, dr \\
&= \frac{1}{\sqrt{2\pi}} \int_{-\infty}^{\infty} e^{ixw} dx \int_{x^2}^{\infty} (\rho - x^2)^{-\alpha} f\left(\sqrt{\rho}\right) d\rho \\
&= \frac{1}{\sqrt{2\pi}} \int_0^{\infty} f\left(\sqrt{\rho}\right) d\rho \int_{-\sqrt{\rho}}^{\sqrt{\rho}} \frac{e^{ixw}}{(\rho - x^2)^\alpha} dx
\end{aligned}$$

$$= \frac{\Gamma(1-\alpha)}{2^{\alpha}} \int\limits_0^{\infty} f\left(\sqrt{\rho}\right) w^{\alpha-1/2} \rho^{1/4-a/2} J_{1/2-\alpha}(\sqrt{\rho}w)\, d\rho$$

$$= 2^{1-\alpha}\Gamma(1-\alpha) w^{\alpha-1} \int\limits_0^{\infty} \left(y^{1-\alpha} f(y)\right) \sqrt{yw} J_{1/2-\alpha}(yw)\, dy$$

$$= 2^{1-\alpha}\Gamma(1-\alpha) w^{\alpha-1} \mathcal{H}_{1/2-\alpha}\left[y^{1-\alpha} f(y)\right](w),$$

where J_ν is the Bessel function of the first kind and order ν. In particular,

$$\mathcal{F}\left(\tilde{A}_{1/2}[f]\right)(w) = \sqrt{\frac{2\pi}{w}} \mathcal{H}_0\left[\sqrt{y} f(y)\right](w).$$

19.5 Inversion Formula

Relation (19.2) can be put in the form

$$A_\alpha[g](y) = G_\alpha(y) = \int\limits_{-\infty}^{\infty} k(y-\rho) g(\rho) d\rho$$

with

$$k(z) = \begin{cases} (-z)^{-\alpha}, & z < 0 \\ 0, & z \geq 0 \end{cases}.$$

Having reduced the modified Abel transform of the first kind to a convolution equation, we may use the convolution theorem for the Fourier transform to obtain

$$(\widehat{A_\alpha[g]})(w) = \hat{G}_\alpha(w) = \sqrt{2\pi} \hat{k}(w) \hat{g}(w),$$

in which

$$\hat{k}(w) = \frac{1}{\sqrt{2\pi}} (iw)^{\alpha-1} \Gamma(1-\alpha),$$

where $(iw)^{\alpha-1}$ is real on the positive imaginary axis. Thus, we have the relation

$$\hat{g}(w) = \frac{1}{(i)^{\alpha-1}\Gamma(1-\alpha)} w^{-\alpha} \left(w \hat{G}_\alpha(w)\right).$$

If we apply the inverse Fourier transform to this equation, it will yield the inversion formula for the modified Abel transform of the first kind

$$g(x) = \frac{-\sin \alpha\pi}{\pi} \frac{d}{dx} \int_x^{\infty} (y-x)^{\alpha-1} \left(G_\alpha\right)(y)\, dy. \tag{19.3}$$

If G_α is differentiable, it may be possible to take the derivative inside the integral sign to obtain

$$g(x) = \frac{-\sin\alpha\pi}{\pi}\int_x^\infty (y-x)^{\alpha-1}\frac{d}{dy}(G_\alpha)(y)\,dy,$$

provided that the integral exists. In terms of the Abel transform defined in (19.1), the inversion formula may take the following form: if

$$\tilde{A}_\alpha[f](x) = 2\int_x^\infty (r^2-x^2)^{-\alpha} f(r) r\,dr,$$

then

$$f(r) = \frac{\sin\alpha\pi}{\pi}\frac{d}{dr}\int_r^\infty (x^2-r^2)^{\alpha-1}\left(\tilde{A}_\alpha[f]\right)(x)dx.$$

These last two equations are known, in particular when $\alpha = 1/2$, as the *Abel transform pair*. A similar argument can be used to derive the following more general formulas: if

$$g(x) = \int_x^b \frac{f(t)\,dt}{(t-x)^\alpha}, \quad 0 < x < b \le \infty,\ 0 < \alpha < 1,\ g(b) = 0, \tag{19.4}$$

then

$$f(x) = \frac{-\sin\alpha\pi}{\pi}\frac{d}{dx}\int_x^b (t-x)^{\alpha-1} g(t)\,dt. \tag{19.5}$$

The inversion formula for the modified Abel transform of the second kind can be written as follows: If

$$g(x) = \int_0^x \frac{f(t)\,dt}{(x-t)^\alpha}, \quad 0 < x,\ 0 < \alpha < 1,\ g(0) = 0, \tag{19.6}$$

then

$$f(x) = \frac{\sin\alpha\pi}{\pi}\frac{d}{dx}\int_0^x (x-t)^{\alpha-1} g(t)\,dt. \tag{19.7}$$

By using a change of variables, the inversion formulas of the Abel transforms can be generalized to the following. Let $\phi(t)$ be a monotone increasing, differentiable function with $\phi'(t) \neq 0$. If

$$g(x) = \int_x^b \frac{f(t)\,dt}{[\phi(t)-\phi(x)]^\alpha}, \quad a < x < b \le \infty, \tag{19.8}$$

where $0 < \alpha < 1$, $g(b) = 0$, then

$$f(x) = \frac{-\sin\alpha\pi}{\pi}\frac{d}{dx}\int_x^b \frac{\phi'(t)g(t)\,dt}{[\phi(t)-\phi(x)]^{1-\alpha}}, \tag{19.9}$$

and if

$$g(x) = \int_a^x \frac{f(t)\,dt}{[\phi(x) - \phi(t)]^\alpha}\,, \quad a < x < b\,, \tag{19.10}$$

where $0 < \alpha < 1$, $g(a) = 0$, then

$$f(x) = \frac{\sin \alpha\pi}{\pi} \frac{d}{dx} \int_a^x \frac{\phi'(t) g(t)\,dt}{[\phi(x) - \phi(t)]^{1-\alpha}}\,. \tag{19.11}$$

The conditions imposed on ϕ ensure that its inverse function exists, which is needed in the proof [216, p. 210, 318].

19.6 The Abel Transform of Generalized Functions

The Abel transform of generalized functions can be derived as a special case of the fractional derivatives and integrals of generalized functions. The latter is treated in the chapter on the Riemann–Liouville and Weyl fractional integrals, Chapter 20, and to which the discussion of the Abel transform of generalized functions is relegated.

19.7 Applications

The Abel transformation is used in many problems where circularly symmetrical distributions in two dimensions are projected on one dimension. A typical example is the electrical response of a television camera as it scans across a narrow line; see [23, p. 202]. The Abel transform is also closely related to the fractional order derivatives, which arise in problems of transmissions of electrical signals through cables. For the definition of a fractional order derivative, see Section 20.2.

Another application of the Abel transform can be found in the field of fluid dynamics, where it is used in determining the form of a symmetrical weir when the quantity of the flow per unit time is proportional to a given power of the depth of the stream [28].

The Abel transform also arises in connection with the Radon transform when circular symmetry is assumed. If a function f of two variables is given by

$$f(r, \theta) = \sum_{n=-\infty}^{\infty} g_n(r) e^{in\theta}, \ 0 \le r < \infty$$

then its Radon transform is given by (see Section 33.5.3)

$$f^{\dagger}(\rho,\phi) = \sum_{n=-\infty}^{\infty} \breve{g}_n(\rho)e^{in\phi}, \quad 0 \le \rho < \infty,$$

where g_n and $\breve{g}_n$ are related by the *Tchebycheff transform pair*

$$g_n(r) = -\frac{1}{\pi}\int_r^{\infty} \breve{g}_n(\rho)T_n\left(\frac{\rho}{r}\right)(\rho^2 - r^2)^{-1/2}d\rho,$$

and

$$\breve{g}_n(\rho) = 2\int_{\rho}^{\infty} g_n(r)T_n\left(\frac{\rho}{r}\right)(r^2 - \rho^2)^{-1/2}r\,dr.$$

These equations reduce to the Abel transform pair when $n = 0$.

20

The Riemann–Liouville and Weyl Fractional Integrals

20.1 Introduction

The concept of a non-integer differentiation and integration of a function is almost as old as calculus itself. The notion of the one-half derivative of a function was mentioned by G.W. Leibniz in a letter to L'Hospital in 1695 [141]. Many famous mathematicians, including Euler, Lagrange, Laplace, Fourier, and Abel, either pondered over the idea or made some contributions to it. But it was not until Liouville and Riemann that the idea became more precisely formulated.

S. Liouville [150] was the first to conduct a major study on fractional calculus. First, he studied the one-half derivative of exponential functions, then he extended it and defined the qth derivative (q not necessarily a non-negative integer) of a function $f(x)$ expansible as a series of exponentials. He defined the qth derivative of

$$f(x) = \sum_k c_k \exp(b_k x),$$

as

$$\frac{d^q f}{dx^q} = \sum_k c_k b_k^q \exp(b_k x).$$

More than one hundred years later, this definition was extended by Civin [41] to functions expansible as integrals of exponentials. His definition goes as follows: if

$$f(x) = \int_{-\pi}^{\pi} \exp(ixt) d\alpha(t),$$

then

$$\frac{d^q f(x)}{dx^q} = \int_{-\pi}^{\pi} (it)^q \exp(itx) d\alpha(t).$$

Noting that the nth derivative of x^p (n is a non-negative integer and $p > 0$) is

$$\frac{d^n x^p}{dx^n} = p(p-1)\ldots(p-n+1)x^{p-n} = \frac{\Gamma(p+1)}{\Gamma(p-n+1)}x^{p-n}, \tag{20.1}$$

it is natural to extend this result and define the qth derivative of x^p as

$$\frac{d^q x^p}{dx^q} = \frac{\Gamma(p+1)}{\Gamma(p-q+1)}x^{p-q}, \qquad q > 0. \tag{20.2}$$

But when p is negative, say $p = -r$, the analogue of (20.1) is

$$\frac{d^n}{dx^n}x^{-r} = (-1)^n \frac{\Gamma(r+n)}{\Gamma(r)}x^{-n-r},$$

and hence we may define the qth derivative of x^{-r} as

$$\frac{d^q}{dx^q}x^{-r} = (-1)^q \frac{\Gamma(r+q)}{\Gamma(r)}x^{-q-r}, \qquad q > 0, \qquad r + q > 0. \tag{20.3}$$

Deducing the fractional derivative of the constant function $f(x) = x^0 = 1$, by setting $p = 0$ in (20.2) and $r = 0$ in (20.3) leads to the contradiction that

$$\frac{d^q}{dx^q}1 = \frac{1}{\Gamma(1-q)}x^{-q}, \qquad q > 0,$$

and

$$\frac{d^q}{dx^q}1 = 0,$$

which was the center of a controversy that lasted for a number of years.

Seeking a generalization of a Taylor's series expansion, Riemann [203] proposed a different approach that involved definite integrals and was applicable to power series with non-integer exponents. One of Riemann's main contributions in this paper was his definition of the qth integral of a function f as

$$\frac{d^{-q} f}{dx^{-q}} = \frac{1}{\Gamma(q)} \int_a^x (x-t)^{q-1} f(t)dt.$$

Because integration can be viewed as an inverse operation to differentiation, it is natural to associate with it the symbol

$$\frac{d^{-1}}{dx^{-1}}.$$

It is also customary to denote the definite integral of f by

$$\frac{d^{-1} f}{dx^{-1}} = \int_a^x f(t)dt.$$

Notwithstanding its popularity, this notation is rather ambiguous because it does not display the dependence of the definite integral on the lower limit of integration, which in some instances can be crucial. Some authors suggest the notation

$$\frac{d^{-1}f}{d(x-a)^{-1}}$$

to denote the definite integral

$$\int_a^x f(x)dt.$$

Although this notation is more accurate, it will not be used here because the role that the lower limit of integration plays in our investigation is minor.

Riemann's definition can now be justified. From the identity

$$\frac{d^{-1}f}{dx^{-1}} = \int_a^x f(y)dy,$$

we obtain by another integration

$$\frac{d^{-2}f}{dx^{-2}} = \int_a^x dx_1 \int_a^{x_1} f(y)dy = \frac{1}{1!}\int_a^x (x-y)f(y)dy\,.$$

An $n-1$-fold integration produces

$$\frac{d^{-n}f}{dx^{-n}} = \frac{1}{(n-1)!}\int_a^x (x-y)^{n-1}f(y)dy.$$

Thus, if we replace n by a non-negative number q, we obtain Riemann's definition.

For example, if $a=0$, the qth integral ($q>0$) of x^p, where $p>-1$, can be evaluated with the aid of the Beta function (cf. Section 2.1) to yield

$$\begin{aligned}\frac{d^{-q}}{dx^{-q}}x^p &= \frac{1}{\Gamma(q)}\int_0^x (x-t)^{q-1}t^p dt \\ &= \frac{x^{q+1}}{\Gamma(q)}\int_0^1 (1-t)^{q-1}t^p dt = \frac{\Gamma(p+1)}{\Gamma(q+p+1)}x^{q+p}.\end{aligned} \tag{20.4}$$

Using Riemann's definition of a fractional integral, we can now define the fractional derivative as follows: If $q \geq 0$, choose m such that $m-1 \leq q < m$ and define

$$\frac{d^q}{dx^q}f(x) = \frac{d^m}{dx^m}\frac{d^{-(m-q)}}{dx^{-(m-q)}}f(x) = \frac{d^m}{dx^m}\frac{1}{\Gamma(m-q)}\int_a^x (x-t)^{m-q-1}f(t)dt.$$

In particular, if $f(x)=x^p$, we obtain (20.2) as a special case.

Working with periodic functions, Weyl [243] observed that Riemann's definition was not appropriate for a large class of functions because it was dependent on the lower limit of integration, a. He found that it was more appropriate to replace the lower limit of integration by $-\infty$ and define the qth integral of f as

$$\frac{d^{-q}}{dx^{-q}} f = \frac{1}{\Gamma(q)} \int_{-\infty}^{x} (x-t)^{q-1} f(t)dt.$$

20.2 Definitions

DEFINITION 20.1 *The Riemann–Liouville and Weyl fractional integrals of order q of a function $f(t)$ are defined, respectively, by*

$$\begin{aligned}(I_q f)(x) &= \frac{d^{-q}}{dx^{-q}} f(t) = I_q[f(t)](x) \\ &= \frac{1}{\Gamma(q)} \int_a^x (x-t)^{q-1} f(t)dt, \qquad q > 0,\end{aligned}$$

and

$$(W_q f)(x) = \frac{1}{\Gamma(q)} \int_{-\infty}^{x} (x-t)^{q-1} f(t)dt.$$

Without loss of generality, we may take a to be zero. Another version of the Weyl fractional integral can be obtained by a change of variables:

$$(W_q f)(x) = \frac{1}{\Gamma(q)} \int_{x}^{\infty} (t-x)^{q-1} f(t)dt.$$

It is convenient to introduce a notation that unifies differentiation and integration.

DEFINITION 20.2 *The Riemann–Liouville fractional derivative operator of order q is defined by*

$$D_x^q f(x) = \frac{d^q}{dx^q} f(x) = \begin{cases} 1/\Gamma(-q) \int_a^x (x-t)^{-q-1} f(t)dt\,, & q < 0 \\ \frac{d^m}{dx^m} (D_x^{q-m} f(x))\,, & q \geq 0, \end{cases}$$

where

$$\frac{d^m}{dx^m} (D_x^{q-m} f(x)) = \frac{1}{\Gamma(m-q)} \frac{d^m}{dx^m} \int_a^x (x-t)^{m-q-1} f(t)dt,$$

and $m - 1 \leq q < m\,, m = 1, 2, 3, \ldots$.

In particular, if $0 < q < 1$,

$$D_x^q f(x) = \frac{d}{dx}(I_{1-q} f)(x).$$

Because the operator D^q embraces both differentiation and integration within a single formula, some authors call it a differintegration operator [180].

DEFINITION 20.3 *The integral operators*

$$I^{\alpha,\eta}[f](x) = \frac{1}{\Gamma(\alpha)} x^{-\alpha-\eta} \int_0^x (x-t)^{\alpha-1} t^{\eta} f(t) dt,$$

$$K^{\alpha,\eta}[f](x) = \frac{1}{\Gamma(\alpha)} x^{\eta} \int_x^\infty (t-x)^{\alpha-1} t^{-\alpha-\eta} f(t) dt,$$

are called the Erdélyi–Kober fractional integral operators, and

$$I_m^{\alpha,\eta}[f](x) = \frac{m}{\Gamma(\alpha)} x^{-m\eta - m\alpha} \int_0^x (x^m - t^m)^{\alpha-1} t^{m+m\eta-1} f(t) dt, \tag{20.5}$$

$$K_m^{\alpha,\eta}[f](x) = \frac{m}{\Gamma(\alpha)} x^{m\eta} \int_x^\infty (t^m - x^m)^{\alpha-1} t^{m-m\alpha-m\eta-1} f(t) dt, \tag{20.6}$$

are called Erdélyi–Kober–type fractional integral operators, where $m > 0$, $\operatorname{Re}\alpha > 0$ *and* η *is a suitable complex number.*

Clearly, when $m = 1$, $I_m^{\alpha,\eta}$, $K_m^{\alpha,\eta}$ reduce to $I^{\alpha,\eta}$ and $K^{\alpha,\eta}$, respectively, and when $\eta = 0$, $I^{\alpha,\eta}$ and $K^{\alpha,\eta}$ become related to the Riemann–Liouville and Weyl fractional integrals; see [56, 57, 123] for details. For special values of η, the operator $K_m^{\alpha,\eta}$ becomes the adjoint of $I_m^{\alpha,\eta}$. In fact,

$$\langle I_m^{\alpha,\eta}[f]\,,\, g\rangle = \langle f\,,\, K_m^{\alpha,\eta+1-1/m}[g]\rangle\,, \tag{20.7}$$

where the inner product is taken as

$$\langle f\,,\, g\rangle = \int_0^\infty f(x) g(x)\, dx.$$

In particular, we have

$$\langle I^{\alpha,\eta}[f]\,,\, g\rangle = \langle f\,,\, K^{\alpha,\eta}[g]\rangle\,.$$

Other definitions of fractional derivatives and integrals (differintegrals) have been obtained using the Cauchy integral formula for analytic functions; see Section 1.3.

DEFINITION 20.4 *Let q be a non-negative integer. We define the q th derivative of an analytic function $f(z)$ as*

$$\frac{d^q f}{dz^q} = \frac{\Gamma(q+1)}{2\pi i} \int_\gamma \frac{f(\zeta)\, d\zeta}{[\zeta - z]^{q+1}},$$

where γ is a contour that lies entirely inside the region of analyticity of f. The contour γ begins and ends at some point $\zeta = a$ enclosing the point z in the positive direction.

When q is not an integer, the function $(\zeta - z)^{q+1}$ no longer has a pole at $\zeta = z$, but a branch point. Since the integral depends on the location of the point at which γ crosses the branch line for $(\zeta - z)^{q+1}$, we choose the branch line to be the straight line joining a and z and continuing indefinitely in the quadrant $\mathrm{Re}\,(\zeta - a) \le 0$, $\mathrm{Im}\,(\zeta - a) \le 0$. To determine the denominator of the integrand uniquely, we define

$$(\zeta - z)^{q+1} = \exp\left[(q+1)\ln(\zeta - z)\right],$$

where $\ln(\zeta - z)$ is real when $(\zeta - z)$ is positive.

To relate this definition to the Riemann–Liouville definition, we first deform the contour into a contour γ' lying on both sides of the branch line. Then use the relations,

$$(-1)^{-q-1} = \exp[i\pi(q+1)],$$

$$(\zeta - z)^{q+1} = \exp[(q+1)\ln(\zeta - z)]$$

with $\ln(\zeta - z)$ being real when $\zeta - z > 0$, and

$$\Gamma(x)\Gamma(1-x) = \pi \csc(\pi x) = \frac{2\pi i}{e^{i\pi x} - e^{-i\pi x}}$$

to obtain

$$\begin{aligned}\frac{d^q f}{dz^q} &= \frac{\Gamma(q+1)}{2\pi i} \int_{\gamma'} \frac{f(\zeta)\, d\zeta}{(\zeta - z)^{q+1}} \\ &= \frac{\Gamma(q+1)}{2\pi i}[1 - \exp(-2\pi i(q+1))] \int_a^z \frac{f(\zeta)\, d\zeta}{(\zeta - z)^{q+1}} \\ &= \frac{-e^{-i\pi q}}{\Gamma(-q)} \int_a^z \frac{f(\zeta)\, d\zeta}{(z - \zeta)^{q+1}} = \frac{1}{\Gamma(-q)} \int_a^z \frac{f(\zeta) d\zeta}{(z - \zeta)^{q+1}}.\end{aligned}$$

Another definition of fractional derivatives based on Cauchy's formula was introduced by Erdélyi.

DEFINITION 20.5 *The Erdélyi qth-order derivative of an analytic function $f(z)$ with respect to the function z^n is defined by*

$$D^q_{z^n} f(z) = \frac{\Gamma(q+1)}{2\pi i} \int_a^z \frac{f(\zeta) n\zeta^{n-1} d\zeta}{[\zeta^n - z^n]^{q+1}}, \tag{20.8}$$

where a and z are in the region of analyticity of f.

Erdélyi's definition of the qth fractional derivative has been generalized by Osler [181] to the following.

DEFINITION 20.6 *Let $f(z)$ be analytic in the simply connected region Ω. Let $g(z)$ be analytic and univalent in Ω, and let $g^{-1}(0)$ be an interior or a boundary point of Ω. Assume that $\oint_\gamma f(z)dz = 0$ for any simple closed contour γ in $\Omega \cup \{g^{-1}(0)\}$ through $g^{-1}(0)$. Then if q is not a negative integer and z is in Ω, the qth-order derivative of $f(z)$ with respect to $g(z)$ is defined to be*

$$D^q_{g(z)} f(z) = \frac{\Gamma(q+1)}{2\pi i} \int_{g^{-1}(0)}^{z^+} \frac{f(\zeta) g'(\zeta) d\zeta}{[g(\zeta) - g(z)]^{q+1}}. \tag{20.9}$$

For non-integral q, the integral has a branch line which begins at $\zeta = z$ and passes through $\zeta = g^{-1}(0)$. The contour of integration starts at $g^{-1}(0)$, encloses z once in the positive sense and returns to $g^{-1}(0)$ without cutting the branch line or leaving $\Omega \cup \{g^{-1}(0)\}$.

Cases of particular interest are:

1) $g(z) = z^n - a^n$, which reduces (20.9) to (20.8),
2) $g(z) = z - a$, which yields

$$D^q_{z-a} f(z) = \frac{\Gamma(q+1)}{2\pi i} \int_a^z \frac{f(\zeta) d\zeta}{(\zeta - z)^{q+1}}. \tag{20.10}$$

While ordinary derivatives with respect to z and $z - a$ are equal, formula (20.10) shows that this is not the case for fractional derivatives since the value of the contour integral depends on the point $\zeta = a$ at which the contour crosses the branch line.

If f is analytic in a simply connected region containing the origin, then the fractional integral of order q of f can be defined, as in the real-valued case, by

$$D_z^{-q} f(z) = \frac{d^{-q}}{dz^{-q}} f(z) = \frac{1}{\Gamma(q)} \int_0^z \frac{f(\zeta)}{(z-\zeta)^{1-q}} d\zeta, \quad q > 0,$$

where $\ln(z - \zeta)$ is required to be real when $z - \zeta > 0$. Likewise, the fractional derivative of order q of f can be defined by

$$D_z^q f(z) = \frac{d^q}{dz^q} f(z) = \frac{1}{\Gamma(1-q)} \frac{d}{dz} \int_0^z \frac{f(\zeta)}{(z-\zeta)^q} d\zeta, \quad (0 \le q < 1),$$

It now follows that for $f(z) = z^n$,

$$D_z^{-q} z^n = \frac{1}{\Gamma(q)} \int_0^z \frac{\zeta^n}{(z-\zeta)^{1-q}} d\zeta = \frac{z^{n+q}}{\Gamma(q)} \int_0^1 u^n (1-u)^{q-1} du$$

$$= \frac{\Gamma(n+1)}{\Gamma(n+1+q)} z^{n+q}, \qquad q > 0$$

and

$$D_z^q z^n = \frac{1}{\Gamma(1-q)} \frac{d}{dz} \int_0^z \frac{\zeta^n}{(z-\zeta)^q} d\zeta = \frac{\Gamma(n+1)}{\Gamma(n-q+1)} z^{n-q}.$$

Therefore, if

$$f(z) = \sum_{n=0}^{\infty} a_n z^n$$

is analytic in the unit disc, then

$$D_z^{-q} f(z) = \sum_{n=0}^{\infty} a_n \frac{\Gamma(n+1)}{\Gamma(n+1+q)} z^{n+q}, \qquad q > 0,$$

and

$$D_z^q z^n = \sum_{n=0}^{\infty} a_n \frac{\Gamma(n+1)}{\Gamma(n-q+1)} z^{n-q}, \qquad 0 \leq q < 1. \qquad (20.11)$$

For $q = 0, 1$, we have

$$D_z^0 f(z) = \sum_{n=0}^{\infty} a_n z^n = f(z),$$

and

$$D_z^1 f(z) = \sum_{n=0}^{\infty} n a_n z^{n-1} = \frac{df}{dz},$$

which shows that (20.11) is valid also for $q = 1$.

20.3 Existence

If $f \in L^1(\mathcal{R}^+)$, then its Riemann–Liouville fractional integral $(I_\alpha f)(x)$ exists almost everywhere for $x > 0$ and defines an integrable function as well. The following theorems are due to Hardy and Littlewood [89].

THEOREM 20.1 **(Hardy and Littlewood)**
If $f \in L^p(a,b)$, $p > 1$, $-\infty < a < b \leq \infty$, $0 < \alpha < 1/p$, then $(I_\alpha f) \in L^q(a,b)$, where $q = p/(1-\alpha p)$ with

$$\left(\int_a^b |(I_\alpha f)(x)|^q dx\right)^{1/q} \leq c\left(\int_a^b |f(x)|^p dx\right)^{1/p}.$$

The result is not true for $p = 1$.

The theorem is valid for $a = 0$, $b = 2\pi$, f is periodic with period 2π and has mean value zero, i.e., $\int_0^{2\pi} f(x)dx = 0$,

If $0 < \beta \leq k \leq 1$, and $f \in Lip_k(\mathcal{R}^+)$, then the fractional derivative, $D^\beta f$, in the sense of Definition 20.2 exists and defines a continuous function for which the βth fractional integral in the Riemann–Liouville sense is $f(x)$.

THEOREM 20.2
If $p \geq 1$, $k \geq 0$, $\alpha > 0$, $k + \alpha < 1$, $f \in L^p(\mathcal{R}^+)$ and

$$\int_a^b |f(x+h) - f(x)|^p dx = O(h^{pk})$$

for every finite interval (a,b), then

$$\int_a^b |(I_\alpha f)(x+h) - (I_\alpha f)(x)|^p dx = O(h^{p(k+\alpha)}).$$

Also if $p > 1$, $0 < k \leq 1$, $f \in L^p(\mathcal{R})$, and

$$\int_a^b |f(x+h) - f(x)|^p dx = O(h^{pk})$$

for every finite interval (a,b), then $\frac{d^\beta f}{dx^\beta}$ exists almost everywhere and belongs to $L^p(\mathcal{R})$ for all $\beta < k$ and

$$\int_a^b |(\frac{d^\beta f}{dx^\beta})(x+h) - (\frac{d^\beta f}{dx^\beta})(x)|^p dx = O(h^{p(k-\beta)}).$$

Similar results hold for the Weyl fractional integral.

THEOREM 20.3
If $f \in L^1_{loc}(\mathcal{R}^+)$ is of bounded variation in a neighborhood of x_0 ($x_0 \geq 0$), then $(I_\alpha f)(x)$ is also of bounded variation in a neighborhood of x_0 and satisfies Lipschitz condition of order α at x_0,

$$|(I_\alpha f)(x_0+h) - (I_\alpha f)(x_0)| = O(|h|^\alpha) \quad \text{as} \quad |h| \longrightarrow 0, x_0 + h > 0.$$

20.4 Elementary Properties

20.4.1 Elementary Properties of I_α and W_α

1. **Linearity:**
   Both I_α and W_α are linear transformations.

2. **Translation:**
$$W_\alpha[T_a f(t)](x) = T_a(W_\alpha f)(x),$$
   and if $f = 0$ on $(-\infty, 0)$, then
$$I_\alpha[T_a f(t)](x) = T_a(I_\alpha f)(x), \qquad a > 0.$$

3. **Dilation:**
$$I_\alpha[D_{1/a} f(t)](x) = a^{-\alpha} D_{1/a}(I_\alpha f)(x), \qquad a > 0,$$
$$W_\alpha[D_{1/a} f(t)](x) = a^{-\alpha} D_{1/a}(W_\alpha f)(x), \qquad a > 0.$$

4. **Differentiation:**
$$I_\alpha[f'(t)](x) = I_{\alpha-1}[f](x) - \frac{1}{\Gamma(\alpha)} f(0) x^{\alpha-1},$$
$$W_\alpha[f'(t)](x) = W_{\alpha-1}[f(t)](x).$$

5. **Integration:**
$$I_\alpha\left[\int_0^t f(u)\,du\right](x) = I_{\alpha+1}[f(t)](x),$$
$$W_\alpha\left[\int_{-\infty}^t f(u)du\right](x) = W_{\alpha+1}[f(t)](x).$$

6. **Iteration:**
$$I_\alpha[I_\beta f](x) = I_{\alpha+\beta}[f](x), \tag{20.12}$$
$$W_\alpha\left[W_\beta[f]\right](x) = W_{\alpha+\beta}[f](x), \tag{20.13}$$

The proof of (1)–(5) is straightforward, but that of (6) is slightly more difficult. In fact, we have

$$I_\alpha(I_\beta f)(x) = \frac{1}{\Gamma(\alpha)} \int_0^x (I_\beta f)(t)(x-t)^{\alpha-1} dt$$

$$= \frac{1}{\Gamma(\alpha)} \int_0^x (x-t)^{\alpha-1} dt \{ \frac{1}{\Gamma(\beta)} \int_0^t f(y)(t-y)^{\beta-1} dy \}$$

$$= \frac{1}{\Gamma(\alpha)\Gamma(\beta)} \int_0^x f(y) dy \int_y^x (x-t)^{\alpha-1}(t-y)^{\beta-1} dt.$$

By substituting

$$u = \frac{t-y}{x-y}$$

in the inside integral, we have

$$\int_y^x (x-t)^{\alpha-1}(t-y)^{\beta-1} dt = (x-y)^{\alpha+\beta-1} \int_0^1 (1-u)^{\alpha-1} u^{\beta-1} du$$

$$= \frac{\Gamma(\alpha)\Gamma(\beta)}{\Gamma(\alpha+\beta)} (x-y)^{\alpha+\beta-1}.$$

Therefore

$$I_\alpha(I_\beta f)(x) = \frac{1}{\Gamma(\alpha+\beta)} \int_0^x f(y)(x-y)^{\alpha+\beta-1} dy = (I_{\alpha+\beta} f)(x).$$

20.4.2 Elementary Properties of the Erdélyi–Kober Fractional Integrals

1. **Dilation:**

$$I_m^{\alpha,\eta}[D_a f(t)](x) = D_a(I_m^{\alpha,\eta} f)(x), \quad a > 0, \tag{20.14}$$

$$K_m^{\alpha,\eta}[D_a f(t)](x) = D_a(K_m^{\alpha,\eta} f)(x), \quad a > 0. \tag{20.15}$$

2. **Differentiation:**

$$R[I_m^{\alpha,\eta} f] = I_m^{\alpha,\eta}[Rf], \tag{20.16}$$

$$R[K_m^{\alpha,\eta} f] = K_m^{\alpha,\eta}[Rf], \tag{20.17}$$

where

$$R = x\frac{d}{dx}.$$

3. **Recurrence Relations:**

$$I_m^{\alpha+1,\eta}[Rf] = mI_m^{\alpha,\eta+1}[f] - (m\eta + m)I_m^{\alpha+1,\eta}[f], \tag{20.18}$$

$$mI_m^{\alpha,\eta}[f] = (m\eta + m\alpha + m)I_m^{\alpha+1,\eta}[f] + I_m^{\alpha+1,\eta}[Rf], \tag{20.19}$$

$$mK_m^{\alpha,\eta}[f] = (m\eta + m\alpha)K_m^{\alpha+1,\eta}[f] - K_m^{\alpha+1,\eta}[Rf], \tag{20.20}$$

provided that $\operatorname{Re}\alpha > 0$, $\operatorname{Re}(m\eta + m) > 0$ and f is a suitably restricted function. For such restrictions on f, see Section 20.7.2.

4. **Iteration:**

$$I_m^{\beta,\zeta} I_m^{\alpha,\eta} = I_m^{\alpha+\beta,\eta}, \quad \zeta = \alpha + \eta, \tag{20.21}$$

$$K_m^{\alpha,\eta} K_m^{\beta,\zeta} = K_m^{\alpha+\beta,\eta} \quad \zeta = \alpha + \eta. \tag{20.22}$$

For $I_m^{\alpha,\eta}$, properties (1) and (2) follow immediately from the relation

$$(I_m^{\alpha,\eta} f)(x) = \frac{m}{\Gamma(\alpha)} \int_0^1 (1 - u^m)^{\alpha-1} u^{m\eta+m-1} f(xu) du,$$

which is obtained from the definition of $I_m^{\alpha,\eta}$ (20.5) by the change of variable $t = xu$. Equation (20.18) follows from

$$I_m^{\alpha+1,\eta}(Rf)(x) = \frac{m}{\Gamma(\alpha+1)} \int_0^1 (1 - t^m)^{\alpha} t^{m\eta+m} x f'(xt) dt,$$

by integration by parts. Straightforward calculations show that

$$I_m^{\alpha,\eta+1} = I_m^{\alpha,\eta} - \alpha I_m^{\alpha+1,\eta},$$

and by substituting this into (20.18), we obtain (20.19). As for (20.21), we have

$$I_m^{\beta,\zeta}[I_m^{\alpha,\eta} f](x) =$$

$$\frac{m^2 x^{-m\zeta-m\beta}}{\Gamma(\alpha)\Gamma(\beta)} \int_0^x (x^m - t^m)^{\beta-1} t^{m(\zeta-\eta-\alpha+1)-1} dt \int_0^t (t^m - u^m)^{\alpha-1} u^{m+m\eta-1} f(u) du$$

$$= \frac{m^2}{\Gamma(\alpha)\Gamma(\beta)} x^{-m\zeta-m\beta} \int_0^x u^{m+m\eta-1} f(u) du \int_u^x (x^m - t^m)^{\beta-1} (t^m - u^m)^{\alpha-1} t^{m-1} dt$$

The result now follows upon making the change of variable

$$y = \frac{t^m - u^m}{x^m - u^m}$$

in the inside integral.

The properties of $K_m^{\alpha,\eta}$ can be proved similarly.

20.5 Inversion Formulas

For the Riemann–Liouville fractional integral, we have

$$D_x^q (I_q f)(x) = \frac{d^q}{dx^q} \left(\frac{d^{-q}}{dx^{-q}} f(x) \right) = f(x), \quad q > 0.$$

Before we prove this formula, let us recall that $D^{\alpha}D^{-\alpha}f = f$ if α is a non-negative integer, but $D^{-\alpha}D^{\alpha}f$ is not always equal to f, even if α is a positive integer, as can be seen from the relation

$$D^{-1}Df = \int_0^x f'(t)dt = f(x) - f(0).$$

The left-hand side is equal to f only if $f(0) = 0$. The next theorem extends these results to non-integer values of α. But first, let us note that Equation (20.12) can be rewritten in the form $D^{-\alpha}D^{-\beta}f = D^{-(\alpha+\beta)}f$, where $\alpha, \beta \geq 0$.

Let $f \in C^{(n)}(I)$, where $I = [0, \sigma]$. Using (20.4) and the Maclaurin formula

$$f(x) = \sum_{k=0}^{n-1} f^{(k)}(0)\frac{x^k}{\Gamma(k+1)} + D^{-n}\left[f^{(n)}(x)\right], \quad x \in I, \tag{20.23}$$

we obtain

$$D^{-\alpha}f(x) = \sum_{k=0}^{n-1} f^{(k)}(0)\frac{x^{k+\alpha}}{\Gamma(k+\alpha+1)} + D^{-(n+\alpha)}\left[f^{(n)}(x)\right], \quad \alpha \geq 0. \tag{20.24}$$

From Definition 20.2 we have for $\alpha \geq 0$

$$D^{\alpha}f = D^{p}g,$$

where $p - 1 \leq \alpha < p$, $p = 1, 2, \cdots$, and $g = D^{-(p-\alpha)}f$.

THEOREM 20.4
Let $\alpha \geq 0$ and p, f, g be as given above. Then

$$D^{\alpha}D^{-\alpha}f = f. \tag{20.25}$$

If in addition $g \in C^{(p)}(I)$, then

$$D^{-\alpha}D^{\alpha}f(x) = f(x) - \sum_{k=1}^{p} g^{(p-k)}(0)\frac{x^{\alpha-k}}{\Gamma(1+\alpha-k)}, \tag{20.26}$$

if α is not an integer, but if α is an integer, say $\alpha = m$, then

$$D^{-m}D^{m}f(x) = f(x) - \sum_{k=0}^{m-1} f^{(k)}(0)\frac{x^k}{\Gamma(k+1)}.$$

Proof: To prove (20.25), let $h = D^{-\alpha} f$. Then by Definition 20.2 and (20.12) we have

$$\begin{aligned} D^{\alpha} D^{-\alpha} f &= D^{\alpha} h = D^{p} \left(D^{-(p-\alpha)} h \right) \\ &= D^{p} \left(D^{-(p-\alpha)} D^{-\alpha} f \right) = D^{p} D^{-p} = f. \end{aligned}$$

As for (20.26), first we replace n by p, α by $\alpha - p + 1$ and f by g in (20.24) to obtain

$$D^{-(\alpha-p+1)} g = \sum_{k=0}^{p-1} g^{(k)}(0) \frac{x^{\alpha-p+k+1}}{\Gamma(\alpha - p + k + 2)} + D^{-(\alpha+1)} D^{p} g \, .$$

On the other hand,

$$D^{-\alpha} D^{\alpha} f = D(D^{-1} D^{-\alpha} D^{\alpha} f) = D(D^{-(\alpha+1)} D^{p} g).$$

Eliminating $D^{-(\alpha+1)} D^{p} g$ from the last two equations leads to

$$\begin{aligned} D^{-\alpha} D^{\alpha} f &= D \left(D^{-(\alpha-p+1)} g - \sum_{k=0}^{p-1} g^{(k)}(0) \frac{x^{\alpha-p+k+1}}{\Gamma(\alpha - p + k + 2)} \right) \\ &= D \left(D^{-(\alpha-p+1)} D^{-(p-\alpha)} f \right) - \sum_{k=0}^{p-1} g^{(k)}(0) \frac{x^{\alpha-p+k}}{\Gamma(\alpha - p + k + 1)} \\ &= D \left(D^{-1} f \right) - \sum_{k=0}^{p-1} g^{(k)}(0) \frac{x^{\alpha-p+k}}{\Gamma(\alpha - p + k + 1)}, \end{aligned}$$

which is (20.26). The case, $\alpha = m$ is trivial. ∎

THEOREM 20.5
Let f satisfy the same conditions as in the last theorem. Then

$$\begin{aligned} D^{\alpha} D^{-\beta} f &= D^{-(\beta-\alpha)} f \quad \text{if } \beta \geq \alpha \geq 0 \\ D^{\alpha} D^{-\beta} f &= D^{(\alpha-\beta)} f \quad \text{if } \alpha \geq \beta \geq 0. \end{aligned}$$

Proof: The first relation follows easily from (20.12) and (20.25) since

$$D^{-(\beta-\alpha)} f = D^{\alpha} D^{-\alpha} D^{-(\beta-\alpha)} f = D^{\alpha} D^{-\beta} f.$$

To prove the second, let p and q be positive integers such that $p - 1 \leq \alpha < p$ and $q - 1 \leq \alpha - \beta < q$. Then because $q \leq p$, it follows from Definition 2 that

$$\begin{aligned} D^{\alpha} D^{-\beta} f &= D^{p} \left(D^{-(p-\alpha)} D^{-\beta} f \right) = D^{p} \left(D^{-(p+\beta-\alpha)} f \right) \\ & D^{q} D^{p-q} \left(D^{-(p-q)} D^{-(q+\beta-\alpha)} f \right) = D^{q} \left(D^{-(q+\beta-\alpha)} f \right) = D^{\alpha-\beta} f, \end{aligned}$$

because $D^p = D^q D^{p-q}$ when p and q are positive integers with $q \leq p$. ∎

In fact, if the derivatives are taken in the sense of generalized functions (see Section 20.7.1), we have

$$\frac{d^\alpha}{dx^\alpha}\left(\frac{d^\beta}{dx^\beta} f(x)\right) = \frac{d^{\alpha+\beta}}{dx^{\alpha+\beta}} f(x),$$

for all α, β.

THEOREM 20.6

The inverses of the Erdélyi–Kober-type fractional integral operators $I_m^{\alpha,\eta}$ and $K_m^{\alpha,\eta}$ are

$$(I_m^{\alpha,\eta})^{-1} = I_m^{-\alpha,\alpha+\eta}, \tag{20.27}$$

and

$$(K_m^{\alpha,\eta})^{-1} = K_m^{-\alpha,\alpha+\eta}. \tag{20.28}$$

Proof: It is easily shown that $I_m^{\alpha,\eta}(f)$ is analytic in α for $\operatorname{Re}\alpha > 0$ and by using Equation (20.19) we can continue it analytically to the region $-1 < \operatorname{Re}\alpha \leq 0$, and step by step, to the whole complex α-plane . Therefore, for $\alpha = 0$, we have from (20.19)

$$mI_m^{0,\eta}(f) = (m\eta + m)I_m^{1,\eta}(f) + I_m^{1,\eta}(Rf). \tag{20.29}$$

But

$$I_m^{1,\eta}(Rf)(x) = m\int_0^1 t^{m\eta+m} x f'(xt)dt$$

$$= m\left\{t^{m\eta+m} f(xt)|_0^1 - (m\eta+m)\int_0^1 t^{m\eta+m-1} f(xt)\, dt\right\}$$

$$= mf(x) - (m\eta+m)I_m^{1,\eta}(f)(x),$$

which, when combined with (20.29), yields

$$I_m^{0,\eta}(f)(x) = f(x).$$

Finally, in view of (20.21), we obtain

$$I_m^{-\alpha,\alpha+\eta} I_m^{\alpha,\eta}(f) = I_m^{0,\eta}(f) = f,$$

and

$$I_m^{\alpha,\eta} I_m^{-\alpha,\alpha+\eta}(f) = I_m^{0,\alpha+\eta}(f) = f,$$

which establishes (20.27). The proof of (20.28) is similar. ∎

20.6 Relationships with Other Transforms

20.6.1 The Laplace Transform

Let us recall formulas (5.9) and (5.10)

$$\mathcal{L}\left[\frac{d^q f}{dt^q}\right](s) = s^q \mathcal{L}[f](s) - \sum_{k=0}^{q-1} s^{q-1-k} \frac{d^k f}{dx^k}(0), \quad q = 1, 2, 3, \ldots,$$

$$\mathcal{L}\left[\frac{d^{-q} f}{dt^{-q}}\right](s) = s^{-q} \mathcal{L}[f](s), \quad q = 0, 1, 2, \ldots\ ,$$

where $\mathcal{L}$ is the Laplace transformation operator. Both formulas can be combined in a single formula

$$\mathcal{L}\left[\frac{d^q f}{dt^q}\right](s) = s^q \mathcal{L}(f)(s) - \sum_{k=0}^{q-1} s^k \frac{d^{q-1-k} f}{dt^{q-1-k}}(0), \quad q = 0, \pm 1, \pm 2, \ldots,$$

where the summation is zero when q is non-positive. We now show that this formula remains valid if q is replaced by a non-integer value. That is

$$\mathcal{L}\left[\frac{d^q f}{dt^q}\right](s) = s^q \mathcal{L}[f](s) - \sum_{k=0}^{n-1} s^k \frac{d^{q-1-k} f}{dt^{q-1-k}}(0), \quad \text{for all } q, \tag{20.30}$$

where n is an integer such that $n - 1 < q \leq n$. First, consider q negative so that

$$\begin{aligned} \frac{d^q f}{dt^q} &= \frac{1}{\Gamma(-q)} \int_0^t (t - x)^{-q-1} f(x) dx, \quad q < 0 \\ &= (f * \Phi_q)(t), \end{aligned}$$

where $*$ stands for the convolution operation associated with the Laplace transformation (see Section 5.9.1) and

$$\Phi_q(t) = \frac{1}{\Gamma(-q)} t^{-q-1}.$$

From the convolution theorem for the Laplace transform (Theorem 5.9), we obtain

$$\mathcal{L}[\frac{d^q f}{dt^q}](s) = s^q \mathcal{L}[f](s), \quad q < 0, \tag{20.31}$$

which is (20.30) since the summation is zero. For non-integer positive q, we use (5.9) together with

$$\frac{d^q f}{dt^q} = \frac{d^n}{dt^n} \frac{d^{q-n} f}{dt^{q-n}}, \quad n - 1 < q < n,$$

to obtain

$$\mathcal{L}\left[\frac{d^q f}{dt^q}\right](s) = \mathcal{L}\left[\frac{d^n}{dt^n}\frac{d^{q-n} f}{dt^{q-n}}\right](s)$$

$$= s^n \mathcal{L}\left[\frac{d^{q-n}}{dt^{q-n}} f\right](s) - \sum_{k=0}^{n-1} s^k \frac{d^{n-1-k}}{dt^{n-1-k}}\left[\frac{d^{q-n} f}{dt^{q-n}}\right](0).$$

Since $q-n < 0$, we use (20.31) to evaluate the first term on the right-hand side and obtain

$$\mathcal{L}\left[\frac{d^q f}{dt^q}\right](s) = s^q \mathcal{L}[f](s) - \sum_{k=0}^{n-1} s^k \left[\frac{d^{q-1-k} f}{dt^{q-1-k}}\right](0),$$

where $0 < q \neq 1, 2, \ldots$. Here we have used the fact that

$$\frac{d^\alpha}{dt^\alpha}\left(\frac{d^{-\beta} f}{dt^{-\beta}}\right) = \frac{d^{\alpha-\beta} f}{dt^{\alpha-\beta}}, \quad \alpha, \beta \geq 0,$$

to simplify the second term on the right-hand side (cf. Theorem 20.5)

20.6.2 The Mellin Transform

The fractional derivative of order α associated with the Weyl fractional integral is defined by

$$\begin{aligned}\tilde{D}^\alpha[f](x) &= \frac{1}{\Gamma(n-\alpha)} \frac{d^n}{dx^n} \int_x^\infty f(t)(t-x)^{n-\alpha-1} dt \\ &= \frac{d^n}{dx^n}(h_{n-\alpha}(x)), \quad n-1 < \operatorname{Re}\alpha < n,\end{aligned}$$

where

$$h_q(x) = \frac{1}{\Gamma(q)} \int_x^\infty f(t)(t-x)^{q-1} dt.$$

THEOREM 20.7
Let $\mathcal{M}$ denote the Mellin transformation operator. Then

$$\mathcal{M}\left[\tilde{D}^\alpha f\right](s) = \frac{(-1)^n \Gamma(s)}{\Gamma(s-\alpha)} (\mathcal{M} f)(s-\alpha), \tag{20.32}$$

$$\mathcal{M}\left[I_m^{\alpha,\eta} f\right](s) = \frac{\Gamma(\eta - \frac{s}{m} + 1)}{\Gamma(\alpha + \eta - \frac{s}{m} + 1)} \mathcal{M}(f)(s), \tag{20.33}$$

$$\mathcal{M}\left[K_m^{\alpha,\eta} f\right](s) = \frac{\Gamma(\eta + \frac{s}{m})}{\Gamma(\alpha + \eta + \frac{s}{m})} \mathcal{M}(f)(s). \tag{20.34}$$

Proof: From item 4 of Section 10.3

$$\mathcal{M}(\tilde{D}^{\alpha} f)(s) = \frac{(-1)^n \Gamma(s)}{\Gamma(s-n)} \mathcal{M}(h_{n-\alpha})(s-n).$$

Upon observing that $h_{n-\alpha}(x)$ can be written in the form

$$h_{n-\alpha}(x) = \int_0^{\infty} f_1(u) g(\frac{x}{u}) \frac{du}{u},$$

where

$$f_1(u) = u^{n-\alpha} f(u) \,, \quad g(u) = \frac{1}{\Gamma(n-\alpha)} (1-u)^{n-\alpha-1} H(1-u),$$

and $H(u)$ is the Heaviside function, it follows from the convolution theorem for the Mellin transform (Theorem 10.2) that

$$\mathcal{M}(h_{n-\alpha})(s) = \mathcal{M}(f_1)(s) \mathcal{M}(g)(s) = \frac{\Gamma(s)}{\Gamma(s+n-\alpha)} \mathcal{M}(f)(s+n-\alpha);$$

thus

$$\mathcal{M}(\tilde{D}^{\alpha} f)(s) = \frac{(-1)^n \Gamma(s)}{\Gamma(s-\alpha)} \mathcal{M}(f)(s-\alpha),$$

which is (20.32).

To prove (20.33), make the change of variable, $t = xu$, in (20.5), to obtain

$$(I_m^{\alpha,\eta} f)(x) = \frac{m}{\Gamma(\alpha)} \int_0^1 (1-u^m)^{\alpha-1} u^{m+m\eta-1} f(xu)\, du,$$

then apply the Mellin transformation to both sides of this equation. This leads to

$$\begin{aligned} \mathcal{M}\left[I_m^{\alpha,\eta} f\right](s) &= \mathcal{M}(f)(s) \frac{m}{\Gamma(\alpha)} \int_0^1 (1-u^m)^{\alpha-1} u^{m+m\eta-1-s}\, du, \\ &= \frac{\Gamma(\eta - \frac{s}{m} + 1)}{\Gamma(\alpha + \eta - \frac{s}{m} + 1)} \mathcal{M}(f)(s)\,, \end{aligned}$$

which is (20.33). The proof of (20.34) is similar. ∎

20.7 Fractional Differentiation and Integration of Generalized Functions

20.7.1 Riemann–Liouville Integrals of Generalized Functions

Extending the Riemann–Liouville and Weyl fractional integrals to generalized functions is straightforward once we realize that these integrals can be written as convolution integrals. For example, the Riemann–Liouville integral can be written as

$$(I_\lambda f)(x) = \frac{1}{\Gamma(\lambda)} \int_0^x (x-t)^{\lambda-1} f(t)dt = (f * \Phi_\lambda)(x), \tag{20.35}$$

where

$$\Phi_\lambda(x) = \frac{x^{\lambda-1}}{\Gamma(\lambda)} \chi_{(0,\infty)}(x) = \frac{1}{\Gamma(\lambda)} x_+^{\lambda-1},$$

$$(f * g)(x) = \int_{-\infty}^{\infty} f(t)g(x-t)dt,$$

and f has been extended to the whole real line by setting $f = 0$ on $(-\infty, 0)$.

Equation (20.35) remains valid if f is a generalized function concentrated on $[0, \infty)$. For $\mathrm{Re}\,\lambda > 0$, the generalized function Φ_λ is well defined and for other values of λ it is interpreted as in [70, Vol. 1, p. 48], namely,

$$< x_+^\lambda, \phi >= \int_0^\infty x^\lambda [\phi(x) - \phi(0) - x\phi'(0) - \cdots - \frac{x^{n-1}}{(n-1)!}\phi^{(n-1)}(0)]dx,$$

$-n-1 < \mathrm{Re}\lambda < -n$. At $\lambda = -k$, $k = 1, 2, \ldots$, x_+^λ has a simple pole with residue $(-1)^{k-1}\delta^{(k-1)}(x)/(k-1)!$, $k = 1, 2, \ldots$. Thus, since

$$\lim_{\lambda \to -k} (\lambda + k)\Gamma(\lambda + 1) = \frac{(-1)^{k-1}}{(k-1)!},$$

it follows that

$$\Phi_{-k}(x) = \delta^{(k)}(x), \quad k = 0, 1, 2, \ldots,$$

and therefore

$$(I_0 f)(x) = (f * \Phi_0)(x) = (f * \delta)(x) = f(x),$$

$$(I_{-k} f)(x) = (f * \Phi_{-k})(x) = (f * \delta^{(k)})(x) = f^{(k)}(x).$$

Having extended $\Phi_\lambda(x)$ analytically to the whole λ-plane, we can extend relation (20.35) to any α and β to yield

$$\frac{d^\alpha}{dx^\alpha}(\frac{d^\beta f}{dx^\beta}) = \frac{d^{\alpha+\beta} f}{dt^{\alpha+\beta}},$$

for any α, β,and any generalized function f concentrated in $[0, \infty)$ [70, Vol. 1, p. 116].

20.7.2 The Erdélyi–Kober–type Integrals of Generalized Functions

To extend the fractional integral operators $I_m^{\alpha,\eta}$ and $K_m^{\alpha,\eta}$ to generalized functions, we use the adjoint method suggested by A. McBride [157].

For $1 \leq p \leq \infty$, and $I = (0, \infty)$, let

$$F_p = \{\phi \in C^\infty(I) : x^k \phi^{(k)}(x) \in L^p(I), k = 0, 1, 2, \ldots\}.$$

Define the family of seminorms $M_p = \{\gamma_k^p\}_{k=0}^\infty$ on F_p by

$$\gamma_k^p(\phi) = ||x^k \phi^{(k)}(x)||_p. \tag{20.36}$$

It is readily seen that M_p is a countable family of separating seminorms (norms) and that (F_p, M_p) is a Fréchet space. In fact, it is a testing function space (see Section 3.3.1). If we denote the operator $x\frac{d}{dx}$ by R, then one can easily show that $\phi \in F_p$ if and only if $R^k \phi \in L^p(I)$ and that $\phi_n \longrightarrow 0$ in F_p if and only if $R^k \phi_n \longrightarrow 0$ in $L^p(I)$ for every $k = 0, 1, 2, \ldots$. From this it follows that R is a continuous linear operator on F_p. For, if $\phi \in F_p$, then $R^k(R\phi) = R^{k+1}\phi \in L^p(I)$; hence $R\phi \in F_p$. Moreover, if $\phi_n \longrightarrow 0$ in F_p as $n \longrightarrow \infty$, then $R^k(R\phi_n) = R^{k+1}\phi_n \longrightarrow 0$ in $L^p(I)$ as $n \longrightarrow \infty$, hence $R\phi_n \longrightarrow 0$ in F_p as $n \longrightarrow \infty$.

Any function $f \in L^q(I)$ generates an element in F_p^* by means of the formula

$$< f, \phi >= \int_0^\infty f(x)\phi(x)dx, \quad \phi \in F_p$$

where $(1/p) + (1/q) = 1$.

It is clear that $\mathcal{D}(I) \subset F_p \subset \mathcal{E}(I)$ and that the topology of $\mathcal{D}(I)$ is stronger than the one it inherits from F_p. For $1 \leq p < \infty$, $\mathcal{D}(I)$ is dense in F_p; hence, $E^*(I) \subset F_p^* \subset \mathcal{D}^*(I)$.

In order to perform multiplication by power of x and differentiation we introduce the space

$$F_{p,\mu} = \{\phi : x^{-\mu}\phi(x) \in F_p\},$$

where μ is a complex number.

The topology of $F_{p,\mu}$ is generated by the family $M_{p,\mu} = \{\gamma_k^{p,\mu}\}_{k=0}^\infty$ of seminorms defined by

$$\gamma_k^{p,\mu}(\phi) = \gamma_k^p(x^{-\mu}\phi), \quad \phi \in F_{p,\mu},$$

where γ_k^p is given by (20.36). It is evident that $\phi \longrightarrow x^\mu \phi$ is an isomorphism of F_p onto $F_{p,\mu}$ and therefore $(F_{p,\mu}, M_{p,\mu})$ is also a Fréchet space.

The following facts can be easily established:

1) $\phi \longrightarrow x^\lambda \phi$ is an isomorphism from $F_{p,\mu}$ onto $F_{p,\mu+\lambda}$,

2) the operator R is a continuous linear operator from $F_{p,\mu}$ into itself,

3) the operators R and $I_m^{\alpha,\eta}$ commute (cf.(20.16)) and

$$x^k \frac{d^k}{dx^k}(I_m^{\alpha,\eta}\phi)(x) = I_m^{\alpha,\eta}\left(t^k \frac{d^k\phi}{dt^k}\right)(x). \tag{20.37}$$

Equation (20.37) can be proved in the same way as (20.16).

For $m\mathrm{Re}\,\eta + m > 1/p$, $I_m^{\alpha,\eta}$ is a continuous linear operator on $L^p(I)$ into itself ([157]). Using this fact, together with (20.37), one can show that

$$\gamma_k^p(I_m^{\alpha,\eta}\phi) \leq C\gamma_k^p(\phi), \quad k = 0, 1, 2, \dots\, ; \; \phi \in F_p\,, \tag{20.38}$$

for some constant C that depends on α, η and m. From the relation

$$(I_m^{\alpha,\eta}\phi)(x) = x^\mu I_m^{\alpha,\eta+(\frac{\mu}{m})}\left(t^{-\mu}\phi(t)\right)(x)$$

we will show that $I_m^{\alpha,\eta}$ is a continuous linear operator on $F_{p,\mu}$ into itself, provided that $\mathrm{Re}\,(m\eta + \mu) + m > 1/p$, $1 \leq p \leq \infty$, $\mathrm{Re}\,\alpha > 0$. To this end, let

$$\psi(x) = (I_m^{\alpha,\eta}\phi)\,(x)\,.$$

Then, by (20.38), we obtain

$$\gamma_k^{p,\mu}(\psi) = \gamma_k^p\left(I_m^{\alpha,\eta+\mu/m}\left[t^{-\mu}\phi(t)\right]\right) \leq C\gamma_k^p\left(t^{-\mu}\phi(t)\right) = C\gamma_k^{p,\mu}(\phi).$$

Similar results hold for the operator $K_m^{\alpha,\eta}$. In fact, we have the following stronger result.

THEOREM 20.8 (**McBride** [157])
Let $\mathrm{Re}\,(m\eta + \mu) + m > \frac{1}{p}$, $1 \leq p \leq \infty$. *Then for any complex number* α, $I_m^{\alpha,\eta}$ *is a continuous linear operator on* $F_{p,\mu}$ *into itself and for fixed* η, $I_m^{\alpha,\eta}$ *is entire with respect to* α *on* $F_{p,\mu}$.

Similarly, for $\mathrm{Re}\,(m\eta - \mu) + 1 > \frac{1}{q}$, *and any complex number* α, $K_m^{\alpha,\eta}$ *is a continuous linear operator on* $F_{p,\mu}$ *into itself and for any fixed* η, $K_m^{\alpha,\eta}$ *is entire with respect to* α *on* $F_{p,\mu}$, *where* $\frac{1}{p} + \frac{1}{q} = 1$.

This theorem, together with (20.7), enables us to extend the operators $I_m^{\alpha,\eta}$ and $K_m^{\alpha,\eta}$ to generalized functions.

DEFINITION 20.7 *Let* $f \in F_{p,\mu}^*$. *Then the fractional integrals of Erdélyi–Kober type of* f, $I_m^{\alpha,\eta}(f)$ *and* $K_m^{\alpha,\eta}(f)$ *are defined by*

$$< I_m^{\alpha,\eta}f, \phi > = < f, K_m^{\alpha,\eta+1-(\frac{1}{m})}\phi >, \qquad \mathrm{Re}\,(m\eta - \mu) + m > \frac{1}{q}$$

and

$$< K_m^{\alpha,\eta}f, \phi > = < f, I_m^{\alpha,\eta-1+(\frac{1}{m})}\phi >, \qquad \mathrm{Re}\,(m\eta + \mu) > -\frac{1}{q},$$

where $\phi \in F_{p,\mu}$.

From Theorem 20.8 and item (3) of Section 3.4.1, it follows that $I_m^{\alpha,\eta}$ is a continuous linear transformation on $F_{p,\mu}^*$ for $\mathrm{Re}(m\eta - \mu) + m > 1/q$ and likewise $K_m^{\alpha,\eta}$ is a continuous linear transformation on $F_{p,\mu}^*$ for $\mathrm{Re}(m\eta + \mu) > -1/q$.

20.8 Applications

20.8.1 Ordinary Differential Equations with Fractional Orders

Some of the techniques explained in this chapter can be used to solve ordinary differential equations with fractional orders. Such equations are sometimes called *extraordinary differential equations*, the simplest of which is

$$\frac{d^q y}{dx^q} = F(x),$$

where q is arbitrary and F is a known function. The obvious solution

$$y = \frac{d^{-q}}{dx^{-q}} F(x)$$

is not the most general solution, but it can be verified that

$$y = \frac{d^{-q}}{dx^{-q}} F(x) + C_1 x^{q-1} + c_2 x^{q-2} + \cdots + C_n x^{q-n}, \tag{20.39}$$

is, where for $q > 0$, $q \leq n < q + 1$, and for $q \leq 0$, $n = 0$, and $C_1, \ldots, C_n$ are arbitrary constants.

Example A: The general solution of the differential equation

$$\frac{d^{\frac{5}{2}} y}{dx^{\frac{5}{2}}} = x^2,$$

is given by

$$y = \frac{d^{-\frac{5}{2}}}{dx^{-\frac{5}{2}}} x^2 + C_1 x^{\frac{3}{2}} + C_2 x^{\frac{1}{2}} + C_3 x^{-\frac{1}{2}} = \frac{2}{\Gamma(\frac{11}{2})} x^{\frac{9}{2}} + C_1 x^{\frac{3}{2}} + C_2 x^{\frac{1}{2}} + C_3 x^{-\frac{1}{2}}.$$

This solution can be obtained by applying the operator $D^{-5/2}$ to the differential equation and then using (20.26).

Analogously, the differential equation

$$\frac{d^q y}{dx^q} + a\frac{d^{q-1}y}{dx^{q-1}} = F(x),$$

can be reduced to the ordinary differential equation

$$\frac{dy}{dx} + ay = \frac{d^{1-q}F}{dx^{1-q}} + C_1 x^{q-2} + C_2 x^{q-3} + \cdots + c_n x^{q-n-1}\,,\ q-1 \le n < q$$

by applying the operator D^{1-q} and using techniques similar to those discussed above. The latter differential equation is of first order and can then be solved by standard methods. Similar arguments can be used to solve the differential equation

$$\frac{d^q y}{dx^q} + \frac{d^p y}{dx^p} = F(x),$$

where $q - p$ is an integer or half-integer; for more details, see [180, Ch. 8].

20.8.2 Partial Differential Equations

Consider the diffusion equations

$$\frac{\partial}{\partial t}F(x,y,z,t) = \kappa\nabla^2 F(x,y,z,t)$$

in which F is some intensive scalar quantity (temperature or concentration), κ is a constant that depends on the medium and ∇^2 is the Laplacian differential operator. If we consider spherical or cylindrical surfaces, we may convert this equation to

$$\frac{\partial}{\partial t}F(r,t) - \kappa\frac{\partial^2}{\partial r^2}F(r,t) - \frac{2g\kappa}{r+R}\frac{\partial}{\partial r}F(r,t) = 0,$$

where r is the spatial coordinate normal to the boundary and having its origin at the boundary surface, R denotes the radius of curvature of the surface and g is a number that depends on the geometry of the surface [180, p. 198].

The initial boundary conditions are

$$F(r,t) = F_0, \quad \text{a constant, for } t < 0,\ r \ge 0,$$

and

$$F(r,t) = F_0, \quad 0 \le t \le \tau\,,$$

for r in regions remote from the boundary $r = 0$. Normally, r is equal to ∞ or R depending on the boundary of the surface. Oldham and Spanier have shown that this problem can be described equally well by the partial differential equation with fractional order derivatives

$$\frac{\partial}{\partial t}F + \frac{1}{\sqrt{\kappa}}\frac{\partial^{\frac{1}{2}}}{\partial t^{\frac{1}{2}}}(F - F_0) + \frac{g}{r+R}(F - F_0) = 0.$$

For the solution of this equation and the like, see [180, Ch. 11].

20.8.3 Integral Equations

A prototype of integral equations that can be solved by using fractional derivatives is Abel's integral equation

$$g(x) = \frac{1}{\Gamma(1-\alpha)}\int_0^x \frac{f(y)dy}{(x-y)^\alpha},$$

where g is a known function and f is unknown; see Chapter 19. In the classical theory, α is assumed less than 1. But in view of the results of Section 7.1 of this chapter, this assumption is no longer needed if f and g are generalized functions. In the language of fractional calculus, Abel's integral equation reads

$$g(x) = \frac{d^{\alpha-1}f}{dx^{\alpha-1}}.$$

Using the notation of Section 7.1 , we may write Abel's integral equation in the form

$$g(x) = (f * \Phi_\lambda)(x), \quad \lambda = -\alpha + 1.$$

By taking the convolution with $\Phi_{-\lambda}$, we obtain

$$\begin{aligned}(g * \Phi_{-\lambda})(x) &= (f * (\Phi_\lambda * \Phi_{-\lambda}))(x)\\ &= (f * \Phi_0)(x) = (f * \delta)(x) = f(x).\end{aligned}$$

If $0 < \alpha < 1$, and g is differentiable, we obtain

$$f = g * \Phi_{\alpha-1} = g * \frac{d}{dx}\Phi_\alpha = \frac{d}{dx}g * \Phi_\alpha = \frac{1}{\Gamma(\alpha)}\int_0^x \frac{g'(y)dy}{(x-y)^{1-\alpha}},$$

which is the classical solution of Abel's integral equation.

20.8.4 Special Functions

The fractional derivatives and integrals are used to derive and simplify many formulas involving special functions. We begin with the integral representation of the Gauss hypergeometric function $F(a, b; c; x)$ (cf. Section 2.3.2)

$$F(a, b; c; x) = \frac{\Gamma(c)}{\Gamma(b)\Gamma(c-b)}\int_0^1 t^{b-1}(1-t)^{c-b-1}(1-tx)^{-a}dt, \qquad (20.40)$$

where $\operatorname{Re} a > \operatorname{Re} b > 0$, $c \neq 0, -1, -2, \ldots$, $|x| < 1$. Evidently, in terms of Erdélyi's integral operator we can write

$$(I_1^{c-b,b-1}(1-t)^{-a})(x) = (I^{c-b,b-1}(1-t)^{-a})(x) = \frac{\Gamma(b)}{\Gamma(c)} F(a,b;c;x),$$

or equivalently

$$\begin{aligned}\frac{x^{c-1}}{\Gamma(c)} F(a,b;c;x) &= \frac{1}{\Gamma(c-b)} \int_0^x (x-u)^{c-b-1} \frac{u^{b-1}}{\Gamma(b)} (1-u)^{-a} du \\ &= \frac{d^{b-c}}{dx^{b-c}} \Big(\frac{x_+^{b-1}(1-x)_+^{-a}}{\Gamma(b)} \Big) \end{aligned} \tag{20.41}$$

where x_+^λ is interpreted as in Section 20.7.1. On the other hand, if we put

$$y = \frac{(1-t)x}{(1-tx)}$$

in (20.40), we obtain

$$F(a,b;c;x) = \frac{\Gamma(c)}{\Gamma(b)\Gamma(c-b)} \frac{(1-x)^{c-b-a}}{x^{c-1}} \int_0^x (x-y)^{b-1}(1-y)^{a-c} y^{c-b-1} dy;$$

hence

$$\frac{x^{c-1}(1-x)^{a+b-c}}{\Gamma(c)} F(a,b;c;x) = \frac{1}{\Gamma(c-b)} \frac{d^{-b}}{dx^{-b}} [(1-x)_+^{a-c} x_+^{c-b-1}].$$

If we replace a by $c-a$ and b by $c-b$ in the last equation, we obtain

$$\frac{x^{c-1}(1-x)^{c-a-b}}{\Gamma(c)} F(c-a, c-b; c; x) = \frac{d^{b-c}}{dx^{b-c}} [\frac{x_+^{b-1}(1-x)_+^{-a}}{\Gamma(b)}],$$

which, when compared with (20.41), yields the well-known relation

$$F(a,b;c;x) = (1-x)^{c-a-b} F(c-a, c-b; c; x).$$

Other types of special function relations can be obtained by straightforward differentiation as shown in the next example.

Example B:

$$\frac{d^q}{dx^q} \sin\sqrt{x} = \frac{\sqrt{\pi}}{2} (2\sqrt{x})^{\frac{1}{2}-q} J_{\frac{1}{2}-q}(\sqrt{x}), \tag{20.42}$$

To show this, we write the Taylor series of $\sin\sqrt{x}$ and differentiate term by term using (20.2) to obtain

$$\frac{d^q}{dx^q} \sin\sqrt{x} = \frac{d^q}{dx^q} \sum_{n=0}^{\infty} \frac{(-1)^n (\sqrt{x})^{2n+1}}{\Gamma(2n+2)}$$

$$= \sum_{n=0}^{\infty} \frac{(-1)^n \Gamma(n+\frac{3}{2})}{\Gamma(2n+2)\Gamma(n+\frac{3}{2}-q)} x^{n+\frac{1}{2}-q}.$$

The result now follows from Legendre's duplication formula (2.5). Many similar results and representations of other special functions can be found in [180, Ch. 6,7].

20.8.5 Evaluation of Definite Integrals

The fractional calculus provides a useful tool for the evaluation of a large number of definite integrals and infinite series. We shall give a single example to illustrate the idea, but the experienced reader can come up with many more. We recall from (20.42) that

$$\frac{d^q}{dx^q} \sin\sqrt{x} = \frac{\sqrt{\pi}}{2}(2\sqrt{x})^{\frac{1}{2}-q} J_{\frac{1}{2}-q}(\sqrt{x}), \quad x > 0$$

which, for $q = -\frac{1}{2}$, gives us

$$\frac{1}{\Gamma(\frac{1}{2})} \int_0^x \frac{\sin\sqrt{t}}{\sqrt{x-t}} dt = \sqrt{\pi x} J_1(\sqrt{x}), \quad x > 0.$$

This equation can be simplified to the well-known formula [79, p. 419]

$$\int_0^1 \frac{\sin\sqrt{xt}}{\sqrt{1-t}} \, dt = \pi J_1(\sqrt{x});$$

a result that is not easy to establish by using standard methods. Other applications can be found in [179].

21

The Hankel and Hankel-Type Transforms

The Hankel transform is named after the German mathematician, Hermann Hankel [1839-1873], who was a student of the great mathematician Riemann, and later became a professor of mathematics at Leibzig.

The Hankel transformation of order ν is a linear integral transformation whose kernel is essentially the Bessel function of the first kind and order ν. The Bessel function is named after the German astronomer, F. W. Bessel [1784-1846]. But its earliest appearance was in a paper by John Bernoulli in 1694 on planer curves in which he studied a differential equation of Riccati's type. Bernoulli, however, was interested only in the first few terms of the Bessel series expansion.

The Bessel function of order zero appeared for the first time in the work of Daniel Bernoulli [1738] on the oscillations of heavy chains. It was later investigated by Euler who, in his studies of the vibrations of a stretched membrane in 1764, also discovered the Bessel functions of integer orders. But it was Bessel himself who studied these functions in great details and pointed out their importance.

We shall call a linear integral transformation a *Hankel-type transformation* if its kernel is one of the Bessel functions, whatever its type. Under this classification, the known $Y_\nu, I_\nu, \mathbf{H}_\nu$, and K_ν transformations are considered Hankel-type transformations.

Most properties of the Hankel-type transformations, with some exceptions for the K_ν transformation, can be derived in a similar way to those of the Hankel transformation, with some minor modifications. Therefore, we shall focus our attention in this chapter on the latter. Because the modified Bessel function of the third kind, K_ν, behaves differently from the other Bessel functions, the K_ν transformation will be treated separately in another chapter (Chapter 23).

The Hankel transform arises naturally in solving boundary-value problems formulated in cylindrical coordinates. It also arises in connection with the n-dimensional Fourier transform when circular symmetry is assumed.

For example, for $n = 2$, the Fourier transform of f is given by

$$F(u,v) = \frac{1}{2\pi} \int_{-\infty}^{\infty} \int_{-\infty}^{\infty} f(x,y) e^{i(ux+vy)} dxdy\,,$$

so that

$$f(x,y) = \frac{1}{2\pi} \int_{-\infty}^{\infty} \int_{-\infty}^{\infty} F(u,v) e^{-i(ux+vy)} dudv\,.$$

If we set

$$x = r\cos\theta \quad , \quad y = r\sin\theta,$$

and

$$u = \rho\cos\phi \quad , \quad v = \rho\sin\phi,$$

we obtain

$$F(\rho,\phi) = \frac{1}{2\pi} \int_0^{\infty} \int_0^{2\pi} f(r,\theta) e^{ir\rho\cos(\theta-\phi)} r dr d\theta\,.$$

But since

$$J_0(r\rho) = \frac{1}{2\pi} \int_0^{2\pi} e^{ir\rho\cos(\theta-\phi)}\, d\theta\,, \qquad \text{cf. (2.18)}$$

it follows that if f is independent of θ, i.e., $f(r,\theta) = f(r)$, then

$$F(\rho) = F(\rho,\phi) = \int_0^{\infty} r f(r) J_0(\rho r)\, dr\,,$$

and similarly

$$f(r) = \int_0^{\infty} \rho F(\rho) J_0(r\rho)\, d\rho\,.$$

The functions f and F are sometimes called a *Hankel transform pair*; however, our definition of the Hankel transform pair will differ slightly from that.

21.1 Definitions

DEFINITION 21.1 *The Hankel transform of order ν of a function $f(x)$ is defined as*

$$\mathcal{H}_\nu\left[f\right](y) = F_\nu(y) = \int_0^{\infty} f(x)\sqrt{xy}\, J_\nu(yx) dx\,, \quad y > 0\,, \quad \nu > -1/2,$$

where $J_\nu(z)$ *is the Bessel function of the first kind of order* ν*, which is defined by*

$$J_\nu(z) = \sum_{n=0}^{\infty} \frac{(-1)^n (z/2)^{2n+\nu}}{n!\Gamma(n+\nu+1)}.$$

The Hankel transforms of order 1/2 and −1/2 are equal to the Fourier sine and cosine transforms, respectively, because

$$J_{1/2}(x) = \sqrt{\frac{2}{\pi x}} \sin x \quad , \quad J_{-1/2}(x) = \sqrt{\frac{2}{\pi x}} \cos x.$$

As with the Fourier transform, there seems to be no agreement on the definition of the Hankel transform. Some authors define it as

$$G_\nu(y) = \int_0^\infty x g(x)\, J_\nu(yx) dx\,;$$

however, the two definitions are equivalent, as can be seen by replacing $f(x)$ by $\sqrt{x}g(x)$ and $F_\nu(y)$ by $\sqrt{y}G_\nu(y)$.

Some Hankel–type transforms are given in the next definition.

DEFINITION 21.2 *The* Y, I, H, *and* K *transforms of order* ν *of a function* $f(x)$ *are defined as*

$$\mathcal{Y}_\nu\,[f]\,(y) = \int_0^\infty f(x)\sqrt{xy}\, Y_\nu(yx) dx\,, \quad y > 0\,, \quad \nu > -1/2,$$

$$\mathcal{I}_\nu\,[f]\,(y) = \int_0^\infty f(x)\sqrt{xy}\, I_\nu(yx) dx\,, \quad y > 0\,, \quad \nu > -1/2,$$

$$H_\nu\,[f]\,(y) = \int_0^\infty f(x)\sqrt{xy}\, \mathbf{H}_\nu(yx) dx\,, \quad y > 0\,, \quad \nu > -1/2,$$

$$\mathcal{K}_\nu\,[f]\,(y) = \int_0^\infty f(x)\sqrt{xy}\, K_\nu(yx) dx\,, \quad y > 0\,, \quad \nu > -1/2,$$

respectively, where Y_ν *is the Bessel function of the second kind or Neumann's function,* I_ν *is the modified Bessel function of the first kind,* $\mathbf{H}_\nu$ *is Struve's function, and* K_ν *is the modified Bessel function of the third kind or Machdonald's function.* For the definition of these functions, see Section 2.5.

All the above transforms are of the form

$$F(y) = \int_0^\infty f(x) k(yx) dx,$$

for some appropriate kernel function k, and most of them have inversion formulas of the form

$$f(x) = \int_0^\infty F(y)h(yx)dy,$$

for some kernel function h. An integral transform whose kernel is a function of xy is sometimes called an integral transform of *Fourier–type* or an integral transform with *Fourier–type kernel.*

DEFINITION 21.3 *The Hankel potential transform of a function f is defined as*

$$F_\nu(x) = \int_0^\infty \frac{t}{(x^2+t^2)^{\nu+1}} f(t)\, d\mu_\nu(t)\,, \quad \nu > 0,$$

where

$$d\mu_\nu(t) = \frac{t^{2\nu}}{2^{\nu-1/2}\Gamma(\nu+1/2)} dt.$$

The Hankel potential transform is closely related to the Hankel transform. In fact, its kernel can be obtained from the Hankel transform since

$$\frac{t}{(x^2+t^2)^{\nu+1}} = \frac{\sqrt{\pi}}{2^{\nu+1/2}\Gamma(\nu+1)} \int_0^\infty \mathcal{J}_\nu(xy)e^{-ty} d\mu_\nu(y)\,,$$

where

$$\mathcal{J}_\nu(z) = 2^{\nu-1/2}\Gamma(\nu+1/2)z^{-\nu+1/2}J_{\nu-1/2}(z)\,, \tag{21.1}$$

or

$$\frac{t}{(x^2+t^2)^{\nu+1}} = \frac{\sqrt{\pi}x^{-\nu}}{2^{\nu+1/2}\Gamma(\nu+1)} \mathcal{H}_{\nu-1/2}\left[y^\nu e^{-ty}\right](x)\,.$$

21.2 Existence

Since $\sqrt{x}J_\nu(x)$ is bounded on the positive real-axis, it follows that the Hankel transform of f exists if $f \in L^1(0,\infty)$. The following is the analogue of the Riemann–Lebesgue lemma for the Hankel transform.

THEOREM 21.1 [216, p. 304]
Let $\nu > -1/2$. If $f \in L^1(a,b)$ is piecewise continuous for $0 \le a < b \le \infty$, then

$$\int_a^b f(x)\sqrt{x}J_\nu(\lambda x)\, dx = O\left(\frac{1}{\sqrt{\lambda}}\right), \quad \textit{as } \lambda \to \infty.$$

21.3 Elementary Properties

1. **Multiplication by x^m:**

$$\mathcal{H}_\nu\left[x^m f(x)\right](y) = y^{1/2-\nu}\left(\frac{1}{y}\frac{d}{dy}\right)^m \left[y^{\nu+m-1/2}F_{\nu+m}(y)\right].$$

2. **Division by x:**

$$\mathcal{H}_\nu\left[\frac{2\nu}{x}f(x)\right](y) = y\left[F_{\nu-1}(y) + F_{\nu+1}(y)\right].$$

and also

$$\mathcal{H}_\nu\left[\frac{f(x)}{x}\right](y) = y^{1/2-\nu}\int_0^y t^{\nu-1/2}F_{\nu-1}(t)dt\,.$$

3. **Differentiation:**

$$\mathcal{H}_\nu\left[2\nu f'(x)\right](y) = (\nu - 1/2)yF_{\nu+1}(y) - (\nu + 1/2)yF_{\nu-1}(y).$$

4. **Differentiation and Multiplication by Powers of x:**

$$\mathcal{H}_\nu\left[x^{1/2-\nu}\left(\frac{1}{x}\frac{d}{dx}\right)^m\left(x^{\nu+m-1/2}f(x)\right)\right](y) = y^m F_{\nu+m}(y).$$

Property (1) follows from (2.12), the first part of (2) from (2.16) and the second part from the integral of (2.10). Property (3) is established by integrating its left-hand side by parts and using (2.14) and (2.16). Similarly, property (4) is proved by integrating its left-hand side by parts and using (2.13).

21.4 Inversion Formulas

21.4.1 The Hankel, Y, and H Transforms

Since the Hankel and Hankel–type transforms are of Fourier–type, i.e., they are of the form

$$F(y) = \int_0^\infty f(x)k(yx)dx, \tag{21.2}$$

and their inversion formulas are also of the form

$$f(x)=\int_0^\infty F(y)h(xy)dy, \tag{21.3}$$

for some appropriate kernel functions k and h, we shall first derive some general properties of integral transforms of Fourier–type.

Multiplying (21.2) by y^{s-1} and integrating from 0 to ∞, we obtain

$$\mathcal{M}_F(s)=\int_0^\infty y^{s-1}F(y)dy=\int_0^\infty f(x)\left(\int_0^\infty y^{s-1}k(yx)dy\right)dx$$

$$=\int_0^\infty x^{-s}f(x)dx\int_0^\infty u^{s-1}k(u)du=\mathcal{M}_f(1-s)\mathcal{M}_k(s), \tag{21.4}$$

where $\mathcal{M}$ denotes the Mellin transformation. Similarly, from (21.3) we obtain

$$\mathcal{M}_f(s)=\mathcal{M}_F(1-s)\mathcal{M}_h(s). \tag{21.5}$$

Combining (21.4) and (21.5) leads to

$$\mathcal{M}_k(1-s)\mathcal{M}_h(s)=1 \quad \text{or} \quad \mathcal{M}_k(\frac{1}{2}-s)\mathcal{M}_h(s+\frac{1}{2})=1. \tag{21.6}$$

The kernel functions k and h are called *unsymmetrical Fourier kernels.* One can verify that if $k(x)$ and $h(x)$ are unsymmetrical Fourier kernels, then so are $\sqrt{b}k(bx)$ and $\sqrt{b}h(bx)$, and also $cx^{(c-1)/2}k(x^c)$ and $cx^{(c-1)/2}h(x^c)$.

Suppose we know the kernel, k, of the forward transform. The kernel of the inverse transform can be determined by

$$h(x)=\mathcal{M}^{-1}\left(\mathcal{M}_h(s)\right)(x)=\mathcal{M}^{-1}\left(\frac{1}{\mathcal{M}_k(1-s)}\right)(x)$$

$$=\frac{1}{2\pi i}\int_{c-i\infty}^{c+i\infty}\frac{1}{\mathcal{M}_k(1-s)}x^{-s}ds. \tag{21.7}$$

The following are consequences of (21.6):

A) Since

$$\mathcal{M}\left[\sqrt{x}J_\nu(x)\right](s)=\frac{2^{s-1/2}\Gamma(\frac{s}{2}+\frac{1}{4}+\frac{\nu}{2})}{\Gamma(\frac{\nu}{2}-\frac{s}{2}+\frac{3}{4})},$$

it is easy to see that

$$\mathcal{M}\left[\sqrt{x}J_\nu(x)\right](1-s)=\frac{1}{\mathcal{M}\left[\sqrt{x}J_\nu(x)\right](s)},$$

and hence the kernels of the Hankel transform and its inverse are the same. That is, if

$$F(y)=\int_0^\infty f(x)\sqrt{yx}J_\nu(yx)dx\,,$$

then

$$f(x) = \int_0^\infty F(y)\sqrt{yx}J_\nu(yx)\,dy\,.$$

The following theorem gives a more precise statement for the inversion of the Hankel transform.

THEOREM 21.2

If f is absolutely integrable on $(0,\infty)$ and of bounded variation in a neighborhood of the point x, then

$$\int_0^\infty F_\nu(y)\sqrt{xy}\,J_\nu(yx)dy = \frac{f(x+0)+f(x-0)}{2}$$

whenever the expression on the right-hand side has a meaning and to $f(x)$ whenever $f(x)$ is continuous.

B) Since

$$\mathcal{M}\left[\sqrt{x}Y_\nu(x)\right](s) = -\frac{2^{s-1/2}}{\pi}\Gamma(\frac{s}{2}+\frac{1}{4}+\frac{\nu}{2})\Gamma(\frac{s}{2}+\frac{1}{4}-\frac{\nu}{2})\cos[(s+\frac{1}{2}-\nu)\pi/2],$$

and

$$\mathcal{M}\left[\sqrt{x}\mathbf{H}_\nu(x)\right](1-s) = \frac{2^{1/2-s}\pi}{\Gamma(\frac{s}{2}+\frac{1}{4}+\frac{\nu}{2})\Gamma(\frac{s}{2}+\frac{1}{4}-\frac{\nu}{2})\cos[(\frac{3}{2}-s+\nu)\pi/2]},$$

it follows that the Y and H transforms are the inverse of each other.

More generally, let F and f be given by (21.2) and (21.3) so that

$$f(x) = \int_0^\infty h(xy)\,dy\int_0^\infty f(u)k(yu)\,du. \tag{21.8}$$

We shall call F the k transform of f and f the h transform of F. Let us denote the antiderivative

$$\int_0^x f(t)dt$$

of f by $f^{(-1)}(x)$.

THEOREM 21.3

Let k and h be given as above. Then

$$\int_0^\infty h(xy)k^{(-1)}(yz)\frac{dy}{y} = \chi_{[0,z]}(x) \tag{21.9}$$

if and only if (21.8) holds. Moreover,

$$\int_0^\infty h^{(-1)}(uy)k^{(-1)}(yz)\frac{dy}{y^2} = \begin{cases} u, \text{ if } u \le z \\ z, \text{ if } z \le u \end{cases} = \min(u, z).$$

Proof: Let (21.8) hold. If we set $f(x) = \chi_{[0,z]}(x)$, we obtain

$$\chi_{[0,z]}(x) = \int_0^\infty h(xy)dy \int_0^z k(yu)du = \int_0^\infty h(xy)k^{(-1)}(zy)\frac{dy}{y},$$

which is (21.9). Conversely, let (21.9) hold. Then

$$\int_0^x f(y)\,dy = \int_0^\infty f(y)\chi_{[0,x]}(y)\,dy$$

$$= \int_0^\infty f(y)\,dy \int_0^\infty h(yu)k^{(-1)}(xu)\frac{du}{u} = \int_0^\infty k^{(-1)}(xu)\frac{du}{u}\int_0^\infty f(y)h(uy)dy,$$

which, when we differentiate it with respect to x, gives us (21.8).

If we integrate (21.9), we obtain

$$\int_0^\infty h^{(-1)}(uy)k^{(-1)}(yz)\frac{dy}{y^2} = \int_0^u \chi_{[0,z]}(x)dx = \begin{cases} u, \text{ if } u \le z \\ z, \text{ if } z \le u \end{cases} = \min(u, z).$$

∎

Let us denote the Mellin transform of g by $\mathcal{M}_g(s)$. From Section 10.3, item 5, we have

$$\mathcal{M}_{f^{(-1)}(x)}(s) = \mathcal{M}_{f^{(-1)}(x)/x}(s+1) = \frac{\mathcal{M}_{f(x)}(s+1)}{-s},$$

which implies that for each fixed y

$$\mathcal{M}_{f^{(-1)}(yx)/(yx)}(s) = \frac{\mathcal{M}_f(s)}{(1-s)y^s}. \tag{21.10}$$

Let

$$g_k(x) = \frac{d}{dx}\int_0^\infty \frac{k^{(-1)}(xy)}{y}f(y)\,dy, \tag{21.11}$$

and

$$g_h(x) = \frac{d}{dx}\int_0^\infty \frac{h^{(-1)}(xy)}{y}f(y)\,dy. \tag{21.12}$$

If $h^{(-1)}(x)/x$ and $k^{(-1)}(x)/x$ are in $L^2(0,\infty)$, then, in view of Parseval's relation of the Mellin transform (cf. (10.10)) and (21.10), we have for $f \in L^2(0,\infty)$,

$$\int_0^x g_k(u)\,du = \int_0^\infty \frac{k^{(-1)}(xy)}{y}f(y)\,dy$$

$$= \frac{1}{2\pi i}\int_{c-i\infty}^{c+i\infty} x^{1-s}\frac{\mathcal{M}_k(s)}{1-s}\mathcal{M}_f(1-s)\,ds.$$

Thus,

$$g_k(x) = \frac{1}{2\pi i}\int_{c-i\infty}^{c+i\infty} x^{-s}\mathcal{M}_k(s)\mathcal{M}_f(1-s)\,ds\,,$$

and similarly

$$g_h(x) = \frac{1}{2\pi i}\int_{d-i\infty}^{d+i\infty} x^{-s}\mathcal{M}_h(s)\mathcal{M}_f(1-s)\,ds\,.$$

By the inversion formula for the Mellin transform (Section 10.4), we infer that

$$\mathcal{M}_{g_k}(s) = \mathcal{M}_k(s)\mathcal{M}_f(1-s) \text{ and } \mathcal{M}_{g_h}(s) = \mathcal{M}_h(s)\mathcal{M}_f(1-s)\,.$$

Let η be the k transform of g_h. Then by (21.5)

$$\mathcal{M}_\eta(s) = \mathcal{M}_{g_h}(1-s)\mathcal{M}_k(s) = \mathcal{M}_h(1-s)\mathcal{M}_k(s)\mathcal{M}_f(s) = \mathcal{M}_f(s)\,.$$

This means that the k transform of g_h is f, and so is the h transform of g_k.

21.4.2 The I Transform

The following inversion formula for the I transform has been obtained by Koh and Zemanian [126] using a different technique: If

$$F(s) = \int_0^\infty f(x)\sqrt{sx}I_\nu(sx)\,dx \quad ,\nu \geq -1/2\,,$$

then

$$f(x) = \lim_{r\to\infty}\frac{1}{\pi i}\int_{c-ir}^{c+ir} F(s)\sqrt{xs}K_\nu(xs)\,ds\,,$$

for some appropriate c.

21.4.3 The Hankel Potential Transform

To obtain an inversion formula for the Hankel potential transform, let $D = \frac{d}{dx}$, and define

$$L_1[f] = \frac{\sqrt{\pi}\,\Gamma(2\nu+1)}{2^{5/2+\nu}[\Gamma(2+\nu/2)]^2 x^{1+2\nu}} Dx^{5+2\nu} D\frac{1}{x}D[f]\,,$$

and

$$L_n[f] = \frac{(-1)^{n+1}\sqrt{\pi}\,\Gamma(2\nu+1)}{2^{2n+\nu+1/2}[\Gamma(n+1+\nu/2)]^2 x^{2n-1+2\nu}} Dx^{4n+1+2\nu}$$

$$\times\left[\prod_{k=1}^{n-1} D\frac{1}{x^{4k+2\nu+1}}Dx^{4k+2\nu+1}\right] D\frac{1}{x}D[f]\,, n = 2, 3, \cdots.$$

THEOREM 21.4 [36]
Let $d\mu_\nu$ be as given in Definition 21.3 and let f be locally integrable with respect to $d\mu_\nu$. If the Hankel potential transform, F_ν, of f exists for $x \neq 0$, then

$$\lim_{n\to\infty} L_n[F_\nu(x)] = f(x).$$

21.5 Parseval's Relation

Let F_ν and G_ν denote the Hankel transforms of order ν of f and g, respectively. Then

$$\int_0^\infty F_\nu(y)G_\nu(y)dy = \int_0^\infty f(x)g(x)dx.$$

In particular,

$$\int_0^\infty |F_\nu(y)|^2\, dy = \int_0^\infty |f(x)|^2 dx.$$

More generally, if g_k and g_h are defined by (21.11) and (21.12), then by Parseval's relation for the Mellin transform (cf. 10.10), we have

$$\int_0^\infty g_k(x)g_h(x)\,dx = \frac{1}{2\pi i}\int_{c-i\infty}^{c+i\infty} \mathcal{M}_k(s)\mathcal{M}_f(1-s)\mathcal{M}_h(1-s)\mathcal{M}_f(s)\,ds$$

$$= \frac{1}{2\pi i}\int_{c-i\infty}^{c+i\infty} \mathcal{M}_f(s)\mathcal{M}_f(1-s)\,ds = \int_0^\infty f^2(x)\,dx\,.$$

21.6 Convolution

Let

$$d\rho_\nu(x) = \frac{x^{2\nu+1}dx}{2^\nu\Gamma(\nu+1)}\,,$$

$$\mathbf{J}_\nu(x) = 2^\nu\Gamma(\nu+1)x^{-\nu}J_\nu(x)\,,$$

and define

$$(\tilde{\mathcal{H}}_\nu f)(t) = \tilde{F}_\nu(t) = \int_0^\infty f(x)\mathbf{J}_\nu(tx)\,d\rho_\nu(x)\,. \tag{21.13}$$

It is easy to see that,

$$\mathcal{H}_\nu\left(x^{\nu+1/2}f(x)\right)(t) = t^{\nu+1/2}\tilde{F}_\nu(t)\,.$$

It follows from the inversion formula for the Hankel transform that if $\tilde{F}_\nu \in L^1(0,\infty)$, then

$$f(x) = \int_0^\infty \tilde{F}_\nu(t)\mathbf{J}_\nu(xt)\,d\rho_\nu(t)\,.$$

The convolution operation, $\otimes$, associated with the Hankel transform is defined as

$$(f \otimes g)(x) = \int_0^\infty \int_0^\infty f(y)g(z)D_\nu(x,y,z)\,d\rho_\nu(y)\,d\rho_\nu(z)\,, \tag{21.14}$$

where

$$D_\nu(x,y,z) = \frac{2^{3\nu-1/2}\Gamma^2(\nu+1)A^{2\nu-1}(x,y,z)}{\sqrt{\pi}\,\Gamma(\nu+1/2)(xyz)^{2\nu}}\,,$$

and $A(x,y,z)$ is the area of the triangle whose sides are x, y, z if there is such a triangle and zero otherwise. An important property of the convolution operation is given in the following theorem.

THEOREM 21.5 [102]
Let $1 \le r, s \le \infty$, and let p be such that

$$\frac{1}{p} = \frac{1}{r} + \frac{1}{s} - 1.$$

If $f \in L^r(0,\infty), g \in L^s(0,\infty)$, then $h(x) = (f \otimes g)(x)$ is defined for all $0 < x < \infty$ and

$$\|h\|_p \le \|f\|_r \|g\|_s\,.$$

Moreover, if $p = \infty$, then h is defined for $0 < x < \infty$ and is continuous.

Using the relation

$$\int_0^\infty \mathbf{J}_\nu(tz)D_\nu(x,y,z)\,d\rho_\nu(z) = \mathbf{J}_\nu(tx)\mathbf{J}_\nu(ty) \qquad [242,\ \text{p. } 411]$$

one can prove the following convolution theorem for the Hankel transform; see [102].

THEOREM 21.6 **(Convolution Theorem)**
Let $h(t) = (f \otimes g)(t)$. If $f, g \in L^1(0,\infty)$, then

$$\tilde{H}_\nu(t) = \tilde{F}_\nu(t)\tilde{G}_\nu(t)\,,$$

where $\tilde{H}_\nu\,, \tilde{F}_\nu\,, \tilde{G}_\nu$ are the Hankel transforms of h, f, g as defined by (21.13).

21.7 Generalized Functions

21.7.1 The Hankel Transform

The Hankel transform can be extended to generalized functions by using the embedding or the adjoint method. We shall outline the adjoint method as introduced by Zemanian [276, Ch. 5].

Let $I = (0, \infty)$ and define A_μ as the space of all infinitely differentiable functions ϕ on I that satisfy

$$\gamma^{\mu}_{m,k}(\phi) = \sup_{0<x<\infty} \left| x^m \left(x^{-1}D\right)^k \left[x^{-\mu-1/2}\phi(x)\right]\right| < \infty ,$$

where

$$D = \frac{d}{dx}, \ \left(x^{-1}D\right)^k = \left(x^{-1}D\right)^{k-1}\left(x^{-1}D\right); \quad k, m = 0, 1, 2, \cdots, \mu \geq -1/2.$$

It is easy to see that $\left(A_\mu, \left\{\gamma^{\mu}_{m,k}\right\}\right)$ is a testing-function space and that $\mathcal{D}(I)$ is a subspace of A_μ for every μ, but it is not a dense subspace. Moreover, A_μ is a subspace of $\mathcal{E}(I)$ and $A_{\mu+2q}$ is a subspace of A_μ for any $q = 0, 1, 2, \cdots$. The latter claim requires a little bit more work to establish. First, we show by induction that

$$R^k x^{-\mu-1/2}\phi(x) = 2kR^{k-1}\eta(x) + x^2 R^k \eta(x) , \tag{21.15}$$

where $R = x^{-1}D$ and $\eta(x) = x^{-\mu-5/2}\phi(x)$. Since

$$\begin{aligned} Rx^{-\mu-1/2}\phi(x) &= R\left(x^2\eta(x)\right) \\ &= x^{-1}\left[2x\eta(x) + x^2 D\eta(x)\right] = 2\eta(x) + x^2 R\eta(x) , \end{aligned}$$

(21.15) is true for $k = 1$. Now assume that it is true for k, we shall show that it is also true for $k+1$. From the induction assumption, it follows

$$\begin{aligned} R^{k+1}x^{-\mu-1/2}\phi(x) &= R\left(R^k x^{-\mu-1/2}\phi(x)\right) \\ &= R\left(2kR^{k-1}\eta(x) + x^2R^k\eta(x)\right) \\ &= 2kR^k\eta(x) + x^{-1}\left[2xR^k\eta(x) + x^2 D(R^k\eta(x)\right] \\ &= 2(k+1)R^k\eta(x) + x^2R^{k+1}\eta(x) , \end{aligned}$$

which is (21.15) when k is replaced by $k+1$. If we multiply (21.15) by x^m and then take the sup-norm, we immediately obtain

$$\gamma^{\mu}_{m,k}(\phi) \leq 2k\gamma^{\mu+2}_{m,k-1}(\phi) + \gamma^{\mu+2}_{m+2,k}(\phi) ,$$

which, together with $\gamma^{\mu}_{m,0}(\phi) = \gamma^{\mu+2}_{m+2,0}(\phi)$, implies that $A_{\mu+2}$ is a subspace of A_μ. The general case can be proved by induction over q.

Functions in the space A_μ are of rapid descent as $x \to \infty$. If f is a locally integrable function on I such that $f(x) = O(x^p)$ as $x \to \infty$ for some $p \geq 0$, and $x^{\mu+1/2} f(x)$ is integrable over $(0,1)$, then it defines a regular member of the dual space A^*_μ. For,

$$\langle f, \phi \rangle = \int_0^1 x^{\mu+1/2} f(x) x^{-\mu-1/2}\phi(x)dx + \int_1^\infty x^{-m+\mu+1/2} f(x) x^{m-\mu-1/2}\phi(x)dx.$$

Thus,

$$|\langle f, \phi\rangle| \leq \gamma^{\mu}_{0,0}(\phi) \int_0^1 x^{\mu+1/2}|f(x)|dx + \gamma^{\mu}_{m,0}(\phi) \int_1^\infty x^{-m+\mu+1/2}|f(x)|dx < \infty,$$

for $m > \mu + p + 3/2$.

LEMMA 21.1

Let us define the operator N_μ as

$$N_\mu \phi(x) = x^{\mu+1/2} D x^{-\mu-1/2}\phi.$$

Then

1) N_μ is a continuous linear mapping from A_μ into $A_{\mu+1}$.

2) The following relations hold:

$$N_{\mu+k} N_{\mu+k-1} \cdots N_\mu f(x) = x^{\mu+k+3/2} \left(x^{-1}D\right)^{k+1} x^{-\mu-1/2} f(x), \tag{21.16}$$

$$N_{\mu+k+m-1} \cdots N_{\mu+k+1} N_{\mu+k} x^k f(x) = x^k N_{\mu+m-1} \cdots N_{\mu+1} N_\mu f(x). \tag{21.17}$$

3) For any $\phi \in A_\mu$,

$$N_\mu \left(\mathcal{H}_\mu \phi(x)\right)(y) = -\mathcal{H}_{\mu+1}\left[x\phi(x)\right](y), \tag{21.18}$$

and

$$\mathcal{H}_{\mu+1}\left[N_\mu \phi\right](y) = -y\mathcal{H}_\mu\left[\phi\right](y). \tag{21.19}$$

Proof: 1) For any $\phi \in A_\mu$, we have

$$\begin{aligned}\gamma^{\mu+1}_{m,k}(N_\mu\phi) &= \sup_{0<x<\infty} \left| x^m (x^{-1}D)^k [x^{-\mu-3/2} x^{\mu+1/2} D x^{-\mu-1/2}\phi(x)]\right| \\ &= \sup_{0<x<\infty} \left| x^m (x^{-1}D)^{k+1} [x^{-\mu-1/2}\phi(x)]\right| = \gamma^{\mu}_{m,k+1}(\phi).\end{aligned}$$

2) Equation (21.16) is proved by induction on k and (21.17) by induction on m.

3) Since $\phi \in A_\mu$ is of rapid descent at infinity, we have

$$N_\mu\left(\mathcal{H}_\mu \phi\right)(y) = y^{\mu+1/2} D_y y^{-\mu-1/2} \int_0^\infty \phi(x)\sqrt{yx}J_\mu(yx)\,dx\,,$$

but by (2.19)

$$D_y y^{-\mu} J_\mu(yx) = -xy^{-\mu} J_{\mu+1}(yx)\,;$$

hence

$$N_\mu\left(\mathcal{H}_\mu \phi\right)(y) = -\int_0^\infty x\phi(x)\sqrt{yx}J_{\mu+1}(yx)\,dx = -\mathcal{H}_{\mu+1}\left[x\phi(x)\right](y)\,,$$

which is (21.18). As for (21.19), we have

$$\begin{aligned}\mathcal{H}_{\mu+1}\left[N_\mu \phi\right](y) &= \int_0^\infty \left(x^{\mu+1/2} D x^{-\mu-1/2}\phi(x)\right)\sqrt{yx}J_{\mu+1}(yx)\,dx \\ &= \sqrt{y}\int_0^\infty x^{\mu+1} J_{\mu+1}(yx) D\left(x^{-\mu-1/2}\phi(x)\right)\,dx.\end{aligned}$$

By integration by parts and with the aid of the relation (cf. 2.18)

$$D_x x^{\mu+1} J_{\mu+1}(yx) = yx^{\mu+1} J_\mu(yx)\,,$$

we obtain

$$\mathcal{H}_{\mu+1}\left[N_\mu \phi\right](y) = -y\int_0^\infty \phi(x)\sqrt{yx}J_\mu(yx)\,dx = -y\mathcal{H}_\mu[\phi](y)\,.$$

∎

THEOREM 21.7
For $\mu \geq -1/2$, the Hankel transformation is an automorphism on A_μ.

Proof: Let us denote the Hankel transform of order μ of $\phi(x)$ by $\Phi_\mu(y) = \mathcal{H}_\mu[\phi](y)$. By repeated applications of (21.18), we obtain

$$N_{\mu+k-1}\cdots N_{\mu+1}N_\mu \Phi_\mu(y) = (-1)^k \mathcal{H}_{\mu+k}\left[x^k\phi(x)\right](y)\,;$$

hence

$$(-y)^m N_{\mu+k-1}\cdots N_{\mu+1}N_\mu \Phi_\mu(y) = (-1)^{k+m}(y)^m \mathcal{H}_{\mu+k}\left[x^k\phi(x)\right](y)\,, \tag{21.20}$$

and by repeated applications of (21.19), it follows that

$$\mathcal{H}_{\nu+l}\left[N_{\nu+l-1}\cdots N_{\nu+1}N_\nu \phi\right](y) = (-y)^l \mathcal{H}_\nu[\phi](y)\,.$$

If we replace ν by $\mu+k$, l by m, and ϕ by $x^k\phi$ we obtain

$$\mathcal{H}_{\mu+k+m}\left[N_{\mu+k+m-1}\cdots N_{\mu+k+1}N_{\mu+k}x^k\phi(x)\right](y) = (-y)^m\mathcal{H}_{\mu+k}\left[x^k\phi(x)\right](y). \tag{21.21}$$

But from (21.17), we have

$$\mathcal{H}_{\mu+k+m}\left[N_{\mu+k+m-1}\cdots N_{\mu+k}x^k\phi(x)\right] = \mathcal{H}_{\mu+k+m}\left[x^k N_{\mu+m-1}\cdots N_{\mu+1}N_\mu\phi\right],$$

which, when we combine it with (21.21), gives us

$$\mathcal{H}_{\mu+k+m}\left[x^k N_{\mu+m-1}\cdots N_{\mu+1}N_\mu\phi\right](y) = (-1)^m y^m\mathcal{H}_{\mu+k}\left[x^k\phi(x)\right](y). \tag{21.22}$$

Putting (21.20) and (21.22) together, we find that

$$(-y)^m\left(N_{\mu+k-1}\cdots N_{\mu+1}N_\mu\Phi_\mu\right)(y) = (-1)^k\mathcal{H}_{\mu+k+m}\left[x^k N_{\mu+m-1}\cdots N_{\mu+1}N_\mu\phi\right](y),$$

which, in view of (21.16), can be reduced to

$$(-)^{k+m}y^{\mu+k+m+1/2}(R_y)^k y^{-1/2-\mu}\Phi_\mu(y) = \mathcal{H}_{\mu+k+m}\left[x^{\mu+k+m+1/2}(R_x)^m x^{-\mu-1/2}\phi(x)\right](y),$$

where $R_x = x\frac{d}{dx}$.

Therefore,

$$(-y)^m(R_y)^k y^{-1/2-\mu}\Phi_\mu(y) = \int_0^\infty x^{2k+2\mu+m+1}\left[(R_x)^m x^{-\mu-1/2}\phi(x)\right]\frac{J_{\mu+k+m}(yx)}{(yx)^{\mu+k}}dx.$$

If we choose n large enough, say $n > 2\mu+2k+m+1$, and take the sup-norm, we obtain

$$\begin{aligned}\gamma^\mu_{m,k}(\Phi_\mu) &\le C\int_0^\infty (1+x^2)^{n+1}\left|(x^{-1}D)^m x^{-\mu-1/2}\phi(x)\right|\frac{dx}{(1+x^2)}\\ &\le \frac{\pi}{2}C\sum_{j=0}^{n+1}\binom{n+1}{j}\gamma^\mu_{2j,m}(\phi),\end{aligned}$$

where C is a constant that depends on k and m.

Thus, $\Phi_\mu \in A_\mu$ and the mapping $\mathcal{H}_\mu$ is continuous on A_μ. By the inversion theorem for the Hankel transformation and the fact that $\mathcal{H}_\mu^{-1} = \mathcal{H}_\mu$, we infer that $\mathcal{H}_\mu$ is a one-to-one transformation from A_μ onto itself. ∎

Now we are able to extend the Hankel transform to generalized functions.

DEFINITION 21.4 *The Hankel transform of a generalized function $f \in A_\mu^*$ is defined as*

$$\langle\mathcal{H}_\mu f,\phi\rangle = \langle f,\mathcal{H}_\mu\phi\rangle \quad \text{for all } \phi \in A_\mu.$$

It follows from Theorem 21.7 that the Hankel transformation is an automorphism on A_μ^*. If f is a regular member of A_μ^* that is integrable over $(0, \infty)$, then its generalized Hankel transform as given by Definition 21.4 coincides with that given by Definition 21.1. For,

$$\begin{aligned}\langle f, \mathcal{H}_\mu \phi\rangle &= \int_0^\infty f(y)\left(\int_0^\infty \phi(x)\sqrt{yx}J_\mu(yx)\,dx\right)dy\\ &= \int_0^\infty \phi(x)\left(\int_0^\infty f(y)\sqrt{yx}J_\mu(yx)\,dy\right)dx = \langle \mathcal{H}_\mu f, \phi\rangle\,,\end{aligned}$$

hence

$$(\mathcal{H}_\mu f)(y) = \int_0^\infty f(x)\sqrt{yx}J_\mu(yx)\,dx = \langle f(x), \sqrt{yx}J_\mu(yx)\rangle.$$

Similarly, if $f \in \mathcal{E}^*(I)$, one can show that

$$(\mathcal{H}_\mu f)(y) = \langle f, \sqrt{yx}J_\mu(yx)\rangle\,, \quad \mu \geq -1/2.$$

21.7.2 The I transform

Koh and Zemanian [126] extended the I transform to generalized functions by using the embedding method. They introduced the space $\mathbf{I}_{\mu,a}$, where $a > 0$, $\mu \geq -1/2$, as the space of all complex-valued, infinitely differentiable functions $\phi(x)$ on $I = (0, \infty)$ such that

$$\gamma_k^{\mu,a}(\phi) = \sup_{x\in I}\left|e^{-ax}x^{-\mu-1/2}S_\mu^k(\phi(x))\right| < \infty\,, \quad k = 0, 1, 2, \cdots,$$

where

$$S_\mu = x^{-\mu-1/2}Dx^{2\mu+1}Dx^{-\mu-1/2}.$$

Then,they showed that for every fixed complex number s such that $|\mathrm{Re}\, s| < a$, $\sqrt{sx}I_\mu(sx) \in \mathbf{I}_{\mu,a}$. Therefore, the I transform of any $f \in \mathbf{I}_{\mu,a}^*$ can be defined as

$$F_\mu(s) = \langle f(x)\,, \sqrt{sx}I_\mu(sx)\rangle \quad |\mathrm{Re}\, s| < a.$$

21.8 Applications

The Hankel transform is particularly useful in solving initial-value problems in cylindrical coordinates. We illustrate this by an example. More examples of a similar nature can be found in [216].

21.8.1 Axisymmetric Dirichlet Problem for a Half-Space

Consider the problem

$$\frac{\partial^2 v}{\partial r^2} + \frac{1}{r}\frac{\partial v}{\partial r} + \frac{\partial^2 v}{\partial z^2} = 0\,, \quad r, z > 0\,,$$

with

$$v(r,0) = f(r)\,, \tag{21.23}$$

$$v(r,z) \to 0 \quad \text{as } \sqrt{r^2+z^2} \to \infty. \tag{21.24}$$

By using the change of variables

$$u(r,z) = \sqrt{r}v(r,z) \quad , \quad g(r) = \sqrt{r}f(r)\,,$$

the differential equation is converted into

$$\frac{\partial^2 u}{\partial r^2} + \frac{1}{4r^2}u + \frac{\partial^2 u}{\partial z^2} = 0\,.$$

If we apply the Hankel transform of order zero, taken with respect to r, to this equation, we obtain

$$-\rho^2 U(\rho,z) + \frac{\partial^2 U(\rho,z)}{\partial z^2} = 0\,,$$

where

$$U(\rho,z) = \int_0^\infty u(r,z)\sqrt{\rho r}J_0(\rho r)\,dr\,.$$

Solving for U and using (21.24), we find

$$U(\rho,z) = A(\rho)e^{-\rho z}\,.$$

Because of (21.23)

$$U(\rho,0) = A(\rho) = G_0(\rho)\,,$$

where G_0 is the Hankel transform of order zero of g, and it follows that

$$U(\rho,z) = G_0(\rho)e^{-\rho z}\,.$$

Therefore, in view of the inversion formula for the Hankel transform, we have the solution $v(r,z) = r^{-1/2}u(r,z)$, where

$$u(r,z) = \int_0^\infty G_0(\rho)e^{-\rho z}\sqrt{r\rho}J_0(r\rho)\,d\rho\,.$$

With some obvious modifications, the above solution remains valid if f is a suitably chosen generalized function.

21.8.2 Hankel Harmonic Functions

A function $u(x,t)$ is said to be Hankel harmonic in a region R if

$$\Delta_x u + \frac{\partial^2 u}{\partial t^2} = 0$$

in R where

$$\Delta_x = \frac{\partial^2}{\partial x^2} + \frac{2\nu}{x}\frac{\partial}{\partial x}.$$

Since $\Delta_x \mathcal{J}_\nu(yx) = -y^2 \mathcal{J}_\nu(yx)$, where $\mathcal{J}_\nu(x)$ is given by (21.1), it can be readily verified that

$$\frac{t}{(x^2+t^2)^{\nu+1}}$$

is Hankel harmonic in the half plane $t > 0$. Therefore, the Hankel potential transform is an integral transform whose kernel is a Hankel harmonic function. This makes the Hankel potential transform useful in solving problems involving the Hankel harmonic partial differential equation.

22

The Hardy Transform

The Hardy transform was first discussed by the great English mathematician G.H. Hardy in 1925 in a paper [90] in which he was interested in simplifying the proof of a formula involving Bessel functions that was proved earlier by Titchmarsh. Hardy introduced an integral transform, which is now called the Hardy transform, and reported on its inversion formula, but provided no proofs for it. It was R. Cooke [46] who established the validity of Hardy's inversion formula, together with another inversion formula by Titchmarsh.

The Hardy transform is related to the Hankel and the Y transforms in much the same way as the Hartley transform is related to the cosine and sine Fourier transforms.

22.1 Definitions

DEFINITION 22.1 *The Hardy transform of order ν of a function $f(t)$ is defined as*

$$F_\nu(x) = F(x) = \int_0^\infty [\cos a\pi J_\nu(xt) + \sin a\pi Y_\nu(xt)]\, tf(t)dt, \quad Re\ \nu > -1/2, \tag{22.1}$$

where J_ν and Y_ν are the Bessel functions of the first and second kind, respectively (c.f., Section 2.5).

The Hardy transform reduces essentially to:

1. the Hankel transform when $a = 0$,

2. the Y transform when $a = 1/2$,

3. the Hankel transform with kernel $J_{-\nu}$ when $a = -\nu$.

The last one follows from the observation that

$$Y_\nu(z) = \frac{J_\nu(z)\cos\nu\pi - J_{-\nu}(z)}{\sin\nu\pi}.$$

To get the exact forms of the Hankel and Y transforms from the Hardy transform, one needs to replace $\sqrt{t}f(t)$ by $g(t)$ and $\sqrt{x}F(x)$ by $G(x)$. Another interesting case is when $\nu = 1/2$. It yields

$$F(x) = \sqrt{\frac{2}{\pi x}} \int_0^\infty \sin(xt - a\pi)\, f(t)\sqrt{t}\, dt\,,$$

or

$$G(x) = \sqrt{\frac{2}{\pi}} \int_0^\infty \sin(xt - a\pi) g(t) dt, \tag{22.2}$$

where $G(x) = \sqrt{x}F(x)$ and $g(t) = \sqrt{t}f(t)$.

This last version of the Hardy transform makes it easy to relate it to the Hartley transform, which is defined by

$$G_1(x) = \frac{1}{\sqrt{2\pi}} \int_{-\infty}^{\infty} (\cos xt + \sin xt) g(t) dt\,, \qquad \text{see Chapter 13.}$$

As we pointed out in Chapter 13, the Hartley transform is used in solving some electrical engineering problems; for more details see [24]. Functions that vanish on the negative real axis represent what is known as *causal signals* in electric engineering. For such functions, the Hartley transform coincides (up to a factor of $1/\sqrt{2}$) with the Hardy transform (22.2) when $a = -1/4$.

22.2 Existence

A sufficient condition for the Hardy transform of $f(t)$ to exist is that

1. $t^\sigma f(t) \in L^1(0, \delta)$ for some $\delta > 0$, where $\sigma = 1 - |\nu|$,

2. $\sqrt{t}f(t) \in L^1(\delta, \infty)$.

22.3 Elementary Properties

Since the Hardy transform is essentially a combination of the Hankel and the Y transforms, any common property of these two transforms is also a property of the Hardy transform. For example, if we denote the Hardy transform (22.1) by $F_\nu(x) = \mathbf{T}_\nu\left(f(t)\right)(x)$, we then have

1. **Dilation:**

$$D_{1/b}\mathbf{T}_\nu\left[f(t)\right](x) = \frac{1}{b}\mathbf{T}_\nu\left[D_b f(t)\right](x)\ b > 0.$$

2. **Differentiation:**

$$x^{-\nu}\left(\frac{1}{x}\frac{d}{dx}\right)^n\left(x^{n+\nu}\mathbf{T}_{n+\nu}[f](x)\right) = \mathbf{T}_\nu\left[t^n f(t)\right](x).$$

22.4 Asymptotic Behavior and Abelian Theorems

Let f be a locally integrable function such that $t^\sigma f(t) \in L^1(\delta, \infty)$, where $3/2 < \sigma < 2 - |Re\ \nu|$, and $\delta > 0$.

If

$$\lim_{t \to 0+} t^\sigma f(t) = 0,$$

then

$$\lim_{x \to \infty} x^{2-\sigma} F_\nu(x) = 0.$$

Similarly, if f is a locally integrable function such that $t^{1-|\nu|} f \in L^1(0, \delta)$ for some $\delta > 0$ and $3/2 < \sigma < 2 - |Re\ \nu|$, then

$$\lim_{t \to \infty} t^\sigma f(t) = 0,$$

implies that

$$\lim_{x \to 0+} x^{2-\sigma} F_\nu(x) - 0.$$

22.5 Inversion Formula

Let $a > -1, a + \nu > -1$ and $|2a + \nu| < 3/2$. Under the assumptions that

1. $t^{\sigma} f(t) \in L^1(0, \delta)$ for some $\delta > 0$, and $\sigma = \min(1 - |\nu|, 1 - 2a - \nu, 1/2)$,
2. $\sqrt{t} f(t) \in L^1(\delta, \infty)$,
3. f is of bounded variation in a neighborhood of the point t,

we have

$$f(t) = \int_0^\infty G_\nu(xt) x F(x) dx, \tag{22.3}$$

where

$$G_\nu(z) = \sum_{k=0}^\infty \frac{(-1)^k (z/2)^{\nu+2a+2k}}{\Gamma(a+1+k)\Gamma(a+1+\nu+k)}$$
$$= \frac{2^{1-\mu} s_{\mu,\nu}(z)}{\Gamma\left(\frac{\mu+\nu+1}{2}\right)\Gamma\left(\frac{\mu-\nu+1}{2}\right)},$$

and $s_{\mu,\nu}(z)$ is Lommel's function (cf. Section 2.5.10) with $\mu = \nu + 2a - 1$.

The inversion formula (22.3) is the one that Hardy reported in [90] and Cooke proved in [46]. By combining (22.1) and (22.3) we obtain

$$f(x) = \int_0^\infty G_\nu(xu) u du \int_0^\infty C_\nu(tu) t f(t) dt,$$

where

$$C_\nu(t) = \cos a\pi J_\nu(t) + \sin a\pi Y_\nu(t). \tag{22.4}$$

If $a > -1, a + \nu > -1, \nu + 2a < 3/2$ and f is such that

1. $t^{\sigma} f(t) \in L^1(0, \delta)$, where $\sigma = \min(1 + \nu + 2a, 1/2), \delta > 0$,
2. $\sqrt{t} f(t) \in L^1(\delta, \infty)$,
3. f is of bounded variation,

then

$$f(x) = \int_0^\infty C_\nu(xu) u du \int_0^\infty G_\nu(tu) t f(t) dt.$$

As special cases of (22.1) and (22.3) we obtain:

1. For $a = 0$, the Hankel transform pair

$$F(x) = \int_0^\infty J_\nu(xt) t f(t) dt,$$

$$f(t) = \int_0^\infty J_\nu(xt)xF(x)dx.$$

2. For $a = 1/2$, the Y transform pair

$$F(x) = \int_0^\infty Y_\nu(xt)tf(t)dt,$$

$$f(t) = \int_0^\infty \mathbf{H}_\nu(xt)xF(x)dx,$$

where $\mathbf{H}_\nu(z)$ is the Struve function (see Section 2.5.10), which is related to Lommel's function by

$$\mathbf{H}_\nu(z) = \frac{2^{1-\nu}s_{\nu,\nu}(z)}{\sqrt{\pi}\Gamma(\nu+1/2)}$$

3. For $\nu = 1/2$, we have that if

$$F(x) = \sqrt{\frac{2}{\pi x}} \int_0^\infty \sin(xt - a\pi)\sqrt{t}f(t)dt,$$

then

$$f(t) = \frac{1}{\Gamma(2a)}\sqrt{\frac{2}{\pi}} \int_0^\infty s_{2a-1/2,1/2}(xt)xF(x)dx.$$

The last formula can be written in terms of Young's function [252]

$$\mathbf{Y}_v(z) = \sum_{k=0}^\infty \frac{(-1)^k z^{v+2k}}{\Gamma(v+2k+1)} = \frac{\sqrt{z}}{\Gamma(v-1)} s_{v-3/2,1/2}(z),$$

as

$$f(t) = \sqrt{\frac{2}{\pi t}} \int_0^\infty \mathbf{Y}_{2a+1}(xt)\sqrt{x}F(x)dx.$$

A case of particular interest because of its connection with the Hartley transform is when $a = -1/4$. There we have the pair

$$F(x) = \frac{1}{\sqrt{\pi x}} \int_0^\infty (\sin xt + \cos xt)\sqrt{t}f(t)dt,$$

and

$$f(t) = \sqrt{\frac{2}{\pi t}} \int_0^\infty \mathbf{Y}_{1/2}(xt)\sqrt{x}F(x)dx.$$

Young's function, $\mathbf{Y}_v(z)$, coincides with $\cos z$ when $v = 0$ and with $\sin z$ when $v = 1$. For intermediate values of v, it is no longer periodic, but fills up the analytical gap between the sine and cosine functions.

22.6 The Hardy Transform of Generalized Functions

The Hardy transform of generalized functions can be defined by using the embedding method (cf. Section 4.5.1). In [190], R. Pathak and J. Pandey constructed two testing-function spaces to extend the Hardy transform to generalized functions, in the first of which they embedded the function $C_v(xt)$ and the second is obtained by dividing members of the first by t. A more direct approach, of course, is to regard the kernel of the transform as $xtC_\nu(xt)$, then embed it in a suitably concocted testing-function space; however, we shall not pursue that here since the final outcome is the same.

22.6.1 Definition

Let $I = (0, \infty)$,

$$\eta(x, a, b) = \eta(x, \alpha) = \begin{cases} x^\alpha, & 0 < x \leq 1 \\ x^{\alpha-2}, & x > 1 \end{cases},$$

and

$$R = \frac{d^2}{dt^2} + \frac{1}{t}\frac{d}{dt} - \frac{\nu^2}{t^2}.$$

Since η depends on one parameter, we shall denote this space by $G_\alpha(I)$. The seminorms

$$\gamma_k(\phi) = \sup_{0<t<\infty} \left|\eta(t, \alpha)R^k\phi(t)\right| < \infty,$$

$k = 0, 1, 2, \ldots$, form a separating family of seminorms on $G_\alpha(I)$ making $(G_\alpha(I), \{\gamma_k\}_0^\infty)$ a Fréchet space.

In view of the asymptotic behavior of J_ν and Y_ν near zero and infinity (see Section 2.5.8) and the fact that

$$R_t^k C_\nu(t) = (-1)^k C_\nu(t), \quad k = 0, 1, 2, \ldots,$$

where $C_\nu(t)$ is given by (22.4), it follows that $C_\nu(xt) \in G_\alpha(I)$ for $|\nu| \leq \alpha \leq 1/2$.

Now define the space $H_\alpha(I)$ as follows: a C^∞-function $\phi(t)$ belongs to $H_\alpha(I)$ if and only if $\phi(t)/t$ belongs to $G_\alpha(I)$. The seminorms on $H_\alpha(I)$ are defined by

$$\tilde{\gamma}_k(\phi) = \gamma_k\left(\frac{\phi(t)}{t}\right), \quad k = 0, 1, \ldots.$$

Clearly, we have that $f \in H_\alpha^*(I)$ if and only if $tf(t) \in G_\alpha^*(I)$. For,

$$\langle tf(t), \phi\rangle = \langle f, t\phi\rangle$$

for all $\phi \in G_\alpha(I)$.

DEFINITION 22.2 *The Hardy transform of a generalized function $f \in H_\alpha^*(I)$ is defined as*

$$F_\nu(x) = \langle tf(t), C_\nu(xt)\rangle = \langle f(t), tC_\nu(xt)\rangle, \tag{22.5}$$

for $|\nu| \leq \alpha \leq 1/2$. For fixed ν, we may suppress the subscript and write $F(x)$ instead.

22.6.2 Inversion Formula

Let $f \in H_\alpha^*(t)$ and $F(x)$ be its Hardy transform as defined by (22.5). Then in the sense of $\mathcal{D}^*(I)$,

$$\lim_{N\to\infty} \int_0^N F(x)C_\nu(xt)x\,dx = f(t) - \langle yf(y), Q(t,y)\rangle,$$

where

$$Q(t,y) = \frac{2\sin(\alpha\pi)\sin(a+\nu)\pi}{\pi\sin(v\pi)}\,\frac{(t^{2v} - y^{2v})}{t^v y^v (t^2 - y^2)}.$$

This formula gives $f(t)$ in terms of $F(x)$; yet it does not correspond exactly to (22.3). To obtain an analogue of (22.3), a new testing-function space containing $G_\nu(t)$ is needed. This was done by the same authors in [188].

22.7 Applications

The importance of the Hardy transform lies in the fact that it generalizes the Hankel and the Y transforms, each of which has numerous applications, see Chapter 21. Moreover, the Hardy transform is also related to the

Hartley transform which, as we pointed out in Chapter 13, has many applications in electrical engineering [23, p. 179]. In addition, formula (22.2) is related to an integral that was used earlier by W.H. Young [252] in his study of Fourier integrals, which in turn have many applications in physics and engineering.

23

The K (Meijer) Transform

The K transform, also known as the Meijer transform [124, 125], was first introduced in 1940 by C. S. Meijer [158] who also derived its representation theorem and inversion formula. It was further investigated by Greenwood [80], Erdélyi [58] and Boas [16, 17] who obtained a sufficient condition for a function to be represented by a K transform. The K transformation is a generalization of the Laplace transformation and it has its own operational calculus which can be used to solve certain initial and boundary-value problems.

Meijer obtained an inversion formula for the K transform that generalizes the complex inversion formula for the Laplace transform, whereas Boas obtained an inversion formula that uses the Post-Widder inversion operator.

It should be pointed out that Meijer investigated other integral transformations, some of which are also referred to in the literature as Meijer's transformations [130, 131, 132] . Worse yet, some generalizations of the Meijer transformation are also called Meijer transformation [199] and [33, Section 7.2].

23.1 Definitions

DEFINITION 23.1 *The K transform or the Meijer transform of order μ of a function $f(t)$ is defined as*

$$\mathcal{K}_\mu[f(t)](s) = F_\mu(s) = \int_0^\infty f(t)\sqrt{st}\,K_\mu(st)\,dt\,, \quad \mu \quad \textit{complex} \tag{23.1}$$

where $K_\mu(z)$ is the modified Bessel function of the third kind and order μ and s is a complex variable.

Because $K_\mu(z) = K_{-\mu}(z)$, there is no loss of generality in assuming that

Re $\mu \geq 0$. In view of the relation (cf. Section 2.5, item 6)

$$K_{\frac{1}{2}}(z) = \sqrt{\pi/2z}\, e^{-z} ,$$

it is evident that the K transformation is a generalization of the Laplace transformation to which it reduces when $\mu = \frac{1}{2}$.

Another integral transformation that is sometimes also called the generalized Meijer transformation [124, 125] is the following:

$$G_\mu(y) = \frac{2y}{\Gamma(\mu+1)} \int_0^\infty g(x)(yx)^{\mu/2} K_\mu(2\sqrt{yx})\, dx .$$

However, this can be transformed back into (23.1) by replacing y by $s^2/2$, x by $t^2/2$, and setting

$$f(t) = t^{\mu+\frac{1}{2}} g\left(\frac{t^2}{2}\right) \quad , \quad F_\mu(s) = \frac{2^\mu \Gamma(\mu+1)}{s^{\mu+\frac{3}{2}}} G\left(\frac{s^2}{2}\right) .$$

23.2 Existence

If f is a locally integrable function on $(0,\infty)$ such that $f(t) = O(t^\alpha)$ as $t \to 0^+$ and $O(e^{at})$ as $t \to \infty$, where $\alpha \geq \mu - \frac{3}{2}$, then its K transform of order μ ($\mu > 0$) exists almost everywhere for Re $s > a$. For $\mu = 0$, we require that $\alpha > -1$.

In general, if f is a suitably restricted function, its K transform $F_\mu(s)$ is analytic in Re $s > 0$. Because of the singularity of the function $K_\mu(z)$ at $z = 0$, Theorem 1.1 cannot be applied directly, but if we assume that $t^{-p} f(t)$ is integrable in a neighborhood of $t = 0$ for every $p \geq 0$, in particular, if f vanishes in a neighborhood of $t = 0$, then we may apply Theorem 1.1 to conclude that F_μ is analytic in Re $s > 0$.

23.3 Elementary Properties

1) **Linearity:** The K transformation is linear.

2) **Dilation:**

$$\mathcal{K}_\mu[f(at)](s) = \frac{1}{a} \mathcal{K}_\mu[f(t)]\left(\frac{s}{a}\right) , \quad a > 0$$

3) **Multiplication by powers of t:**

$$\mathcal{K}_\mu[t^m f(t)](s) = s^{\frac{1}{2}-\mu}\left(-\frac{1}{s}\frac{d}{ds}\right)^m \left\{s^{m+\mu-\frac{1}{2}}\mathcal{K}_{\mu+m}(f)(s)\right\}, \quad m = 0, 1, 2, \ldots$$

4) **Division by t:**

$$\mathcal{K}_\mu\left[\frac{2\mu}{t}f(t)\right](s) = s\left(\mathcal{K}_{\mu+1}[f(t)](s) - \mathcal{K}_{\mu-1}[f(t)](s)\right) .$$

5) **Differentiation:**

$$\mathcal{K}_\mu\left[t^{\frac{1}{2}-\mu}\left(\frac{1}{t}\frac{d}{dt}\right)^m \left(t^{m+\mu-\frac{1}{2}}f(t)\right)\right](s) = s^m \mathcal{K}_{\mu+m}[f(t)](s) .$$

Formulas (3)–(5) follow from the identities

$$\left(\frac{-1}{z}\frac{d}{dz}\right)^m (z^\mu K_\mu(z)) = z^{\mu-m}K_{\mu-m}(z) ,$$

$$K_{\mu+1}(z) - K_{\mu-1}(z) = \frac{2\mu}{z}K_\mu(z) ,$$

and

$$\left(\frac{-1}{z}\frac{d}{dz}\right)^m \left(\frac{K_\mu(z)}{z^\mu}\right) = \frac{K_{\mu+m}(z)}{z^{\mu+m}} ,$$

respectively.

23.4 Relationships with Other Transforms

23.4.1 The Laplace Transform

Let $\mathcal{L}$ and $\mathcal{K}_\mu$ denote the Laplace and K-transformation of order μ, respectively. It was shown in Section 1 of this chapter that

$$\mathcal{K}_{\pm\frac{1}{2}}[f](s) = \sqrt{\frac{\pi}{2}}\mathcal{L}[f](s) .$$

Moreover, in view of the relation (2.19)

$$K_\mu(z) = \frac{\sqrt{\pi}}{\Gamma(\mu+\frac{1}{2})}\left(\frac{z}{2}\right)^\mu \int_0^\infty e^{-z\cosh t}(\sinh t)^{2\mu}dt ,$$

where $\operatorname{Re} z > 0$, $\operatorname{Re} \mu > -\frac{1}{2}$, it can be easily shown that

$$\mathcal{K}_\mu[f](s) = \frac{\sqrt{\pi}\, s^{\mu+\frac{1}{2}}}{\Gamma\left(\mu+\frac{1}{2}\right) 2^\mu} \mathcal{L}[g](s) \,, \quad \operatorname{Re} \mu > -\frac{1}{2}$$

where

$$g(x) = \int_0^x (x^2 - t^2)^{\mu-\frac{1}{2}} t^{\frac{1}{2}-\mu} f(t) dt \,,$$

and also

$$\mathcal{K}_\mu[f](s) = \frac{\sqrt{\pi}\, s^{\frac{1}{2}-\mu}}{2^\mu \Gamma(\mu+\frac{1}{2})} \int_s^\infty (t^2 - s^2)^{\mu-\frac{1}{2}} h(t) dt \,,$$

where

$$h(t) = \mathcal{L}[x^{\frac{1}{2}+\mu} f(x)](t) \,.$$

23.5 Inversion Formula

The following inversion formula was obtained by Meijer for $|\operatorname{Re} \mu| \leq \frac{1}{2}$ and extended to $\operatorname{Re} \mu \geq -\frac{1}{2}$ by Zemanian [276, p. 173].

THEOREM 23.1
Let μ be a complex number such that Re $\mu \geq -\frac{1}{2}$. Assume that

$$F(s) = \int_0^\infty f(t) \sqrt{st}\, K_\mu(st)\, dt$$

is analytic in Re $s > b \geq 0$ and bounded thereon according to $|F(s)| < M|s|^{-q}$, where M is a positive constant and $q >$ Re $\mu + \frac{3}{2}$. Then for any real number $c > b$

$$f(t) = \frac{1}{\pi i} \int_{c-i\infty}^{c+i\infty} F(s) \sqrt{ts}\, I_\mu(ts)\, ds \,, \quad \text{Re } s > c,$$

where $I_\mu(z)$ is the modified Bessel function of the first kind and order μ.

23.6 The K transform of L^2-Functions

Since $K_\mu(z) \in L^2(0,\infty)$ for $|\operatorname{Re} \mu| < \frac{1}{2}$, it follows from the Cauchy-Schwarz inequality that $F_\mu(s)$ exists whenever $f \in L^2(0,\infty)$. Moreover,

$F_\mu(s)$ is analytic in Re $s > 0$. Some kind of a converse statement was established by Boas [16, 17] who gave the following characterization for a function $F(s)$ that is analytic in Re $s > 0$ to be the K transform of a function $f \in L^2(0, \infty)$. First, let us define the space $A(c)$ as the linear space of all functions $f(z)(z = x + iy)$ which are analytic for Re $z > c$ and satisfy

$$\int_{-\infty}^{\infty} |f(x+iy)|^2 dy \leq M \ , \quad x > c \ ,$$

where M depends on f. It can be shown that for any such function, $\lim_{x \to c} f(x + iy) = f(c + iy)$ exists almost everywhere. Moreover, it is not hard to show that $A(c)$ is a Hilbert space with inner product

$$\langle f, g \rangle = \int_{-\infty}^{\infty} f(c+iy)\overline{g}(c+iy) dy \ .$$

THEOREM 23.2
Let $|Re\, \mu| < \frac{1}{2}$. Then for every $F(s) \in A(0)$ there exists a unique $f \in L^2(0, \infty)$ such that

$$F(s) = \int_0^{\infty} f(t)\sqrt{st}\, K_\mu(st)\, dt \ , \quad s > 0 \ .$$

23.7 Generalized Functions

23.7.1 Definitions

The K transform has been extended to generalized functions by Zemanian [276, Ch. 6]. Below we outline his method. Let μ be a fixed complex number and define

$$S_\mu = t^{-\mu-\frac{1}{2}} D t^{2\mu+1} D t^{-\mu-\frac{1}{2}} \ ,$$

where $D = \frac{d}{dt}$. It is routine to show that

$$S_\mu f(t) = D^2 f + \left(\frac{\frac{1}{4} - \mu^2}{t^2} \right) f \ , \tag{23.2}$$

and

$$S_\mu \sqrt{st}\, K_\mu(st) = s^2 \sqrt{st}\, K_\mu(st) \ . \tag{23.3}$$

Throughout this section we shall assume that μ is either 0 or a complex number satisfying Re $\mu > 0$ and $I = (0, \infty)$. Let

$$h(t) = \begin{cases} \ln t, \ 0 < t < 1/e \\ -1, \ 1/e \leq t < \infty \ , \end{cases}$$

and

$$\eta_\mu(t) = \begin{cases} t^{\mu-\frac{1}{2}}, & \text{when Re } \mu > 0 \\ \left(\sqrt{t}\, h(t)\right)^{-1}, & \text{when } \mu = 0\,. \end{cases}$$

We define the space $\mathbf{K}_\mu(a)$ (a real) as the linear space of all infinitely differentiable functions ϕ satisfying

$$\gamma_{\mu,a,k}(\phi) = \sup_{0<t<\infty} \left| e^{at}\eta_\mu(t) S_\mu^k \phi(t)\right| < \infty\,, \quad k = 0, 1, 2, \ldots\,.$$

It can be shown that $(\mathbf{K}_\mu(a), \{\gamma_{\mu,a,k}\}_{k=0}^\infty)$ is a testing-function space which is a dense subspace of $\mathcal{E}(I)$ whatever the choices of μ and a. Although $\mathcal{D}(I)$ is a subspace of $\mathbf{K}_\mu(a)$ and convergence in $\mathcal{D}(I)$ implies convergence in $\mathbf{K}_\mu(a)$, $\mathcal{D}(I)$ is not a dense subspace of $\mathbf{K}_\mu(a)$. Moreover, if $a < b$, $\mathbf{K}_\mu(b) \subset \mathbf{K}_\mu(a)$ and the topology of $\mathbf{K}_\mu(b)$ is stronger than the topology induced on it by $\mathbf{K}_\mu(a)$ since $\gamma_{\mu,a,k}(\phi) \le \gamma_{\mu,b,k}(\phi)$, for any $\phi \in \mathbf{K}_\mu(b)$.

A generalized function f is said to be a K-transformable generalized function if it belongs to $\mathbf{K}_\mu^*(a)$ for some a. Since $\mathbf{K}_\mu^*(b) \supset \mathbf{K}_\mu^*(a)$ if $a < b$, it follows that there exists a real number σ_f (possibly $-\infty$) such that $f \in \mathbf{K}_\mu^*(a)$ for every $a > \sigma_f$ and $f \notin \mathbf{K}_\mu^*(a)$ for any $a < \sigma_f$. Let

$$\Omega_f = \{s : \text{Re } s > \sigma_f, s \neq 0, -\pi < \arg s < \pi\}\,.$$

Equation (23.3) shows that $\sqrt{st}\, K_\mu(st) \in \mathbf{K}_\mu(a)$ for every s such that $s \neq 0$ and Re $s > a$. Hence, we have the following definition.

DEFINITION 23.2 *Let f be a K transformable generalized function. We define the K transform of order μ of f as*

$$F_\mu(s) = \langle f(t), \sqrt{st}\, K_\mu(st)\rangle\,; \quad s \in \Omega_f\,. \tag{23.4}$$

If f is a locally integrable function such that for some $a, e^{-at} f(t)/\eta_\mu(t)$ is absolutely integrable on I, then it generates a regular member of $\mathbf{K}_\mu^*(a)$ whose generalized and conventional K transforms given by (23.1) and (23.4) coincide.

23.7.2 Characterization

It can be easily shown that $F_\mu(s)$ given by (23.4) is analytic in the region Ω_f and is of polynomial growth thereon as $|s| \to \infty$. More precisely, let $b > \sigma_f$ and define

$$\Lambda_b = \{s : s \in \Omega_f, \text{Re } s \ge b, |s| \ge b - \sigma_f\}\,.$$

The function $F(s)$ is analytic and bounded by a polynomial as $|s| \to \infty$ in Λ_b. In fact, the converse is also true.

THEOREM 23.3 (Zemanian [276, p. 195])
A necessary and sufficient condition for a function $F(s)$ to be the μth-order K-transform of some generalized function as defined by (23.4) is that there exists a half-plane $\{s : Re\ s \geq b > 0\}$ on which $F(s)$ is analytic and bounded according to $|F(s)| \leq P_b(|s|)$, where $P_b(|s|)$ is a polynomial in $|s|$.

23.7.3 Operational Calculus

The operator S_μ is continuous on $\mathbf{K}_\mu(a)$ as can be seen from the relation

$$\gamma_{\mu,a,k}(S_\mu\phi) = \gamma_{\mu,a,k+1}(\phi) , \quad \phi \in \mathbf{K}_\mu(a) .$$

Moreover, S_μ is formally self-adjoint, hence it is continuous on $\mathbf{K}^*_\mu(a)$ with

$$\langle S_\mu f, \phi\rangle = \langle f, S_\mu\phi\rangle , \quad \phi \in \mathbf{K}_\mu(a) , \quad f \in \mathbf{K}^*_\mu(a) .$$

From Equation (23.3) it is easy to see that if $P(s)$ is a polynomial in s, then

$$P(S_\mu)k_\mu(st) = P(s^2)k_\mu(st) ,$$

where $k_\mu(st) = \sqrt{st}\ K_\mu(st)$ is the kernel of the K transform of order μ. Thus,

$$\langle P(S_\mu)f(t), k_\mu(st)\rangle = \langle f(t), P(S_\mu)k_\mu(st)\rangle = P(s^2)\langle f(t), k_\mu(st)\rangle$$

or

$$\mathcal{K}_\mu[P(S_\mu)f](s) = P(s^2)\mathcal{K}_\mu[f](s) . \tag{23.5}$$

23.7.4 Inversion Formula

THEOREM 23.4
Let F_μ be the K transform of order μ of a K-transformable generalized function f. Then, in the sense of $\mathcal{D}^*(I)$

$$f(t) = \lim_{R\to\infty} \frac{1}{\pi i}\int_{\sigma-iR}^{\sigma+iR} F(s)\sqrt{st}\ I_\mu(st)\, ds ,$$

where σ is any real positive number in Ω_f.

23.8 Applications

The operational calculus developed in the last section can be used to solve differential equations involving the operator S_μ. For example, to solve the differential equation

$$P(S_\mu)u = f\ ,$$

where P is a given polynomial and f is a K-transformable function or generalized function, we apply the K transformation of order μ to both sides of the equation to obtain, with the aid of (23.5),

$$P(s^2)U_\mu(s) = F_\mu(s)\ ,$$

where U_μ and F_μ stand for the K transforms of u and f, respectively. Thus, we have formally

$$u(t) = \mathcal{K}_\mu^{-1}\left[\frac{F_\mu(s)}{P(s^2)}\right](t)\ .$$

To apply the inversion theorem for the K transform, we take Re $s > \max(0, \sigma_f, \sigma_P)$, where σ_P is the largest of the real parts of the roots of $P(s^2)$.

Example A: Consider the differential equation

$$ay'' + \frac{b}{t}y' + cy = f(t)\ , \quad a \neq 0\ .$$

Setting $y = t^{\mu-\frac{1}{2}}u$, $g(t) = \frac{1}{a}t^{-\mu+\frac{1}{2}}f(t)$, and using (23.2), we convert this equation into

$$u'' + \frac{\left(\frac{1}{4} - \mu^2\right)}{t^2}u + \frac{c}{a}u = S_\mu u + \frac{c}{a}u = g(t)\ ,$$

where $b/a = 1 - 2\mu$. By applying the K transformation to the last equation and using (23.5), we find that

$$u(t) = \mathcal{K}_\mu^{-1}\left[\frac{G_\mu(s)}{\left(s^2 + \frac{c}{a}\right)}\right](t)\ .$$

24

The Kontorovich–Lebedev Transform

The Kontorovich–Lebedev transform was introduced by the Soviet mathematicians M. J. Kontorovich and N. N. Lebedev in 1938-1939 [128] to solve certain boundary-value problems. The mathematical theory and the inversion formulas were later introduced by Lebedev in 1946 and 1949 [139, 140]; for a generalization and some related material see [116, 137, 138]. The Kontorovich–Lebedev transform arises naturally when one uses the method of separation of variables to solve boundary-value problems formulated in terms of cylindrical coordinate systems. Therefore, it comes as no surprise that many more applications of it have been discovered in recent years.

24.1 Definitions

DEFINITION 24.1 *The Kontorovich–Lebedev transform of a function $f(t)$ is defined as*

$$(\mathcal{K}f)(t) = F(t) = \int_0^\infty f(x) K_{it}(x)\, dx\,, \quad t \text{ real} \tag{24.1}$$

where $K_{it}(x)$ is the Macdonald function (see Section 2.5) *Another version of the transform is*

$$(\mathcal{K}_1 f)(t) = F_1(t) = \int_0^\infty f(x) K_{ix}(t)\, dx\,, \quad t, x \text{ real}.$$

In the sequel, we shall derive properties of the first version only, since the corresponding ones for the second version can be obtained similarly. Also because $K_t(x) = K_{-t}(x)$, we shall assume without loss of generality that $t \geq 0$.

24.2 Existence

In view of the relation (24.4) below, we have

$$|(\mathcal{K}f)(t)| \leq \int_0^\infty |f(x)|K_0(x)dx\,,$$

which shows that if the above integral converges, then the integral (24.1) defining $(\mathcal{K}f)(t)$ converges absolutely and uniformly with respect to t and thus, by Theorem 1.1, $(\mathcal{K}f)(t)$ is a continuous function. From the asymptotic expansions of $K_0(x)$ (see Section 2.5, item 8), it follows that a sufficient condition for the above integral to converge is that $f(x)$ be a locally integrable function on $(0,\infty)$, with $f(x) = O\left(e^{cx}\right)$ as $x \to \infty$, for some $0 \leq c < 1$. Moreover, under this condition, $(\mathcal{K}f)(t) \to 0$ as $t \to \infty$. For asymptotic expansions of the Kontorovich–Lebedev transform, see [250].

24.3 Relationship with Other Transforms

24.3.1 The Fourier Transform

In the relation

$$K_t(x) = \frac{1}{2}\int_{-\infty}^\infty e^{-x\cosh w + tw}dw\,, \quad x \geq 0 \quad [60,\text{ Vol. 2, 7.4.2 (15)}]$$

if we replace t by it, we obtain

$$\mathcal{F}\left(\sqrt{\pi/2}e^{-x\cosh w}\right)(t) = K_{it}(x)\,, \quad x,t \geq 0\,,$$

where $\mathcal{F}$ denotes the Fourier transformation, which is taken with respect to w. In particular, we have

$$K_{it}(x) = \int_0^\infty e^{-x\cosh w}\cos(tw)\,dw\,, \quad x,t \geq 0\,, \tag{24.2}$$

and

$$K_0(x) = \int_0^\infty e^{-x\cosh w}dw\,. \tag{24.3}$$

From (24.2) and (24.3), it follows that

$$|K_{it}(x)| \leq K_0(x)\,; \tag{24.4}$$

see Section 24.2. In (24.3) if we replace x by ax and then take the Fourier cosine transform, we obtain

$$\int_0^\infty K_0(ax)\cos(ux)\,dx = \int_0^\infty \frac{a\cosh w}{a^2\cosh^2 w + u^2}dw = \frac{\pi}{2}\frac{1}{\sqrt{a^2+u^2}}\,.$$

The inversion formula of the Fourier cosine transformation (see Section 12.8) now yields

$$\int_0^\infty K_{it}(x)\cos(wt)\,dt = \frac{\pi}{2}e^{-x\cosh w}\,, \tag{24.5}$$

which, when we differentiate it with respect to w, gives us

$$\int_0^\infty tK_{it}(x)\sin(wt)\,dt = \frac{\pi}{2}x(\sinh w)e^{-x\cosh w}\,.$$

If we multiply (24.5) by $G(w)$ and integrate with respect to w from zero to infinity, we obtain

$$\int_0^\infty K_{it}(x)g(t)\,dt = \frac{\pi}{2}\int_0^\infty G(w)e^{-x\cosh w}dw\,,$$

where

$$g(t) = \sqrt{\frac{\pi}{2}}\mathcal{F}_c(G)(t)\,,$$

and $\mathcal{F}_c$ is the Fourier cosine transformation.

Similarly, by multiplying (24.1) by $\cos tw$, integrating with respect to t from zero to infinity and interchanging the integration signs, we find, in view of (24.5), that

$$\int_0^\infty F(t)\cos(wt)\,dt = \frac{\pi}{2}\int_0^\infty f(x)e^{-x\cosh w}dx\,,$$

or symbolically

$$\mathcal{F}_c\left(\mathcal{K}f\right)(w) = \sqrt{\pi/2}\int_0^\infty f(x)e^{-x\cosh w}dx\,. \tag{24.6}$$

From (24.5), it follows that the Fourier cosine transform of $e^{-x\cosh u}$ and $e^{-y\cosh u}$ taken with respect to u are $\sqrt{2/\pi}K_{it}(x)$ and $\sqrt{2/\pi}K_{it}(y)$, respectively. By invoking the convolution theorem for the Fourier cosine transform (see Section 12.9.2), we obtain

$$\begin{aligned}
&\frac{2}{\pi}\int_0^\infty K_{it}(x)K_{it}(y)\cos(wt)\,dt \\
&= \frac{1}{2}\int_0^\infty e^{-x\cosh u}\left[e^{-y\cosh(u+w)} + e^{-y\cosh(u-w)}\right]du
\end{aligned}$$

$$= \frac{1}{2}\int_{-\infty}^{\infty} e^{-(x\cosh u + y\cosh u\cosh w + y\sinh u\sinh w)}\,du$$

$$= \frac{1}{2}\int_{-\infty}^{\infty} e^{-R\cosh(u+\xi)}\,du\,,$$

where

$$R\cosh\xi = x + y\cosh w \quad\text{and}\quad R\sinh\xi = y\sinh w\,.$$

Hence, by (24.3)

$$\frac{2}{\pi}\int_0^{\infty} K_{it}(x)K_{it}(y)\cos(wt)\,dt = \int_0^{\infty} e^{-R\cosh v}\,dv = K_0(R)\,, \tag{24.7}$$

where

$$R = \sqrt{x^2 + 2xy\cosh w + y^2}\,.$$

In particular,

$$\frac{2}{\pi}\int_0^{\infty} K_{it}(x)K_{it}(y)\,dt = K_0(x+y)\,.$$

Differentiating both sides of (24.7) with respect to w, we obtain

$$\frac{2}{\pi}\int_0^{\infty} tK_{it}(x)K_{it}(y)\sin(wt)\,dt = \frac{xy\sinh w}{R}K_1(R)\,,$$

and, upon replacing w by iw, we immediately have

$$\frac{2}{\pi}\int_0^{\infty} tK_{it}(x)K_{it}(y)\sinh(wt)\,dt =$$

$$\frac{xy\sin w}{\sqrt{x^2+2xy\cos w+y^2}}K_1\left(\sqrt{x^2+2xy\cos w+y^2}\right)\,.$$

24.3.2 The Laplace Transform

Equation (24.6) can be rewritten in the form

$$\mathcal{L}\left[\sqrt{\pi/2}f\right](\cosh w) = F_c(\mathcal{K}f)(w)\,,$$

where $\mathcal{L}$ denotes the Laplace transformation. If we replace w by $\cosh^{-1} s$, we have

$$\mathcal{L}(f)(s) = \frac{2}{\pi}\int_0^{\infty} (\mathcal{K}f)(t)\cos[t\cosh^{-1}s]\,dt, \tag{24.8}$$

which, in view of item 8 of Section 5.5, yields

$$\mathcal{L}(xf(x))(s) = \frac{2}{\pi}\int_0^{\infty} (\mathcal{K}f)(t)\frac{t\sin[t\cosh^{-1}s]}{\sqrt{s^2-1}}\,dt\,. \tag{24.9}$$

The Laplace transform of $K_{it}(x)$ taken with respect to x can be derived directly from (24.2) as follows,

$$\begin{aligned}\mathcal{L}(K_{it}(x))(s) &= \int_0^\infty e^{-sx} K_{it}(x)dx \\ &= \int_0^\infty e^{-sx}\left(\int_0^\infty e^{-x\cosh w}\cos(tw)\,dw\right)dx \\ &= \int_0^\infty \cos tw\left(\int_0^\infty e^{-x(s+\cosh w)}dx\right)dw \\ &= \int_0^\infty \frac{\cos tw}{s+\cosh w}dw\,.\end{aligned}$$

If we replace s by $\cosh s$ and make use of formula (7) in [62, Vol.1, p. 30], we find that

$$\mathcal{L}_x\left(K_{it}(x)\right)(\cosh s) = \frac{\pi \sin st}{\sinh s \sinh \pi t}\,,$$

in which the subscript x indicates that the Laplace transform is taken with respect to x. Again, replacing s by $\cosh^{-1} s$, we obtain

$$\mathcal{L}_x\left(\frac{1}{\pi}\sinh(\pi t)K_{it}(x)\right)(s) = \frac{\sin[t\cosh^{-1} s]}{\sqrt{s^2-1}}\,. \tag{24.10}$$

24.4 Inversion Formulas

The following inversion formulas for the Kontorovich–Lebedev transform hold. If

$$(\mathcal{K}f)(t) = \int_0^\infty f(x)K_{it}(x)\,dx\,,$$

then

$$f(x) = \frac{2}{\pi^2 x}\int_0^\infty t\sinh(\pi t)K_{it}(x)(\mathcal{K}f)(t)\,dt\,, \tag{24.11}$$

and if

$$(\mathcal{K}_1 f)(t) = \int_0^\infty f(x)K_{ix}(t)\,dx\,,$$

then

$$f(x) = \frac{2x}{\pi^2}\sinh(\pi x)\int_0^\infty (\mathcal{K}_1 f)(t)K_{ix}(t)\frac{dt}{t}\,. \tag{24.12}$$

Formally, if we apply the inverse Laplace transformation to (24.9) and assume that it commutes with the integration operation, we find that

$$f(x) = \frac{2}{\pi x}\int_0^\infty t(\mathcal{K}f)(t)\left[\mathcal{L}_s^{-1}\left(\frac{\sin[t\cosh^{-1} s]}{\sqrt{s^2-1}}\right)(x)\right]dt\,,$$

which, in view of (24.10), yields (24.11). To justify these steps, further assumptions are needed. For example, it suffices to assume that f is continuously differentiable and $x^2 f$, $xf'(x)$ are absolutely integrable on $[0, \infty)$. For a more rigorous proof see [216, pp. 359-361].

24.5 Parseval's Relation

Using the inversion formula (24.11), we have

$$\begin{aligned}
&\frac{2}{\pi^2}\int_0^\infty t\sinh(\pi t)(\mathcal{K}f)(t)(\mathcal{K}g)(t)dt \\
&\quad = \frac{2}{\pi^2}\int_0^\infty t\sinh(\pi t)(\mathcal{K}f)(t)\,dt \int_0^\infty g(x)K_{it}(x)\,dx \\
&\quad = \frac{2}{\pi^2}\int_0^\infty g(x)\left(\int_0^\infty t\sinh(\pi t)K_{it}(x)(\mathcal{K}f)(t)\,dt\right)dx \\
&\quad = \int_0^\infty xf(x)g(x)\,dx\,.
\end{aligned}$$

This equation can be written as

$$\langle \mathcal{K}f, \mathcal{K}g\rangle = \langle f, g\rangle\,, \tag{24.13}$$

where the inner product on the left-hand side is taken with respect to the measure, $d\mu(t) = (2/\pi^2)t\sinh(\pi t)\chi_{[0,\infty)}(t)dt$, and the one on the right-hand side is taken with respect to the measure $d\nu(x) = x\chi_{[0,\infty)}(x)dx$. As a special case of (24.13) we obtain

$$\int_0^\infty f(x)e^{-x}dx = \frac{2}{\pi}\int_0^\infty (\mathcal{K}f)(t)\,dt\,,$$

which is also a special case of (24.6) for $w = 0$ and of (24.8) for $s = 1$.

24.6 Generalized Functions

The Kontorovich–Lebedev transform was first extended to generalized functions by A. Zemanian [277] using the embedding method. His result was later extended to a larger class of generalized functions by R. Pathak and J. Pandey [189]; see also [34, 221, 73]. Below is an outline of their results.

Zemanian defined the Kontorovich–Lebedev transform of a generalized function f with compact support as

$$(\mathcal{K}f)(t) = \langle f(x), K_{it}(x) \rangle \,, \quad f \in \mathcal{E}^*(I) \,,$$

where $I = (0, \infty)$. The definition makes sense because $K_{it}(x)$ is analytic in x for every fixed t. The inversion formula for this transform is

$$\langle f, \phi \rangle = \lim_{T \to \infty} \langle \frac{2}{\pi^2 x} \int_0^T t \sinh(\pi t) K_{it}(x) (\mathcal{K}f)(t) \, dt, \phi(x) \rangle \,,$$

for every $\phi \in \mathcal{D}(I)$, where $f \in \mathcal{E}^*(I)$.

Pathak and Pandey first defined the Kontorovich–Lebedev transform for a function f as

$$(\mathcal{K}f)(t) = \int_0^\infty f(x) K_{it}(x) \frac{dx}{\sqrt{x}} \,,$$

so that its inverse transform is

$$f(x) = \frac{2}{\pi^2 \sqrt{x}} \int_0^\infty t \sinh(\pi t) K_{it}(x) (\mathcal{K}f)(t) \, dt \,.$$

Then, they introduced the space $\mathbf{K}_\alpha(I)$ consisting of all complex-valued, infinitely differentiable functions ϕ on $I = (0, \infty)$, satisfying

$$\gamma_k(\phi) = \sup_{x \in I} \mid \eta(x) \Delta^k \phi(x) \mid < \infty \,, \quad k = 0, 1, 2, ...,$$

where

$$\eta(x) = \begin{cases} \sqrt{x} & , 0 < x \leq 1 \\ x^{3/2 - \alpha} \ln x & , 1 < x \,, \end{cases}$$

for some positive real number α and

$$\Delta = x^2 \frac{d}{dx^2} + 2x \frac{d}{dx} - x^2 \,.$$

It is easy to see that $(\mathbf{K}_\alpha(I), \{\gamma_k\}_{k=0}^\infty)$ is a Fréchet space with $\mathcal{D}(I) \subset \mathbf{K}_\alpha(I)$ and that the topology of $\mathcal{D}(I)$ is stronger than the one it inherits from $\mathbf{K}_\alpha(I)$. The function $K_{it}(x)/\sqrt{x}$ belongs to $\mathbf{K}_\alpha(I)$ and

$$\Delta_x [K_{it}(x)/\sqrt{x}] = -\left(\frac{1}{4} + t^2 \right) [K_{it}(x)/\sqrt{x}] \,.$$

Therefore, for any generalized function $f \in \mathbf{K}_\alpha^*(I)$, the Kontorovich–Lebedev transform can be defined by

$$(\mathcal{K}f)(t) = \langle f(x), \frac{K_{it}(x)}{\sqrt{x}} \rangle \,.$$

The inversion formula for this transform is shown to be

$$\langle f(x), \phi(x) \rangle = \lim_{T \to \infty} \langle \frac{2}{\pi^2} \int_0^T t \sinh(\pi t) K_{it}(x) (\mathcal{K}f)(t) \frac{dt}{\sqrt{t}}, \phi(x) \rangle$$

for every $\phi \in \mathcal{D}(I)$, where $f \in \mathbf{K}_\alpha^*(I)$.

24.7 Applications

The Kontorovich–Lebedev transform has been used in many applications, including fluid mechanics [13] and quantum optics [210].The transform arises naturally when one uses the method of separation of variables to solve boundary-value problems formulated in terms of cylindrical coordinate systems. To illustrate the idea, consider Laplace's equation in cylindrical coordinates

$$\frac{\partial^2 u}{\partial r^2} + \frac{1}{r}\frac{\partial u}{\partial r} + \frac{1}{r^2}\frac{\partial^2 u}{\partial \theta^2} + \frac{\partial^2 u}{\partial z^2} = 0 .$$

If we seek a solution of this equation in the form

$$u(r,\theta,z) = e^{\pm t\theta \pm i\sigma z} R(r) ,$$

then it can be verified that R satisfies the differential equation

$$R'' + \frac{1}{r}R' + \left(\frac{t^2}{r^2} - \sigma^2\right) R = 0 .$$

One solution of this equation is $K_{it}(|\sigma| r)$; hence

$$u(r,\theta,z) = e^{\pm t\theta \pm i\sigma z} K_{it}(|\sigma| r) .$$

Superimposing these solutions leads to the more general solution

$$U(r,\theta,z) = \int_{-\infty}^{\infty} e^{-i\sigma z} d\sigma \int_0^{\infty} f(\sigma,t) \begin{Bmatrix} \cosh(t\theta) \\ \sinh(t\theta) \end{Bmatrix} K_{it}(|\sigma| r)\, dt$$

for some suitable function $f(\sigma,t)$. This solution can be written symbolically as

$$U(r,\theta,z) = \mathcal{F}_\sigma^{-1}\left[\mathcal{K}_t^{-1}(g)(|\sigma| r,\theta)\right](z) ,$$

where

$$g(\sigma,t,r,\theta) = \sqrt{2\pi}\,\frac{\pi^2 |\sigma| r f(\sigma,t)}{2t \sinh \pi t} \begin{cases} \cosh(t\theta) \\ \sinh(t\theta) \end{cases} .$$

Example A: Consider Laplace's equation in cylindrical coordinates

$$\frac{\partial^2 u}{\partial r^2} + \frac{1}{r}\frac{\partial u}{\partial r} + \frac{1}{r^2}\frac{\partial^2 u}{\partial \theta^2} + \frac{\partial^2 u}{\partial z^2} = 0 ,$$

and let us seek a solution of the form $u(r,\theta,z) = e^{i\sigma z} u(r,\theta)$ in the wedge $\{(r,\theta) : 0 < r < \infty, 0 \le \theta \le \alpha\}$ subject to the conditions

$$u(r,0) = 0 \quad \text{and} \quad u(r,\alpha) = r f(r) ,$$

for some suitable function or generalized function f. It can be easily shown that u must satisfy the equation

$$r^2\frac{\partial^2 u}{\partial r^2}+r\frac{\partial u}{\partial r}-r^2\sigma^2 u+\frac{\partial^2 u}{\partial \theta^2}=0\,.$$

Without loss of generality, we assume that $\sigma > 0$, and set $\rho = \sigma r$; hence

$$\rho^2\frac{\partial^2 u}{\partial \rho^2}+\rho\frac{\partial u}{\partial \rho}-\rho^2 u+\frac{\partial^2 u}{\partial \theta^2}=0\,.$$

If we put $u(\rho,\theta)=\rho v(\rho,\theta)$, we obtain after some easy calculations that

$$\rho^2\frac{\partial^2 v}{\partial \rho^2}+3\rho\frac{\partial v}{\partial \rho}+(1-\rho^2)v+\frac{\partial^2 v}{\partial \theta^2}=0\,,$$

or

$$A_\rho^* v+\frac{\partial^2 v}{\partial \theta^2}=0\,, \tag{24.14}$$

where the differential operator A^* is given by

$$A_\rho^*=\rho^2\frac{\partial^2}{\partial \rho^2}+3\rho\frac{\partial}{\partial \rho}+(1-\rho^2)\,.$$

Applying the Kontorovich–Lebedev transform to (24.14), we have

$$\langle K_{it}(\rho), A_\rho^* v\rangle+\langle K_{it}(\rho), \frac{\partial^2 v}{\partial \theta^2}\rangle=0\,,$$

and hence

$$\langle A_\rho K_{it}(\rho), v\rangle+\frac{\partial^2}{\partial \theta^2}\langle K_{it}(\rho), v(\rho,\theta)\rangle=0\,,$$

where

$$A_\rho=\rho^2\frac{\partial^2}{\partial \rho^2}+\rho\frac{\partial}{\partial \rho}-\rho^2\,.$$

But since $A_\rho K_{it}(\rho)=-t^2K_{it}(\rho)$, we have

$$\frac{\partial^2 V}{\partial \theta^2}-t^2V=0\,, \tag{24.15}$$

where $V(t,\theta)$ is the Kontorovich–Lebedev transform of $v(\rho,\theta)$, i.e.,

$$V(t,\theta)=\int_0^\infty K_{it}(\rho)v(\rho,\theta)d\rho\,.$$

Upon solving (24.15) and employing the boundary conditions, we obtain

$$V(t,\theta)=\frac{F(t)}{\sinh \alpha t}\sinh(\theta t)\,,$$

where $F(t)$ is the Kontorovich–Lebedev transform of $f(\rho)$. An application of the inversion formula (24.11) gives us the solution

$$u(r,\theta)=\frac{2}{\pi^2}\int_0^\infty \frac{t\sinh(\pi t)}{\sinh(\alpha t)}K_{it}(r)F(t)\sinh(\theta t)dt\,.$$

25

The Mehler–Fock Transform

The Mehler–Fock transformation, which was introduced by V. Fock in 1943 [64], is particularly useful in solving many problems in elasticity theory, where axial symmetry is present or toroidal coordinates are used. It has also been used in mechanics and fluid dynamics [30] to solve some problems with mixed boundary conditions. One of its drawbacks, however, is that calculating it explicitly is rather difficult; the transforms of only a few functions can be calculated in closed form.

Recall from Section 2.3.2, item 7(c) that the associated Legendre function $P_\nu^m(z)$ may be defined by

$$P_\nu^m(z) = \frac{\Gamma(\nu+m+1)}{2^m\Gamma(m+1)\Gamma(\nu-m+1)}(z^2-1)^{m/2} \\ {}_2F_1\left(m-\nu, m+\nu+1; m+1; \frac{1-z}{2}\right),$$

where ν is an arbitrary complex number, $m = 0, 1, 2, ...,$ $|z-1| < 2$, $|\arg(z-1)| < \pi$, and is known to satisfy the differential equation

$$(1-x^2)y'' - 2xy' + \left\{\nu(\nu+1) - \frac{m^2}{1-x^2}\right\} y = 0\,.$$

Clearly, $P_\nu^{-m}(z)$ is also a solution of the same differential equation. These two solutions can be related as follows

$$P_\nu^{-m}(z) = \frac{\Gamma(\nu-m+1)}{\Gamma(\nu+m+1)}P_\nu^m(z) \\ = \frac{1}{m!}\left(\frac{z-1}{z+1}\right)^{m/2} {}_2F_1\left(-\nu, \nu+1; m+1; \frac{1-z}{2}\right)\,. \quad (25.1)$$

25.1 Definitions

DEFINITION 25.1 *We define The Mehler–Fock transform of order m of a function f(x) by*

$$\tilde{\Phi}_m[f(x)](t) = \tilde{F}_m(t) = \int_0^\infty f(x) P^{-m}_{-\frac{1}{2}+it}(\cosh x) \sinh x \, dx \, .$$

The Mehler–Fock transform of order zero will be simply called The Mehler–Fock transform and will be denoted by $\tilde{F}$. *Another version of the transform is*

$$\Phi_m[g(x)](t) = G_m(t) = \int_1^\infty g(y) P^{-m}_{-\frac{1}{2}+it}(y) \, dy \, .$$

The two versions are easily seen to be equivalent upon setting $g(\cosh x) = f(x)$.

The following generalization of the Mehler–Fock transform was introduced in [26].

DEFINITION 25.2 *The Mehler–Fock transform of order* (k, n) *of a function f is defined by*

$$F_{k,n}(t) = \int_0^\infty f(x) P^{k,n}_{-\frac{1}{2}+it}(\cosh x) \sinh x \, dx \, ,$$

where

$$P^{k,n}_\nu(z) = \frac{(z+1)^{n/2}}{\Gamma(1-k)(z-1)^{k/2}} \, {}_2F_1\left(-\nu + \frac{n-k}{2}, \nu + \frac{n-k}{2} + 1; 1-k; \frac{1-z}{2}\right) .$$

The Mehler–Fock transform of order (k, n) reduces to the Mehler–Fock transform of order m when $k = n = -m$. For other generalizations of the Mehler–Fock transform see [33, p. 79] and [155, 138].

25.2 Existence

Since $P^m_{-\frac{1}{2}+it}(y) = O(y^{-\frac{1}{2}})$ as $y \to \infty$, it follows that the Mehler–Fock transform F_m of f exists if $f(y)/\sqrt{y}$ is absolutely integrable over $[1, \infty)$.

In particular, if f is locally integrable over $[1,\infty)$ and of order $O(y^\alpha)$ as $y \to \infty$ for some $\alpha < -1/2$.

25.3 Elementary Properties

1) **Linearity:** The Mehler–Fock transformation is linear.

2) **Transform with negative order:**

$$F_{-m}(t) = \frac{\Gamma(it + m + \frac{1}{2})}{\Gamma(it - m + \frac{1}{2})} F_m(t) \; .$$

3) **Symmetry:**

$$F_m(t) = F_m(-t)$$

Properties (2) and (3) follow from (25.1).

25.4 Relationship with Other Transforms

25.4.1 The Fourier and Weyl Fractional Integrals

From the relation (2.9) (cf. [216, p. 381])

$$P^{-m}_{-\frac{1}{2}+it}(\cosh\alpha) = \sqrt{\frac{2}{\pi}} \frac{(\sinh\alpha)^{-m}}{\Gamma(m+\frac{1}{2})} \int_0^\alpha \cos(ty)[\cosh\alpha - \cosh y]^{m-\frac{1}{2}} dy \; , \tag{25.2}$$

we obtain

$$\Gamma\left(m+\frac{1}{2}\right) \int_0^\infty f(\cosh\alpha) P^{-m}_{-\frac{1}{2}+it}(\cosh\alpha)(\sinh\alpha)^{m+1} d\alpha$$
$$= \sqrt{\frac{2}{\pi}} \int_0^\infty \cos(ty) dy \int_y^\infty f(\cosh\alpha)\sinh\alpha[\cosh\alpha - \cosh y]^{m-\frac{1}{2}} d\alpha \; ,$$

or

$$\int_1^\infty f(x) P^{-m}_{-\frac{1}{2}+it}(x)(x^2-1)^{m/2} dx$$
$$= \sqrt{\frac{2}{\pi}} \frac{1}{\Gamma\left(m+\frac{1}{2}\right)} \int_0^\infty \cos(ty) dy \int_{\cosh y}^\infty f(x)[x - \cosh y]^{m-\frac{1}{2}} dx \; .$$

Recognizing the inside integral on the right-hand side as the Weyl fractional integral of f evaluated at $\cosh y$, we may rephrase this equation as follows: The Mehler–Fock transform of order m of $f(x)(x^2-1)^{m/2}$ is the Fourier cosine transform (taken with respect to y) of the Weyl fractional integral of order $m+1/2$ of f evaluated at $\cosh y$. Or symbolically

$$\Phi_m\left[(x^2-1)^{m/2}f(x)\right](t) = \mathcal{F}_c\left[W_{m+1/2}(f)(\cosh y)\right](t).$$

25.4.2 The Riemann–Liouville Fractional Integral

From the relation [216, p. 381]

$$P_{-\frac{1}{2}+it}(\cosh\alpha) = \frac{\sqrt{2}}{\pi}\coth(\pi t)\int_\alpha^\infty \frac{\sin(ty)}{\sqrt{\cosh y-\cosh\alpha}}dy\,, \tag{25.3}$$

we obtain

$$\begin{aligned}&\int_0^\infty f(\cosh\alpha)P_{-\frac{1}{2}+it}(\cosh\alpha)\sinh\alpha\, d\alpha\\ &\quad= \frac{\sqrt{2}}{\pi}\coth(\pi t)\int_0^\infty \sin(ty)dy\int_0^y \frac{f(\cosh\alpha)\sinh\alpha d\alpha}{\sqrt{\cosh y-\cosh\alpha}}\end{aligned}$$

or

$$\int_1^\infty f(x)P_{-\frac{1}{2}+it}(x)dx = \frac{\sqrt{2}}{\pi}\coth(\pi t)\int_0^\infty \sin(ty)dy\int_1^{\cosh y}\frac{f(x)dx}{\sqrt{\cosh y - x}}\,.$$

In other words, the Mehler–Fock transform of f is equal to $\coth(\pi t)$ times the Fourier sine transform (taken with respect to y) of the Riemann–Liouville Fractional integral of order $1/2$ of f evaluated at $\cosh y$. Symbolically

$$F_0(t) = \coth(\pi t)\mathcal{F}_s\left[I_{1/2}(f)(\cosh y)\right](t).$$

25.5 Inversion Formula

First, we begin by finding an inversion formula for the Mehler–Fock transform of order zero. We recall from Equations (19.8) and (19.9) that the solution of the Abel-type integral equation

$$g(t) = \int_t^\infty \frac{f(x)dx}{[\phi(x)-\phi(t)]^\alpha}\,, \quad 0<\alpha<1\,,$$

is

$$f(x) = \frac{-\sin(\pi\alpha)}{\pi}\frac{d}{dx}\int_x^\infty \frac{g(t)\phi'(t)dt}{[\phi(t)-\phi(x)]^{1-\alpha}}\,, \tag{25.4}$$

where ϕ is a monotone increasing function with $\phi' > 0$; see Equations (19.8) and (19.9).

Now consider the integral

$$I(\alpha) = \int_0^\infty t\tanh(\pi t)P_{-\frac{1}{2}+it}(\cosh\alpha)\tilde{F}_0(t)dt\,.$$

Since $P_\nu^m(z) = O(\nu^{m-\frac{1}{2}}e^{(\nu+\frac{1}{2})z})$ as $\nu \to \infty$ for fixed z, it is easy to see that $I(\alpha)$ exists if $\sqrt{t}\tilde{F}_0(t)$ is absolutely integrable on $[0,\infty)$.

Substituting the right-hand side of (25.3) into I and interchanging the integration signs, we obtain

$$I(\alpha) = \frac{1}{\sqrt{\pi}}\int_\alpha^\infty \frac{\mathcal{F}_s(t\tilde{F}_0(t))(x)}{\sqrt{\cosh x - \cosh\alpha}}dx\,,$$

which, with the aid of (25.4), yields

$$\mathcal{F}_s(t\tilde{F}_0(t))(x) = -\frac{1}{\sqrt{\pi}}\frac{d}{dx}\int_x^\infty \frac{I(\alpha)\sinh\alpha}{\sqrt{\cosh\alpha - \cosh x}}d\alpha\,.$$

By the inversion formula for the Fourier sine transform and integration by parts, we obtain

$$\begin{aligned} t\tilde{F}_0(t) &= -\frac{\sqrt{2}}{\pi}\int_0^\infty \sin(tx)\left(\frac{d}{dx}\int_x^\infty \frac{I(\alpha)\sinh\alpha}{\sqrt{\cosh\alpha-\cosh x}}\,d\alpha\right)dx \\ &= \frac{\sqrt{2}}{\pi}t\int_0^\infty \cos(tx)\left(\int_x^\infty \frac{I(\alpha)\sinh\alpha}{\sqrt{\cosh\alpha-\cosh x}}d\alpha\right)dx \\ &= \frac{\sqrt{2}}{\pi}t\int_0^\infty I(\alpha)\sinh\alpha\left(\int_0^\alpha \frac{\cos tx}{\sqrt{\cosh\alpha-\cosh x}}dx\right)d\alpha \\ &= t\int_0^\infty I(\alpha)\sinh\alpha\, P_{-\frac{1}{2}+it}(\cosh\alpha)\,d\alpha\,, \end{aligned}$$

where the last equality follows from (25.2) with $m = 0$. Thus,

$$\tilde{F}_0(t) = \int_0^\infty I(\alpha)P_{-\frac{1}{2}+it}(\cosh\alpha)\sinh\alpha\,d\alpha\,,$$

which implies that $I(\alpha) = f(\alpha)$ almost everywhere. Again, since $P^m_{-\frac{1}{2}+it}(z) = O(z^{-\frac{1}{2}})$ as $z \to \infty$, it is easy to see that the last integral exists whenever $f(\alpha)$ is locally integrable and $O\left(e^{-(\frac{1}{2}+\epsilon)\alpha}\right)$ as $\alpha \to \infty$ for some $\epsilon > 0$. This leads to the following inversion formula.

THEOREM 25.1
Let f be a locally integrable function such that $f(x)/\sqrt{x}$ is absolutely integrable on $[1,\infty)$ and let

$$F(t)=\int_1^\infty f(x)P_{-\frac{1}{2}+it}(x)dx$$

be such that $\sqrt{t}\,F(t)$ is absolutely integrable on $[0,\infty)$. If f is of bounded variation in a neighborhood of a point x, then

$$\int_0^\infty t\tanh(\pi t)P_{-\frac{1}{2}+it}(x)F(t)dt=\frac{1}{2}[f(x+0)+f(x-0)]\ .$$

The conclusion of the theorem also holds under weaker hypothesis on f; see [216]. The inversion formula of the Mehler–Fock transform of order m can be proved similarly [216, pp. 414-417] and put in the form: If

$$F_m(t)=\int_1^\infty f(x)P^{-m}_{-\frac{1}{2}+it}(x)dx\ ,$$

then

$$f(x)=(-1)^m\int_0^\infty t\tanh(\pi t)F_m(t)P^m_{-\frac{1}{2}+it}(x)dt\ . \tag{25.5}$$

For these integrals to exist it is sufficient that $f(x)/\sqrt{x}$ be absolutely integrable on $[1,\infty)$ and $t^{m+\frac{1}{2}}F_m$ be absolutely integrable on $[0,\infty)$.

25.6 Parseval-Type Relation

Let F_m and G_m denote the Mehler-Folk transforms of order m of f and g, respectively. Then

$$(-1)^m\int_0^\infty t\tanh(\pi t)\frac{\Gamma\left(\frac{1}{2}+m+it\right)}{\Gamma\left(\frac{1}{2}-m+it\right)}F_m(t)G_m(t)\,dt=\int_1^\infty f(x)g(x)\,dx\ .$$

For, by (25.1) and (25.5) we have

$$\begin{aligned}&(-1)^m\int_0^\infty t\tanh(\pi t)\frac{\Gamma(\frac{1}{2}+m+it)}{\Gamma(\frac{1}{2}-m+it)}F_m(t)\,dt\int_1^\infty g(x)P^{-m}_{-\frac{1}{2}+it}(x)\,dx\\&=(-1)^m\int_1^\infty g(x)\,dx\int_0^\infty t\tanh(\pi t)\frac{\Gamma(\frac{1}{2}+m+it)}{\Gamma(\frac{1}{2}-m+it)}F_m(t)P^{-m}_{-\frac{1}{2}+it}(x)\,dt\\&=(-1)^m\int_1^\infty g(x)\,dx\int_0^\infty t\tanh(\pi t)F_m(t)P^m_{-\frac{1}{2}+it}(x)\,dt=\int_1^\infty f(x)g(x)\,dx.\end{aligned}$$

In particular, for $m = 0$, we have

$$\int_0^\infty t\tanh(\pi t)F_0(t)G_0(t)\,dt = \int_1^\infty f(x)g(x)\,dx\,.$$

25.7 Generalized Functions

25.7.1 Definition

The Mehler–Fock transformation has been extended to generalized functions by Buggle [34] and Tiwari and Pandey [231]; see also [72]. We introduce Buggle's definition. Let $I = (1, \infty)$

$$R_{m,x} = (x^2-1)\frac{d^2}{dx^2} + 2x\frac{d}{dx} + \left(\frac{1}{4} - \frac{m^2}{x^2-1}\right),$$

and

$$\eta(x,a) = \begin{cases} 1, & 1 < t \le 2 \\ (x-2)^{(1-a)/2}, & 2 < t < \infty\,, \end{cases}$$

where $0 < a \le 1$. We define $\mathcal{L}_{a,m}$ as the space of all infinitely differentiable functions ϕ on I such that

$$\gamma_k(\phi) = \sup_{x\in I} |\eta(x,a)R^k_{m,x}\phi(x)| < \infty\,, \quad k = 0, 1, 2, \ldots\,.$$

It can be shown that $(\mathcal{L}_{a,m}, \{\gamma_k\}_{k=0}^\infty)$ is a testing-function space and that $\mathcal{D}(I) \subset \mathcal{L}_{a,m} \subset \mathcal{E}(I)$. Moreover, $P^{-m}_{-\frac{1}{2}+it}(x) \in \mathcal{L}_{a,m}$.

DEFINITION 25.3 *The Mehler–Fock transform of order m of a generalized function $f \in \mathcal{L}^*_{a,m}$ is defined by*

$$F_m(t) = \langle f(x), P^{-m}_{-\frac{1}{2}+it}(x)\rangle\,, \quad t > 0\,.$$

If f is a locally integrable function such that $f(x)/\eta(x,a) \in L^1(I)$, then f generates a regular member of $\mathcal{L}^*_{a,m}$ and its conventional and generalized Mehler–Fock transforms coincide. The inversion formula for the generalized Mehler–Fock transform is given in the next theorem.

THEOREM 25.2
Let $f \in \mathcal{L}^*_{a,m}$ and $F_m(t)$ be its Mehler–Fock transform of order m, and let r be a real number. Then in the sense of $\mathcal{D}^*(I)$

$$f(x) = \lim_{r\to\infty}(-1)^m \int_0^r F_m(t)P^m_{-\frac{1}{2}+it}(x)t\tanh(\pi t)dt\,.$$

25.7.2 Convolution

Using the relation

$$P^{-m}_{-\frac{1}{2}+it}(x)P^{-m}_{-\frac{1}{2}+it}(y) = \frac{1}{\pi(2m-1)!!}[(x^2-1)(y^2-1)]^{m/2} \times \int_0^{\pi} (z^2-1)^{-m/2} P^{-m}_{-\frac{1}{2}+it}(z) \sin^{2m} v \, dv$$

where

$$z = xy + [(x^2-1)(y^2-1)]^{1/2}\cos v \quad \text{and } k!! = k(k-2)(k-4)\cdots,$$

Glaeske and Hess [72] defined a convolution operation for the Mehler–Fock transform involving the generalized translation operation

$$T_x\phi(y) = \frac{1}{\pi(2m-1)!!}[(x^2-1)(y^2-1)]^{m/2} \int_0^{\pi} (z^2-1)^{-m/2} \sin^{2m} v \, \phi(z)\, dv,$$

where $x, y > 1$. They defined the convolution, $h = f \star g$, of f and g in $\mathcal{L}^*_{a,m}$. by

$$\langle f \star g, \phi\rangle = \langle f(x), \langle g(y), T_x\phi(y)\rangle\rangle \quad \text{for all } \phi \in \mathcal{L}_{a,m},$$

then they showed that

$$H_m(t) = F_m(t)G_m(t).$$

25.8 Applications

25.8.1 Solutions of Laplace's Equation in Spheroidal Coordinates

Consider the prolate spheroidal coordinates α, β, θ defined by

$$x = c\sinh\alpha\sin\beta\cos\theta, \quad y = c\sinh\alpha\sin\beta\sin\theta, \quad z = c\cosh\alpha\cos\beta,$$

where $0 \le \alpha < \infty$, $0 \le \beta \le \pi$, $-\pi < \theta \le \pi$, and $c > 0$ is a scale factor. If (r, θ, z) represents the cylindrical coordinates of a point, then $z + ir = c\cosh(\alpha + i\beta)$. The surface $\beta =$ constant is the double-sheeted hyperboloid of revolution

$$\frac{z^2}{b^2} - \frac{x^2}{a^2} - \frac{y^2}{a^2} = 1\,,$$

where $a = c\sin\beta$, $b = c\cos\beta$. Laplace's equation in spheroidal coordinates takes the form

$$\nabla^2 u = \frac{1}{c^2(\sinh^2\alpha + \sin^2\beta)}\left[\frac{1}{\sinh\alpha}\frac{\partial}{\partial\alpha}(\sinh\alpha\frac{\partial u}{\partial\alpha})+\right.$$

$$\left.\frac{1}{\sin\beta}\frac{\partial}{\partial\beta}(\sin\beta\frac{\partial u}{\partial\beta}) + (\frac{1}{\sinh^2\alpha} + \frac{1}{\sin^2\beta})\frac{\partial^2 u}{\partial\theta^2}\right] = 0 \qquad (25.6)$$

If we assume that $u(\alpha,\beta,\theta) = A(\alpha)B(\beta)C(\theta)$, it is easy to see that A, B and C satisfy the ordinary differential equations

$$\frac{d^2C}{d\theta^2} + \mu^2 C = 0$$

$$\frac{1}{\sin\beta}\frac{d}{d\beta}\left(\sin\beta\frac{dB}{d\beta}\right) + \left[\nu(\nu+1) - \frac{\mu^2}{\sin^2\beta}\right]B = 0$$

$$\frac{1}{\sin\alpha}\frac{d}{d\alpha}\left(\sinh\alpha\frac{dA}{d\alpha}\right) - \left[\nu(\nu+1) + \frac{\mu^2}{\sinh^2\alpha}\right]A = 0$$

where μ and ν are the separation parameters. If the solution u is periodic in θ, then $\mu = m(m = 0, 1, 2, ...)$. A particular solution of (25.6) that is bounded for $0 \le \beta \le \pi$ and $0 \le \alpha \le d\,(d > 0)$ is readily seen to be

$$P_\nu^m(\cos\beta)P_\nu^m(\cosh\alpha)e^{im\theta}\ .$$

Hence, it follows from the principle of superposition that

$$\sum_{m=-\infty}^{\infty} e^{im\theta}\int_0^\infty A_m(\nu)P_\nu^m(\cos\beta)P_\nu^m(\cosh\alpha)d\nu$$

is a solution of (25.6). If the solution is axially symmetric, i.e., independent of θ, then

$$u(\alpha,\beta) = \int_0^\infty A_0(\nu)P_\nu(\cos\beta)P_\nu(\cosh\alpha)d\nu\ .$$

Let us solve the following Dirichlet problem for the domain bounded by the hyperboloid of revolution $\beta = \beta_0$:

i) $\nabla^2 = 0$ in $0 \le \beta < \beta_0$,

ii) u is axially symmetric,

iii) u satisfies $u(\alpha,\beta_0) = f(\alpha)$ for some continuous function f on $[0,\infty)$ with $\lim_{\alpha\to\infty} f(\alpha) = 0$.

Clearly,

$$u(\alpha,\beta) = \int_0^\infty A(t)P_{-\frac{1}{2}+it}(\cos\beta)P_{-\frac{1}{2}+it}(\cosh\alpha)dt$$

satisfies (i) and (ii). To satisfy (iii) we must have

$$u(\alpha, \beta_0) = f(\alpha) = \int_0^\infty A(t) P_{-\frac{1}{2}+it}(\cos\beta_0) P_{-\frac{1}{2}+it}(\cosh\alpha) dt \ .$$

From the inversion formula of the Mehler-Foch transform, we immediately obtain that

$$A(t) = \frac{t\tanh(\pi t)}{P_{-\frac{1}{2}+it}(\cos\beta_0)} \int_0^\infty f(\alpha) P_{-\frac{1}{2}+it}(\cosh\alpha) \sinh\alpha d\alpha \ ,$$

and hence

$$u(\alpha, \beta) = \int_0^\infty F(t) \frac{P_{-\frac{1}{2}+it}(\cos\beta)}{P_{-\frac{1}{2}+it}(\cos\beta_0)} P_{-\frac{1}{2}+it}(\cosh\alpha) dt \ ,$$

where

$$F(t) = t\tanh(\pi t) \int_0^\infty f(\alpha) P_{-\frac{1}{2}+it}(\cosh\alpha) \sinh\alpha d\alpha \ .$$

In a similar fashion one can show that if the condition of axial symmetry is dropped, then the solution will involve the inversion formula for the Mehler–Fock transform of order m.

25.8.2 Solutions of Laplace's Equation in Toroidal Coordinates

Consider the toroidal coordinates (α, β, θ) defined by

$$x = c\frac{\sinh\alpha\cos\theta}{\cosh\alpha + \cos\beta} \ , \quad y = c\frac{\sinh\alpha\sin\theta}{\cosh\alpha + \cos\beta} \ , \quad z = c\frac{\sin\beta}{\cosh\alpha + \cos\beta} \ ,$$

where $0 \le \alpha < \infty$, $-\pi < \beta \le \pi$, $-\pi < \theta \le \pi$, and $c > 0$ is a scale factor. If (r, θ, z) represents the cylindrical coordinates of a point, then $r + iz = c\tanh\frac{1}{2}(\alpha + i\beta)$.

The surface $\beta = \beta_0$ has equation

$$r^2 + (z + c\cot\beta_0)^2 = c^2\csc^2\beta_0 \ ;$$

hence if $0 < \beta_0 < \pi$, the surface $\beta = \beta_0$ is that part of the sphere with center $(0, 0, -c\cot\beta_0)$ and radius $c/\sin\beta_0$ which lies above the xy-plane.

The surface $\alpha = \alpha_0$ has equation

$$(r - c\coth\alpha_0)^2 + z^2 = c^2 \text{cosech}^2\alpha_0 \ ,$$

which is an equation of a torus.

Laplace's equation in toroidal coordinates takes the form

$$\nabla^2 u = \frac{\partial}{\partial \alpha}\left[\frac{\sinh\alpha}{\cosh\alpha + \cos\beta}\frac{\partial u}{\partial \alpha}\right] +$$

$$\frac{\partial}{\partial \beta}\left[\frac{\sinh\alpha}{\cosh\alpha + \cos\beta}\frac{\partial u}{\partial \beta}\right] + \frac{1}{\sinh\alpha(\cosh\alpha + \cos\beta)}\frac{\partial^2 u}{\partial \theta^2} = 0\ . \quad (25.7)$$

Unlike the previous example, here we cannot separate the variables. However, if we introduce a new function υ by

$$u(\alpha,\beta,\theta) = \sqrt{\cosh\alpha + \cos\beta}\ \upsilon(\alpha,\beta,\theta)\ ,$$

we find that υ satisfies the equation

$$\frac{\partial^2 \upsilon}{\partial \alpha^2} + \frac{\partial^2 \upsilon}{\partial \beta^2} + \coth\alpha\frac{\partial \upsilon}{\partial \alpha} + \frac{1}{4}\upsilon + \frac{1}{\sinh^2\alpha}\frac{\partial^2 \upsilon}{\partial \theta^2} = 0\ , \quad (25.8)$$

which can be solved by the separation of variables technique. Thus, by letting

$$\upsilon(\alpha,\beta,\theta) = A(\alpha)B(\beta)C(\theta)$$

in (25.8), we obtain

$$C'' + \mu^2 C = 0\ ,$$

$$B'' - t^2 B = 0\ ,$$

$$A'' + \coth\alpha A' + \left(\frac{1}{4} + t^2 - \frac{\mu^2}{\sinh^2\alpha}\right)A = 0\ .$$

To have a periodic solution in θ, μ must be an integer $m (m = 0, \pm1, \pm2, ...)$. If we make the change of variable, $x = \cosh\alpha$ and set $\nu = -\frac{1}{2} + it$, the last differential equation becomes

$$(1 - x^2)\frac{d^2 A}{dx^2} - 2x\frac{dA}{dx} + \left[\nu(\nu+1) - \frac{m^2}{1 - x^2}\right]A = 0\ .$$

Thus, equation (25.7) has solutions of the form

$$\sqrt{\cosh\alpha + \cos\beta}\ P^m_{-\frac{1}{2}+it}(\cosh\alpha)e^{\pm t\beta}e^{\pm im\theta}\ ,$$

and by the superposition principle

$$u(\alpha,\beta,\theta) = \sqrt{\cosh\alpha + \cos\beta}\ \sum_{m=-\infty}^{\infty} e^{im\theta}u_m(\alpha,\beta)\ ,$$

is a solution of Laplace's equation (25.7), where

$$u_m(\alpha,\beta) = \int_0^\infty P^m_{-\frac{1}{2}+it}(\cosh\alpha)\left[A_m(t)\cosh\beta t + B_m(t)\sinh\beta t\right] dt\ ,$$

and $A_m(t), B_m(t)$ are arbitrary functions of t. If we further assume that $u(\alpha, 0, \theta) = f(\alpha, \theta)$ and $\frac{\partial u}{\partial \beta}|_{\beta=0} = 0$, then it is easy to see that

$$u(\alpha, \beta, \theta) = \sqrt{\cosh\alpha + \cos\beta} \sum_{m=-\infty}^{\infty} e^{im\theta} v_m(\alpha, \beta) ,$$

where

$$v_m(\alpha, \beta) = \int_0^\infty A_m(t) \cosh(\beta t) P^m_{-\frac{1}{2}+it}(\cosh\alpha)\, dt ,$$

and

$$f(\alpha, \theta) = \sqrt{\cosh\alpha + 1} \sum_{m=-\infty}^{\infty} e^{im\theta} \int_0^\infty A_m(t) P^m_{-\frac{1}{2}+it}(\cosh\alpha) dt .$$

Hence

$$\int_0^\infty A_m(t) P^m_{-\frac{1}{2}+it}(\cosh\alpha) dt = \frac{1}{2\pi} \int_0^{2\pi} \frac{f(\alpha, \theta)}{\sqrt{\cosh\alpha + 1}} e^{-im\theta}\, d\theta = g(\alpha) .$$

To determine A_m, we invoke the inversion formula for the Mehler–Fock transform of order m to obtain

$$A_m(t) = (-1)^m t \tanh(\pi t) \int_0^\infty g(\alpha) P^{-m}_{-\frac{1}{2}+it}(\cosh\alpha) \sinh\alpha\, d\alpha .$$

26

The Sturm-Liouville-Type Transforms

The kernels of many integral transformations are solutions of Sturm-Liouville differential equations. Since the spectral theory of the Sturm-Liouville boundary-value problems is very well developed, it is natural to use it to study many function transformations in a unified way. By a *Sturm-Liouville transformation* we mean an integral transformation whose kernel is a solution of a Sturm-Liouville differential equation. The transform and its inverse form what we call a *Sturm-Liouville pair.*

If the spectrum of a Sturm-Liouville boundary-value problem is discrete, the associated transformation will also be discrete in the sense that the transform of a function will be a sequence of complex numbers. In this chapter, we shall show, among other things, that there exists also a continuous transformation associated with the discrete spectrum such that when the continuous transform variable is replaced by the eigenvlaues, the continuous transformation reduces to the discrete one.

Consider the Sturm-Liouville differential equation

$$\frac{d}{dx}\left\{p(x)\frac{dy}{dx}\right\} + \ell(x)y = -\lambda r(x)y\ , \quad a \le x \le b\ , \tag{26.1}$$

where $p(x), r(x) > 0$ in $[a, b]$. If $p(x)$ has a continuous first derivative and $p(x)r(x)$ has a continuous second derivative, then by setting

$$z = \frac{1}{c}\int_a^x \sqrt{r(t)/p(t)}\ dt\ , \quad u(x) = \sqrt[4]{r(x)p(x)}\ y\ , \quad \mu = c^2\lambda\ ,$$

where

$$c = \frac{1}{\pi}\int_a^b \sqrt{r(x)/p(x)}\ dx\ ,$$

we can convert Equation (26.1) into the form

$$u'' - q(z)u = -\mu u\ , \tag{26.2}$$

where

$$q(z) = \frac{\theta''(z)}{\theta(z)} - c^2 \frac{\ell(x)}{r(x)} \quad \text{and} \quad \theta(z) = (r(x)p(x))^{\frac{1}{4}} .$$

In the remainder of this chapter we shall be mainly concerned with the differential operator (26.2).

DEFINITION 26.1 *Let $[a,b]$ be a bounded closed interval and $q(x)$ be a continuous function on (a,b) such that $\lim_{x\to a^+} q(x)$ and $\lim_{x\to b^-} q(x)$ are finite. The boundary-value problem (BVP)*

$$y'' - q(x)y = -\lambda y\ , \quad -\infty < a \le x \le b < \infty\ , \tag{26.3}$$

$$y(a)\cos\alpha + y'(a)\sin\alpha = 0 \tag{26.4}$$

$$y(b)\cos\beta + y'(b)\sin\beta = 0\ , \tag{26.5}$$

is called a regular Sturm-Liouville boundary-value problem. The operator $-\frac{d^2}{dx^2} + q(x)$ will be denoted by L so that Equation (26.3) can be written as $Ly = \lambda y$.

If q has a singularity at, at least, one of the end points of the interval or if the interval extends to infinity in one direction or in both, the problem is called a singular Sturm-Liouville BVP or a singular Sturm-Liouville problem for short.

The problem in the case of a finite interval (a,b) with q having a singularity at one of the end points of the interval is similar to that in which q is continuous and the interval extends to infinity in one direction, and likewise the problem in which (a,b) is finite and q has singularities at the end points of the interval is similar to that in which q is continuous and the interval extends to infinity in both directions. Therefore, without loss of generality we shall restrict ourselves to singular problems on a half-line and the whole line. For more information on the Sturm–Liouville theory, consult [226, 147].

26.1 The Regular Case

26.1.1 The Discrete Transform

The regular Sturm-Liouville problem (26.3)–(26.5), possesses a set of real numbers, called the eigenvalues,$\{\lambda_n\}_{n=0}^{\infty}$, having no finite limit point and a corresponding set of functions $\{\phi_n\}_{n=0}^{\infty}$, called the eigenfunctions such that

$L\phi_n = \lambda_n \phi_n$. The eigenfunctions form a complete orthogonal set in $L^2[a,b]$ in the sense that

$$\langle \phi_m, \phi_n \rangle = \int_a^b \phi_m(x)\overline{\phi}_n(x)dx = 0 \quad \text{if} \quad m \neq n ,$$

and if $f \in L^2[a,b]$ is such that $\langle f, \phi_n \rangle = 0$ for all n, then $f = 0$. Moreover, if we normalize the eigenfunctions so that $\langle \phi_n, \phi_n \rangle = 1$ for all n, then for any $f \in L^2[a,b]$, we have

$$f(x) = \sum_{n=0}^{\infty} \langle f, \phi_n \rangle \phi_n(x) \tag{26.6}$$

in the sense of $L^2[a,b]$. If $f \in C^2[a,b]$ satisfies the boundary conditions (26.4)–(26.5), then the series (26.6) converges absolutely and uniformly on $[a,b]$.

Every regular Sturm-Liouville problem gives rise to a discrete function transformation on $L^2[a,b]$, $f \to \left\{\hat{f}(n)\right\}_{n=0}^{\infty}$ defined by

$$\hat{f}(n) = \int_a^b f(x)\overline{\phi}_n(x)dx .$$

The inversion formula is given by

$$f(x) = \sum_{n=0}^{\infty} \hat{f}(n)\phi_n(x) ,$$

and the Parseval relation is

$$\int_a^b f(x)\overline{g}(x)\, dx = \sum_{n=0}^{\infty} \hat{f}(n)\overline{\hat{g}}(n), \tag{26.7}$$

in particular,

$$\int_a^b |f(x)|^2 dx = \sum_{n=0}^{\infty} |\hat{f}(n)|^2 .$$

26.1.2 The Continuous Transform

The next theorem shows that every regular Sturm-Liouville problem gives rise to a continuous transformation as well.

THEOREM 26.1 **(Zayed [259, 266])**
Consider the regular Sturm-Liouville problem (26.3)-(26.5). Let $\phi(x,\lambda), \psi(x,\lambda)$ be the solutions of (26.3) such that

$$\phi(a,\lambda) = \sin\alpha , \quad \phi'(a,\lambda) = -\cos\alpha$$
$$\psi(b,\lambda) = \sin\beta , \quad \psi'(b,\lambda) = -\cos\beta .$$

Then there exists a linear integral transformation T from $L^2[a,b]$ into the space of all entire functions of order $\frac{1}{2}$ and type at most $(b-a)$ defined by

$$(Tf)(\lambda) = F(\lambda) = \int_a^b f(x)\phi(x,\lambda)dx, \quad f \in L^2[a,b]\ , \lambda \text{ complex}, \tag{26.8}$$

such that $F(\lambda_n) = \hat{f}(n)$. Moreover, F is completely determined by its values at the eigenvalues $\{\lambda_n\}_{n=0}^{\infty}$ and can be reconstructed from these values according to the formula

$$F(\lambda) = \sum_{n=0}^{\infty} F(\lambda_n)\frac{P(\lambda)}{(\lambda-\lambda_n)P'(\lambda_n)} \tag{26.9}$$

where $P(\lambda)$ is the Wronskian of ϕ and ψ, which without loss of generality can be written in the form

$$P(\lambda) = \begin{cases} \prod_{n=1}^{\infty}\left(1-\dfrac{\lambda}{\lambda_n}\right) & , \text{ if zero is not an eigenvalue} \\ \lambda\prod_{n=1}^{\infty}\left(1-\dfrac{\lambda}{\lambda_n}\right) & , \text{ if zero is an eigenvalue.} \end{cases}$$

The series (26.9) converges uniformly on compact subsets of the λ-plane.

COROLLARY 26.1

Let $\tilde{P}(t) = P(\lambda)$, $\tilde{\phi}(x,t) = \phi(x,\lambda)$, where $\lambda = t^2$, $\lambda_n = t_n^2$, with $t_{-n} = -t_n$, $n = 0, 1, \ldots$. If $f \in L^2[a,b]$ and

$$F(t) = \int_a^b f(x)\tilde{\phi}(x,t)dx\ ,$$

then $F(t)$ is an entire function of order 1 (of exponential type) with type at most $(b-a)$ that admits the expansion

$$F(t) = \sum_{n=1}^{\infty} F(t_n)\frac{\tilde{P}(t)(2t_n)}{(t^2-t_n^2)\tilde{P}'(t_n)}$$

if zero is not an eigenvalue, and

$$F(t) = F(0)\frac{\tilde{P}(t)}{t^2} + \sum_{n=1}^{\infty} F(t_n)\frac{\tilde{P}(t)(2t_n)}{(t^2-t_n^2)\tilde{P}'(t_n)}$$

if zero is an eigenvalue.

Example A: The Finite Continuous Cosine-Transform

Consider the regular Sturm-Liouville problem

$$\begin{aligned} y'' &= -\lambda y \qquad , \quad 0 \le x \le \pi \\ y'(0) &= 0 = y'(\pi) \ . \end{aligned}$$

It is easy to see that $\phi(x, \lambda) = \cos\sqrt{\lambda}\, x$ and that the eigenvalues are $\lambda_n = n^2$, $n = 0, 1, 2, \dots$; hence

$$P(\lambda) = \lambda \prod_{n=1}^{\infty} \left(1 - \frac{\lambda}{n^2}\right) = \frac{\sqrt{\lambda}\, \sin\sqrt{\lambda}\, \pi}{\pi} \ .$$

The finite continuous cosine transform is defined by

$$F(\lambda) = \int_0^{\pi} f(x) \cos\sqrt{\lambda}\, x dx \ , \quad f \in L^2[0, \pi].$$

It reduces to the discrete cosine transform when $\lambda = n^2$. We also have

$$F(\lambda) = F(0) \frac{\sin \pi\sqrt{\lambda}}{\pi\sqrt{\lambda}} + 2 \sum_{n=1}^{\infty} F(n^2) \frac{\sqrt{\lambda}\, \sin \pi(\sqrt{\lambda} - n)}{\pi(\lambda - n^2)} \ .$$

Many more examples can be obtained similarly; see [266].

26.1.3 Inversion of the Continuous Transform

Finding an inversion formula for the continuous transform developed in the preceding section requires more work. We shall state the main result and refer the interested reader to [274] for details.

Since the substitution $x = \pi(x - a)/(x - b)$ transforms the interval $[a, b]$ into the interval $[0, \pi]$ without changing the form of the boundary-value problem, we shall assume throughout this section that $a = 0$, $b = \pi$. We shall also write $\phi(x, t)$ for the kernel of the continuous Sturm-Liouville transform (26.8), where $t^2 = \lambda$, so that the transform takes the form

$$\Phi(f)(t) = F(t) = \int_0^{\pi} f(x) \phi(x, t) dx \ .$$

Let H_0^m be the Sobolev space on $(0, \pi)$, consisting of the completion of $C_0^m[0, \pi]$ with respect to the norm $\|f\|_m = \sum_{j=0}^{m} \|f^{(j)}\|$, where $\|.\|$ stands for the L^2-norm. Integration by parts, together with the asymptotic expansion of ϕ, implies that if $f \in H_0^m$, then $F(t) = 0(|t|^{-m})$ as $|t| \to \infty$.

THEOREM 26.2 (**Zayed and Walter** [274])
Consider the regular Sturm-Liouville boundary-value problems (26.3)-(26.5). Then the continuous Sturm-Liouville transform (26.8) is a linear transformation that maps $L^2[0, \pi]$ into a subspace of the Paley-Wiener space B_π^2

consisting of even entire functions of exponential type $\leq \pi$. *Moreover, if*

$$F(t) = \int_0^\pi f(x)\phi(x,t)dx\ , \quad \textit{for some} \quad f \in H_0^m\ , \quad m > 1\ ,$$

then

$$f(x) = \int_{-\infty}^{\infty} F(t)k(x,t)dt\ ,$$

where

$$k(x,t) = \begin{cases} \frac{1}{2}\sum_{n=-\infty}^{\infty}{}' \dfrac{\phi_n(x)}{\|\phi_n\|^2}\, B_n(t) & , \textit{if zero is not an eigenvalue} \\ \frac{\phi_0(x)}{\|\phi_0\|^2}\, B_0(t) + \frac{1}{2}\sum_{n=-\infty}^{\infty}{}' \dfrac{\phi_n(x)}{\|\phi_n\|^2}\, B_n(t) & , \textit{if zero is an eigenvalue}\ , \end{cases}$$

in which

$$B_n(t) = \frac{\sin \pi(t-t_n)}{\pi(t-t_n)}\ ,$$

ϕ_n *are the eigenfunctions of the problem and* $\sum'$ *means that the term* $n = 0$ *is excluded.*

Another inversion formula based on the Poisson summation formula (cf. Equation (12.24)) has also been derived in [274].

Example B: Consider the boundary-value problem of Example A. One can easily find that $k(x,t) = \cos tx/\pi$ and hence if

$$F(t) = \int_0^\pi f(x)\cos(tx)\, dx\ , \quad f \in H_0^2(0,\pi)\ ,$$

then

$$f(x) = \frac{2}{\pi}\int_0^\infty F(t)\cos(xt)\, dt\ ,$$

which is in agreement with the inversion formula for the Fourier cosine-transform.

26.2 The Singular Case on a Half-Line

Consider the Sturm-Liouville differential equation

$$y'' - q(x)y = -\lambda y\ , \quad 0 \leq x < \infty\ . \tag{26.10}$$

where q is continuous on $(0, \infty)$. Let $\phi(x, \lambda)$ be the solution of (26.10) that satisfies the initial conditions

$$\phi(0, \lambda) = \sin \alpha\ , \quad \phi'(0, \lambda) = -\cos \alpha\ .$$

and let $f \in L^2(0, \infty)$. Then there exists a nondecreasing function $\rho(\lambda)$ which does not depend on f such that

$$F(\lambda) = \int_0^\infty f(x)\phi(x, \lambda)dx \tag{26.11}$$

exists in the mean and

$$\int_0^\infty |f(x)|^2 dx = \int_{-\infty}^\infty |F(\lambda)|^2 d\rho(\lambda)\ .$$

More generally, we have Parseval's relation

$$\int_0^\infty f(x)\overline{g}(x)dx = \int_{-\infty}^\infty F(\lambda)\overline{G}(\lambda)d\rho(\lambda)\ ,$$

where g is related to G as f is related to F.

Let f be a continuous function on $[0, \infty)$ and assume that the integral

$$\int_{-\infty}^\infty F(\lambda)\phi(x, \lambda)d\rho(\lambda)$$

converges absolutely and uniformly on any finite interval. Then, we can invert the integral transform (26.11) to obtain

$$f(x) = \int_{-\infty}^\infty F(\lambda)\phi(x, \lambda)d\rho(\lambda)\ .$$

This inversion formula can be obtained by substituting $\int_0^\infty g(x)\phi(x, \lambda)dx$ for $G(\lambda)$ in Parseval's relation, then interchanging the integrals. The two functions f and F are called a *Sturm–Liouville transform pair.*

Let $\theta(x, \lambda)$ be the solution of (26.10) that satisfies

$$\theta(0, \lambda) = \cos \alpha\ , \quad \theta'(0, \lambda) = \sin \alpha\ .$$

It is known [226, p. 25] that there exists a function $m(\lambda)$ analytic in the upper and lower halves of the complex λ-plane ($Im\,\lambda \neq 0$) such that for every non-real λ, Equation (26.10) has a solution

$$\psi(x, \lambda) = \theta(x, \lambda) + m(\lambda)\phi(x, \lambda)$$

belonging to $L^2[0, \infty)$. The function $m(\lambda)$ takes conjugate values for conjugate values of λ. If it has poles on the real axis, they are all simple. The functions $m(\lambda)$ and $\rho(\lambda)$ are related by

$$m(\lambda) = -\cot \alpha + \int_{-\infty}^\infty \frac{d\rho(t)}{\lambda - t}\ ,$$

$$\lim_{\delta\to 0}\frac{1}{\pi}\int_0^\lambda -\operatorname{Im} m(u+i\delta)du=\rho(\lambda)\ .$$

The following theorem is an analogue of Theorem 26.1. It shows that even for a singular Sturm-Liouville problem on a half-line if the spectrum is discrete, we can define a continuous Sturm-Liouville transform that reduces to the discrete Sturm-Liouville transform when the parameter λ is replaced by the eigenvalues.

THEOREM 26.3 **(Zayed [259])**
Consider the singular Sturm-Liouville problem

$$y''-q(x)y=-\lambda y\ ,\quad 0\le x<\infty\ ,$$

$$y(0)\cos\alpha+y'(0)\sin\alpha=0\ ,$$

where $q(x)$ is continuous on $[0,\infty)$. Assume that $m(\lambda)$ is a meromorphic function that is real-valued on the real axis and whose only singularities are simple poles $\{\lambda_n\}_{n=0}^\infty$ on the non-negative real axis; λ_0 will be reserved for the eigenvalue zero. Let p be the smallest integer for which the series $\sum_{n=1}^\infty 1/(\lambda_n)^{p+1}$ converges.

1. *If none of the λ_n's is zero, set*

$$G(\lambda)=\begin{cases}\prod_{n=0}^\infty\left(1-\frac{\lambda}{\lambda_n}\right)\exp\left[\left(\frac{\lambda}{\lambda_n}\right)+\cdots+\frac{1}{p}\left(\frac{\lambda}{\lambda_n}\right)^p\right] & \text{if } p=1,2,\ldots,\\ \prod_{n=0}^\infty\left(1-\frac{\lambda}{\lambda_n}\right) & \text{if } p=0\,.\end{cases}\tag{26.12}$$

2. *If one of the λ_n's is zero, say $\lambda_0=0$, set*

$$G(\lambda)=\begin{cases}\lambda\prod_{n=0}^\infty\left(1-\frac{\lambda}{\lambda_n}\right)\exp\left[\left(\frac{\lambda}{\lambda_n}\right)+\cdots+\frac{1}{p}\left(\frac{\lambda}{\lambda_n}\right)^p\right] & \text{if } p=1,2,\ldots\\ \lambda\prod_{n=0}^\infty\left(1-\frac{\lambda}{\lambda_n}\right) & \text{if } p=0\,.\end{cases}\tag{26.13}$$

Let $\Phi(x,\lambda)=G(\lambda)\psi(x,\lambda)$, $f(x)\in L^2(0,\infty)$ and

$$F(\lambda)=\int_0^\infty f(x)\Phi(x,\lambda)dx\ .\tag{26.14}$$

Then $F(\lambda)$ is an entire function that admits the following sampling representation:

$$F(\lambda)=\sum_{n=0}^\infty F(\lambda_n)\frac{G(\lambda)}{(\lambda-\lambda_n)G'(\lambda_n)}\ ,\tag{26.15}$$

where the series converges uniformly on any compact subset of the complex λ-plane.

COROLLARY 26.2

Let $\tilde{G}(t) = G(\lambda)$ and $\tilde{\Phi}(x,t) = \Phi(x,\lambda)$ where $\lambda = t^2$, $\lambda_n = t_n^2$ and $t_{-n} = -t_n$, $n = 0, 1, 2, \ldots$. If $f(x) \in L^2(0,\infty)$ and

$$F(t) = \int_0^\infty f(x)\tilde{\Phi}(x,t)dx \,,$$

then, $F(x)$ is an entire function that admits the following sampling representations:

$$F(t) = \begin{cases} \displaystyle\sum_{n=0}^{\infty} F(t_n)\frac{\tilde{G}(t)(2t_n)}{\tilde{G}'(t_n)(t^2 - t_n^2)} & \text{if } t_n \neq 0 \text{ for all } n\,, \\ F(0)\frac{\tilde{G}(t)}{t^2} + \displaystyle\sum_{n=1}^{\infty} F(t_n)\frac{\tilde{G}(t)(2t_n)}{\tilde{G}'(t_n)(t^2 - t_n^2)} & \text{if one of the } t_n\text{'s say } t_0 = 0\,. \end{cases}$$

26.3 The Singular Case on the Whole Line

Now we consider the case where either the function $q(x)$ has singularities at both end points of the interval (a,b) or the interval (a,b) extends to infinity in both directions, or it extends to infinity in one direction and $q(x)$ has a singularity at the other finite end point. Since the analysis in all these cases is similar we shall, without loss of generality, confine our attention to the case where (a,b) extends to infinity in both directions and $q(x)$ is continuous.

Consider

$$y'' - q(x)y = -\lambda y \,, \quad -\infty < x < \infty \,, \tag{26.16}$$

where $q(x)$ is continuous on $(-\infty,\infty)$.

Let $\phi(x) = \phi(x,\lambda)$ and $\theta(x) = \theta(x,\lambda)$ be the solutions of (26.16) such that

$$\phi(0) = 0 \,, \quad \phi'(0) = -1 \,,$$
$$\theta(0) = 1 \,, \quad \theta'(0) = 0 \,.$$

We infer, from the results mentioned at the beginning of the preceding section, that there exist two functions $m_1(\lambda)$ and $m_2(\lambda)$, analytic in the upper and lower half-planes such that

$$\psi_1(x,\lambda) = \theta(x,\lambda) + m_1(\lambda)\phi(x,\lambda)$$

is in $L^2(0,\infty)$ and

$$\psi_2(x,\lambda) = \theta(x,\lambda) + m_2(\lambda)\phi(x,\lambda)$$

is in $L^2(-\infty,0)$.

It is easy to see that the Wronskian of ψ_1, ψ_2

$$W(\psi_1,\psi_2) = m_1(\lambda) - m_2(\lambda)\,,$$

and

$$\int_0^\infty \psi_1(x,\lambda)\psi_1(x,\lambda')dx = \frac{m_1(\lambda) - m_1(\lambda')}{\lambda' - \lambda}\,,$$

$$\int_{-\infty}^0 \psi_2(x,\lambda)\psi_2(x,\lambda')dx = \frac{m_2(\lambda) - m_2(\lambda')}{\lambda - \lambda'}\,.$$

Throughout this section we shall assume that both $m_1(\lambda)$ and $m_2(\lambda)$ are meromorphic functions satisfying the same conditions as $m(\lambda)$ given in the preceding section. A sufficient condition for this to hold, for example, is that $\lim_{x\to\pm\infty} q(x) = \infty$ (cf. [226, p. 127]). It is known [226, p. 43] that the eigenvalues $\{\lambda_n\}_{n=0}^\infty$ are the points where:

i) $m_1(\lambda)$, $m_2(\lambda)$ have poles of the form $m_1(\lambda) \sim a_1/(\lambda - \lambda_n)$, $m_2(\lambda) \sim a_2/(\lambda - \lambda_n)$, $a_1 \neq a_2$, or

ii) $m_1(\lambda_n) = m_2(\lambda_n) \neq 0$, or

iii) $m_1(\lambda) \sim a_1(\lambda - \lambda_n)$, $m_2(\lambda) \sim a_2(\lambda - \lambda_n)$, $a_1 \neq a_2$.

Now we are able to state our main theorem for this section.

THEOREM 26.4 (**Zayed** [259])
Consider the differential equation

$$y'' - q(x)y = -\lambda y\,, \quad -\infty < x < \infty\,,$$

where $q(x)$ is an even continuous function on $(-\infty,\infty)$. Let $m_i(\lambda)$, $\psi_i(x,\lambda)$, $i = 1, 2$ have the same meaning as above. Denote the zeros and the poles of $m_i(\lambda)$ by $\{\lambda_{2n}\}_{n=0}^\infty$ and $\{\lambda_{2n+1}\}_{n=0}^\infty$, respectively, and the corresponding residues at the poles by $\{r_{2n+1}\}_{n=0}^\infty$. Let $G(\lambda)$ be the canonical product of $\{\lambda_{2n+1}\}_{n=0}^\infty$ as defined in Theorem 26.3. If $f(x) \in L^2(-\infty,\infty)$ and

$$F(\lambda) = \int_0^\infty f(x)\Phi_1(x,\lambda)dx + \int_{-\infty}^0 f(x)\Phi_2(x,\lambda)dx\,,$$

where

$$\Phi_i(x,\lambda) = G(\lambda)\psi_i(x,\lambda)\,, \quad i = 1, 2,\,,$$

then

$$F(\lambda) = \sum_{n=0}^{\infty} \frac{F(\lambda_{2n})G(\lambda)m_1(\lambda)}{G(\lambda_{2n})(\lambda - \lambda_{2n})m_1'(\lambda_{2n})} + \sum_{n=0}^{\infty} \frac{F(\lambda_{2n+1})G(\lambda)}{(\lambda - \lambda_{2n+1})G'(\lambda_{2n+1})} ,$$

where the two series converge uniformly on compact subsets of the complex plane.

26.4 Summability

Suppose that the spectrum of the Sturm-Liouville problem is discrete with eigenvalues $\lambda_0, \lambda_1, \ldots$, tending to ∞. If $f \in L^2(0, \infty)$, and

$$\int_0^{\eta} |f(x+u) - f(x)| du = 0(\eta) \quad \text{as} \quad \eta \to 0 ,$$

then the eigenfunction expansion of f is Riesz $(R, \lambda, 1)$ summable to f, i.e.,

$$\lim_{\lambda \to \infty} \sum_{\lambda_n < \lambda} \left(1 - \frac{\lambda_n}{\lambda}\right) c_n \psi_n(x) = f(x) ,$$

where $c_n = \int_0^\infty f(x)\psi_n(x)\, dx$; see Section 1.11.1.

In particular, it is summable $(R, \lambda, 1)$ to $f(x)$ almost everywhere and to $f(x)$ wherever f is continuous.

If the spectrum is continuous but bounded below by λ_0 set $g(\lambda) = \int_0^\lambda \int_0^\infty f(x)\phi(x, u)\, dx d\rho(u)$. If g is of bounded variation, then

$$f(x) = \lim_{\lambda \to \infty} \int_{\lambda_0}^{\lambda} \left(1 - \frac{u}{\lambda}\right) \phi(x, u) dg(u) ,$$

if the spectrum extends to $-\infty$, then

$$f(x) = - \lim_{\lambda \to \infty} \int_{-\lambda}^{\lambda} \left(1 - \frac{u^2}{\lambda^2}\right) \phi(x, u) dg(u) .$$

26.5 Examples

26.5.1 Sturm–Liouville Transform Pairs

1. **Continuous Spectrum**

(a) **The Cosine and Sine Transforms:**
Consider the differential equation (26.10) with $q(x) = 0$. Then,

$$\phi(x,\lambda) = \sin\alpha\cos(\sqrt{\lambda}\,x) - \frac{1}{\sqrt{\lambda}}\cos\alpha\sin(\sqrt{\lambda}\,x)$$
$$\theta(x,\lambda) = \cos\alpha\cos(\sqrt{\lambda}\,x) + \frac{1}{\sqrt{\lambda}}\sin\alpha\sin(\sqrt{\lambda}\,x)\,.$$

For $\operatorname{Im}\lambda > 0$, the function $\psi(x,\lambda) = \theta(x,\lambda) + m(\lambda)\phi(x,\lambda)$ can only differ from $e^{i\sqrt{\lambda}x}$ by a constant factor, as $e^{-i\sqrt{\lambda}x} \notin L^2(0,\infty)$. Consequently,

$$m(\lambda) = \frac{\sin\alpha - i\sqrt{\lambda}\,\cos\alpha}{\cos\alpha + i\sqrt{\lambda}\,\sin\alpha}\,,$$

and hence

$$-\operatorname{Im}\,m(\lambda) = \begin{cases} \sqrt{\lambda}/(\cos^2\alpha + \lambda\sin^2\alpha)\,, & \lambda > 0 \\ 0 & ,\ \lambda < 0\,. \end{cases}$$

The Sturm-Liouville transform pair takes the form

$$F(\lambda) = \int_0^\infty f(x)\left[\sin\alpha\cos(\sqrt{\lambda}\,x) - \frac{1}{\sqrt{\lambda}}\cos\alpha\sin(\sqrt{\lambda}\,x)\right]dx\,,$$

and

$$f(x) = \frac{1}{\pi}\int_0^\infty F(\lambda)\left[\sin\alpha\cos(\sqrt{\lambda}\,x) - \frac{1}{\sqrt{\lambda}}\cos\alpha\sin(\sqrt{\lambda}\,x)\right]$$
$$\times\frac{\sqrt{\lambda}}{\cos^2\alpha + \lambda\sin^2\alpha}d\lambda\,.$$

In particular, if $\alpha = \frac{\pi}{2}$, we have the Fourier cosine-transform pair

$$F(\lambda) = \int_0^\infty f(x)\cos(\sqrt{\lambda}\,x)dx\,,$$
$$f(x) = \frac{1}{\pi}\int_0^\infty F(\lambda)\cos(\sqrt{\lambda}\,x)\frac{d\lambda}{\sqrt{\lambda}}\,.$$

The Fourier sine-transform pair can be obtained similarly by putting $\alpha = 0$.

(b) **The Hankel Transform:**
Consider the differential equation (26.10) on $(0,\infty)$ with $q(x) = (\nu^2 - \frac{1}{4})/x^2$, $\lambda = t^2\,, \nu > 1$. The Sturm-Liouville transform pair

becomes

$$F(t) = \int_0^\infty f(x)\sqrt{x}\, J_\nu(tx)dx \; ,$$
$$f(x) = \int_0^\infty \sqrt{x}\, F(t)J_\nu(xt)tdt \; .$$

For $0 < \nu < 1$, we have

$$F(t) = \int_0^\infty \sqrt{x}\left[cJ_\nu(tx) - t^{2\nu}J_{-\nu}(tx)\right] f(x)dx \; ,$$
$$f(x) = \int_0^\infty \sqrt{x}\left[\frac{cJ_\nu(xt) - t^{2\nu}J_{-\nu}(xt)}{c^2 - 2ct^{2\nu}\cos(\nu\pi) + t^{4\nu}}\right] tdt \; ,$$

for $c < 0$.

(c) **The Weber Transform:**
Consider the differential equation (26.10) on (a, ∞) with $a > 0$ and $q(x)$ as in the preceding example, but with $\lambda = t^2$ and $\alpha = 0$ in the boundary condition at a. Then the Sturm-Liouville transform pair takes the form

$$F(t) = \int_a^\infty \sqrt{x}\, f(x)\left[J_\nu(tx)Y_\nu(at) - Y_\nu(tx)J_\nu(at)\right] dx \; ,$$
$$f(x) = \int_0^\infty \sqrt{x}\left(\frac{J_\nu(xt)Y_\nu(at) - Y_\nu(xt)J_\nu(at)}{J_\nu^2(at) + Y_\nu^2(at)}\right) tdt \; .$$

(d) **The Kontorovich-Lebedev Transform:**
If $q(x) = e^{2x}$ on $(-\infty, \infty)$, then

$$F(t) = \int_{-\infty}^\infty \frac{f(x)}{x} K_{it}(x)dx \; ,$$

and

$$f(x) = \frac{2}{\pi^2}\int_0^\infty t\sinh(\pi t)F(t)K_{it}(x)dt \; .$$

2. **Discrete Spectrum**

(a) **The Discrete Jacobi Transform:**
If

$$q(x) = \frac{\frac{1}{4} - \alpha^2}{4\sin^2\frac{x}{2}} + \frac{\frac{1}{4} - \beta^2}{4\cos^2\frac{x}{2}} \; , \qquad \alpha, \beta > -1$$

on $0 < x < \pi$, then there exists eigenvalues $\lambda_n = (n + \gamma)^2$, $n = 0, 1, 2, \ldots$, where $2\gamma = \alpha + \beta + 1$. The corresponding eigenfunctions are

$$\psi_n(x) = \left(\sin\frac{x}{2}\right)^{\alpha+\frac{1}{2}}\left(\cos\frac{x}{2}\right)^{\beta+\frac{1}{2}} P_n^{(\alpha,\beta)}(\cos x) \; ,$$

where $P_n^{(\alpha,\beta)}(x)$ is the Jacobi polynomial of degree n.

The Sturm-Liouville transform pair reduces to the Jacobi series expansion.

(b) **The Discrete Laguerre Transform:**

If $q(x) = x^2 + (\alpha^2 - \frac{1}{4})/x^2$ on $(0, \infty)$, then the eigenvalues are $\lambda_n = 4n + 2\alpha + 2$ and the corresponding eigenfunctions are the Laguerre functions

$$\psi_n(x) = x^{\alpha+\frac{1}{2}} e^{-x^2/2} L_n^\alpha(x^2) ,$$

where $L_n^\alpha(x)$ is the generalized Laguerre polynomial. The Sturm-Liouville transform pair is reduced to the Laguerre series expansion.

(c) **The Discrete Hermite Transform:**

If $q(x) = x^2$, the eigenvalues are $\lambda_n = 2n + 1$ and the eigenfunctions are

$$\psi_n(x) = e^{-x^2/2} H_n(x) ,$$

where $H_n(x)$ is the Hermite polynomial of degree n. The Sturm-Liouville transform pair is the Hermite series expansion.

26.5.2 Continuous Transformations Associated with Discrete Spectra

The aforementioned discrete transformations arise from singular Sturm–Liouville boundary-value problems; therefore, they may also be generated from their corresponding continuous transformations as suggested by Theorems 26.3 and 26.4.

1. **The Finite Continuous Jacobi Transform:**

Let

$$R_t^{(\alpha,\beta)}(z) = {}_2F_1\left(-t, t + 2\gamma; \alpha + 1; \frac{1-z}{2}\right) .$$

The function $R_t^{(\alpha,\beta)}$ is called the Jacobi function of the first kind. It reduces to a multiple of the Jacobi polynomial of degree n when $\lambda = n$.

The continuous Jacobi transform of a function $f(x)$ is defined by

$$F^{(\alpha,\beta}(\lambda) = \int_0^\pi f(x) \left(\sin\frac{x}{2}\right)^{\alpha+\frac{1}{2}} \left(\cos\frac{x}{2}\right)^{\beta+\frac{1}{2}} R_{\sqrt{\lambda}-\gamma}^{(\alpha,\beta)}(\cos x) dx . \tag{26.17}$$

If the integral converges absolutely and uniformly in λ, then

$$F(\lambda) = \begin{cases} \sum_{n=0}^{\infty} F\left((n+\gamma)^2\right) \dfrac{(-1)^{n+1} 2(n+\gamma)\Gamma(n+2\gamma)}{\Gamma(\gamma+\sqrt{\lambda})\Gamma(\gamma-\sqrt{\lambda})\left[\lambda-(n+\gamma)\right]^2 \Gamma(1+n)}, \\ F(0)\frac{\sin \pi\sqrt{\lambda}}{\pi\sqrt{\lambda}} + \sum_{n=1}^{\infty} F(n^2) \dfrac{2\sqrt{\lambda}\, \sin \pi(\sqrt{\lambda}-n)}{\pi(\lambda-n^2)}, \end{cases}$$

depending on whether $\gamma \neq 0$ or $\gamma = 0$.

To invert the continuous Jacobi transform, let us introduce the normalized Jacobi function of the first kind

$$\phi_t^{(\alpha,\beta)}(x) = \frac{\Gamma(t+2\gamma)}{\Gamma(\alpha+1)\Gamma(\beta+1)\Gamma(t+1)}\, {}_2F_1\left(-t, t+2\gamma; \alpha+1; \frac{1-z}{2}\right) .$$

By a change of variable and a scalar multiplication we may convert (26.17) into the modified continuous Jacobi transform

$$F(t) = F^{(\alpha,\beta)}(t) = \frac{1}{2^{2\gamma}} \int_{-1}^{1} (1-x)^{\alpha}(1+x)^{\beta} f(x) \phi_t^{(\alpha,\beta)}(x) dx \,. \quad (26.18)$$

THEOREM 26.5 (**Walter–Zayed** [241])

Let f be a C^{∞}-function with support in $(-1,1)$, and 2γ be a positive integer, $\beta > -1$. Let F be its modified Jacobi transform (26.18). Then

1) F is an entire function in t satisfying $F(-t-\gamma) = F(t-\gamma)(-1)^{2\gamma+1}$,

2)

$$F(t) = \sum_{n=0}^{\infty} F(n) S_n^{\gamma}(t) \,,$$

where

$$S_n^{\gamma}(t-\gamma) = \frac{\sin \pi(t-n-\gamma)(2n+2\gamma)}{\pi\left(t^2-(n+\gamma)^2\right)} \,.$$

3) The inverse transform is given by

$$f(x) = \begin{cases} C_{\alpha,\beta} \int_0^{\infty} F(t-\gamma)\phi_{t-\gamma}^{(\beta,\alpha)}(-x) \frac{t \sin \pi t}{(t^2-(1/2)^2)\ldots(t^2-(\gamma-1)^2)} dt \\ C_{\alpha,\beta} \int_0^{\infty} F(t-\gamma)\phi_{t-\gamma}^{(\beta,\alpha)}(-x) \frac{\cos \pi t}{(t^2-1)\ldots(t^2-(\gamma-1)^2)} dt \,, \end{cases}$$

depending on whether 2γ is odd or even, where

$$C_{\alpha,\beta} = \begin{cases} (-1)^{\gamma-\frac{1}{2}} 4\Gamma(\alpha+1)\Gamma(\beta+1) & \text{if } 2\gamma \text{ is odd} \\ (-1)^{\gamma} 4\Gamma(\alpha+1)\Gamma(\beta+1) & \text{if } 2\gamma \text{ is even.} \end{cases}$$

The integrals converge uniformly for x in the interior of $(-1, 1)$.

The inversion formula for the finite continuous Jacobi transform was extended to general α and β by Koornwinder and Walter [129].

The finite continuous Jacobi transform was extended to generalized functions by A. Zayed [261] using the embedding method. A characterization and inversion formula for it (when $\alpha+\beta=0$) are also given in [52]. Later A. Zayed and E. Deeba [267] extended Koornwinder and Walter's result to generalized functions as well.

2. **The Continuous Laguerre Transform with a Squared Argument:**

Consider the Whittaker function $W_{k,\mu}(x)$ defined by

$$W_{k,\mu}(x) = e^{-x/2}x^{c/2}\Psi(a, c; x) ,$$

where Ψ is the confluent hypergeometric function, $k = -a + \frac{c}{2}, \mu = \frac{c}{2} - \frac{1}{2}$. The function $\phi_0(x, \lambda) = x^{-\frac{1}{2}}W_{\frac{\lambda}{4},\frac{\alpha}{2}}(x^2)$ reduces to the Laguerre functions $\psi_n(x) = x^{\alpha+\frac{1}{2}}e^{-x^2/2}L_n^\alpha(x^2)$, when $\lambda = 4n + 2\alpha + 2$. The continuous Laguerre transform is defined by

$$F(\lambda) = \int_0^\infty f(x)x^{-\frac{1}{2}}W_{\frac{\lambda}{4},\frac{\alpha}{2}}(x^2)dx .$$

It can be shown that if $f \in L^2(\mathcal{R})$ and $f(x) = 0(x^\delta)$ as $x \to 0^+$ where $\delta > \alpha - \frac{3}{2}$, then F is an entire function that admits the following sampling expansion

$$F(\lambda) = \sum_{n=0}^{\infty} F(4n+2\alpha+2)\frac{4(-1)^{n+1}\exp\left[\frac{1}{4}\psi\left(\frac{\alpha+1}{2}\right)(4n+2\alpha+2)-\lambda\right]}{\Gamma\left(\frac{\alpha+1}{2}-\frac{\lambda}{4}\right)\Gamma(n+1)\left[\lambda-(4n+2\alpha+2)\right]}$$

or

$$\tilde{F}(\lambda) = \sum_{n=0}^{\infty} \tilde{F}(n)\frac{\sin\pi(\lambda-n)}{\pi(\lambda-n)}$$

where

$$\tilde{F}(\lambda) = \frac{F(4\lambda+2\alpha+2)}{\Gamma(1+\lambda)}\exp\left(\lambda\psi\left(\frac{\alpha+1}{2}\right)\right) ,$$

and ψ is the logarithmic derivative of the Γ-function.

3. **The Continuous Laguerre Transform:**

Consider the differential equation

$$xy'' + y' - \left(\frac{x}{4} + \frac{\alpha^2}{4x}\right)y = -\lambda y , \quad 0 < x < \infty , \quad \alpha > -1 .$$

with

$$y(0) = 0 \,, \quad |y(\infty)| < \infty \,.$$

The eigenvalues are $\lambda_n = n + (\alpha+1)/2$ and the eigenfunctions are $\psi_n(x) = x^{\alpha/2} e^{-x/2} L_n^\alpha(x)$. The continuous Laguerre transform is defined in terms of the function

$$\phi_0(x,\lambda) = x^{\frac{\alpha}{2}} e^{-\frac{x}{2}} \Psi\left(\frac{\alpha+1}{2} - \lambda, \alpha+1; x\right)$$

which reduces to the eigenfunctions when $\lambda = \lambda_n$. The continuous Laguerre transform is then defined by

$$F(\lambda) = \int_0^\infty f(x) x^{\frac{\alpha}{2}} e^{-\frac{x}{2}} \Psi\left(\frac{\alpha+1}{2} - \lambda, \alpha+1; x\right) dx \,.$$

If $f \in L^2(\mathcal{R})$ and of order x^δ as $x \to 0^+$ where $\delta > \frac{\alpha}{2} - 1$, then F is an entire function that admits the sampling expansion

$$F(\lambda) = \sum_{n=0}^\infty F\left(n + \frac{\alpha+1}{2}\right) \frac{(-1)^{n+1} \exp\left\{\psi\left(\frac{\alpha+1}{2}\right)\left[n + \frac{\alpha+1}{2} - \lambda\right]\right\}}{\Gamma\left(\frac{\alpha+1}{2} - \lambda\right)\Gamma(1+n)\left[\lambda - \left(n + \frac{\alpha+1}{2}\right)\right]}$$

or

$$\tilde{F}(\lambda) = \sum_{n=0}^\infty \tilde{F}(n) \frac{\sin \pi(\lambda - n)}{\pi(\lambda - n)} \,,$$

where

$$\tilde{F}(\lambda) = \frac{F\left(\lambda + \frac{\alpha+1}{2}\right) \exp\left(\lambda x^{(\alpha+1)/2}\right)}{\Gamma(1+\lambda)} \,.$$

4. **The Continuous Hermite Transform:**

   Consider the differential equation

$$\begin{aligned} & y'' - x^2 y = -\lambda y \,, \; -\infty < x < \infty \\ & |y(\pm\infty)| < \infty \,. \end{aligned}$$

   The eigenvalues are $\lambda_n = 2n+1$, and the eigenfunctions are $\psi_n(x) = e^{-x^2/2} H_n(x)$, $n = 0, 1, 2, \ldots$, where $H_n(x)$ is the Hermite polynomial of degree n.

   Consider the parabolic cylindrical function

$$D_\lambda(x) = 2^{\frac{1}{4}+\frac{\lambda}{2}} x^{-\frac{1}{2}} W_{\frac{1}{4}+\frac{\lambda}{2}, \frac{1}{4}}\left(\frac{x^2}{2}\right) \,,$$

   and set

$$\phi_0(x,\lambda) = 2^{(1-\lambda)/4} D_{(\lambda-1)/2}(\sqrt{2}\, x) \,.$$

When $\lambda = \lambda_n$, the function $\phi_0(x, \lambda)$ reduces to a multiple of the eigenfunction, namely $2^{-n}\psi_n(x)$. The continuous Hermite transform is defined by

$$F(\lambda) = 2^{(1-\lambda)/4} \int_0^\infty f(x) D_{(\lambda-1)/2}(\sqrt{2}\, x)\, dx .$$

Unlike the aforementioned transforms, the Hermite transform reduces to the discrete Hermite transform when $\lambda = \lambda_n$ only for square-integrable functions vanishing on $(-\infty, 0)$. The reason is the function $\phi_0(x, \lambda)$ is not in $L^2(\mathcal{R})$, unless $\lambda = \lambda_n$. For more details, see [257, p. 145-146].

If $f \in L^2(\mathcal{R}^+)$, where $\mathcal{R}^+ = [0, \infty)$, and the integral converges uniformly, then $F(\lambda)$ is entire and can be written in the form

$$F(\lambda) = \sum_{n=0}^{\infty} F(4n+3) \frac{4(-1)^{n+1} \exp\left\{\frac{1}{4}\psi\left(\frac{3}{4}\right)[(4n+3) - \lambda]\right\}}{\Gamma\left((3-\lambda)/4\right)\Gamma(n+1)[\lambda - (4n+3)]}$$

or

$$\tilde{F}(\lambda) = \sum_{n=0}^{\infty} \tilde{F}(n) \frac{\sin \pi(\lambda - n)}{\pi(\lambda - n)}$$

where

$$\tilde{F}(\lambda) = \frac{F(4\lambda + 3) e^{\lambda\psi(3/4)}}{\Gamma(1+\lambda)} .$$

26.6 Applications

26.6.1 The Discrete Sturm–Liouville Transform

1. Perturbed Diffusion Equation: Consider the perturbed diffusion equation

$$Lu(x, t) = \frac{\partial u(x, t)}{\partial t}, \quad a \le x \le b, \quad t \ge 0,$$

subject to the conditions

$$\begin{aligned} \cos\alpha u(a, t) + \sin\alpha u_x(a, t) &= 0 \\ \cos\beta u(b, t) + \sin\beta u_x(b, t) &= 0 \\ u(x, 0) &= f(x), \end{aligned}$$

where L is the Sturm–Liouville differential operator. It is easy to see that

$$u(x, t) = \sum_{n=0}^{\infty} a_n \phi_n(x) e^{-\lambda_n t}$$

is a solution of the above problem, where λ_n , $\phi_n(x)$ are the eigenvalues and eigenfuncions of the Sturm–Liouville boundary-value problem (26.3)–(26.5) and

$$a_n = \frac{1}{\|\phi_n\|^2} \int_a^b f(x)\phi_n(x)\,dx.$$

2. Vibration of a Variable String: Consider the equation of a vibrating string

$$u_{xx} - q(x)u = u_{tt}\,, \quad a \le x \le b\,,\ 0 \le t$$

subject to the conditions

$$\begin{aligned} u(a,t)\cos\alpha + \frac{\partial u(a,t)}{\partial x}\sin\alpha &= 0 \\ u(b,t)\cos\beta + \frac{\partial u(b,t)}{\partial x}\sin\beta &= 0\,, \end{aligned}$$

$$\begin{aligned} u(x,0) &= f(x), \\ \frac{\partial u(x,0)}{\partial t} &= 0, \end{aligned}$$

where $q(x)$ represents some external force.

By writing $u(x,t) = X(x)T(t)$ and using the method of separation of variables, we find that X and T satisfy

$$X'' - q(x)X = -\lambda X$$

$$\begin{aligned} X(a)\cos\alpha + X'(x)\sin\alpha &= 0 \\ X(b)\cos\beta + X'(b)\sin\beta &= 0\,, \end{aligned}$$

and

$$T'' = -\lambda T\,, \qquad \text{with } T'(0) = 0.$$

Solving these two problems leads to the solution

$$u(x,t) = \sum_{n=0}^{\infty} a_n\phi_n(x)\cos(\sqrt{\lambda_n}t),$$

where

$$a_n = \int_a^b f(x)\phi_n(x)\,dx/\|\phi_n\|^2,$$

λ_n and ϕ_n are the eigenvalues and eigenfunctions of the boundary-value problem satisfied by $X(x)$.

26.6.2 The Finite Continuous Jacobi Transform

Applications similar to those given in the foregoing section can be easily found for the continuous Sturm–Liouville transform. In this section, we outline some applications of a continuous Sturm–Liouville transform associated with a discrete spectrum, as explained in Sections 26.2 and 26.3. The finite continuous Jacobi transform and its inversion formula, for example, can be used to solve Cauchy's problem for a perturbed heat equation and the generalized biaxially symmetric Schrödinger equation.

Example A: (Perturbed Heat Equation) Consider the partial differential equation

$$\frac{\partial u(\theta,T)}{\partial T} = D_\theta^{(\alpha,\beta)} u(\theta,T) + c^2 u(\theta,T)\,, \quad 0<\theta<\pi, T\geq 0$$

where

$$D_\theta^{(\alpha,\beta)} = \frac{1}{\omega_{\alpha,\beta}(\theta)}\frac{d}{d\theta}\left(\omega_{\alpha,\beta}(\theta)\frac{d}{d\theta}\right),$$

and

$$\omega_{\alpha,\beta}(\theta) = \left(\sin\frac{\theta}{2}\right)^{2\alpha+1}\left(\cos\frac{\theta}{2}\right)^{2\beta+1}.$$

This differential equation can be solved by applying the continuous Jacobi transform and using the relation

$$D_\theta^{(\alpha,\beta)}\left(R_{t-\gamma}^{(\alpha,\beta)}(\cos\theta)\right) = (\gamma^2 - t^2) R_{t-\gamma}^{(\alpha,\beta)}(\cos\theta).$$

For more details, see [267].

Example B: (The Generalized Biaxially Symmetric Schrödinger Equation) The generalized biaxially symmetric Schrödinger equation is given by

$$\frac{\partial^2\psi}{\partial x^2} + \frac{\partial^2\psi}{\partial y^2} + \frac{2\nu}{x}\frac{\partial\psi}{\partial x} + \frac{2\mu}{y}\frac{\partial\psi}{\partial y} + [k^2 - V(r)]\psi = 0\,, \tag{26.19}$$

where $\nu,\mu\geq 0$, $V(r)=\sum_{n=0}^{\infty} v_n r^{n-1}$, $V_0>0\,,\ r=\sqrt{x^2+y^2}$.

To see how the continuous Jacobi transform arises in connection with this problem we introduce polar coordinates to convert the above equation into

$$\frac{\partial^2\psi}{\partial r^2} + \frac{1}{r^2}\frac{\partial^2\psi}{\partial\theta^2} + \left(\frac{2\nu+2\mu+1}{r}\right)\frac{\partial\psi}{\partial r} + \frac{1}{r^2}\left(2\mu\cot\theta - 2\nu\tan\theta\right)\frac{\partial\psi}{\partial\theta}$$
$$+[k^2 - V(r)]\psi = 0.$$

Setting $\psi(r,\theta) = R(r)\Theta(\theta)$ and using the method of separation of variables, we obtain

$$S_{\mu,\nu}(R) = [4t^2 - (\mu+\nu)^2]R\,,$$

and

$$\frac{1}{\omega_{\mu-\frac{1}{2},\nu-\frac{1}{2}}(2\theta)}\frac{d}{d\theta}\omega_{\mu-\frac{1}{2},\nu-\frac{1}{2}}(2\theta)\frac{d\Theta}{d\theta}$$
$$= 4D_{2\theta}^{(\mu-\frac{1}{2},\nu-\frac{1}{2})}\Theta = \left[(\mu+\nu)^2 - 4t^2\right]\Theta\,,$$

where the separation constant has been set equal to $(\mu+\nu)^2 - 4t^2$ and

$$S_{\mu,\nu} = r^2\frac{d^2}{dr^2} + (2\mu+2\nu+1)r\frac{d}{dr} + r^2\left(k - V(r)\right)\,.$$

Note that Equation (26.19) can be written in the form

$$\left(S_{\mu,\nu} + 4D_{2\theta}^{(\mu-1/2\,,\,\nu-1/2)}\right)\psi(r,\theta) = 0\,.$$

The last equation can now be solved as in Example A. For more details, see [267].

27

Miscellaneous Transforms

The number of integral transforms cited in the literature is overwhelming. Every conceivable special function has been or is bound to be the kernel of some integral transformation. But whether that transformation is of any utility is not certain. Fortuitously, one does not need to know all of these transforms because a great number of them are either superficial generalizations, or combinations, or iterations of other known transforms. For example, the so-called *Glasser transform* [74], which is defined by

$$F(y) = \int_{-\infty}^{\infty} \frac{f(x)}{(x^2+y^2)^{\frac{1}{2}}} dx \ ,$$

is nothing more than an iteration of the Fourier cosine and the $\mathcal{K}$- transforms as can be seen from the relation

$$\frac{1}{(x^2+y^2)^{\frac{1}{2}}} = \frac{2}{\pi} \int_{-\infty}^{\infty} K_0(xt) \cos yt \, dt \ .$$

Thus,

$$\begin{aligned} F(y) &= \frac{2}{\pi} \int_{-\infty}^{\infty} f(x) \left(\int_{-\infty}^{\infty} K_0(xt) \cos yt \, dt \right) dx \\ &= \frac{2}{\pi} \int_{-\infty}^{\infty} \cos yt \left(\int_{-\infty}^{\infty} f(x) K_0(xt) dx \right) dt \\ &= \sqrt{2/\pi} \, \mathcal{F}_c \left[t^{-\frac{1}{2}} \mathcal{K}_0[g(x)](t) \right] (y) \end{aligned}$$

where $g(x) = x^{-\frac{1}{2}} f(x)$.

In this chapter we collect some transforms that are worth mentioning because of their generality, but are not worth being in separate chapters because of their limited utility. These transforms are of very general nature. The *hypergeometric transform*, for example, is in fact a whole class of integral transforms, and the remarkable E and G transforms are so general that many known integral transforms are included in them as special cases. One disappointing feature of these transforms, however, is that their great generality seems to preclude their applicability.

27.1 The Hypergeometric Transforms

By a hypergeometric transformation we mean any linear integral transformation whose kernel involves the hypergeometric function ${}_pF_q$. This class of transformations encompasses a great number of transformations; some are well known, some are less known, and the rest are either obscure or unnamed. Examples of those that are well known are the Laplace, Stieltjes, continuous Jacobi, and Legendre transforms. Examples of less known transforms are:

1. **The Kummer Transforms:**

   A) **The Kummer–Erdélyi Transform:**

   Erdélyi [59] introduced the following transform which generalizes the Laplace transform. For a suitably restricted function $f(x)$, define the *Kummer transform* as

$$F_\beta^\alpha(u) = \frac{\Gamma(\alpha)}{\Gamma(\beta)} u \int_0^\infty {}_1F_1(\alpha;\beta;-ux)(ux)^\gamma f(x)dx\ ,$$

   where ${}_1F_1$ denotes Kummer's (confluent) hypergeometric function, $\alpha, \beta \neq 0, -1, -2, \ldots$, Re $u > 0$.

   This transform is sometimes referred to as the *Erdélyi transform.* If $\alpha = \beta$ and $\gamma = 0$, we obtain

$$\frac{F_\alpha^\alpha(u)}{u} = \int_0^\infty e^{-ux} f(x)dx = \mathcal{L}[f](u)\ ,$$

   where $\mathcal{L}$ denotes the Laplace transformation. The Kummer–Erdélyi transform was extended to generalized functions by Rao [202].

   B) **The Kummer Transform of Two Variables:**

   This transform was introduced in [237].

DEFINITION 27.1 *The Kummer transform of two variables of a function $f(x, y)$ is defined as*

$$F_{\beta,\delta}^{\alpha,\gamma}(u,v) = \frac{\Gamma(\alpha)\Gamma(\gamma)}{\Gamma(\beta)\Gamma(\delta)} uv \int_0^\infty \int_0^\infty {}_1F_1(\alpha;\beta;-ux)\,{}_1F_1(\gamma;\delta;-vy)f(x,y)d \tag{27.1}$$

where $\alpha, \beta, \gamma, \delta \neq 0, -1, -2, \ldots$, Re $u > 0$ and Re $v > 0$.

When $\alpha = \beta$, $\gamma = \delta$, we obtain

$$\frac{F_{\alpha,\gamma}^{\alpha,\gamma}(u,v)}{uv} = \int_0^\infty \int_0^\infty e^{-(ux+vy)} f(x,y)dxdy = \mathcal{L}[f](u,v) ,$$

where the right-hand side is the two-dimensional Laplace transform of f. If $f(x,y) = g(x)h(y)$, the Kummer transform of two variables is reduced to a product of Kummer transforms of one variable. This transform was also extended to generalized functions by Rao [201].

2. **The Generalized Whittaker Transform:**

   The *generalized Whittaker transform* of a function $f(t)$, as introduced by Srivastava [218], is defined by

   $$F_{k,m,\sigma}^{p,q}(x) = \int_0^\infty (xt)^{\sigma-\frac{1}{2}} \exp\left(-\frac{qxt}{2}\right) W_{k,m}(pxt)f(t)dt ,$$

   where $W_{k,m}$ is the Whittaker function defined by

   $$W_{k,m}(x) = e^{-x/2}x^k {}_2F_0\left(\frac{1}{2} - k + m, \frac{1}{2} - k - m; -\frac{1}{x}\right) .$$

   The generalized Whittaker transform reduces to the so-called *Varma transform* [235] when $\sigma = m\,, q = 1\,, p = 1$; to the *Whittaker transform* when $\sigma = 1/2\,, q = 0\,, p = 1$; and to the Laplace transform when $\sigma = m\,, k = \frac{1}{2} - m\,, p = q = 1$. The generalized Whittaker transform was extended to generalized functions by Tiwari and Ko [230].

3. **A Generalized Stieltjes Transform:**

   This generalization of the Stieltjes transform, which was introduced by Arya [4], is defined by

   $$F(x) = \frac{\Gamma(2\mu+1)}{\Gamma(\mu-\nu+\frac{3}{2})x} \int_0^\infty {}_2F_1\left(2\mu+1\,,1\,;\mu-\nu+\frac{3}{2}\,;-\frac{t}{x}\right) f(t)dt ,$$

   where $\mathrm{Re}(2\mu+1) > 0$. It reduces to the Stieltjes transform when $\nu = \frac{1}{2} - \mu$. An extension of this transform to generalized functions was considered in [200].

   More transforms of the above types and their extension to generalized functions can be found in [33].

4. **The Index ${}_2F_1$-Transform:**

   Another type of hypergeometric transformations is the one in which the transform variable is one of the parameters of the hypergeometric function. A case in mind is the so-called *Index* ${}_2F_1$*-transform.* It is defined by

$$F(\tau) = \int_0^\infty f(x)\mathbf{F}(\mu, \alpha, \tau, x)dx \ ,$$

where

$$\mathbf{F}(\mu, \alpha, \tau, x) ={}_2 F_1\left(\mu + \frac{1}{2} + i\tau, \mu + \frac{1}{2} - i\tau; \mu + 1; -x\right) x^\alpha \ ,$$

where α, μ are complex parameters and $\tau > 0$. For its inversion formula and extension to generalized functions, see [94]. The inversion formula is given by

$$f(x) = \int_0^\infty S(\mu, \tau)G(\mu, \alpha, \tau, x)F(\tau)d\tau \ ,$$

with

$$S(\mu, \tau) = \frac{2}{\pi\Gamma^2(\mu + 1)}\tau \sinh(\pi\tau)\Gamma(\mu + \frac{1}{2} + i\tau)\Gamma(\mu + \frac{1}{2} - i\tau) \ ,$$

and

$$G(\mu, \alpha, \tau, x) = x^{\mu-\alpha}\,{}_2F_1\left(\frac{1}{2} + i\tau, \frac{1}{2} - i\tau; \mu + 1; -x\right) \ ,$$

where $\operatorname{Re}\alpha > 0, \operatorname{Re}\mu > 0, 1/8 < \operatorname{Re}(\mu - \alpha) < 1/4$ and $\operatorname{Re}(\mu - 2\alpha) < -1$, for some suitably restricted function f.

One of the most common techniques used to invert hypergeometric transformations is based on the inversion formula for the Mellin transformation. We might also add that the inversion formulas of the E and the G transformations are both based on the inversion formula for the Mellin transformation. To demonstrate this technique, we shall close this section with a derivation of the inversion formula for the Kummer transformation of two variables.

Assume that $u^{-r-1}v^{-s-1}F^{\alpha,\gamma}_{\beta,\delta}(u, v) \in L^1(\mathcal{R}^2)$ and consider

$$\begin{aligned}
I &= \int_0^\infty \int_0^\infty u^{-r-1}v^{-s-1}F^{\alpha,\gamma}_{\beta,\delta}(u, v)dudv \\
&= \frac{\Gamma(\alpha)\Gamma(\gamma)}{\Gamma(\beta)\Gamma(\delta)} \int_0^\infty \int_0^\infty u^{-r}v^{-s}du\,dv \\
&\int_0^\infty \int_0^\infty {}_1F_1(\alpha; \beta; -ux){}_1F_1(\gamma; \delta; -vy)f(x, y)dxdy \\
&= \frac{\Gamma(\alpha)\Gamma(\gamma)}{\Gamma(\beta)\gamma(\delta)} \int_0^\infty \int_0^\infty x^{r-1}y^{s-1}f(x, y)dxdy \\
&\int_0^\infty p^{-r}\,{}_1F_1(\alpha; \beta; -p)dp \int_0^\infty q^{-s}\,{}_1F_1(\gamma; \delta; -q)dq \ ,
\end{aligned}$$

which, in view of the formula (cf. [79, p. 285])

$$\int_0^\infty t^{b-1}\,{}_1F_1(a;c;-t)dt = \frac{\Gamma(b)\Gamma(c)\Gamma(a-b)}{\Gamma(a)\Gamma(c-b)}\,, \qquad 0 < \mathrm{Re}\, b < \mathrm{Re}\, a\,,$$

leads to

$$I = \frac{\Gamma(1-r)\Gamma(\alpha-1+r)}{\Gamma(\beta-1+r)}\,\frac{\Gamma(1-s)\Gamma(\gamma-1+s)}{\Gamma(\delta-1+s)}\int_0^\infty\int_0^\infty x^{r-1}y^{s-1}f(x,y)dxdy\,.$$

Therefore

$$\begin{aligned}&\int_0^\infty\int_0^\infty x^{r-1}y^{s-1}f(x,y)dxdy\\ &= \int_0^\infty\int_0^\infty u^{-r-1}v^{-s-1}\frac{F_{\beta,\delta}^{\alpha,\gamma}(u,v)}{H_1(1-r)H_2(1-s)}dudv = G(r,s)\,,\end{aligned}$$

provided that $\mathrm{Re}\, r > 1-\alpha\,, \mathrm{Re}\, s > 1-\gamma$, where

$$H_1(r) = \frac{\Gamma(r)\Gamma(\alpha-r)}{\Gamma(\beta-r)}\,, \qquad H_2(\delta) = \frac{\Gamma(s)\Gamma(\gamma-s)}{\Gamma(\delta-s)}\,.$$

From the inversion formula of the Mellin transform, we have

$$\begin{aligned}f(x,y) &= \frac{1}{(2\pi i)^2}\int_{c-i\infty}^{c+i\infty}\int_{d-i\infty}^{d+i\infty} x^{-r}y^{-s}G(r,s)drds\\ &= \frac{1}{(2\pi i)^2}\int_{c-i\infty}^{c+i\infty}\int_{d-i\infty}^{d+i\infty} x^{-r}y^{-s}drds\int_0^\infty\int_0^\infty\\ &\qquad u^{-r-1}v^{-s-1}\frac{F_{\beta,\delta}^{\alpha,\gamma}(u,v)}{H_1(1-r)H_2(1-s)}dudv\,.\end{aligned}$$

By interchanging the integration signs, we obtain the following inversion formula of (27.1): if $F_{\beta,\delta}^{\alpha,\gamma}$ is given by (27.1), then

$$f(x,y) = \int_0^\infty\int_0^\infty F_{\beta,\delta}^{\alpha,\gamma}(u,v)g(xu,yv)\frac{du}{u}\frac{dv}{v}\,,$$

where

$$g(x,y) = \frac{1}{(2\pi i)^2}\int_{c-i\infty}^{c+i\infty}\int_{d-i\infty}^{d+i\infty}\frac{x^{-r}y^{-s}}{H_1(1-r)H_2(1-s)}drds\,.$$

Parseval's relation for the Kummer transform of two variables takes the form

$$\int_0^\infty\int_0^\infty F_{\beta,\delta}^{\alpha,\gamma}(u,v)g(u,v)\frac{du}{u}\frac{dv}{v} = \int_0^\infty\int_0^\infty G_{\beta,\delta}^{\alpha,\gamma}(u,v)f(u,v)\frac{du}{u}\frac{dv}{v}\,,$$

where $G_{\beta,\delta}^{\alpha,\gamma}$ is related to g as $F_{\beta,\delta}^{\alpha,\gamma}$ is related to f. Parseval's relation can be established directly by substituting $F_{\beta,\delta}^{\alpha,\gamma}$ into the left-hand side, then interchanging the integral signs.

27.2 The E transform

The *E transform* was introduced by the Egyptian mathematician F. M. Ragab in 1967 [198]. He obtained an inversion formula for it which involves the generalized MacRobert E function. The transform and its inversion formula form what is called the *E transform pair.* It turns out that the E transform pair generalizes very nicely many known and important integral transform pairs. It is somewhat surprising that this transform has not received much attention. This may be attributed to the complexity of the generalized MacRobert E function and to the fact that the E transform is superseded by the G transform.

Although the generalized MacRobert E function can be written in terms of Meijer's G function, we shall introduce it exactly as Ragab did, but we shall use different notations for simplicity.

DEFINITION 27.2 *The generalized MacRobert E-function is defined as*

$$E_{q,l+1}^{p,m}\begin{pmatrix}\alpha_r;\, \rho_{q+s} & : z \\ \rho_s;\, \alpha_{p+r}; 1 & \end{pmatrix} =$$
$$\frac{1}{2\pi i}\int_\gamma \frac{\prod_{r=1}^{p}\Gamma(\alpha_r-\xi)\prod_{s=1}^{m}\Gamma(\xi-\rho_{q+s}+1)}{\prod_{s=1}^{q}\Gamma(\rho_s-\xi)\prod_{r=1}^{l}\Gamma(\xi-\alpha_{p+r}+1)} z^\xi\, d\xi\,,$$

where l, m, p, q are non-negative integers and the contour, γ, is along the imaginary ξ axis from $-\infty$ to ∞ with loops, if necessary, to ensure that the poles of the integrand at the origin and at $\rho_{q+1}-1, \cdots, \rho_{q+m}-1$ lie to the left and the poles at $\alpha_1, \cdots, \alpha_p$ to the right of the contour. When necessary, the contour is bent to the left or to the right at both ends without crossing any poles of the integrand so that it is ultimately parallel to the real ξ axis. Zero and negative integral values of the α's and the ρ's are excluded; the parameters $\alpha_1, \cdots, \alpha_p$ must not differ by integral values, and likewise the parameters $\rho_{q+1}, \cdots, \rho_{q+m}$ must not differ by integral values.

It is easily seen that the generalized MacRobert E-function reduces to the ordinary E function when $l = 0\,, m = 1$ and $\rho_{q+1} = 1$. For the definition of the E function, see Section 2.6. By using the relation

$$\Gamma(z) = \pi[\Gamma(1-z)\sin(\pi z)]^{-1}$$

and summing the residues at the poles of the integrand, one can show that

$$E_{q,l+1}^{p,m}\begin{pmatrix}\alpha_r;\, \rho_{q+s} & : z \\ \rho_s;\, \alpha_{p+r}; 1 & \end{pmatrix}$$

$$= \pi^{m-l-1} \sum_{s=1}^{m} \frac{\prod_{r=1}^{l} \sin(\rho_{q+s} - \alpha_{p+r})\pi}{\prod'^{m}_{t=1} \sin(\rho_{q+s} - \rho_{q+t})\pi} z^{\rho_{q+s}-1}$$

$$\times E \begin{pmatrix} p+l\,; \alpha_r - \rho_{q+s} + 1 & : \omega z \\ \rho_1 - \rho_{q+s} + 1, \cdots, *, \cdots, \rho_{q+m} - \rho_{q+s} + 1 & \end{pmatrix},$$

where ω is equal to $e^{\pm i\pi}$ or 1 according to $l+m$ is even or odd and the prime in $\prod'$ indicates the omission of the factor $\sin(\rho_{q+s} - \rho_{q+s})$; for more details, see [152].

DEFINITION 27.3 *The E transform of a function $f(x)$ is defined as*

$$F(y) = \int_0^\infty f(x)(yx)^k E(p; \alpha_r : q; \rho_t : (yx)^{\pm n}) dx \, ,$$

where n is any positive integer, $q \neq 0$.

Using symbolic manipulation, Ragab derived the following inversion formula for the E transform without stating the conditions that ensure its validity.

THEOREM 27.1 **(Ragab [198])**
Let $F(y)$ be the E transform of $f(x)$. Then,

$$f(x) = n^2 \int_0^\infty F(y)(xy)^{-k} E^{0,q}_{1,p+1} \begin{pmatrix} ; & 1 - \rho_t \mp 1/n & : (xy)^{\pm n} \\ \mp 1/n; & 1 - \alpha_r \mp 1/n, 1 & \end{pmatrix} dy.$$

The functions $f(x)$ *and* $F(y)$ *are called an E transform pair.* The following are special cases of the E transform pair:

1. **Generalizations of the Hankel Transform Pair:**

   A. For $n = 2, p = 0, q = 1, \rho_1 = \nu + 1$, we have after scaling by a constant that

   $$F(y) = \int_0^\infty f(x)(yx)^{k-\nu} J_\nu(yx) dx \, ,$$

   and

   $$f(x) = \int_0^\infty F(y)(xy)^{-k+\nu+1} J_\nu(xy) dy \, .$$

   In particular, if $k = \nu + \frac{1}{2}$, we obtain the Hankel transform pair which in turn leads to the K-transform pair

   $$F(y) = \int_0^\infty f(x)\sqrt{yx}\, K_\nu(yx) dx \, ,$$

and

$$f(x) = \frac{1}{\pi i} \int_{c-i\infty}^{c+i\infty} F(y)\sqrt{xy}\, I_\nu(xy) dy \, .$$

B. For $n = 2m (m = 1, 2, ...)$, we obtain, in view of the formula

$$E(\alpha; ., x) = \Gamma(\alpha) \left(1 + \frac{1}{x}\right)^{-\alpha} ,$$

another generalization of the Hankel transform pair, namely

$$F(y) = \int_0^\infty f(x)(yx)^{k-\nu m} J_\nu(2y^m x^m) dx \, ,$$

and

$$f(x) = 4m^2 \int_0^\infty F(y)(xy)^{m\nu+2m-k-1} J_\nu(2x^m y^m) dy \, .$$

These two relations can be reduced to the Hankel transform pair by a change of variable.

2. **A Generalization of the Y Transform Pair:** For $n = -2, p = 1, q = 2, \alpha_1 = 1, \rho_1 = \frac{3}{2}$, and $\rho_2 = \frac{3}{2} + \nu$, we obtain the pair

$$F(y) = \int_0^\infty f(x)(yx)^{k-\nu-1} \mathbf{H}_\nu(xy) dx \, ,$$

and

$$f(x) = \int_0^\infty F(y)(xy)^{-k+\nu+2} Y_\nu(xy) dy \, ,$$

which reduces to the Y transform pair when $k = \nu + \frac{3}{2}$.

3. **The KJ Transform Pair:** For $n = 4, p = 0, q = 3, \rho_1 = 1 + \nu, \rho_2 = 1 + \frac{\nu}{2}$, and $\rho_3 = \frac{3}{2} + \frac{\nu}{2}$, we obtain the KJ transform pair

$$F(y) = \frac{1}{\Gamma(1+\nu)\Gamma(1+\frac{\nu}{2})\Gamma(\frac{3}{2}+\frac{\nu}{2})} \int_0^\infty f(x)(yx)^k \, {}_0F_3\left(1+\nu, 1+\frac{\nu}{2}, \frac{3}{2}+\frac{\nu}{2}; -(yx)^4\right) dx \, ,$$

and

$$f(x) = 16\sqrt{2\pi} \int_0^\infty F(y)(xy)^{-k+2\nu+3} K_\nu(2\sqrt{2}\, xy) J_\nu(2\sqrt{2}\, xy) dy \, .$$

4. **The YJ Transform Pair:** For $n = 1, p = 1, q = 2, \alpha_1 = \frac{3}{2} + \nu, \rho_1 = 1 + \nu, \rho_2 = 1 + 2\nu$, we obtain the YJ transform pair

$$F(y) = \frac{\Gamma(\frac{3}{2} + \nu)}{\Gamma(1 + \nu)\Gamma(1 + 2\nu)} \int_0^\infty f(x)(yx)^k {}_1F_2\left(\frac{3}{2} + \nu; 1 + \nu, 1 + 2\nu; -yx\right) dx ,$$

and

$$f(x) = -\sqrt{\pi} \int_0^\infty F(y)(xy)^{-k+\nu} J_\nu(\sqrt{xy}) Y_\nu(\sqrt{xy}) dy .$$

27.3 The G transform

Following Hardy and Titchmarsh's treatment of Fourier-type integrals, R. Narain [174, 175] introduced the G *transform* and its inversion formula as an example of a Fourier-type integral transform pair with unsymmetrical Fourier kernels. The G transform is an integral transform whose kernel is essentially Meijer's G function, which can be considered a generalization of MacRobert's E function. As such, the G transform can be regarded as a generalization of the E transform. The G transform and its inversion formula form the G transform pair. Analogous to the E transform, the G transform pair includes many known integral transform pairs as special cases.

DEFINITION 27.4 *The G transform of a function $f(t)$ is defined as*

$$F(y) = 2\beta\gamma A \int_0^\infty f(x)(yx)^{\gamma - \frac{1}{2}} G^{m,p}_{p+q,m+n}\left(\beta^2 (yx)^{2\gamma} \,\middle|\, \begin{matrix} a_1, ..., a_p, b_1, ..., b_q \\ c_1, ..., c_m, d_1, ..., d_n \end{matrix}\right) dx$$

where $A \neq 0$ is real, $\beta, \gamma > 0$, and $n - p = m - q > 0$.

The inversion formula for the G transform is given in the next theorem.

THEOREM 27.2
If

1. $\gamma > 0, 0 < n - p = m - q = \eta/2,$
2. $\sum_{i=1}^p a_i + \sum_{i=1}^q b_i = \sum_{i=1}^m c_i + \sum_{i=1}^n d_i ,$
3. $Re(\frac{1}{2} - a_i), Re(\frac{1}{2} + b_j) > \eta/2 , \quad i = 1, ..., p , j = 1, ..., q ,$

   $Re(\frac{1}{2} + c_i), Re(\frac{1}{2} - d_j) \geq \eta/2 , \quad i = 1, ..., m , j = 1, ..., n .$

4. $y^{\eta/\gamma-1/2}f(y) \in L^1(0,\infty)$ *and* f *is of bounded variation near* $y = x$ $(x > 0)$, *then*

$$f(x) = 2\beta\gamma A^{-1}\int_0^\infty F(y)(xy)^{\gamma-\frac{1}{2}}$$

$$G^{n,q}_{p+q,m+n}\left(\beta^2(yx)^{2\gamma}\left|\begin{matrix}-b_1,...,-b_q,-a_1,...,-a_p\\-d_1,...,-d_n,-c_1,...,-c_m\end{matrix}\right.\right)dy.$$

The functions f and F are called a G transform pair. Let us denote the kernel of the G transform by $k(x)$ and the kernel of the inverse transform by $h(x)$ so that

$$F(y) = \int_0^\infty f(x)k(yx)dx\ ,$$

$$f(x) = \int_0^\infty F(y)h(xy)dy\ .$$

The following are special cases of the G transform pair:

1. **The Y-transform Pair:**

   The Y-transform pair can be obtained from the G transform pair by choosing

$$k(x) = x^{\frac{1}{2}}G^{1,1}_{1,3}\left(x^2/4\left|\begin{matrix}1/2+\nu/2\\1/2+\nu/2,\nu/2,-\nu/2\end{matrix}\right.\right) = x^{\frac{1}{2}}\mathbf{H}_\nu(x)\ ;$$

   hence

$$h(x) = x^{\frac{1}{2}}G^{2,0}_{1,3}\left(x^2/4\left|\begin{matrix}-1/2-\nu/2\\-\nu/2,\nu/2,-1/2-\nu/2\end{matrix}\right.\right) = x^{\frac{1}{2}}Y_\nu(x)\ .$$

2. **The Hartley-Laplace Transform Pair:**

$$k(x) = \left(\frac{1}{4}\right)x^{\frac{3}{2}}G^{2,0}_{0,4}\left(\left(\frac{x}{4}\right)^4\left|-\frac{3}{8},\frac{3}{8},\frac{1}{8},-\frac{1}{8}\right.\right) = \pi^{-\frac{1}{2}}[\cos x+\sin x+e^{-x}]\ ,$$

$$h(x) = \left(\frac{1}{4}\right)x^{\frac{3}{2}}G^{2,0}_{0,4}\left((\frac{x}{4})^4\left|-\frac{1}{8},\frac{1}{8},\frac{3}{8},-\frac{3}{8}\right.\right) = \pi^{-\frac{1}{2}}[\cos+\sin x-e^{-x}]\ .$$

   This transform pair resembles the Hartley transform pair.

3. **A Hypergeometric Transform Pair:**

   Let

$$k(x) = x^{\frac{1}{2}}\left(\frac{x}{2}\right)^{\frac{3}{2}-\gamma}G^{1,1}_{1,3}\left(x^2/4\left|\begin{matrix}-\alpha_1\\-1,-\rho_1,-\rho_2\end{matrix}\right.\right)$$

$$= x^{\frac{1}{2}}(\frac{x}{2})^{-\gamma-\frac{1}{2}}\,{}_1F_2\left(\alpha_1;\rho_1,\rho_2;-\frac{x^2}{4}\right)$$

then

$$h(x) = x^{\frac{1}{2}} \left(\frac{x}{2}\right)^{\gamma-\frac{3}{2}} G^{2,0}_{1,3}\left(\frac{x^2}{4}\middle| \begin{matrix}\alpha_1\\ \rho_1,\rho_2,1\end{matrix}\right)$$
$$= x^{\frac{1}{2}} \sum_{\rho_1,\rho_2} \frac{\sin(\alpha_1-\rho_1)\pi}{\sin(\rho_2-\rho_1)\pi}\left(\frac{x}{2}\right)^{2\rho_1+\gamma-\frac{3}{2}}$$
$${}_1F_2\left(1-\alpha_1+\rho_1; 1-\rho_2+\rho_1,\rho_1; -\frac{x^2}{4}\right),$$

where $\gamma = \alpha_1 - \rho_1 - \rho_2 + 1/2$ and $\sum_{\rho_1,\rho_2}$ means that to the expression following it a similar expression, in which ρ_1 and ρ_2 are interchanged, is to be added.

4. **The Hardy Transform Pair:**

This is a special case of the hypergeometric transform pair obtained by setting $\alpha_1 = 1, \rho_1 = \alpha+1, \rho_2 = \nu+\alpha+1$ so that $\gamma = -2\alpha-\nu-1/2$. Thus,

$$k(x) = x^{\frac{1}{2}} \left(\frac{x}{2}\right)^{2\alpha+\nu} {}_1F_2\left(1; \alpha+1, \nu+\alpha+1; -\frac{x^2}{4}\right)$$

and

$$h(x) = x^{\frac{1}{2}} \left(\frac{x}{2}\right)^{-2\alpha-\nu-2} G^{2,0}_{1,3}\left(\frac{x^2}{4}\middle| \begin{matrix}1\\ \alpha+1,\nu+\alpha+1,1\end{matrix}\right)$$
$$= x^{\frac{1}{2}} G^{2,0}_{1,3}\left(\frac{x^2}{4}\middle| \begin{matrix}-\alpha-\nu/2\\ -\nu/2,\nu/2,-\alpha-\nu/2\end{matrix}\right)$$
$$= x^{\frac{1}{2}} \left[\cos\alpha\pi J_\nu(x) + \sin\alpha\pi Y_\nu(x)\right],$$

see Chapter 22.

Other integral transform pairs involving the hypergeometric function ${}_pF_q$ can also be obtained, see [33, Ch. 7]. The G transform was extended to generalized functions by Misra [166, 167] and Pathak and Pandy [187].

28

The Bargmann Transform

The states of a quantum mechanical system of n degrees of freedom can be represented by functions of $2n$ variables. One such representation is the phase-space in which n of the variables are the spatial coordinates, $q_1, \cdots, q_n$ and the rest are the momentum coordinates, $p_1, \cdots, p_n$. The representation of a system by functions of n complex variables is also plausible. The use of the complex variables

$$\eta_k = \frac{1}{\sqrt{2}}(q_k - ip_k)\,, \quad \zeta_k = \frac{1}{\sqrt{2}}(q_k + ip_k)\,, \quad k = 1, 2, ..., n\,.$$

has proved to be very fruitful in quantum mechanics and field theory.

If q_k and p_k are self-adjoint operators satisfying the commutation relations

$$[q_k, p_l] = i\delta_{k,l}\,,\ [q_k, q_l] = 0\,,\ [p_k, p_l] = 0\,,$$

then

$$[\zeta_k, \eta_l] = \delta_{k,l}\,,\ [\zeta_k, \zeta_l] = 0\,,\ [\eta_k, \eta_l] = 0\,,$$

and $\zeta_k = \eta_k^*\,,\ \eta_k = \zeta_k^*$.

In analogy to Schrödinger's solution, $p_k = -i\partial/\partial q_k$, of the commutation relation $[q_k, p_k] = i$, V. Foch introduced in 1928 [65] the operator solution $\zeta_k = \partial/\partial\eta_k$ of the commutation relation $[\zeta_k, \eta_k] = 1$, and used it in quantum field theory.

In 1961, V. Bargmann [8] investigated in greater detail the function space **S** on which Foch's solution is realized. He also studied the relationship between the space **S**, which is called the Bargmann-Segal-Foch space, and the Hilbert space of square integrable functions $L^2(\mathcal{R})$. That connection was established using an integral transform that is now known as the Bargmann transform. The Bargmann transform resurfaced again in recent years because of its connection with other important transforms such as the Gabor and Zak transforms [112].

28.1 Definitions

Before we give a definition of the Bargmann transform, we should point out that Bargmann introduced his definition in n dimensions; however, to simplify the notation, we shall restrict ourselves to the case $n = 1$. Except for some obvious modifications, the extension to n dimensions is very similar.

DEFINITION 28.1 *The Bargmann transform $\mathcal{A}[f]$ of a function f is defined by*

$$\mathcal{A}[f](z) = F(z) = \frac{1}{\sqrt[4]{\pi}} \int_{-\infty}^{\infty} f(q) e^{-\frac{1}{2}(z^2+q^2)+\sqrt{2}\, zq} dq = \langle k(z,q), \overline{f}(q) \rangle_{L^2(\mathcal{R})}$$

where $z = x + iy$, and

$$k(z,q) = \frac{1}{\sqrt[4]{\pi}} \exp\left(-\frac{1}{2}(z^2+q^2) + \sqrt{2}\, zq\right), \tag{28.1}$$

q real and z complex.

DEFINITION 28.2 *The Bargmann-Segal-Foch space, $\mathbf{S}$, consists of all entire functions $F(z)$ such that*

$$\int_{-\infty}^{\infty} \int_{-\infty}^{\infty} |F(x+iy)|^2 d\mu(z) < \infty ,$$

where $d\mu(z) = (1/\pi) e^{-|z|^2} dx dy$.

The Bargmann-Segal-Foch space is a Hilbert space in which the inner product is defined by

$$\langle F, G \rangle_{\mathbf{S}} = \langle F, G \rangle = \int_{-\infty}^{\infty} \int_{-\infty}^{\infty} F(z) \overline{G}(z) d\mu(z).$$

28.2 Existence

If f is a locally integrable function such that $f(q) = O(\exp(aq^2))$ as $|q| \to \infty$, for some $a < 1/2$, in particular, if $f \in L^2(\mathcal{R})$, then its Bargmann transform exists almost everywhere.

28.3 Elementary Properties

1. **Linearity:** The Bargmann transformation is linear.

2. **Translation:**

$$\mathcal{A}[f(q+a)](z) = e^{-a^2/4 - az/\sqrt{2}} \mathcal{A}[f(q)](z + \frac{a}{\sqrt{2}}).$$

3. **Modulation**

$$\mathcal{A}[e^{-bq} f(q)](z) = e^{b^2/4 - bz/\sqrt{2}} \mathcal{A}[f(q)](z - \frac{b}{\sqrt{2}}).$$

28.4 Relationship with Other Transforms

1. **The Fourier Transform:**
   If we denote the Fourier transform of f by $\hat{f}$, then

$$\mathcal{A}[\hat{f}](z) = \mathcal{A}[f](iz).$$

2. **The Gabor Transform:**
   Let $\mathcal{G}$ denote the Gabor transformation (cf. Chapter 30), then

$$\mathcal{G}[f](w,t) = F_g[f](w,t) = \frac{1}{\sqrt{2\pi}} e^{-(t^2+w^2)/4 + itw/2} \mathcal{A}[f](\frac{t}{\sqrt{2}} + i\frac{w}{\sqrt{2}}),$$

   where $g(x) = (1/\sqrt[4]{\pi}) \exp(-x^2/2)$.

3. **The Weierstrass Transform:**
   If W denotes the Weierstrass transformation, then

$$\mathcal{A}[f(q)](z/\sqrt{2}) = \sqrt{2}\sqrt[4]{\pi} e^{z^2/4} W\left[f(q/\sqrt{2})\right](\sqrt{2}z).$$

Relations (2) and (2) follow easily from the definitions, and (1) follows from Parseval's relation for the Fourier transform and the relation

$$\mathcal{F}\left[e^{-\alpha^2 x^2 + i\beta x}\right](\omega) = \frac{1}{\sqrt{2}\alpha} \exp\left(-\frac{(\beta+\omega)^2}{4\alpha^2}\right).$$

28.5 The Bargmann-Segal-Foch Space

Recall that the Bargmann-Segal-Foch space is a Hilbert space with inner product

$$\langle F, G\rangle = \int_{-\infty}^{\infty}\int_{-\infty}^{\infty} F(z)\overline{G}(z)d\mu(z).$$

The inner product can also be defined in terms of the Taylor series coefficients of F and G. For, if

$$F(z) = \sum_{n=0}^{\infty} a_n z^n \quad \text{and} \quad G(z) = \sum_{m=0}^{\infty} b_m z^m \,, \qquad z = re^{i\theta}$$

then

$$\begin{aligned}\langle F, G\rangle &= \frac{1}{\pi}\sum_{m,n=0}^{\infty} a_n\bar{b}_m \int_0^{\infty}\left(\int_0^{2\pi} e^{i(n-m)\theta}d\theta\right) r^{n+m+1}e^{-r^2}dr \\ &= 2\sum_{n=0}^{\infty} a_n\bar{b}_n \int_0^{\infty} r^{2n+1}e^{-r^2}dr = \sum_{n=0}^{\infty} n!\, a_n\bar{b}_n \,. \end{aligned} \tag{28.2}$$

Interchanging the integration and the summation signs is possible because of the uniform convergence. Therefore, the norm of $F \in \mathbf{S}$ can be defined as

$$\|F\|^2 = \int_{-\infty}^{\infty}\int_{-\infty}^{\infty} |F(z)|^2 d\mu(z)\,,$$

or equivalently

$$\|F\|^2 = \sum_{n=0}^{\infty} n!|a_n|^2 \,.$$

It also follows from (28.2) that $\left\{u_n(z) = z^n/\sqrt{n!}\right\}_{n=0}^{\infty}$ is an orthonormal set in $\mathbf{S}$ and since $\sqrt{n!}\,a_n = \langle F, u_n\rangle$ for all n, it follows that $\langle F, u_n\rangle = 0$ for all n implies that $F = 0$. Hence, $\{u_n(z)\}_{n=0}^{\infty}$ is complete and consequently it is an orthonormal basis of $\mathbf{S}$.

By applying the Cauchy-Schwarz inequality to the Taylor series of F, we obtain

$$|F(z)|^2 \le \left(\sum_{n=0}^{\infty} n!|a_n|^2\right)\left(\sum_{n=0}^{\infty}\frac{|z|^{2n}}{n!}\right) = e^{|z|^2}\|F\|_{\mathbf{S}}^2\,, \tag{28.3}$$

which shows that convergence in $\mathbf{S}$ implies pointwise convergence and uniform convergence on compact sets. Another consequence of (28.3) is that the evaluation map $F \to F(a)$ is continuous; hence, $\mathbf{S}$ is a reproducing-kernel Hilbert space (see Section 1.10.2). Since $\{u_n(z)\}_{n=0}^{\infty}$ is an orthonormal basis of $\mathbf{S}$, the reproducing kernel can be found explicitly. In fact, the reproducing kernel is readily seen to be

$$K(z,w) = \sum_{n=0}^{\infty} \frac{z^n \overline{w}^n}{n!} = e^{z\overline{w}} .$$

Thus, for any $F \in \mathbf{S}$, we have by (28.2)

$$\langle F(z), K(z,w)\rangle_{\mathbf{S}} = \int_{-\infty}^{\infty}\int_{-\infty}^{\infty} F(z) e^{w\overline{z}} d\mu(z) = \sum_{n=0}^{\infty} n! a_n \frac{w^n}{n!} = F(w). \tag{28.4}$$

Let $k(z,q)$ be as in Equation (28.1). Since

$$|k(\alpha z, q)|^2 = \frac{1}{\sqrt{\pi}} \exp\left[-\alpha^2(x^2-y^2) - q^2 + 2\sqrt{2}\,\alpha q x\right] ,$$

it is easy to see that $k(\alpha z, q) \in \mathbf{S}$ if and only if $|\alpha| < 1$. On the other hand, $k(z,q) \in L^2(\mathcal{R})$ as a function of q for any z. In fact,

$$\langle k(z,q), k(w,q)\rangle_{L^2(\mathcal{R})} = \frac{1}{\sqrt{\pi}} e^{-\frac{1}{2}(z^2+\overline{w}^2)} \int_{-\infty}^{\infty} e^{-q^2} e^{\sqrt{2}\, q(z+\overline{w})} dq ,$$

which, in view of the relation

$$\int_{-\infty}^{\infty} e^{-\alpha^2 x^2 \pm \beta x} dx = \frac{\sqrt{\pi}}{\alpha} \exp\left(\frac{\beta^2}{4\alpha^2}\right) , \qquad \alpha > 0$$

yields

$$\langle k(z,q), k(w,q)\rangle = e^{z\overline{w}} . \tag{28.5}$$

The last equation can be viewed as

$$\mathcal{A}\left[\overline{k}_w(q)\right](z) = e^{z\overline{w}} , \quad \text{where} \quad k_w(q) = k(w,q) . \tag{28.6}$$

If F is the Bargmann transform of f, then in view of (28.5) and the Cauchy-Schwarz inequality, we have

$$|F(z)|^2 \le \left(\int_{-\infty}^{\infty} |f(q)|^2 dq\right)\left(\int_{-\infty}^{\infty} |k(z,q)|^2 dq\right) \le e^{|z|^2} \|f\|^2_{L^2(\mathcal{R})} . \tag{28.7}$$

Now, let $\psi_n(x)$ be the normalized Hermite functions defined by

$$\psi_n(x) = \frac{1}{2^{\frac{n}{2}} \sqrt{n!}\, \sqrt[4]{\pi}} e^{-x^2/2} H_n(x) ,$$

where H_n is the Hermite polynomial of degree n. It then follows from the relation

$$\int_{-\infty}^{\infty} e^{-(x-y)^2} H_n(x)dx = \sqrt{\pi}\, 2^n y^n\,, \quad n = 0, 1, \ldots \qquad [79, \text{p. } 837, \text{no. } 16]$$

that the Bargmann transform of the normalized Hermite function ψ_n is $u_n(z) = z^n/\sqrt{n!}\,, n = 0, 1, 2, \ldots$. For,

$$\begin{aligned}\mathcal{A}(\psi_n)(z) &= \frac{1}{2^{\frac{n}{2}}\sqrt{n!}\,\sqrt{\pi}} \int_{-\infty}^{\infty} e^{-q^2/2} H_n(q) e^{-\frac{1}{2}(q^2+z^2)+\sqrt{2}\,zq} dq \\ &= \frac{1}{2^{\frac{n}{2}}\sqrt{n!}\,\sqrt{\pi}} \int_{-\infty}^{\infty} H_n(q) e^{-(q-z/\sqrt{2})^2} dq = \frac{z^n}{\sqrt{n!}}\,.\end{aligned}$$

It also follows from the generating function of the Hermite polynomials

$$\exp(2bx - b^2) = \sum_{n=0}^{\infty} \frac{b^n}{n!} H_n(x)$$

that

$$\begin{aligned}&\sum_{n=0}^{\infty} \psi_n(q) u_n(z) = \\ &\frac{e^{-q^2/2}}{\sqrt[4]{\pi}} \sum_{n=0}^{\infty} \left(\frac{1}{2^{n/2}\sqrt{n!}} H_n(q)\right)\left(\frac{z^n}{\sqrt{n!}}\right) \\ &= \frac{e^{-q^2/2}}{\sqrt[4]{\pi}} \sum_{n=0}^{\infty} \frac{H_n(q)}{n!}\left(\frac{z}{\sqrt{2}}\right)^n \\ &= \frac{1}{\sqrt[4]{\pi}} \exp\left(-\frac{1}{2}(z^2+q^2) + \sqrt{2}zq\right) = k(z,q)\,.\end{aligned} \tag{28.8}$$

Equation (28.8) enables us to obtain another representation for the Bargmann transform of a function $f \in L^2(\mathcal{R})$. First, let

$$\hat{f}_n = \int_{-\infty}^{\infty} f(q)\psi_n(q)\, dq$$

be the nth Hermite coefficient of f when expanded in a series of the normalized Hermite functions ψ_n. Then

$$F(z) = \sum_{n=0}^{\infty} \hat{f}_n u_n(z). \tag{28.9}$$

For,

$$\begin{aligned}F(z) &= \int_{-\infty}^{\infty} k(z,q) f(q)\, dq \\ &= \int_{-\infty}^{\infty} \left[\sum_{n=0}^{\infty} \psi_n(q) u_n(z)\right] f(q)\, dq = \sum_{n=0}^{\infty} \hat{f}_n u_n(z).\end{aligned}$$

Let $F \in \mathbf{S}$ and set $F_\lambda(z) = F(\lambda z), 0 < \lambda < 1$. Then one can easily show that $\|F_\lambda\|^2 = \sum_{n=0}^{\infty} n!|a_n|^2\lambda^{2n}$ and that $F \in \mathbf{S}$ if and only if $F_\lambda \in \mathbf{S}$ for any $0 < \lambda < 1$ and the norms $\|F_\lambda\|$ are uniformly bounded. Furthermore,

$$\|F - F_\lambda\|_{\mathbf{S}}^2 = \sum_{n=0}^{\infty} n!|a_n|^2(1-\lambda^n)^2 \to 0 \quad \text{as} \quad \lambda \to 1 \,.$$

First, by assuming that f is a square integrable function with compact support, Bargmann showed that $F_\lambda \in \mathbf{S}$ for $0 < \lambda < 1$ and that $\|F_\lambda\|$ are uniformly bounded; hence $F \in \mathbf{S}$. He then used the fact that functions with compact support are dense in $L^2(\mathcal{R})$ to extend this result to all $f \in L^2(\mathcal{R})$. Moreover, it is not hard to see that for any $F \in \mathbf{S}$, the following relation is true: $\|F\|_{\mathbf{S}}^2 = \lim_{\lambda\to 1} \|F_\lambda\|_{\mathbf{S}}^2 = \|f\|_{L^2(\mathcal{R})}^2$.

Furthermore, the range of the Bargmann transformation is dense in $\mathbf{S}$. This is a consequence of (28.6) and the fact that finite linear combinations of $\{e_w(z) = e^{\overline{w}z}\}$ are dense in $\mathbf{S}$. Putting all these facts together leads to the following theorem.

THEOREM 28.1 **(Bargmann** [8] **)**
The Bargmann transformation is a unitary transformation from $L^2(\mathcal{R})$ onto $\mathbf{S}$ that maps the normalized Hermite functions $\psi_n(x)$ into $u_n(z) = z^n/\sqrt{n!}\,, n = 0, 1, 2, \ldots$.

28.6 Parseval's Relation

There are several ways to prove Parseval's relation for the Bargmann transform. The one we shall use is based on formula (28.9). Let F and G be the Bargmann transforms of f and g, respectively. Then

$$\begin{aligned}
\langle F\,,\,G\rangle_{\mathbf{S}} &= \int_{-\infty}^{\infty}\int_{-\infty}^{\infty} F(z)\overline{G}(z)d\mu(z) \\
&= \int_{-\infty}^{\infty}\int_{-\infty}^{\infty} \left(\sum_{n=0}^{\infty} \hat{f}_n u_n(z)\right)\left(\sum_{m=0}^{\infty} \overline{\hat{g}}_m \overline{u}_m(z)\right) d\mu(z) \\
&= \sum_{m,n=0}^{\infty} \hat{f}_n \overline{\hat{g}}_m \langle u_n\,,\,u_m\rangle_{\mathbf{S}} = \sum_{m=0}^{\infty} \hat{f}_n \overline{\hat{g}}_n.
\end{aligned}$$

On the other hand, since the Hermite functions are orthonormal, we have

$$\langle f\,,\,g\rangle_{L^2(\mathcal{R})} = \int_{-\infty}^{\infty} \left(\sum_{n=0}^{\infty} \hat{f}_n \psi_n(q)\right)\left(\sum_{m=0}^{\infty} \overline{\hat{g}}_m \overline{\psi}_m(q)\right) dq$$

$$= \sum_{n=0}^{\infty} \hat{f}_n \overline{\hat{g}}_n \, .$$

Therefore, we have the following Parseval's relation for the Bargmann transform

$$\langle F \, , \, G \rangle_{\mathbf{S}} = \langle f \, , \, g \rangle_{L^2(\mathcal{R})} . \tag{28.10}$$

28.7 Inversion Formula

Let $F \in \mathbf{S}$ and define F_λ as in Section 28.5. Let us define the integral transformation $\mathcal{B}$ by

$$\mathcal{B}[F_\lambda](q) = \int_{-\infty}^{\infty} \int_{-\infty}^{\infty} \overline{k}(z,q) F_\lambda(z) d\mu(z) \, . \tag{28.11}$$

One can verify that the integral in (28.11) exists for any q. Now consider the Bargmann transform of $\mathcal{B}[F_\lambda]$,

$$\mathcal{A}[\mathcal{B}[F_\lambda]](w) = \int_{-\infty}^{\infty} \mathcal{B}[F_\lambda](q) k(w,q) dq \, . \tag{28.12}$$

By inserting (28.11) into (28.12), interchanging the integrals and using (28.5), we find that

$$\mathcal{A}[\mathcal{B}[F_\lambda]](w) = \int_{-\infty}^{\infty} \int_{-\infty}^{\infty} F_\lambda(z) e^{w\overline{z}} d\mu(z) = F_\lambda(w) \, ,$$

in which the last equality follows from (28.4). Thus, if $f \in L^2(\mathcal{R})$ is such that its Bargmann transform is of the form

$$F_\lambda(z) = \int_{-\infty}^{\infty} f(q) k(z,q) dq \, , \quad \text{for some} \quad F \in S \, ,$$

then

$$f(q) = \int_{-\infty}^{\infty} \int_{-\infty}^{\infty} F_\lambda(z) \overline{k}(z,q) d\mu(z) \, . \tag{28.13}$$

That is to say, $\mathcal{B} = \mathcal{A}^{-1}$.

The gist of Bargmann's proof of the inversion formula is to extend formula (28.12) to any $f \in L^2(\mathcal{R})$. One of the technical difficulties in the proof is that the kernel $k(z,q)$ is not always in $\mathbf{S}$ (see Section 5); hence, the integral in (28.12) does not converge in general. The general inversion formula of the Bargmann transform is

$$f(q) = \lim_{\lambda \to 1} \int_{-\infty}^{\infty} \int_{-\infty}^{\infty} \overline{k}(z,q) F(\lambda z) d\mu(z) \, .$$

Another version of it is

$$f(q) = \lim_{\sigma \to \infty} \iint_{|z|<\sigma} \overline{k}(z,q) F(z) d\mu(z) .$$

28.8 Generalized Functions

The Bargmann transform can be extended to tempered distributions using the embedding method or the adjoint method. Because the kernel of the Bargmann transform, $A(z,q)$, belongs to the Schwartz space of functions $\mathcal{S}$, the embedding method is routine and, in addition, it does not lead directly to any interesting results. A. Janssen [112], however, has used the embedding method in a different setting to obtain very nice results which he utilized to extend the Bargmann transform not only to tempered distributions, but to another space of generalized functions.

The adjoint method, which was used by Bargmann [7], reveals some interesting properties of the underlying spaces of functions. The basic idea of this method is to extend Parseval's relation (28.10), which is valid for functions in $L^2(\mathcal{R})$, to tempered distributions.

For α, ρ real, let

$$\theta_\rho^\alpha(s) = (1+s^2)^{\rho/2} e^{\alpha s^2/2} , \quad s \geq 0.$$

Let S^k be the set of all functions $f \in C^k$ such that

$$\gamma_k(f) = \max_{0 \leq j \leq k} \left\{ \sup_q \left| f^{(j)}(q) \theta_{k-j}^0(q) \right| \right\} < \infty.$$

It is easy to verify that S^k is a normed space with norm $||f||_k = \gamma_k(f)$, and since $\theta_\rho^\alpha(q) \leq \theta_{\rho'}^\alpha(q)$ if $\rho \leq \rho'$, it follows that $S^{k+1} \subset S^k$. Moreover, $\mathcal{S} = \bigcap_{k=0}^{\infty} S^k$.

Let G^ρ be the set of all entire functions f that satisfy

$$\|f\|_\rho = \sup_z \theta_\rho^{-1}(z) |f(z)| < \infty , \quad \rho \text{ real}.$$

It can be easily shown that G^ρ is a normed vector space with norm $\|f\|_\rho$, and since $\theta_\rho^{-1}(z) \leq \theta_\sigma^{-1}(z)$ if $\rho \leq \sigma$, it follows that $G^\sigma \subset G^\rho$. Let

$$G = \bigcap_{k=0}^{\infty} G^k = \bigcap_{k=-\infty}^{\infty} G^k .$$

The topology of G is the one it inherits from all the G^k's.

It has been shown in [7] that G is complete and that $G^* = \bigcup_{k=-\infty}^{\infty} G^k$. Moreover, the Bargmann transform of any $f \in S^k$ belongs to G^k, or symbolically, $\mathcal{A}(S^k) \subset G^k$. Therefore, $\mathcal{A}(\mathcal{S}) \subset G$. In fact, Bargmann strengthened this result and showed that $\mathcal{A}(\mathcal{S}) = G$. His result is stated more precisely in the next theorem.

THEOREM 28.2 **(Bargmann [7])**
The Bargmann transformation $\mathcal{A}$ is a bijective mapping from $\mathcal{S}$ onto G and the transformation $\mathcal{B}$ given by (28.11) is its inverse. Both $\mathcal{A}$ and $\mathcal{A}^{-1}$ are continuous in the topologies of $\mathcal{S}$ and G. Furthermore, a linear functional L is a continuous linear functional on G if and only if for all $f \in G$

$$L(f) = \langle f\,,\, g\rangle = \int_{-\infty}^{\infty} f(z)\overline{g}(z)d\mu(z)$$

for some fixed $g \in G^{-k}$.

The next theorem, also due to Bargmann, characterizes the Bargmann transform of tempered distributions.

THEOREM 28.3
Let $f \in \mathcal{S}^*$ and $\phi \in \mathcal{S}$. Then, the Bargmann transform $\mathcal{A}[f] = \langle f(q)\,,\, k(\overline{z}, q)\rangle$ of f is in G^{-k} for some k, and

$$\begin{aligned}\langle f\,,\,\phi\rangle &= \langle \mathcal{A}[f]\,,\,\mathcal{A}[\phi]\rangle \\ &= \int_{-\infty}^{\infty} \langle f\,,\, k(\overline{z}, q)\rangle\langle k(\overline{z}, q)\,,\,\phi\rangle\, d\mu(z).\end{aligned}$$

Therefore, the Bargmann transform, $\mathcal{A}[f]$, of $f \in \mathcal{S}^*$ is an element of G^*.

The Bargmann transform of a tempered distribution can also be characterized as in (28.9). But, first let us introduce the Hermite differential operator

$$L = -\frac{d^2}{dx^2} + x^2 - 1\,,$$

so that $L\psi_n = 2n\psi_n$. It follows that for any non-negative integer p and any $\phi \in \mathcal{S}$, we have

$$(2n)^p\langle \phi\,,\,\psi_n\rangle = \langle \phi\,,\, L^p\psi_n\rangle = \langle L^p\phi\,,\,\psi_n\rangle.$$

Thus,

$$(2n)^p|\langle \phi\,,\,\psi_n\rangle| \le ||\psi_n||\,||L^p\phi|| = c < \infty\,,$$

or $\hat{\phi}_n = \langle \phi\,,\,\psi_n\rangle = O(n^{-p})$ as $n \to \infty$. As a corollary, we obtain the well-known result that $\phi = \sum_{n=0}^{\infty} \hat{\phi}_n\psi_n$ is in $\mathcal{S}$ if and only if $\hat{\phi}_n = O(n^{-p})$ for all p, and consequently, $f = \sum_{n=0}^{\infty} \hat{f}_n \psi_n$ is in $\mathcal{S}^*$ if and only if $\hat{f}_n = O(n^p)$ for some p.

With the aid of (28.9), we obtain the following characterization of the Bargmann transforms of $\phi \in \mathcal{S}$ and $f \in \mathcal{S}^*$: $\Phi = \sum_{n=0}^{\infty} a_n u_n$ is the Bargmann transform of some $\phi \in \mathcal{S}$ if and only if $a_n = O(n^{-p})$ for all p, and $F = \sum_{n=0}^{\infty} b_n u_n$ is the Bargmann transform of some $f \in \mathcal{S}^*$ if and only if $b_n = O(n^p)$ for some p. Moreover,

$$\langle F\,,\,\Phi\rangle = \langle f\,,\,\phi\rangle = \sum_{n=0}^{\infty} b_n \overline{a}_n.$$

28.9 Applications

The Bargmann transform is not only intrinsically interesting, but also genuinely useful. Recall that the Bargmann transform was introduced originally to establish the mathematical foundation for some of Foch's work on quantum field theory. It has also appeared in the area of quantum optics [121, 122]. In addition to that, it has other important applications, one of which is its use in the solution of the Gabor representation problem (see Section 30.8) and in the proof of the completeness of the canonical coherent states in quantum mechanics and the Gabor frames in signal analysis [6]. Admittedly, nowadays there are other methods to prove the completeness of these systems, but the Bargmann transform method was one of the earliest. For more details, we refer the reader to [112, 115].

29

The Zak Transform

Unlike most linear function transformations discussed in this book, the Zak transformation is neither an integral transformation nor does it map a function of one variable into another function of one variable. It is a linear series transformation that maps a function of n variables into a function of $2n$ variables.

The Zak transform was discovered by several people in different fields and was called different names depending on the field in which it was discovered. Its first appearance in the literature was in the work of the Russian mathematician, I. M. Gel'fand [71], on eigenfunction expansions associated with Schrödinger operators with periodic potentials. The Zak transform was called the Gel'fand mapping in the Russian literature. In 1967, almost 17 years after the publication of Gel'fand's work, the transform was rediscovered independently by a solid-state physicist, J. Zak, who called it the k-q representation. Zak introduced the k-q representation to construct a quantum mechanical representation for the motion of a Bloch electron in the presence of a magnetic or electric field [254, 253]. It has also been said [213] that some properties of another version of the Zak transform, which is called the Weil–Brezin map in [5, 213], were even known to the great mathematician Gauss.

Although the Gel'fand–Zak transform seems to be a more appropriate name for this transform, as suggested by A. Janssen [111], there seems to be a general consent among experts in the field to call it the Zak transform since Zak was indeed the first to systematically study that transform in a more general setting and recognize its usefulness. Among the different areas in which the Zak transform has been used successfully are signal processing and quantum field theory. A very good review article on the Zak transform and its applications is [111].

29.1 Definitions

DEFINITION 29.1 *The Zak transform $Z_a(f)$ of a function f is defined by*

$$Z_a[f](t,w) = (Z_a f)(t,w) = \sqrt{a} \sum_{k=-\infty}^{\infty} f(at+ak)e^{-2\pi ikw}, \tag{29.1}$$

where $a > 0$, t and w are real.

If f represents a signal, then its Zak transform can be considered as a mixed time-frequency representation of f, and it can also be considered as a generalization of the discrete Fourier transform of f in which an infinite sequence of samples in the form $f(at+ak), k = 0, \pm 1, \pm 2, \cdots$, are used. Without loss of generality, we may take $a = 1$, and denote $Z_a f$ by Zf.

Example A: If $a = 1$ and $f(t) = 0$ outside $[-b, b], 0 < b \leq 1/2$, then $(Zf)(t,w) = f(t), |t| \leq 1/2$.

Example B: The Zak transform of the Gaussian function

$$f(t) = (2\gamma)^{1/4} \exp(-\pi\gamma t^2), \gamma > 0,$$

is easily shown to be

$$(Zf)(t,w) = (2\gamma)^{1/4} e^{-\pi\gamma t^2} \theta_3(w - i\gamma t, e^{-\pi\gamma}),$$

where θ_3 is the third theta function defined by

$$\theta_3(z,q) = \sum_{k=-\infty}^{\infty} q^{k^2} e^{-2\pi ikz}.$$

29.2 Existence

If f is integrable or square integrable, its Zak transform exists almost everywhere. In particular, if f is a continuous function such that $|f(t)| \leq C(1+|t|)^{-(1+\epsilon)}$, for some $\epsilon > 0$, for all t, then its Zak transform exists and defines a continuous function.

29.3 Elementary Properties

1. **Linearity:** The Zak transform is linear.

2. **Translation:**

$$Z[T_a f](t,w) = (Zf)(t-a,w), \quad a \text{ real},$$

and

$$Z[T_{-m} f](t,w) = e^{2\pi imw}(Zf)(t,w), \quad m \text{ is an integer},$$

in particular

$$(Zf)(t+1,w) = e^{2\pi iw}(Zf)(t,w).$$

3. **Modulation:**

$$Z\left[E_a f\right](t,w) = e^{iat}(Zf)(t,w-\frac{a}{2\pi}).$$

4. **Periodicity:** The Zak transform is periodic in w with period one, that is

$$(Zf)(t,w+1) = (Zf)(t,w).$$

5. **Translation and Modulation:** By combining (2) and (3), we obtain

$$Z\left[E_{2\pi m} T_n f\right](t,w) = e^{-2\pi inw} e^{2\pi imt}(Zf)(t,w).$$

6. **Conjugation:**

$$(Z\overline{f})(t,w) = \overline{(Zf)}(t,-w).$$

7. **Symmetry:** If f is even, then

$$(Zf)(t,w) = (Zf)(-t,-w),$$

and if f is odd, then

$$(Zf)(t,w) = -(Zf)(-t,-w).$$

From (6) and (7), it follows that if f is real and even, then

$$(Zf)(t,w) = \overline{(Zf)}(t,-w) = (Zf)(-t,-w).$$

Because of properties (2) and (4), the Zak transform is completely determined by its values on the unit square $Q = [0,1] \times [0,1]$.

Example A: Let $a > 0$ and $e_{m,n;a}(x) = \frac{1}{\sqrt{a}} T_{na} E_{2\pi m/a} \chi_{[0,a)}(x)$. Then,

$$e_{m,n;a}(x) = \frac{1}{\sqrt{a}} e^{2\pi i m(x-na)/a} \chi_{[0,a)}(x - na) = \frac{1}{\sqrt{a}} e^{2\pi i m x/a} \chi_{[na,(n+1)a)}(x).$$

Therefore,

$$(Z_a e_{m,n;a})(t,w) = \sum_{k=-\infty}^{\infty} e^{2\pi i m(at+ak)/a} \chi_{[na,na+a)}(at+ak) e^{-2\pi i k w}$$

$$= \sum_{k=-\infty}^{\infty} e^{2\pi i m t} e^{-2\pi i k w} \chi_{[n-k,n+1-k)}(t) = e^{2\pi i m t} e^{-2\pi i n w}, \qquad (29.2)$$

for $0 \le t, w \le 1$.

29.4 Analytic Properties

If f is a continuous function such that $f(t) = O\left((1+|t|)^{-1-\epsilon}\right)$ as $|t| \to \infty$ for some $\epsilon > 0$, then Zf is continuous on Q. A rather peculiar property of the Zak transform is that if Zf is continuous, it must have a zero in Q. For, if Zf had no zeros in Q, then $\ln(Zf)$ would be single-valued, continuous in Q and can be extended to $\mathcal{R}^2$ by means of properties (2) and (4) of Section 29.3,

$$(Zf)(1,w) = (Zf)(0,w)e^{2\pi i w}, \quad (Zf)(t,1) = (Zf)(t,0).$$

Now let $(Zf) = |(Zf)|e^{2\pi i \phi}$. Then ϕ is continuous and satisfies

$$\phi(1,w) = \phi(0,w) + w + k,$$

in particular

$$\phi(1,1) = \phi(0,1) + 1 + k, \qquad (29.3)$$

and

$$\phi(1,0) = \phi(0,0) + k. \qquad (29.4)$$

On the other hand, we have

$$\phi(t,1) = \phi(t,0) + m,$$

which gives

$$\phi(1,1) = \phi(1,0) + m \qquad (29.5)$$

and

$$\phi(0,1) = \phi(0,0) + m. \tag{29.6}$$

The integers k and m are independent of t and w by the continuity of ϕ. By substituting (29.6) into (29.3), (29.4) into (29.5) and comparing the resulting equations for $\phi(1,1)$, we obtain that $1 = 0$, which is a contradiction.

As special cases, we obtain that if Zf is continuous and f is even, then $(Zf)(\frac{1}{2},\frac{1}{2}) = 0$, and if f is odd, then $(Zf)(0,\frac{1}{2}) = 0 = (Zf)(\frac{1}{2},0)$.

29.5 Inversion Formulas

The following inversion formulas for the Zak transform follow easily from the definition, provided that the series defining the Zak transform converges uniformly,

$$f(t) = \int_0^1 (Zf)(t,w)dw, \quad -\infty < t < \infty,$$

$$\hat{f}(-2\pi w) = \frac{1}{\sqrt{2\pi}} \int_0^1 e^{-2\pi iwt}(Zf)(t,w)dt,$$

and

$$f(2\pi x) = \frac{1}{\sqrt{2\pi}} \int_0^1 e^{-2\pi ixt}(Z\hat{f})(t,x)dt,$$

where $\hat{f}$ is the Fourier transform of f.

29.6 Relationships With Other Transforms

The following formula relates the Zak transform of a function to the Zak transform of its Fourier transform.

$$\frac{1}{\sqrt{2\pi}} (Zf)(t,w) = e^{2\pi itw} (Z\hat{g})(-w,t), \tag{29.7}$$

where $\hat{g}(w) = \hat{f}(2\pi w)$. Had we defined the Fourier transform of f as

$$\hat{f}(w) = \int_{-\infty}^{\infty} e^{2\pi itw} f(t)dt,$$

we would have obtained the more aesthetic formula

$$(Zf)(t,w) = e^{2\pi itw} \left(Z\hat{f}\right)(-w,t).$$

To prove (29.7), let us fix t and w and define

$$h(x) = f(x+t)e^{-2\pi iwx}.$$

Then

$$\begin{aligned}\hat{h}(\gamma) &= \frac{1}{\sqrt{2\pi}}\int_{-\infty}^{\infty} f(x+t)e^{-2\pi iwx}e^{i\gamma x}dx \\ &= \frac{1}{\sqrt{2\pi}}\int_{-\infty}^{\infty} f(y)e^{i(y-t)(\gamma-2\pi w)}dy = e^{-it(\gamma-2\pi w)}\hat{f}(\gamma-2\pi w).\end{aligned}$$

An application of the Poisson summation formula (12.24) with $\sigma = \pi$ and $t = 0$ yields

$$\frac{1}{\sqrt{2\pi}}\sum_{k=-\infty}^{\infty} h(k) = \sum_{k=-\infty}^{\infty} \hat{h}(2\pi k),$$

or

$$\begin{aligned}\frac{1}{\sqrt{2\pi}}\sum_{k=-\infty}^{\infty} f(t+k)e^{-2\pi ikw} &= \sum_{k=-\infty}^{\infty} \hat{f}(2\pi k-2\pi w)e^{-2\pi it(k-w)} \\ = e^{2\pi iwt}\sum_{k=-\infty}^{\infty} \hat{f}(2\pi k-2\pi w)e^{-2\pi ikt} &= e^{2\pi iwt}\sum_{k=-\infty}^{\infty} \hat{g}(k-w)e^{-2\pi ikt},\end{aligned}$$

which can be written as

$$\frac{1}{\sqrt{2\pi}}\left(Zf\right)(t,w) = e^{2\pi iwt}\left(Z\hat{g}\right)(-w,t).$$

As a special case, if $\hat{f}(w) = 0\,,\ |w| \le \pi$, then

$$\frac{1}{\sqrt{2\pi}}\left(Zf\right)(t,w) = e^{2\pi iwt}\hat{f}(-2\pi w).$$

29.7 Product and Convolution of Zak transforms

Although the Zak transform is not periodic in the two variables, t and w, the product of two Zak transforms is. This can be seen from properties (2) and (4) of Section 3, since

$$(Zf)(t+1,w) = e^{2\pi iw}(Zf)(t,w)\,,\ \overline{(Zg)}(t+1,w) = e^{-2\pi iw}\overline{(Zg)}(t,w),$$

and, therefore, if we set

$$H(t,w) = (Zf)(t,w)\overline{(Zg)}(t,w),$$

it follows that

$$H(t+1,w) = (Zf)(t,w)\overline{(Zg)}(t,w) = H(t,w),$$

and

$$H(t,w+1) = (Zf)(t,w)\overline{(Zg)}(t,w) = H(t,w).$$

Hence H is periodic in t and w, and as a result it can be expanded in a Fourier series of the form

$$H(t,w) = \sum_{m,n=-\infty}^{\infty} c_{m,n} e^{2\pi imt} e^{2\pi inw},$$

where

$$c_{m,n} = \int_0^1 \int_0^1 H(t,w) e^{-2\pi imt} e^{-2\pi inw} dt dw.$$

Assuming that the series involved are uniformly convergent, we can interchange the summation and integration signs to obtain

$$\begin{aligned}
c_{m,n} &= \int_0^1 \int_0^1 \left(\sum_{k=-\infty}^{\infty} f(t+k) e^{-2\pi ikw} \right) \left(\sum_{j=-\infty}^{\infty} \overline{g}(t+j) e^{2\pi ijw} \right) e^{-2\pi imt} e^{-2\pi inw} dt dw \\
&= \int_0^1 \left(\sum_{k=-\infty}^{\infty} f(t+k) \right) \left(\sum_{j=-\infty}^{\infty} \overline{g}(t+j) \right) e^{-2\pi imt} dt \int_0^1 e^{2\pi iw(j-n-k)} dw \\
&= \int_0^1 \left(\sum_{k=-\infty}^{\infty} f(t+k) \overline{g}(t+n+k) \right) e^{-2\pi imt} dt \\
&= \sum_{k=-\infty}^{\infty} \int_k^{k+1} f(x) \overline{g}(x+n) e^{-2\pi im(x-k)} dx = \int_{-\infty}^{\infty} f(x) \overline{g}(x+n) e^{-2\pi imx} dx,
\end{aligned}$$

or equivalently

$$c_{m,n} = \langle f(x), e^{2\pi imx} g(x+n) \rangle = \langle f, E_{2\pi m} T_{-n} g \rangle.$$

Therefore, we have

$$(Zf)(t,w)\overline{(Zg)}(t,w) = \sum_{m,n=-\infty}^{\infty} \langle f, E_{2\pi m} T_{-n} g \rangle e^{2\pi imt} e^{2\pi inw} \tag{29.8}$$

The convolution theorem for the Zak transform is obtained as a special case of the following theorem.

THEOREM 29.1

Let $R(t,s)$ be a function satisfying

$$R(t+1, s+1) = R(t,s), \quad -\infty < t, s < \infty, \tag{29.9}$$

and let

$$h(t) = \int_{-\infty}^{\infty} R(t,s) f(s) ds,$$

where the integral converges absolutely and uniformly.

Then

$$(Zf)(t,w) = \int_0^1 (Zf)(s,w)\Phi(t,s,w) ds, \tag{29.10}$$

where

$$\Phi(t,s,w) = \sum_{n=-\infty}^{\infty} R(t+n,s) e^{-2\pi i n w}, \; 0 \le t,s,w \le 1.$$

Proof: From the definition of h and its Zak transform, we have

$$(Zh)(t,w) = \sum_{k=-\infty}^{\infty} h(t+k) e^{-2\pi i k w} = \sum_{k=-\infty}^{\infty} e^{-2\pi i k w} \int_{-\infty}^{\infty} R(t+k,s) f(s) ds$$

$$= \sum_{k=-\infty}^{\infty} e^{-2\pi i k w} \sum_{m=-\infty}^{\infty} \int_m^{m+1} R(t+k,s) f(s) ds$$

$$= \sum_{k=-\infty}^{\infty} e^{-2\pi i k w} \sum_{m=-\infty}^{\infty} \int_0^1 R(t+k,s+m) f(s+m) ds$$

$$= \int_0^1 \left(\sum_{k,m=-\infty}^{\infty} R(t+k,s+m) f(s+m) e^{-2\pi i k w} \right) ds,$$

and because of (29.9), we have

$$(Zh)(t,w) = \int_0^1 \left(\sum_{k,m=-\infty}^{\infty} R(t+k-m,s) f(s+m) e^{-2\pi i k w} \right) ds$$

$$= \int_0^1 \left(\sum_{m,n=-\infty}^{\infty} R(t+n,s) f(s+m) e^{-2\pi i (m+n) w} \right) ds$$

$$= \int_0^1 \left(\sum_{n=-\infty}^{\infty} R(t+n,s) e^{-2\pi i n w} \right) \left(\sum_{m=-\infty}^{\infty} f(s+m) e^{-2\pi i m w} \right) ds$$

$$= \int_0^1 (Zf)(s,w) \Phi(t,s,w) ds,$$

which is (29.10). ∎

In particular, for $R(t,s) = R(t-s)$,

$$\Phi(t,s,w) = \sum_{n=-\infty}^{\infty} R(t-s+n) e^{-2\pi i n w} = (ZR)(t-s,w),$$

and the last theorem takes the following special form.

THEOREM 29.2 **(Convolution Theorem)**
Let

$$h(t) = \int_{-\infty}^{\infty} R(t-s)f(s)ds,$$

then

$$(Zh)(t,w) = \int_0^1 (ZR)(t-s,w)(Zf)(s,w)ds.$$

29.8 The Zak Transform of L^2-Functions

Let Q be the unit square $[0,1) \times [0,1)$, and $L^2(Q)$ be the set of all square integrable complex-valued functions F on Q, i.e.,

$$\int_0^1 \int_0^1 |F(t,w)|^2 dtdw \leq \infty.$$

Provided with the inner product

$$\langle f, G \rangle = \int_0^1 \int_0^1 F(t,w)\overline{G}(t,w)dtdw,$$

$L^2(Q)$ is a Hilbert space with norm

$$||F|| = \left(\int_0^1 \int_0^1 |F(t,w)|^2 dtdw \right)^{1/2}.$$

The set $\left\{E_{m,n} = E_{2\pi m, 2\pi n}(t,w) = e^{2\pi imt}e^{2\pi inw}\right\}_{m,n=-\infty}^{\infty}$ constitutes an orthonormal basis of $L^2(Q)$. This observation simplifies the proof of the next theorem.

THEOREM 29.3
The Zak transform is a unitary transformation from $L^2(\mathcal{R})$ onto $L^2(Q)$.

Proof: From the definition of the inner product in $L^2(Q)$, it follows that the Zak transform is an isometry from $L^2(\mathcal{R})$ into $L^2(Q)$ since

$$\langle Z_a f, Z_a g \rangle =$$

$$a \int_0^1 \int_0^1 \left(\sum_{k=-\infty}^{\infty} f(at+ak)e^{-2\pi ikw} \right) \left(\sum_{m=-\infty}^{\infty} \overline{g}(at+am)e^{2\pi imw} \right) dtdw$$

$$= a\int_0^1 \left(\sum_{k=-\infty}^{\infty} f(at+ak)\overline{g}(at+ak)\right) dt = \sum_{k=-\infty}^{\infty} \int_{ka}^{(k+1)a} f(x)\overline{g}(x)dx$$
$$= \int_{-\infty}^{\infty} f(x)\overline{g}(x)dx,$$

or

$$\langle Z_a f, Z_a g\rangle = \langle f, g\rangle. \tag{29.11}$$

In particular,

$$||Z_a f||^2 = ||f||^2.$$

By (29.2)

$$(Z_a e_{m,n;a})(t,w) = E_{2\pi m,-2\pi n}(t,w) = e^{2\pi imt}e^{-2\pi inw},$$

and since $\left\{e_{m,n;a}(x) = \frac{1}{\sqrt{a}}T_{na}E_{2\pi m/a}\chi_{[0,a)}(x)\right\}_{m,n=-\infty}^{\infty}$ is an orthonormal basis of $L^2(\mathcal{R})$, it follows that the Zak transform maps in a one-to-one fashion an orthonormal basis of $L^2(\mathcal{R})$ onto an orthonormal basis of $L^2(Q)$, and this completes the proof. ∎

For $1 \leq p < 2$, the Zak transform maps $L^p(\mathcal{R})$ into $L^p(Q)$ and the operator norm is ≤ 1. In this case, the Zak transform is injective but not surjective [112].

One of the most important applications of the Zak transform is its use in the study of the Gabor representation problem in signal analysis and the coherent states representation in quantum field theory. This will be discussed in more details in the last section of this chapter, but to facilitate our forthcoming discussion of the crucial role that the Zak transform plays in these problems, we need the following results.

Let $a, b > 0$, $g \in L^2(\mathcal{R})$, and

$$g_{mb,na}(t) = E_{2\pi mb}T_{na}g(t) = e^{2\pi imbt}g(t-na).$$

It is easy to see that if $ab = 1$, then

$$\begin{aligned} Z_a\left[g_{mb,na}\right](t,w) &= e^{2\pi imbt}e^{-2\pi inaw}\left(Z_a g\right)(t,w) \\ &= E_{2\pi mb,-2\pi na}(t,w)\left(Z_a g\right)(t,w). \end{aligned} \tag{29.12}$$

Moreover, if $\{g_{mb,na}(t)\}_{m,n=-\infty}^{\infty}$ is a frame in $L^2(\mathcal{R})$, then the frame operator S is given by

$$S(f) = \sum_{m,n=-\infty}^{\infty} \langle f, g_{mb,na}\rangle g_{mb,na} \quad , \quad f \in L^2(\mathcal{R}).$$

For the definition of a frame, see Section 1.10.1.

THEOREM 29.4
Let $a, b > 0$, $g \in L^2(\mathcal{R})$. If $\{g_{mb,na}\}_{m,n=-\infty}^{\infty}$ is a frame in $L^2(\mathcal{R})$, then its dual frame $\{S^{-1}(g_{mb,na})\}_{m,n=-\infty}^{\infty}$ is also generated by one single function. More precisely,

$$S^{-1}(g_{mb,na}) = \tilde{g}_{mb,na},$$

where $\tilde{g} = S^{-1}(g)$.

Proof: For any $f \in L^2(\mathcal{R})$ and fixed integer k, we have

$$\begin{aligned} S(T_{ka}f)(t) &= \sum_{m,n} \langle T_{ka}f, g_{mb,na}\rangle g_{mb,na}(t) \\ &= \sum_{m,n} e^{-2\pi imbak} \langle f, g_{mb,(n-k)a}\rangle g_{mb,na}(t) \\ &= \sum_{m,n} e^{-2\pi imbak} \langle f, g_{mb,na}\rangle g_{mb,(k+n)a}(t) \\ &= \sum_{m,n} \langle f, g_{mb,na}\rangle e^{2\pi imb(t-ak)} g(t - na - ka) \\ &= \sum_{m,n} \langle f, g_{mb,na}\rangle T_{ka} \left[e^{2\pi imbt} g(t - na)\right] = T_{ka}(S[f])(t), \end{aligned} \tag{29.13}$$

where in $\sum_{m,n}$ both m and n run over the integers. Here we have used the fact that $T_{ka}e^{2\pi imbt} = e^{2\pi imbt}$. Similarly, we can show that S and $E_{2\pi kb}$ commute; hence,

$$S\left(E_{2\pi kb}T_{la}f\right) = E_{2\pi kb}T_{la}\left(S(f)\right), \tag{29.14}$$

and consequently

$$S^{-1}\left(E_{2\pi kb}T_{la}f\right) = E_{2\pi kb}T_{la}\tilde{f}, \tag{29.15}$$

where $\tilde{f} = S^{-1}(f)$. In particular, if $f = g$, we obtain

$$S^{-1}(g_{mb,na}) = S^{-1}(E_{2\pi mb}T_{na}g) = E_{2\pi mb}T_{na}S^{-1}(g) = E_{2\pi mb}T_{na}\tilde{g}.$$

∎

Some important properties of the set $\{g_{mb,na}\}_{m,n=-\infty}^{\infty}$ when $ab - 1$ are given in the next theorem.

THEOREM 29.5
Let $a, b > 0$ be such that $ab = 1$ and let $g \in L^2(\mathcal{R})$. Then the following are equivalent:

1. *there exist constants A and B such that $0 < A \leq |Z_a(g)|^2 \leq B < \infty$,*

2. *the set* $\{g_{mb,na}(t) = e^{2\pi imbt}g(t-na)\}_{m,n=-\infty}^{\infty}$ *is a frame for* $L^2(\mathcal{R})$ *with frame bounds* A *and* B,

3. $\{g_{mb,na}(t)\}_{m,n=-\infty}^{\infty}$ *is an exact frame in* $L^2(\mathcal{R})$ *with frame bounds* A, B.

If any of the above conditions is satisfied, then for any $f \in L^2(\mathcal{R})$, *we have a unique representation of the form* $f = \sum_{m,n} c_{m,n}\, g_{mb,na}$, *where*

$$c_{m,n} = \int_0^1 \int_0^1 \frac{(Z_a f)(t,w)}{(Z_a g)(t,w)} e^{-2\pi imt} e^{2\pi inw} dtdw. \tag{29.16}$$

Proof: 1) $\Rightarrow$ 2): Since by Theorem 29.3 the Zak transform is a unitary transformation from $L^2(\mathcal{R})$ onto $L^2(Q)$, it suffices to show that $\{Z_a g_{mb,na}(t,w)\}_{m,n=-\infty}^{\infty}$ is a frame for $L^2(Q)$. Let $F \in L^2(Q)$. Because $Z_a(g)$ is bounded, $F\overline{(Z_a g)} \in L^2(Q)$ and by (29.12)

$$\langle F, Z_a g_{mb,na}\rangle = \langle F, E_{2\pi m,-2\pi n} Z_a g\rangle = \langle F\overline{(Z_a g)}, E_{2\pi m,-2\pi n}\rangle. \tag{29.17}$$

Since $\{E_{2\pi m,-2\pi n}\}$ is an orthonormal basis of $L^2(Q)$, Parseval's equality implies that

$$\sum_{m,n} |\langle F, Z_a g_{mb,na}\rangle|^2 = ||F\overline{(Z_a g)}||^2.$$

Combining this with the inequalities

$$A||F||^2 \le ||F\overline{(Z_a g)}||^2 \le B||F||^2,$$

we obtain

$$A||F||^2 \le \sum_{m,n} |\langle F, Z_a g_{mb,na}\rangle|^2 \le B||F||^2;$$

hence $\{Z_a g_{mb,na}(t,w)\}$ is a frame in $L^2(Q)$.

2) $\Rightarrow$ 1): If (2) holds, then $\{E_{2\pi m,-2\pi n} Z_a g\}$ is a frame in $L^2(Q)$ with frame bounds A and B. Therefore, for any $F \in L^2(Q)$, we have

$$A||F||^2 \le \sum_{m,n} |\langle F, E_{2\pi m,-2\pi n} Z_a g\rangle|^2 \le B||F||^2.$$

But as we have just shown in (29.17),

$$\sum_{m,n} |\langle F, E_{2\pi m,-2\pi n} Z_a g\rangle|^2 = \sum_{m,n} \left|\langle F\overline{(Z_a g)}, E_{2\pi m,-2\pi n}\rangle\right|^2 = ||F\overline{(Z_a g)}||^2;$$

thus, by putting the last two equations together, we obtain

$$A||F||^2 \le ||F\overline{(Z_a g)}||^2 \le B||F||^2, \quad \text{for all } F \in L^2(Q),$$

which implies (1).

2) $\Rightarrow$ 3): Let (2) hold. Then $\{E_{2\pi m,-2\pi n}Z_a g\}$ is a frame in $L^2(Q)$. But from (1), $Z_a(g)$ is bounded from above and below by positive numbers B and A; hence the mapping $W : L^2(Q) \to L^2(Q)$, defined by

$$W(F) = F(Z_a g), \quad F \in L^2(Q)$$

is a topological isomorphism that maps the orthonormal basis $\{E_{2\pi m,-2\pi n}\}$ onto $\{Z_a(g_{mb,na})(t,w)\}$. Therefore, $\{Z_a(g_{mb,na})(t,w)\}$ is a Riesz basis for $L^2(Q)$, and hence so is $\{g_{mb,na}(t,w)\}$ for $L^2(\mathcal{R})$. As a consequence of Theorem 1.6, $\{g_{mb,na}(t,w)\}$ is an exact frame for $L^2(\mathcal{R})$.

3) $\Rightarrow$ 2) is trivial. To show that (29.16) holds, first, we show that

$$Z_a(Sf) = (Z_a f)|Z_a g|^2, \tag{29.18}$$

where S is the frame operator associated with the frame $\{g_{mb,na}(x)\}$. Since $\{E_{2\pi m,-2\pi n}(t,w)\}$ is an orthonormal basis for $L^2(Q)$, we have in view of (29.11) and (29.12)

$$\begin{aligned}
Z_a(Sf) &= Z_a\left(\sum_{m,n}\langle f, g_{mb,na}\rangle g_{mb,na}\right) = Z_a(g)\sum_{m,n}\langle f, g_{mb,na}\rangle E_{2\pi m,-2\pi n} \\
&= Z_a(g)\sum_{m,n}\langle Z_a f, Z_a g_{mb,na}\rangle E_{2\pi m,-2\pi n} = Z_a(g)\sum_{m,n}\langle Z_a f, Z_a g E_{2\pi m,-2\pi n}\rangle \\
&\times E_{2\pi m,-2\pi n} = Z_a(g)\sum_{m,n}\langle Z_a f\overline{Z_a g}, E_{2\pi m,-2\pi n}\rangle E_{2\pi m,-2\pi n} = Z_a(f)|Z_a(g)|^2,
\end{aligned}$$

which is (29.18). Replacing f by $S^{-1}f$ in (29.18) yields

$$Z_a(S^{-1}f) = \frac{Z_a(f)}{|Z_a(g)|^2}.$$

In particular, if $f = g$, we have

$$Z_a(\tilde{g}) = \frac{1}{\overline{(Z_a g)}}, \tag{29.19}$$

where $\tilde{g} = S^{-1}g$.

Thus, from (29.11), (29.12), (29.19), and Theorem 29.4, we obtain

$$\begin{aligned}
c_{m,n} &= \langle f, S^{-1}(g_{mb,na})\rangle = \langle f, \tilde{g}_{mb,na}\rangle = \langle Z_a f, Z_a(\tilde{g}_{mb,na})\rangle \\
&\langle Z_a f, E_{2\pi m,-2\pi n}Z_a(\tilde{g})\rangle = \langle Z_a f, E_{2\pi m,-2\pi n}\frac{1}{\overline{(Z_a g)}}\rangle\rangle = \langle \frac{Z_a f}{(Z_a g)}, E_{2\pi m,-2\pi n}\rangle,
\end{aligned}$$

which is (29.16). ∎

29.9 The Zak Transform of Generalized Functions

29.9.1 Definition

Since the Zak transform is not an integral transform, the methods described in Section 4.5 do not apply. Two different approaches to extend the Zak transform to generalized functions have been used; one is based on the notion of a tensor product of generalized functions and the other employs the convolution structure of the Zak transform. The latter, which was introduced by A. Zayed and P. Mikusinski [270] requires a new introduction to generalized functions based on Mikusinski's operators. Because this approach to generalized functions is different from the one introduced in Chapter 3, it will not be discussed here. Below we shall outline the former approach, which is based on the tensor product of generalized functions, as introduced by A. Janssen [112].

Let $\mathbf{S}$ be the space of all entire functions f for which there exist $A, B, M > 0$ such that

$$|f(x+iy)| \leq Me^{(-\pi Ax^2+\pi By^2)}, \quad x, y \in \mathcal{R}. \tag{29.20}$$

The topology of $\mathbf{S}$ is defined as follows: a sequence $\{f_n\}_{n=0}^{\infty} \subset \mathbf{S}$ converges to zero in $\mathbf{S}$ if there exist $A, B, M > 0$ such that (29.20) holds for all f_n and $f_n \to 0$ uniformly as $n \to \infty$. The dual space of $\mathbf{S}$ will be denoted as usual by $\mathbf{S}^*$. The set S_α^β $(\alpha, \beta \geq 0)$ is the set of all infinitely differentiable functions f satisfying

$$\left|x^k f^{(q)}(x)\right| \leq CA^k B^q k^{\alpha k} q^{\beta q}, \quad k, q = 0, 1, 2, \cdots,$$

for some positive constants A, B and C. It can be shown that any $f \in S_\alpha^\beta$ satisfies the inequality

$$\left|f^{(q)}(x)\right| \leq CB^q q^{\beta q} \exp(-a|x|^{1/\alpha}), \quad q = 0, 1, 2, \cdots,$$

for some $a > 0$. If $0 \leq \beta < 1$, then f can be continued analytically to the entire complex z-plane, where $z = x + iy$, and in addition, f satisfies

$$\left|x^k f(x+iy)\right| \leq CA^k k^{\alpha k} \exp(b|y|^{1/(1-\beta)}).$$

The space $\mathbf{S}$ can be identified with the space $S_{1/2}^{1/2}$. The space $\mathbf{S}^*$ contains functions F for which $\int_{-\infty}^{\infty} |F(x)|e^{-\epsilon x^2} dx < \infty$ for any $\epsilon > 0$. For, it is easy to show that the integral

$$\langle F, f \rangle = \int_{-\infty}^{\infty} F(x)\overline{f}(x) dx$$

exists for any $f \in \mathbf{S}$ and that if $\lim_{n\to\infty} f_n = 0$ in $\mathbf{S}$, then $\lim_{n\to\infty}\langle F, f_n\rangle = 0$.

To extend the Zak transform to generalized functions, A. Janssen [112] used the notion of a tensor product of generalized functions. He defined the tensor product, $\otimes$, of two generalized functions $F_1, F_2 \in \mathbf{S}^*$ as

$$\langle F_1 \otimes F_2, f_1 \otimes f_2\rangle = \langle F_1, f_1\rangle\langle F_2, f_2\rangle,$$

for $f_1, f_2 \in \mathbf{S}$. In particular, if $F_2 = 1$, we have

$$\langle F_1 \otimes 1, f_1 \otimes f_2\rangle = \langle F_1, f_1\rangle \int_{-\infty}^{\infty} f_2(x)dx.$$

DEFINITION 29.2 *The Zak transform of a generalized function $F \in \mathbf{S}^*$ is defined as*

$$(ZF)(t,w) = \sum_{k=-\infty}^{\infty} T^1_{-k}E^2_{-2\pi k}(F\otimes 1) = \sum_{k=-\infty}^{\infty} (T_{-k}F \otimes E_{-2\pi k}1),$$

where the superscripts 1,2 on the translation and modulation operators indicate that the operator is applied to the first and the second element in the tensor product, respectively.

The definition makes sense because for $f_1, f_2 \in \mathbf{S}$, and $F \in \mathbf{S}^*$ the series

$$\sum_{k=-\infty}^{\infty} \langle T_{-k}F \otimes E_{-2\pi k}1, f_1 \otimes f_2\rangle = \sum_{k=-\infty}^{\infty} \langle T_{-k}F, f_1\rangle \overline{\hat{f_2}}(2\pi k),$$

converges absolutely. Moreover, if $F \in \mathbf{S}$ or $F \in \mathcal{S}$, then Definition 29.2 agrees with Definition 29.1.

Janssen showed that the Zak transform is a continuous linear transformation from $\mathbf{S}^*$ into $\mathbf{S}^* \times \mathbf{S}^*$ and from $\mathcal{S}^*$ into $\mathcal{S}^* \times \mathcal{S}^*$. Furthermore, if F is in $\mathbf{S}^*$ or in $\mathcal{S}^*$ and f is in $\mathbf{S}$ or $\mathcal{S}$, respectively, then an analogue of (29.8) holds, namely

$$\sum_{m,n=-\infty}^{\infty} \langle F, E_{-2\pi k}T_{-n}f\rangle e^{2\pi imt}e^{2\pi inw} = (ZF)(t,w)\overline{(Zf)}(t,w),$$

where the series converges in the sense of $\mathbf{S}^* \times \mathbf{S}^*$ or $\mathcal{S}^* \times \mathcal{S}^*$.

29.9.2 Characterization

The Zak transform of $\mathbf{S}, \mathbf{S}^*, \mathcal{S}$ and $\mathcal{S}^*$ can be characterized as follows: $Z(\mathbf{S})$ is the set of all entire functions f of two variables satisfying

$$f(z+1,w) = e^{2\pi iw}f(z,w), \quad f(z,w+1) = f(z,w),$$

and for which there exist constants $A, B, M > 0$ such that

$$|f(x+iy, u+iv)| \leq M \exp(2\pi xv + \pi Ay^2 + \pi Bv^2),$$

while

$$Z(\mathbf{S}^*) = \{F \in \mathbf{S}^* \times \mathbf{S}^* : T^1_{-1}F = E^2_{2\pi}F \text{ and } T^2_{-1}F = F\}.$$

On the other hand, $Z(\mathcal{S})$ is the set of all functions f of two variables that are infinitely differentiable in each variable and satisfy

$$f(z+1, w) = e^{2\pi i w} f(z, w), \quad f(z, w+1) = f(z, w),$$

while

$$Z(\mathcal{S}^*) = \{F \in \mathcal{S}^* \times \mathcal{S}^* : T^1_{-1}F = E^2_{2\pi}F \text{ and } T^2_{-1}F = F\}.$$

29.10 Applications

As we pointed out in the introduction of this chapter, the Zak transform was brought to light in 1967 by a physicist, J. Zak, in one of his papers on solid-state physics [254]. In that paper, he called the Zak transform the k-q representation and used it to solve Schrödinger's equation for an electron subject to a periodic potential in a constant magnetic field. Thereafter, the Zak transform has been used successfully in various applications in physics, such as in the study of the coherent states representation in quantum field theory [121], and in electrical engineering, such as in time-frequency representation of signals and in digital data transmission. For more references on other applications of the Zak transform, we refer the reader to Janssen's paper [111].

The applications of the Zak transform are not limited only to physics and engineering. More recent applications of it in mathematics have proved to be very useful; in particular, to simplify proofs of some important results. A case in point is the Gabor representation problem, where the Zak transform has been used to study the orthogonality and the completeness of the Gabor frames in the crucial case where $ab = 1$.

In this section we describe briefly the Gabor representation problem in so far as it is related to the Zak transform. A more detailed discussion of this problem can be found in the chapter on the Gabor transform, Chapter 30.

The Gabor representation problem can be stated as follows. Given $g \in L^2(\mathcal{R})$ and two real numbers, a, b, different from zero, is it possible to represent any function $f \in L^2(\mathcal{R})$ in a series of the form

$$f = \sum_{m,n=-\infty}^{\infty} c_{m,n} g_{mb,na},$$

where $g_{mb,na}$ are the Gabor functions defined by

$$g_{mb,na}(x) = e^{2\pi imbx} g(x - na),$$

and $c_{m,n}$ are constants? And under what conditions is the representation unique?

Clearly, the required representation is possible if the functions $\{g_{mb,na}\}$ form an orthonormal basis or a frame for $L^2(\mathcal{R})$ and the uniqueness of the representation depends on whether these functions form a complete set in $L^2(\mathcal{R})$. To answer the above questions, when $ab = 1$, the Zak transform is used. But first, let us observe that because m and n run over the integers, there is no loss of generality in assuming that $a, b > 0$. Second, let us recall from (29.12) that if $ab = 1$, then

$$Z_a\left(g_{mb,na}\right)(t, w) = E_{2\pi m,-2\pi n}(t, w) Z_a\left(g\right)(t, w).$$

THEOREM 29.6

Let $a, b > 0$ with $ab = 1$ and let $g \in L^2(\mathcal{R})$. Then

1. *$\{g_{mb.na}\}$ is an orthonormal basis of $L^2(\mathcal{R})$ if and only if $|Z_a(g)| = 1$ almost everywhere,*

2. *$\{g_{mb.na}\}$ is complete in $L^2(\mathcal{R})$ if and only if $|Z_a(g)| > 0$ almost everywhere.*

Proof: 1) From relations (29.11), (29.12) and Theorem 29.3, we have

$$\begin{aligned}\langle g_{kb,la}, g_{mb,na}\rangle &= \langle Z_a(g_{kb,la}), Z_a(g_{mb,na})\rangle \\ &= \int_0^1 \int_0^1 E_{2\pi k,-2\pi l}(t, w) \overline{E}_{2\pi m,-2\pi n}(t, w) \left|Z_a(g)\right|^2 dt dw,\end{aligned}$$

which implies that the set $\{Z_a(g_{mb,na})\}$ is an orthonormal basis of $L^2(Q)$ if and only if the set $\{g_{mb.na}\}$ is an orthonormal basis of $L^2(\mathcal{R})$ if and only if $|Z_a(g)| = 1$ almost everywhere.

2) Similarly, the relation

$$\langle f, g_{mb,na}\rangle = \langle Z_a(f), Z_a(g_{mb,na})\rangle = \langle Z_a(f)\overline{Z_a(g)}, E_{2\pi m,-2\pi n}\rangle,$$

implies that $\{g_{mb.na}\}$ is complete in $L^2(\mathcal{R})$ if and only if $Z_a g \neq 0$ almost everywhere. ∎

We have seen in Section 29.4 that if $Z_a(g)$ is continuous, it must have a zero in Q; hence $|Z_a g|$ cannot be bounded below away from zero. Consequently, by Theorem 29.6 such a function g cannot generate a frame for $L^2(\mathcal{R})$ when $ab = 1$. However, if $Z_a(g) = 0$, only on a set of measure zero, then the set $\{g_{mb.na}\}$ is complete, though it may not be a frame. An example of such a set is the one generated by the Gaussian function $g(x) = \exp(-cx^2)$ because the Zak transform of g is continuous and has a single zero in Q. Nevertheless, the Gaussian function is known to generate a frame when $0 < ab < 1$.

On the other hand, for $a = 1 = b$, and $g(x) = \chi_{[-1/2,1/2)}(x)$, the Zak transform of g satisfies $|Z(g)| = 1$ on Q; hence by Theorem 29.6, g generates an orthonormal basis for $L^2(\mathcal{R})$, but when $0 < a, b < 1$, it generates only a frame. It is easy to see by Parseval's relation for the Fourier transform that the Fourier transform of a frame of $L^2(\mathcal{R})$ is also a frame for $L^2(\mathcal{R})$. And since the Fourier transform of $\chi_{[-1/2,1/2)}$ is $(1/\sqrt{2\pi})\sin(w/2)/(w/2)$ it follows that $\sin(w/2)/(w/2)$ generates a frame for $L^2(\mathcal{R})$. This is an example of a frame for $L^2(\mathcal{R})$ that is generated by a non-compactly supported function.

We have just seen that in the case where $ab = 1$ no *nice* function (with good localization in the time–frequency plane) can generate a frame of the form $\{g_{mb.na}\}$ for $L^2(\mathcal{R})$. This statement is rephrased more precisely in the Balian–Low theorem, which incidentally can also be proved by using the Zak transform.

THEOREM 29.7 **(Balian–Low)**
Let $a, b > 0$ with $ab = 1$ and let $g \in L^2(\mathcal{R})$. If $\{g_{mb.na}\}$ is a frame in $L^2(\mathcal{R})$, then either $tg(t) \notin L^2(\mathcal{R})$, or $w\hat{g}(w) \notin L^2(\mathcal{R})$.

For a proof of this theorem using the Zak transform, we refer the reader to [49, p. 108].

30

The Gabor Transform

The Gabor and the Zak transformations are intimately related. Both are also linear transformations that map a function of n variables into a function of $2n$ variables. But unlike the Zak transformation, the Gabor transformation is an integral transformation.

The Gabor transformation was introduced in 1946 by Dennis Gabor [67], a Hungarian-British physicist and engineer, who won the Nobel Prize in physics in 1971 for his invention and development of the holographic method. Realizing some of the limitations of the Fourier transform analysis, Gabor introduced a transformation that now bears his name, as a way to overcome some of these limitations. Admittedly, the essence of Gabor's idea had been known to theoretical physicists for some time, the phase-space and the coherent states representations attest to that. Not surprisingly, some of today's jargon pertaining to the Gabor transform has its roots in quantum physics, as we will see shortly.

Recall that if a function of time $f(t)$ represents a signal in the time-domain, then its Fourier transform $F(w)$ represents its frequency spectrum. Therefore, by using the Fourier transform and its inverse, we can go back and forth between the time and frequency representations of the signal, but we cannot view the distribution of the various frequencies in time. This is one of the shortfalls of the Fourier transform analysis since it precludes a very practical notion, namely the notion of frequencies changing with time or equivalently the notion of finding the frequency content of a signal locally in time. For example, when a musician plays a piece of music and passes from a low to a high note, the frequency changes with time. Therefore, to reproduce this musical piece, we need to know which frequency to play at which instant of time.

One of the earliest mathematical techniques used to remedy this deficiency in the Fourier transform analysis is the Gabor transform. It permits a time-frequency representation of a signal, hence making it possible to view the frequency spectrum locally in time. A number of generalizations of Gabor's idea have since then been introduced. One such generalization is the windowed-Fourier transform, also called the short-time Fourier transform. Many authors, however, regard the Gabor and the windowed-Fourier transforms as synonymous.

30.1 Definitions

DEFINITION 30.1 *We define the Gabor transform (windowed-Fourier transform) of a function f with respect to a function g as*

$$\mathcal{G}[f](\mathbf{w},\mathbf{t}) = F_g(\mathbf{w},\mathbf{t}) = \frac{1}{(2\pi)^{n/2}} \int_{-\infty}^{\infty} \cdots \int_{-\infty}^{\infty} f(\mathbf{x}) g(\mathbf{x}-\mathbf{t}) e^{i(\mathbf{w}.\mathbf{x})} d\mathbf{x}\,, \tag{30.1}$$

where $\mathbf{t},\mathbf{w},\mathbf{x} \in \mathcal{R}^n$, $d\mathbf{x} = dx_1...dx_n, \mathbf{x} = (x_1, ..., x_n)$ *and* $\mathbf{w}.\mathbf{x} = \sum_{k=1}^{n} w_k x_k$. *The function g is called the window function or the analyzing function.*

In the remainder of this chapter, we will confine ourselves to the case $n = 1$, and for simplicity if there is only one window function under consideration, we will drop the subscript g. It should be noted that the original Gabor transform was defined with respect to the Gaussian function $g(x) = e^{-x^2}/\sqrt[4]{\pi}$ and was later extended to the above form and called the windowed Fourier transform. That is the reason some authors call the above transform the windowed-Fourier transform and only when the window function is the Gaussian function do they call it the Gabor transform

In practice, f and g represent finite-energy signals; hence $f\,,\ g \in L^2(\mathcal{R})$. Except for this requirement, the choice of g is rather arbitrary. There are, however, some other physical and mathematical considerations that may affect the choice of g. For example, since functions with compact supports in the time domain have infinite supports in the frequency domain and vice versa, there is a tradeoff between time localization and frequency resolution as suggested by the uncertainty principle of the Fourier transform (see Theorem 12.9). Apparently, Gabor's choice for the Gaussian function as a window function was due to the fact that the Gaussian function is optimally localized in both time and frequency in the sense that it attains the equality in the inequality (12.16).

To calculate the Gabor transform of f, first the window function is shifted in time, then the Fourier transform of the product of f and the shifted window is computed. This produces separate Fourier transforms for each location of the window center which, unlike the Fourier transform, allows us to view the signal f in the time and frequency domains simultaneously.

DEFINITION 30.2 *We call the functions $g_{w,t}(x) = g(x-t)e^{iwx}$ the Gabor transform functions or the windowed-Fourier transform functions. When $g(x) = (\pi)^{-\frac{1}{4}}e^{-x^2/2}$, $g_{w,t}(x)$ are called the canonical coherent states and the corresponding Gabor transform of f is called the canonical coherent state representation of f.*

The Gabor transform of f with respect to g can now be written in the form

$$F_g(w,t) = \frac{1}{\sqrt{2\pi}}\langle f\,,\,\overline{g}_{w,t}\rangle\,.$$

The term "coherent states" was first used by R. Glauber [75] in quantum optics, but it was not long before it spilled over to other fields of mathematical physics and now it is used in a more general context. A question of great importance in mathematics and physics is this: Can a function f be reconstructed from the values of its Gabor transform, $F(w,t)$, at a discrete set of sampling points? The standard choice for the sampling points is $w = mw_0$ and $t = nt_0$, where $w_0, t_0 > 0$. To answer this question, the following definitions are needed.

DEFINITION 30.3 *The functions*

$$g_{m,n}(x) = g(x-nt_0)e^{imw_0x} = E_{w_0m}T_{nt_0}g(x)$$

are called the discrete Gabor (windowed-Fourier transform) functions or the Weyl–Heisenberg coherent states. If $\{g_{m,n}(x)\}_{m,n=-\infty}^{\infty}$ form a frame for $L^2(\mathcal{R})$, it is called a Gabor frame. The set of sample points $\{(mw_0, nt_0)\}_{m,n=-\infty}^{\infty}$ is called a Gabor lattice.

DEFINITION 30.4 *The sequence $\{F(m,n)\}_{m,n=-\infty}^{\infty}$, where*

$$F(m,n) = \int_{-\infty}^{\infty} f(x)\overline{g}_{m,n}(x)dx\,,$$

is called the discrete Gabor transform of $f(x)$ and the series $\sum_{m,n=-\infty}^{\infty} F(m,n)g_{m,n}(x)$ is called the Gabor series of f. A general series of the form $\sum_{m,n=-\infty}^{\infty} c_{m,n}g_{m,n}(x)$ is called a Gabor series.

Finding the conditions on t_0, w_0 and g under which the Gabor series of f determines f or converges to it is known as the *Gabor representation problem.* Gabor addressed this problem in the case where g is the Gaussian function and $t_0w_0 = 2\pi$.

The Fourier transform of $g_{m,n}(x)$, provided that it exists, is easily seen to be

$$\hat{g}_{m,n}(w) = e^{int_0(w+mw_0)}\hat{g}(w+mw_0)\,,$$

hence

$$\|\hat{g}_{m,n}\| = \|\hat{g}_{m,n}\|_2 = \|\hat{g}\|_2 \,.$$

If g is centered around $(0,0)$ in the time-frequency domain, i.e.,

$$\int_{-\infty}^{\infty} x|g(x)|^2 dx = 0 = \int_{-\infty}^{\infty} w|g(w)|^2 dw \,, \quad \text{with} \quad \|g\| = 1 \,,$$

then

$$\int_{-\infty}^{\infty} x|g_{m,n}(x)|^2 dx = nt_0 \quad \text{and} \quad \int_{-\infty}^{\infty} w|\hat{g}_{m,n}(w)|^2 dw = -mw_0 \,.$$

30.2 Existence

In most applications, the window function is taken as a bounded function with compact support or of rapid decay at $\pm\infty$. In either case, if f is absolutely integrable, $F(w,t)$ exists almost everywhere in $\mathcal{R}^2$. In many engineering applications the window function represents a signal with finite energy, that is $g \in L^2(\mathcal{R})$. Hence, by the Cauchy-Schwarz inequality F exists almost everywhere whenever $f \in L^2(\mathcal{R})$.

As for the discrete Gabor transform, the following is a consequence of the definition of a frame: If $f \in L^2(\mathcal{R})$, and $\{g_{m,n}\}$ is a frame in $L^2(\mathcal{R})$, then $\sum_{m,n=-\infty}^{\infty} |\langle f, g_{m,n}\rangle|^2 < \infty$. Hence, the Gabor series of f converges in $L^2(\mathcal{R})$, but not necessarily to f. A sufficient condition for $\{g_{m,n}\}$ to be a frame in $L^2(\mathcal{R})$ is given in the next theorem.

THEOREM 30.1
Let $g \in L^2(\mathcal{R})$ be such that

$$0 < A \le G(x) = \sum_n |g(x - na)|^2 \le B < \infty \,, \ a.e. \,, \tag{30.2}$$

and define $g_{m,n}$ as in Definition 30.3, but with $t_0 = a$ and $w_0 = b$. If g has compact support, with supp(g)$\subset I$, where I is an interval of length $1/b$, then $\{g_{m,n}\}$ is a frame in $L^2(\mathcal{R})$, with frame bounds $b^{-1}A$ and $b^{-1}B$. Moreover, for any $f \in L^2(\mathcal{R})$, we have

$$Sf(x) = b^{-1} f(x) G(x),$$

where S is the frame operator. The frame operator is invertible and

$$S^{-1} f(x) = bf(x)/G(x). \tag{30.3}$$

Proof: Let $I_n = I + na = \{x + na : x \in I\}$. For fixed n,

$$f(x)\overline{g(x-na)} \in L^2(I_n)$$

since g is bounded and $f \in L^2(\mathcal{R})$. Let $h_n(x) = f(x)\overline{g(x-na)}$ and $e_m(x) = \sqrt{b}\, E_{2\pi mb}(x) = \sqrt{b}\, e^{2\pi imbx}$. In view of the fact that $\{e_m(x)\}_{m=-\infty}^{\infty}$ is an orthonormal basis of $L^2(I_n)$, we obtain

$$h_n(x) = \sum_m \langle h_n, e_m\rangle e_m(x),$$

which by Parseval's equality for the Fourier series (see Example A, Section 1.5) yields

$$\begin{aligned}\int_{I_n} |h_n(x)|^2 dx &= \sum_m |\langle h_n, e_m\rangle|^2 \\ &= b\sum_m \left| \int_{I_n} f(x)\overline{g(x-na)}\,\overline{E_{2\pi mb}(x)} \right|^2 dx \\ &= b\sum_m |\langle f, E_{2\pi mb}T_{na}g\rangle|^2 .\end{aligned}$$

Therefore, since g has compact support, we have

$$\sum_n \int_{I_n} |h_n(x)|^2\, dx = \sum_n \int_{-\infty}^{\infty} |f(x)|^2\, |g(x-na)|^2\, dx =$$

$$\int_{-\infty}^{\infty} |f(x)|^2\, G(x)\, dx = b\sum_{n,m} |\langle f, E_{2\pi mb}T_{na}g\rangle|^2 . \tag{30.4}$$

From (30.2) and (30.4), it follows that

$$b^{-1}A\,\|f\|^2 \le \sum_{n,m} |\langle f, g_{m,n}\rangle|^2 \le b^{-1}B\,\|f\|^2 ,$$

hence $\{g_{m,n}\}$ is a frame for $L^2(\mathcal{R})$ with the prescribed bounds.

Next, from the definition of the frame operator S, we obtain

$$[Sf](x) = \sum_{m,n} \langle f, g_{m,n}\rangle e^{2\pi imbx} g(x-na) = \sum_n g(x-na)H_n(x), \tag{30.5}$$

where

$$H_n(x) = \sum_m \langle f, g_{m,n}\rangle e^{2\pi imbx}. \tag{30.6}$$

Evidently, $H_n(x)$ is a periodic function with period $1/b$ and hence (30.6) is its Fourier series expansion. Therefore,

$$\begin{aligned}\int_{I_n} H_n(x)e^{-2ximbx}dx &= \frac{1}{b}\int_{-\infty}^{\infty} f(x)\overline{g(x-na)}e^{-2ximbx}dx \\ &= \frac{1}{b}\int_{I_n} f(x)\overline{g(x-na)}e^{-2ximbx}dx,\end{aligned}$$

since supp $g \subset I$. From the completeness of the set $\{E_{2\pi mb}\}_{m=-\infty}^{\infty}$ in $L^2(I_n)$, we conclude that

$$H_n(x) = \frac{1}{b} f(x)\overline{g(x - na)}. \tag{30.7}$$

By substituting (30.7) into (30.5) we obtain

$$[Sf](x) = \frac{1}{b} \sum_n f(x) |g(x - na)|^2 = \frac{1}{b} f(x)G(x). \tag{30.8}$$

By Theorem 1.7, S is invertible and hence by replacing f by $S^{-1}f$ in (30.8) we obtain (30.3). ∎

A more general condition for $\{g_{m,n}\}$ to be a frame in $L^2(\mathcal{R})$, is given in Section 8 of this chapter and for more general results on the Gabor frames, see [95] .

The following is true if g is the Gaussian function [115]:

1. If $f \in L^2(\mathcal{R})$, then

$$\sum_{m,n=-\infty}^{\infty} |\langle f, g_{m,n}\rangle|^2 < \infty ,$$

2. If $f \in L^1(\mathcal{R})$, then the sequence $\{\langle f(x) , g_{m,n}(x)\rangle\}_{m,n}$ is bounded in m and n and tends to zero as $m^2 + n^2 \to \infty$,
3. If $f \in L^p(\mathcal{R})(1 < p < 2)$, the $\sum_{m,n} |\langle f, g_{m,n}\rangle|^q < \infty$ where $\frac{1}{p} + \frac{1}{q} = 1$.

It should be stressed that in (1) nothing has been said to entail that $\{g_{m,n}\}$ is a frame in $L^2(\mathcal{R})$. In fact, it is not.

30.3 Elementary Properties

1. **Linearity:** The Gabor transformation is a linear transformation.
2. **Translation:** If we denote the Gabor transformation by $\mathcal{G}$, then
$$\mathcal{G}[f(x+a)](w,t) = e^{-iaw}\mathcal{G}[f(x)](w,t+a)$$
3. **Modulation:**
$$\mathcal{G}[e^{iax}f(x)](w,t) = \mathcal{G}[f(x)](w+a,t)$$
4. **Conjugation:** If the window function g is real, then
$$\mathcal{G}[\overline{f}(x)](w,t) = \overline{\mathcal{G}}[f(x)](-w,t)$$

30.4 Parseval's Relation

THEOREM 30.2
Let the window function $g \in L^2(\mathcal{R})$ and assume that $f, h \in L^2(\mathcal{R})$. Denote the Gabor transforms of f and h by F and H, respectively. Then Parseval's relation for the Gabor transform takes the form

$$\langle F, H\rangle = \langle F, H\rangle_{L^2(\mathcal{R}^2)} = \|g\|^2\langle f, h\rangle \ , \tag{30.9}$$

where

$$\langle F, H\rangle = \int_{-\infty}^{\infty}\int_{-\infty}^{\infty} F(t,w)\overline{H}(t,w)dtdw \ .$$

In particular, if $\|g\| = 1$, then the Gabor transformation is an isometry from $L^2(\mathcal{R})$ into $L^2(\mathcal{R}^2)$.

Proof: To prove (30.9), note that for a fixed t, $F(w,t)$ is the Fourier transform of the function $f(x)g_t(x)$ where $g_t(x) = g(x-t)$, i.e., $F(w,t) = \mathcal{F}(fg_t)(w)$. Therefore, by Parseval's relation for the Fourier transform, we have

$$\begin{aligned}\int_{-\infty}^{\infty} F(w,t)\overline{H}(w,t)dw &= \langle\mathcal{F}(fg_t), \mathcal{F}(hg_t)\rangle = \langle fg_t, hg_t\rangle \\ &= \int_{-\infty}^{\infty} f(x)\overline{h}(x)|g(x-t)|^2dx \ .\end{aligned}$$

The result now follows by integrating over t from $-\infty$ to ∞. ∎

30.5 Inversion Formula

The inversion formula for the Gabor transform is

$$f(x) = \frac{1}{\sqrt{2\pi}\|g\|^2}\int_{-\infty}^{\infty}\int_{-\infty}^{\infty} F(w,t)\overline{g}(x-t)e^{-ixw}dtdw \ . \tag{30.10}$$

This formula follows from (30.9) since for any fixed f we have

$$\begin{aligned}\langle f, h\rangle &= \frac{1}{\|g\|^2}\int_{-\infty}^{\infty}\int_{-\infty}^{\infty} F(w,t)\left(\frac{1}{\sqrt{2\pi}}\int_{-\infty}^{\infty}\overline{h}(x)\overline{g}(x-t)e^{-iwx}dx\right)dtdw \\ &= \langle\eta, h\rangle \ ,\end{aligned}$$

where

$$\eta(x) = \frac{1}{\sqrt{2\pi}\|g\|^2} \int_{-\infty}^{\infty} \int_{-\infty}^{\infty} F(w,t)\overline{g}(x-t)e^{-iwx}dtdw \,.$$

But since this is true for all $h \in L^2(\mathcal{R})$, it follows immediately that $f = \eta$, which is (30.10).

30.6 L^2-Theory

We have seen that the Gabor transformation is a linear transformation from $L^2(\mathcal{R})$ into $L^2(\mathcal{R}^2)$ that preserves the inner product, provided that the window function, g, is normalized so that $||g|| = 1$. The range of the transformation is a reproducing-kernel Hilbert space with reproducing kernel

$$K(t,w,t',w') = \frac{1}{(2\pi\|g)\|^2} \int_{-\infty}^{\infty} g(x-t)\overline{g}(x-t')e^{iwx-iw'x}dx \,,$$

see Section 1.10.2. For, in view of the inversion formula (30.10), we have

$$\int_{-\infty}^{\infty} \int_{-\infty}^{\infty} F(w',t')K(t,w,t',w')dt'dw'$$

$$= \frac{1}{(2\pi)\|g\|^2} \int_{-\infty}^{\infty} g(x-t)e^{iwx}dx \int_{-\infty}^{\infty} \int_{-\infty}^{\infty} F(w',t')\overline{g}(x-t')e^{-iw'x}dt'dw'$$

$$= \frac{1}{\sqrt{2\pi}} \int_{-\infty}^{\infty} f(x)g(x-t)e^{iwx}dx = F(w,t) \,.$$

If $g(q)$ is the Gaussian function $(1/\sqrt[4]{\pi})e^{-q^2/2}$, we can relate the Gabor transform to the Bargmann transform and the range of the Gabor transformation to that of the Bargmann transformation; see Chapter 28. The latter is called the Bargmann–Segal–Fock space. Recall that the Bargmann transform of a function $f \in L^2(\mathcal{R})$ is a function of the complex variable, $z = x + iy$, defined by

$$\mathcal{A}[f](z) = \frac{1}{\sqrt[4]{\pi}} \int_{-\infty}^{\infty} f(q)e^{-(z^2+q^2)/2+\sqrt{2}zq}dq \,,$$

$$= \frac{1}{\sqrt[4]{\pi}} e^{(x^2+y^2)/2-ixy} \int_{-\infty}^{\infty} f(q)e^{-(q-\sqrt{2}x)^2/2+i\sqrt{2}yq}dq.$$

Now it is easy to show that

$$\mathcal{A}[f](\frac{t}{\sqrt{2}} + i\frac{w}{\sqrt{2}}) = \sqrt{2\pi}e^{(t^2+w^2)/4-itw/2}\mathcal{G}[f](t,w). \tag{30.11}$$

The Bargmann transformation is a unitary transformation from $L^2(\mathcal{R})$ onto the Bargmann–Segal–Fock space, which is the space of all entire functions $F(z)$ satisfying

$$\int_{-\infty}^{\infty}\int_{-\infty}^{\infty} |F(z)|^2 \, d\mu(z) < \infty ,$$

where $d\mu(z) = (1/\pi)e^{-(x^2+y^2)}dxdy$. The Bargmann–Segal–Fock space, denoted by $\mathbf{S}$, is a Hilbert space with inner product

$$\langle F \, , \, H\rangle_{\mathbf{S}} = \int_{-\infty}^{\infty}\int_{-\infty}^{\infty} F(z)\overline{H}(z)d\mu(z).$$

Equation (30.11) defines an isometry from the range of the Gabor transformation onto the Bargmann–Segal–Fock space. In fact, one can easily verify that if f and h are in $L^2(\mathcal{R})$, then

$$\langle \mathcal{A}[f] \, , \, \mathcal{A}[h]\rangle_{\mathbf{S}} = \langle \mathcal{G}[f] \, , \, \mathcal{G}[h]\rangle_{L^2(\mathcal{R}^2)} = \langle f \, , \, h\rangle_{L^2(\mathcal{R})}.$$

Returning to Equation (30.10), we remark that if g is normalized so that $\|g\| = 1$, then formula (30.10) may be written in the form

$$\frac{1}{2\pi}\int_{-\infty}^{\infty}\int_{-\infty}^{\infty} \langle f, \overline{g}_{w,t}\rangle \overline{g}_{w,t}(x)dtdw = f$$

or

$$\frac{1}{2\pi}\int_{-\infty}^{\infty}\int_{-\infty}^{\infty} \langle \cdot \, , \overline{g}_{w,t}\rangle \overline{g}_{w,t}(x)dtdw = I \, ,$$

where I is the identity operator. In particular, if S is a measurable subset of $\mathcal{R}^2$, we define the operator L_S as

$$L_S = \frac{1}{2\pi}\iint_S \langle \cdot \, , \overline{g}_{w,t}\rangle \overline{g}_{w,t}(x)dtdw \, .$$

Thus, in view of (30.9), we have

$$\begin{aligned}\langle L_S f, f\rangle &= \frac{1}{2\pi}\iint_S dtdw|\langle f, \overline{g}_{w,t}\rangle|^2 \\ &\leq \frac{1}{2\pi}\int_{-\infty}^{\infty}\int_{-\infty}^{\infty} |\langle f, \overline{g}_{w,t}\rangle|^2 dtdw = \|f\|^2 \, ,\end{aligned}$$

which implies that $0 \leq L_S \leq I$; see Section 1.5.

A bounded operator A in a Hilbert space $\mathcal{H}$ is called a *trace-class* if $\sum_{n=0}^{\infty} |\langle Ae_n, e_n\rangle| < \infty$ for all orthonormal bases $\{e_n\}_{n=0}^{\infty}$ of $\mathcal{H}$. For such an operator, $\sum_{n=0}^{\infty}\langle Ae_n, e_n\rangle$ is independent of the chosen basis and such a series is called the *trace* of A, i.e., trace $A = \sum_{n=0}^{\infty}\langle Ae_n, e_n\rangle$. If $\mathcal{H}$ has finite dimension, then trace A is the sum of the diagonal elements of any matrix representation of A. To show that a positive operator A is trace-class, it is sufficient to show that $\sum_{n=0}^{\infty}\langle Ae_n, e_n\rangle$ is finite for only one orthonormal basis.

Let $\{\phi_n\}_{n=0}^{\infty}$ be an orthonormal basis of $L^2(\mathcal{R})$ and S be bounded. Then

$$\begin{aligned}\sum_{n=0}^{\infty}|\langle L_S u_n, u_n\rangle| &= \frac{1}{2\pi}\iint_S dtdw\sum_{n=0}^{\infty}|\langle u_n\,, \overline{g}_{w,t}\rangle|^2\\ &= \frac{1}{2\pi}\iint_S dtdw\|g\|^2\\ &= \frac{1}{2\pi}|S| < \infty\,,\end{aligned}$$

where $|S|$ is the measure of S. Hence L_S is a trace-class operator.

Since trace-class operators are compact [45, p. 267-268], it follows that L_S has a complete set of eigenvectors $\{\phi_n\}_{n=0}^{\infty}$ and a decreasing sequence of eigenvalues $\{\lambda_n\}_{n=0}^{\infty}$ satisfying:

$$\begin{aligned}&L_S\phi_n = \lambda_n\phi_n\,,\\ &\lambda_n \geq \lambda_{n+1} \geq 0...\,, \quad \lim_{n\to\infty}\lambda_n = 0\,,\\ &\{\phi_n\}_{n=0}^{\infty} \text{ is an orthonormal basis for } L^2(\mathcal{R})\,.\end{aligned}$$

The eigenvalues and eigenfunctions can be found explicitly when $g(x) = (1/\sqrt[4]{\pi})e^{-x^2/2}$ and $S = \{(w,t) : w^2+t^2 \leq R^2\,, R>0\}$. Since the set S depends on R, we will denote the operator L_S by L_R. It can be shown [49, p.38] that L_R commutes with the differential operator $H = -\frac{d^2}{dx^2}+x^2-1$, and since the eigenfunctions of H are the Hermite functions,

$$\phi_n(x) = c_n H_n(x)e^{-x^2/2}\,,$$

where $c_n = [2^{n/2}\sqrt{n!}\sqrt[4]{\pi}]^{-1}$, and $H_n(x)$ is the Hermite polynomial of degree n, it follows that the Hermite functions are also the eigenfunctions of L_R.

To calculate the corresponding eigenvalues $\lambda_n(R)$, we need to evaluate $\langle\phi_n\,, g_{w,t}\rangle$. To this end, recall that

$$\int_{-\infty}^{\infty} e^{-(x-z)^2}H_n(x)\,dx = \sqrt{\pi}2^n z^n\,, \qquad [79, \text{p.837, no. 6}]\,.$$

Therefore,

$$\begin{aligned}\langle g_{w,t}, \phi_n\rangle &= \frac{c_n}{\sqrt[4]{\pi}}\int_{-\infty}^{\infty} H_n(x)e^{-x^2/2}e^{-(x-t)^2/2+iwx}dx\\ &= \frac{c_n}{\sqrt[4]{\pi}}e^{itw/2-(t^2+w^2)/4}\int_{-\infty}^{\infty} H_n(x)e^{-[x-(t+iw)/2]^2}dx\\ &= \frac{1}{2^{n/2}\sqrt{n!}}(t+iw)^n e^{itw/2-(t^2+w^2)/4}.\end{aligned}$$

From the relation $L_R\phi_n = \lambda_n(R)\phi_n$, we obtain

$$\begin{aligned}\lambda_n(R) = \langle L_R\phi_n\,,\phi_n\rangle &= \frac{1}{2\pi}\int\int_{t^2+w^2\le R} |\langle\phi_n\,,\,g_{w,t}\rangle|^2\,dt\,dw \\ &= \frac{1}{2\pi 2^n n!}\int\int_{t^2+w^2\le R}(t^2+w^2)^n e^{-(t^2+w^2)/2}dt\,dw.\end{aligned}$$

Setting $t + iw = re^{i\theta}$, and integrating over θ lead to

$$\begin{aligned}\lambda_n(R) &= \frac{1}{2^n n!}\int_0^R r^{2n}e^{-r^2/2}r\,dr \\ &= \frac{1}{n!}\int_0^{R^2/2} s^n e^{-s}ds\,.\end{aligned}$$

Clearly, $0 < \lambda_n(R) < 1$ for all n, R, and $\lim_{R\to\infty}\lambda_n(R) = 1$ for all n.

30.7 Relationships with Other Transforms

1. **The Bargmann Transform:**
   If we denote the Bargmann transform of $f \in L^2(\mathcal{R})$ by $\mathcal{A}[f]$, then

   $$\mathcal{G}[f(x)](w,t) = F_g(w,t) = \frac{1}{\sqrt{2\pi}}e^{-(t^2+w^2)/4+itw/2}\mathcal{A}[f](\frac{t}{\sqrt{2}}+i\frac{w}{\sqrt{2}})\,,$$

   where $g(x) = (1/\sqrt[4]{\pi})\exp(-x^2/2)$.

2. **The Convolution Transform:**
   For fixed w

   $$\mathcal{G}[f(x)](w,t) = (e^{iwx}f(x) * g^-(x))(t)\,,$$

   where $g^-(x) = g(-x)$.

3. **The Fourier Transform:**
   For fixed t,

   $$\mathcal{G}[f(x)](w,t) = \mathcal{F}[f(x)g_t(x)](w)\,,$$

   where $g_t(x) = g(x-t)$.

4. **The Weierstrass Transform:**
   For $g(x) = e^{-x^2/4}$,

   $$\mathcal{G}[f(x)](w,t) = \sqrt{2}\,e^{-w^2+itw}(Wf)(t+2iw)\,,$$

   where W denotes the Weierstrass transformation.

30.8 The Gabor Representation Problem

In many physical and engineering applications it is more desirable, at least from a numerical point of view, to work with discrete transforms rather than with continuous ones. Naturally, it is of great interest to know if a function $f \in L^2(\mathcal{R})$ can be determined by its discrete Gabor transform. In other words, is f completely determined by the values

$$F(mw_0, nt_0) = \frac{1}{\sqrt{2\pi}} \int_{-\infty}^{\infty} f(x) g_{m,n}(x)\, dx = \frac{1}{\sqrt{2\pi}} \langle f \,,\, \overline{g}_{m,n} \rangle \quad ,$$

and if so, can it be reconstructed from these values?

The answer to this question, which we call *the Gabor representation problem*, is in the affirmative if the functions $\{g_{m,n}(x)\}_{m,n=-\infty}^{\infty}$ form an orthonormal basis, or more generally if they form a frame for $L^2(\mathcal{R})$. For, in this case,

$$f(x) = \sum_{m,n} \langle f, g_{m,n} \rangle g^*_{m,n}(x) \ ,$$

where $\{g^*_{m,n}(x)\}$ is the dual frame of $\{g_{m,n}(x)\}$.

Therefore, the Gabor representation problem can be rephrased as follows: Under what conditions on the function $g \in L^2(\mathcal{R})$ and the Gabor lattice $\{(mw_0, nt_0)\}$, does the set $\{g_{m,n}\}$ form a frame for $L^2(\mathcal{R})$?

First, let us recall a few facts about this kind of frames:

1) $\{g_{m,n}(x)\}_{-\infty}^{\infty}$ is a frame in $L^2(\mathcal{R})$ if and only if $\{\hat{g}_{m,n}(w)\}_{-\infty}^{\infty}$ is a frame in $L^2(\mathcal{R})$.

2) If $\{g_{m,n}(x)\}$ is the frame generated by g, then its dual frame $\{g^*_{m,n}(x)\}$ is also generated by one single function by translation and modulation.

3) if $\{g_{m,n}\}$ is a frame in $L^2(\mathcal{R})$ with frame bounds A and B, then

$$A \leq \frac{2\pi \|g\|^2}{w_0 t_0} \leq B \ , \tag{30.12}$$

and hence for g to generate a tight frame, the frame bound must be $2\pi\|g\|^2/w_0t_0$. In particular, for $\{g_{m,n}\}$ to be an orthonormal basis, $w_0t_0 = 2\pi$, provided that $\|g\|^2 = 1$.

Fact (1) is an immediate consequence of Parseval's relation for the Fourier transform, (2) is proved in Theorem 29.4, and (3) is proved in [49, p. 81].

The answer to the Gabor representation problem can be summarized as follows:

1) If $0 < t_0 w_0 < 2\pi$, then it is possible for $\{g_{m,n}(x)\}$ to be a frame, provided that g satisfies additional conditions, examples of which are given in the next theorem.

THEOREM 30.3 [49, p. 82]
If g and t_0 are such that

$$0 < \inf_{0 \le x \le t_0} G(x) \le \sup_{0 \le x \le t_0} G(x) < \infty ,$$

where $G(x) = \sum_{n=-\infty}^{\infty} |g(x - nt_0)|^2$, and if

$$\beta(s) = \sup_{0 \le x \le t_0} \sum |g(x - nt_0)||g(x - nt_0 + s)|$$

decays at least as fast as $(1 + |s|)^{-(1+\epsilon)}$ as $s \to \infty$ for $\epsilon > 0$, then there exists $\gamma > 0$ such that $\{g_{m,n}(x)\}$ is a frame whenever $w_0 < \gamma$.

2) If $t_0 w_0 = 2\pi$, which is called the critical case, there is a frame of the form $\{g_{m,n}(x)\}$, but g has bad localization properties either in the time or in the frequency domain. This is made more precise in the Balian-Low theorem (Theorem 29.7), which states that if $\{g_{m,n}(x)\}$ constitutes a frame in $L^2(\mathcal{R})$, then either $\int_{-\infty}^{\infty} x^2 |g(x)|^2 dx = \infty$ or $\int_{-\infty}^{\infty} w^2 |\hat{g}(w)|^2 dw = \infty$.
Since the Gaussian function has very good localization properties, it is impossible for it to generate a frame in this case; however, its associated set $\{g_{m,n}\}$ is complete. This is the original case considered by Gabor.

3) If $w_0 t_0 > 2\pi$, then there is no frame of the form $\{g_{m,n}(x)\}$ whatever the choice of g. In fact, $\{g_{m,n}(x)\}$ is not even complete in the sense that there exists $f \in L^2(\mathcal{R})$ such $\langle f, g_{m,n} \rangle = 0$ for all m, n but $f \neq 0$; see [49, p. 107].

In contrast with the other two cases, in the critical case when $w_0 t_0 = 2\pi$ and only in this case can the family $\{g_{m,n}(x)\}$ be an orthonormal basis. For, by (30.12) it is clear that in order to have a tight frame with frame bounds equal to 1, $w_0 t_0 = 2\pi$.

30.9 Generalized Functions

A. Janssen [114, 115] studied the discrete Gabor transform and the Gabor representation problem of tempered distributions. He considered Gabor series of the form

$$\sum_{m,n=-\infty}^{\infty} c_{n,m} G_{n\alpha,\beta m}(x) \; , \quad \alpha, \beta > 0$$

where

$$G_{a,b}(x) = 2^{\frac{1}{4}} \exp\left[-\pi(x-a)^2 + 2\pi ibx - \pi iab\right] \; .$$

These series are of the form $\sum_{m,n=-\infty}^{\infty} a_{m,n} g_{m,n}(x)$, where

$$a_{m,n} = c_{m,n} e^{-i\pi mn\alpha\beta} \; , \; g(x) = 2^{\frac{1}{4}} e^{-\pi x^2} \; , \; t_0 = \alpha \text{ and } w_0 = 2\pi\beta \; .$$

If $f(x) \in \mathcal{S}$ (Schwarz space of functions), then $h(x,a) = 2^{\frac{1}{4}} f(x) e^{-\pi(x-a)^2} \in \mathcal{S} \times \mathcal{S}$. The Fourier transform of h taken with respect to x is also an element of $\mathcal{S} \times \mathcal{S}$ with $\hat{h}(2\pi b, a) = \frac{1}{\sqrt{2\pi}} e^{-i\pi ab} \langle f(x), \overline{G_{a,b}(x)} \rangle$. Moreover, if $F \in \mathcal{S}^*$, then

$$\langle F(x), G_{a,b}(x) \rangle = e^{-i\pi ab - \pi a^2} \langle F(x), e^{2\pi ix(b-ia)} g(x) \rangle \; .$$

Using Theorem 3.2, we can show that $|\langle F(x), G_{a,b}(x) \rangle| \le C(1+a^2)^N (1+b^2)^N$, for some $C > 0$, and a positive integer N. Conversely, if a sequence of complex numbers, $\{c_{n,m}\}_{m,n=-\infty}^{\infty}$, is such that

$$c_{n,m} = O((1+n^2)^N (1+m^2)^N) \; , \tag{30.13}$$

for some $N > 0$, as $|m|, |n| \to \infty$, then the series $\sum_{m,n} c_{n,m} G_{n\alpha,m\beta}(x)$ for $\alpha > 0, \beta > 0$ is convergent in $\mathcal{S}^*$.

Now let $F \in \mathcal{S}^*$ with $F(x) = \sum_{m,n} c_{n,m} G_{n\alpha,m\beta}(x)$. Clearly, for any integers k and l,

$$\langle F(x), G_{k\alpha,l\beta}(x) \rangle = \sum_{m,n} c_{n,m} \langle G_{n\alpha,m\beta}(x), G_{k\alpha,l\beta}(x) \rangle \; . \tag{30.14}$$

The converse problem, which is intimately related to the Gabor representation problem, is much more subtle. Suppose that $\{c_{n,m}\}_{m,n=-\infty}^{\infty}$, is a sequence of complex numbers satisfying (30.13) so that $\sum_{m,n} c_{n,m} G_{n\alpha,m\beta}(x)$ is in $\mathcal{S}^*$. Let $F \in \mathcal{S}^*$ and suppose that (30.14) holds for all integers k and l, is it true that $F = \sum_{m,n} c_{n,m} G_{n\alpha,m\beta}(x)$?

The answer, in general, depends on α and β. Janssen showed, among other things, that the answer is positive if $\alpha\beta < 1$ and is negative if $\alpha\beta > 1$. In fact, in the latter case there is $F \in \mathcal{S}$ with the property that $F \neq 0$, but $\langle F(x), G_{n\alpha,m\beta}(x) \rangle = 0$ for all m and n. In the critical case where $\alpha\beta = 1$, the answer is more complicated, but can be summarized as follows:

THEOREM 30.4 **(Janssen [115])**
Let $F \in \mathcal{S}^*, \alpha, \beta > 0$, with $\alpha\beta = 1$. Then,

1. *There exists an infinite number of solutions* $\{c_{n,m}\}_{m,n=-\infty}^{\infty}$ *satisfying* (30.13) *such that*

$$\sum_{m,n} c_{n,m}\langle G_{n\alpha,m\beta}(x), G_{k\alpha,l\beta}(x)\rangle = \langle F(x), G_{k\alpha,l\beta}(x)\rangle \,. \qquad (30.15)$$

If $\{c_{n,m}\}, \{d_{n,m}\}$ *are solutions of* (30.15), *there exists a polynomial* P *of two variables such that* $c_{n,m} - d_{n,m} = (-1)^{n+m+nm}P(n,m)$.

2. *If both* F *and* $\sum_{m,n} c_{n,m}G_{n\alpha,m\beta}(x)$ *are regular tempered distributions, then* (30.14) *implies that* $F(x) = \sum_{m,n} c_{n,m}G_{n\alpha,m\beta}(x)$ *in the sense of* $\mathcal{S}^*$.

3. *If* F *is a regular tempered distribution and* $\{c_{n,m}\}$ *satisfies* (30.13), (30.15) *and* $c_{n,m} \to 0$ *as* $m^2+n^2 \to \infty$, *then* $F(x) = \sum_{m,n} c_{n,m}G_{n\alpha,m\beta}(x)$ *in* $\mathcal{S}^*$.

4. *For any* $F \in \mathcal{S}^*$, *there exists a sequence* $\{d_{n,m}\}$ *(not necessarily unique) satisfying* (30.13) *such that* $F(x) = \sum_{m,n} d_{n,m}G_{n\alpha,m\beta}(x)$ *in* $\mathcal{S}^*$.

Part 4 of the theorem shows that any tempered distribution can be represented by a Gabor series. Janssen has generalized this result and showed that any tempered distribution can also be written in the form $\sum_{m,n} a_{m,n}g_{m,n}(x)$, where g is not necessarily a Gaussian function, but generally a nice function, e.g., $g \in \mathcal{S}$.

30.10 Applications

The Gabor transformation provides a representation of a function of n variables by a function of $2n$ variables; an idea that has proved to be very useful in many applications. Examples of its implementation are the time-frequency representation in electrical engineering and the phase-space representation in quantum physics. In fact, many of the ideas presented in this chapter originated in these two fields. For more specific applications in quantum optics, we refer the reader to [6, 121, 122].

31

The Ambiguity Transformation and Wigner Distribution

Analogous to the Zak, the continuous Gabor and wavelet transformations, the ambiguity transformation maps a function of n variables into a function of $2n$ variables, or more precisely it maps a function in $L^2(\mathcal{R}^n)$ into a function in $L^2(\mathcal{R}^{2n})$. But, unlike the Zak and the continuous Gabor and wavelet transformations, the ambiguity transformation is *non-linear.*

The ambiguity transformation is intimately related to what is known in electrical engineering as the radar ambiguity function, which in turn is related to the Wigner distribution function. The Wigner distribution was introduced in 1932 by E. Wigner [247] in the context of quantum mechanics, where he defined the probability function of the simultaneous values of the spatial coordinates, $x_1, ..., x_n$, and the momenta, $p_1, ..., p_n$, as

$$P(x_1, ..., x_n; p_1, ..., p_n) = \frac{1}{(h\pi)^n} \int_{-\infty}^{\infty} \cdots \int_{-\infty}^{\infty} dy_1 ... dy_n \psi(x_1 + y_1, ..., x_n + y_n)$$
$$\overline{\psi}(x_1 - y_1, ..., x_n - y_n) e^{2i(p_1 y_1 + ... + p_n y_n)/h},$$

in which $\psi(x_1, ..., x_n)$ is a given wave function and h is Plank's constant divided by 2π.

Wigner's idea was introduced in signal analysis in 1948 by J. Ville [236], but it did not receive much attention there until 1953 when P. Woodward [249] reformulated it in the context of radar theory.

Woodward proposed treating the question of radar signal ambiguity as a part of the question of target resolution. To do that, he introduced a function that described the correlation between a radar signal and its Doppler-shifted and time-translated version. This function exhibits the interplay between measurement ambiguity and target resolution, and for this reason it is called the radar ambiguity function. Its absolute value is called the uncertainty function as it is related to the uncertainty principle of radar signals. Woodward is also believed to be the first to designate the term *Sinc function* to the function $\sin \pi x / \pi x$.

The study of radar ambiguity functions in the context of pure mathematics and in, particular, in group theory and harmonic analysis was pioneered by C. Wilcox [248], L. Auslander and R. Tolimieri [5], and W. Schempp [213]. But as for the Wigner distribution, it was the Dutch mathematician N. G. de Bruijn [31] who introduced it in mathematics and developed many of its properties. The Wigner distribution was introduced again in signal analysis in 1980 by T. Claasen and W. Mecklenbräuker [42], but in a context different from that of Ville.

31.1 Definitions

Although most of the material presented in this chapter can be easily generalized to functions of n variables, for the sake of simplicity and consistency, we shall restrict ourselves to the case $n = 1$.

DEFINITION 31.1 *The cross-ambiguity function $\mathcal{A}_{f,g}(u,v)$ of two functions f and g is defined by*

$$\begin{aligned} \mathcal{A}_{f,g}(u,v) &= \int_{-\infty}^{\infty} f\left(t+\frac{u}{2}\right)\overline{g}\left(t-\frac{u}{2}\right)e^{-ivt}dt, \\ &= e^{iuv/2}\int_{-\infty}^{\infty} f(x)\overline{g}(x-u)e^{-ivx}dx\,, \end{aligned} \tag{31.1}$$

and the auto-ambiguity function $\mathcal{A}_f(u,v)$ of f is defined as $\mathcal{A}_f(u,v) = \mathcal{A}_{f,f}(u,v)$. The auto-ambiguity function will be also called the ambiguity function for short. The function $R_{f,g}(u) = \mathcal{A}_{f,g}(u,0)$ is called the cross-correlation function of f and g, and $\mathcal{A}_f(u,0)$ is called the auto-correlation function of f.

A closely related expression to the cross–ambiguity function is the function $\mathcal{B}_{f,g}(u,v)$, which is sometimes also called the cross-ambiguity function of f and g.

DEFINITION 31.2 *The cross–ambiguity transformation is defined as a bilinear mapping $\mathcal{B}$ from $L^2(\mathcal{R}) \times L^2(\mathcal{R}) \to L^2(\mathcal{R}^2)$ given by $\mathcal{B}(f,g) = \mathcal{B}_{f,g}(u,v)$, for all f, $g \in L^2(\mathcal{R})$, where*

$$\mathcal{B}_{f,g}(u,v) = \int_{-\infty}^{\infty} f(t)\overline{g}(t-u)e^{-ivt}dt = e^{-iuv/2}\mathcal{A}_{f,g}(u,v). \tag{31.2}$$

The ambiguity transformation $\tilde{\mathcal{B}}$ is defined as a mapping from $L^2(\mathcal{R}) \to L^2(\mathcal{R}^2)$ given by $\tilde{\mathcal{B}}[f] = \mathcal{B}(f, f)$ for all $f \in L^2(\mathcal{R})$.

DEFINITION 31.3 *The cross–Wigner distribution function $W_{f,g}(t, v)$ of two functions f and g is defined by*

$$
\begin{aligned}
W_{f,g}(u,v) &= \int_{-\infty}^{\infty} f\left(u+\frac{x}{2}\right)\overline{g}\left(u-\frac{x}{2}\right)e^{-ivx}dx \\
&= 2e^{2iuv}\int_{-\infty}^{\infty} f(x)\overline{g}(2u-x)e^{-2ivx}dx\,,
\end{aligned}
$$

and the auto-Wigner distribution of a function f is defined as $W_f(u, v) = W_{f,f}(u, v)$. The auto-Wigner distribution may be called Wigner distribution for short.

Although the cross–Wigner distribution is related to the cross–ambiguity function by

$$W_{f,g}(u,v) = 2\mathcal{A}_{f,h}(2u, 2v),$$

where $h(z) = g(-z)$, it is usually easier to derive the properties of $W_{f,g}$ directly instead of deriving them as a special case of those of $\mathcal{A}_{f,g}$.

DEFINITION 31.4 *The generalized Rihaczek distribution function of two functions f and g is defined by*

$$\mathcal{R}^{\alpha}_{f,g}(u,v) = \int_{-\infty}^{\infty} f\left(u+\left(\frac{1}{2}-\alpha\right)x\right)\overline{g}\left(u-\left(\frac{1}{2}+\alpha\right)x\right)e^{-ivx}dx,$$

where α is a constant, $-\infty < \alpha < \infty$.

If $\alpha = 0$, we obtain the cross–Wigner distribution $W_{f,g}$ and if $\alpha = \frac{1}{2}$, we obtain the cross–Rihaczek distribution $R_{f,g}$ defined by

$$R_{f,g}(u,v) = \mathcal{R}^{\frac{1}{2}}_{f,g}(u,-v) = \sqrt{2\pi}\, e^{iuv} f(u)\overline{\hat{g}}(v).$$

31.2 Existence

If for fixed v we set $F(t, u) = f\left(t + \frac{u}{2}\right)\overline{g}\left(t - \frac{u}{2}\right)e^{-ivt}$, then by the translation invariance of Lebesgue measure, it follows that

$$\int_{-\infty}^{\infty} |\mathcal{A}_{f,g}(u,v)|\,du \le \int_{-\infty}^{\infty}\int_{-\infty}^{\infty} |F(t,u)|\,dt\,du \le \|f\|_1\,\|g\|_1 < \infty.$$

Therefore, $\mathcal{A}_{f,g}(u,v)$ exists if f and $g \in L^1(\mathcal{R})$. It also follows from the Cauchy–Schwarz inequality that $\mathcal{A}_{f,g}$ exists if f and $g \in L^2(\mathcal{R})$. Clearly, the same results hold for $W_{f,g}$.

Example A: It is easy to see that for the function $f(t) = e^{-\alpha t^2}$, the ambiguity function is

$$\mathcal{A}_f(u,v) = \sqrt{\pi/2\alpha}\, e^{-\alpha u^2/2 - v^2/(8\alpha)},$$

and the auto-Wigner distribution is

$$W_f(u,v) = \sqrt{2\pi/\alpha}\, e^{-2\alpha u^2 - v^2/(2\alpha)}.$$

Example B: Let $f(t) = \chi_{(-T,T)}(t)$. Then

$$\mathcal{A}_f(u,v) = e^{iuv/2}\int_{-\infty}^{\infty} f(x) f(x-u) e^{-ivx}\,dx = e^{iuv/2}\int_{-T}^{T} f(x-u)e^{-ivx}\,dx.$$

If $u > 0$, then

$$\begin{aligned}\mathcal{A}_f(u,v) &= e^{iuv/2}\int_{-T}^{T} \chi_{(-T+u,T+u)}(x) e^{-ivx}\,dx \\ &= e^{iuv/2}\chi_{[0,2T]}(u)\int_{-T+u}^{T} e^{-ivx}\,dx = \frac{2\sin(vT - uv/2)}{v}\chi_{[0,2T]}(u).\end{aligned}$$

Similarly, if $u < 0$, then

$$\mathcal{A}_f(u,v) = e^{iuv/2}\chi_{[-2T,0]}(u)\int_{-T}^{T+u} e^{-ivx}\,dx = \frac{2\sin(vT + uv/2)}{v}\chi_{[-2T,0]}(u);$$

therefore

$$\mathcal{A}_f(u,v) = \frac{2\sin(vT - |u|v/2)}{v}\chi_{[-2T,2T]}(u).$$

Similar argument shows that

$$W_f(u,v) = \frac{2\sin 2v(T - |u|)}{v}\chi_{[-T,T]}(u)$$

31.3 Elementary Properties

1. **Linearity:** The ambiguity function and the auto-Wigner distribution are non linear. In fact,

$$\mathcal{A}_{f_1+f_2,g_1+g_2}(u,v) = \mathcal{A}_{f_1,g_1}(u,v) + \mathcal{A}_{f_1,g_2}(u,v) + \mathcal{A}_{f_2,g_1}(u,v) + \mathcal{A}_{f_2,g_2}(u,v).$$

In particular, $\mathcal{A}_{cf}(u,v) = |c|^2 \mathcal{A}_f(u,v)$

$$\mathcal{A}_{f+g}(u,v) = \mathcal{A}_f(u,v) + \mathcal{A}_g(u,v) + 2Re\,\mathcal{A}_{f,g}(u,v).$$

The same relations hold for $W_{f,g}$.

2. **Translation:**

$$\mathcal{A}_{T_{-a}f,g}(u,v) = e^{iav}\mathcal{A}_{f,T_a g}(u,v),$$

in particular, $\mathcal{A}_{T_{-a}f}(u,v) = e^{iav}\mathcal{A}_f(u,v)$.

3. **Modulation:**

$$\mathcal{A}_{E_a f,g}(u,v) = e^{iau/2}\mathcal{A}_{f,g}(u,v-a),$$

and

$$W_{E_a f,g}(u,v) = e^{iau}W_{f,g}\left(u, v - \frac{a}{2}\right),$$

in particular,

$$\mathcal{A}_{E_a f}(u,v) = e^{iau}\mathcal{A}_f(u,v)$$

4. **Translation and Modulation:**

$$\mathcal{A}_{E_a T_b f,g}(u,v) = e^{iau/2 - ivb + iab}\mathcal{A}_{f,T_{-b}g}(u,v-a).$$

5. **Complex Conjugation:**

$$\overline{\mathcal{A}}_{f,g}(u,v) = \mathcal{A}_{g,f}(-u,-v),$$

and

$$\overline{W}_{f,g}(u,v) = W_{g,f}(u,v).$$

6. **Differentiation:**

$$iv\mathcal{A}_{f,g}(u,v) = \mathcal{A}_{f',g}(u,v) + \mathcal{A}_{f,g'}(u,v),$$

and

$$2ivW_{f,g}(u,v) = W_{f',g}(u,v) + W_{f,g'}(u,v).$$

7. **Coordinate Transformations:** Let $SL(2,\mathcal{R})$ denote the group of all 2×2 real matrices D of determinant one, acting on $\mathcal{R}^2$ by

$$D\begin{pmatrix} u \\ v \end{pmatrix} = \begin{pmatrix} au + bv \\ cu + dv \end{pmatrix}$$

where

$$D = \begin{pmatrix} a\ b \\ c\ d \end{pmatrix}, \quad ad - bc = 1$$

Let

$$P(\alpha) = \begin{pmatrix} 1 & -\alpha \\ 0 & 1 \end{pmatrix}, \quad Q(\beta) = \begin{pmatrix} 1 & 0 \\ -\beta & 1 \end{pmatrix}, \quad S(\gamma) = \begin{pmatrix} \gamma & 0 \\ 0 & \frac{1}{\gamma} \end{pmatrix}$$

It is easy to verify that

i) if $c \neq 0$, then

$$D = P(\alpha_1)Q(\beta)P(\alpha_2),$$

where

$$\alpha_1 = \frac{1-a}{c}, \quad \beta = -c, \quad \alpha_2 = \frac{1-d}{c},$$

ii) if $b \neq 0$, then

$$D = Q(\beta_1)P(\alpha)Q(\beta_2),$$

where

$$\beta_1 = \frac{1-d}{b}, \quad \alpha = -b, \quad \beta_2 = \frac{1-a}{b},$$

iii) if $b = 0 = c$, then $a = 1/d = \gamma$, and $D = S(\gamma)$. But even in this case D can be written in terms of P and Q as

$$D = S(\gamma) = Q(\beta_1)P(\alpha_1)Q(\beta_2)P(-\alpha),$$

where

$$\beta_1 = \frac{1-1/\gamma}{-\alpha\gamma}, \quad \alpha_1 = \alpha\gamma, \quad \beta_2 = \frac{1-\gamma}{-\alpha\gamma},$$

and α is an arbitrary number $\neq 0$.

Therefore, to evaluate $\mathcal{A}_f(au+bv, cu+vd) = \mathcal{A}_f\left[(D\mathbf{u})^T\right]$ where

$$\mathbf{u} = \begin{pmatrix} u \\ v \end{pmatrix}, \quad D\mathbf{u} \begin{pmatrix} au+bv \\ cu+du \end{pmatrix},$$

and the superscript T stands for the transpose of the matrix, it suffices to calculate

$$\mathcal{A}_f\left[(P(\alpha)\mathbf{u})^T\right] = \mathcal{A}_f(u-\alpha v, v) \quad \text{and} \quad \mathcal{A}_f\left[(Q(\beta)\mathbf{u})^T\right] = \mathcal{A}_f(u, -\beta u+v).$$

The results are as follows:

$$\mathcal{A}_f(u-\alpha v, v) = \mathcal{A}_g(u,v),$$

where g is defined by its Fourier transform G, which is given by

$$G(w) = \sqrt{i}\, F(w) e^{-i\alpha w^2/2},$$

and F is the Fourier transform of f. Moreover,

$$\mathcal{A}_f(u, -\beta u + v) = \mathcal{A}_g(u, v),$$

with

$$g(t) = f(t)e^{i\beta t^2/2}.$$

and

$$\mathcal{A}_f\left[(S(\gamma)\mathbf{u})^T\right] = \mathcal{A}_f\left(\gamma u, \frac{v}{\gamma}\right) = \gamma \mathcal{A}_g(u, v),$$

where $g(t) = f(\gamma t)$.

We only prove (7) since the proofs of (1)–(6) are straightforward. To calculate $\mathcal{A}_f(u - \alpha v, v)$ we use (31.13) to obtain

$$\begin{aligned}\mathcal{A}_f(u - \alpha v, v) &= \mathcal{A}_F(-v, u - \alpha v)\\ &= e^{i(\alpha v^2 - uv)/2} \int_{-\infty}^{\infty} F(w)\overline{F}(w + v)e^{-iw(u - \alpha v)}dw,\end{aligned}$$

where F denotes the Fourier transform of f.

Define $G(w) = cF(w)e^{-i\alpha w^2/2}$ with $|c| = 1$. One can easily verify that $\mathcal{A}_G(-v, u) = \mathcal{A}_F(-v, u - \alpha v)$; hence

$$\mathcal{A}_f(u - \alpha v, v) = \mathcal{A}_g(u, v),$$

where g is the inverse Fourier transform of G. The function g can be written as a convolution of f and a Gaussian function of the form ae^{-bt^2} for some suitable constants a and b.

Similarly, if we define $g(t) = f(t)e^{i\beta t^2/2}$, it follows by direct calculations that

$$\mathcal{A}_f(u, -\beta u + v) = \mathcal{A}_g(u, v).$$

31.4 Relationships with Other Transforms

31.4.1 The Fourier Transform

If we take the Fourier transform of $\mathcal{A}_{f,g}(u, v)$ and $W_{f,g}(u, v)$ with respect to the variable v, we obtain

$$\hat{\mathcal{A}}_{f,g}(u, w) = \sqrt{2\pi}\, f\left(w + \frac{u}{2}\right) \overline{g}\left(w - \frac{u}{2}\right), \tag{31.3}$$

and

$$\hat{W}_{f,g}(u, w) = \sqrt{2\pi}\, f\left(u + \frac{w}{2}\right) \overline{g}\left(u - \frac{w}{2}\right), \tag{31.4}$$

which can be rewritten in the form

$$\frac{1}{2\pi}\int_{-\infty}^{\infty} e^{i(t_1+t_2)v/2}\mathcal{A}_{f,g}(t_1-t_2,v)dv = f(t_1)\overline{g}(t_2), \tag{31.5}$$

with $t_1 = w + u/2$, $t_2 = w - u/2$, and

$$\frac{1}{2\pi}\int_{-\infty}^{\infty} e^{i(\tau_1-\tau_2)v}W_{f,g}\left(\frac{\tau_1+\tau_2}{2},v\right)dv = f(\tau_1)\overline{g}(\tau_2) \tag{31.6}$$

with $\tau_1 = u + w/2$, $\tau_2 = u - w/2$.

In particular, if $t_1 = t_2 = t$ and $\tau_1 = \tau_2 = \tau$, we have

$$\frac{1}{2\pi}\int_{-\infty}^{\infty} e^{itv}\mathcal{A}_{f,g}(0,v)\,dv = f(t)\overline{g}(t),$$

and

$$\frac{1}{2\pi}\int_{-\infty}^{\infty} W_{f,g}(\tau,v)\,dv = f(\tau)\overline{g}(\tau)\,,$$

and from which we obtain the following as special cases:

$$\frac{1}{2\pi}\int_{-\infty}^{\infty} e^{itv}\mathcal{A}_{f}(0,v)\,dv = |f(t)|^2, \tag{31.7}$$

$$\frac{1}{2\pi}\int_{-\infty}^{\infty} W_{f}(\tau,v)\,dv = |f(\tau)|^2, \tag{31.8}$$

$$\frac{1}{2\pi}\int_{-\infty}^{\infty}\int_{-\infty}^{\infty} e^{itv}\mathcal{A}_{f,g}(0,v)\,dv\,dt = \int_{-\infty}^{\infty} f(t)\overline{g}(t)\,dt = \int_{-\infty}^{\infty} F(w)\overline{G}(w)\,dw, \tag{31.9}$$

and

$$\frac{1}{2\pi}\int_{-\infty}^{\infty}\int_{-\infty}^{\infty} W_{f,g}(\tau,v)dv\,d\tau = \int_{-\infty}^{\infty} f(\tau)\overline{g}(\tau)d\tau = \int_{-\infty}^{\infty} F(w)\overline{G}(w)dw, \tag{31.10}$$

where F and G are the Fourier transforms of f and g, respectively.

Moreover, we have

$$\frac{1}{2\pi}\int_{-\infty}^{\infty}\int_{-\infty}^{\infty} e^{itv}\mathcal{A}_{f}(0,v)\,dt\,dv = \int_{-\infty}^{\infty} |f(t)|^2dt = \|f\|_2^2, \tag{31.11}$$

and

$$\frac{1}{2\pi}\int_{-\infty}^{\infty}\int_{-\infty}^{\infty} W_{f}(\tau,v)\,d\tau\,dv = \int_{-\infty}^{\infty} |f(\tau)|^2d\tau = \|f\|_2^2. \tag{31.12}$$

By setting $t_2 = 0$ in (31.5) and $\tau_2 = 0$ in (31.6), we obtain

$$f(t)\overline{g}(0) = \frac{1}{2\pi}\int_{-\infty}^{\infty} e^{itv/2}\mathcal{A}_{f,g}(t,v)\,dv,$$

and

$$f(t)\overline{g}(0) = \frac{1}{2\pi} \int_{-\infty}^{\infty} e^{itv} W_{f,g}\left(\frac{\tau}{2}, v\right) dv.$$

These last two relations show that $f(t)$ can be recovered (up to a constant $\overline{g}(0)$) from $\mathcal{A}_{f,g}$ by taking the Fourier transform with respect to v at time t and from $W_{f,g}$ at time $\tau/2$.

We also have

$$\begin{aligned}
\mathcal{A}_{F,G}(u,v) &= \int_{-\infty}^{\infty} F\left(t+\frac{u}{2}\right) \overline{G}\left(t-\frac{u}{2}\right) e^{-ivt} dt \\
&= \frac{1}{\sqrt{2\pi}} \int_{-\infty}^{\infty} \overline{G}\left(t-\frac{u}{2}\right) e^{-ivt} dt \int_{-\infty}^{\infty} f(x) e^{ix\left(t+\frac{u}{2}\right)} dx \\
&= \frac{1}{\sqrt{2\pi}} \int_{-\infty}^{\infty} f(x) e^{ixu/2} dx \, e^{i(x-v)u/2} \int_{-\infty}^{\infty} \overline{G}(z) e^{iz(x-v)} dz \\
&= e^{-iuv/2} \int_{-\infty}^{\infty} f(x) \overline{g}(x-v) e^{iux} dx = \mathcal{A}_{f,g}(v,-u)
\end{aligned}$$

or

$$\mathcal{A}_{f,g}(u,v) = \mathcal{A}_{F,G}(-v,u) = \int_{-\infty}^{\infty} F\left(t-\frac{v}{2}\right) \overline{G}\left(t+\frac{v}{2}\right) e^{-iut} dt \tag{31.13}$$

Similarly, we can obtain

$$W_{f,g}(u,v) = W_{F,G}(-v,u). \tag{31.14}$$

Now one can easily show that

$$\begin{aligned}
\mathcal{A}_{f,g}(u,0) &= \int_{-\infty}^{\infty} f\left(t+\frac{u}{2}\right) \overline{g}\left(t-\frac{u}{2}\right) dt, \\
\mathcal{A}_{f,g}(0,0) &= \int_{-\infty}^{\infty} f(t)\overline{g}(t)\, dt = \int_{-\infty}^{\infty} F(t)\overline{G}(t)\, dt = \mathcal{A}_{F,G}(0,0), \\
\mathcal{A}_{F,G}(u,0) &= \int_{-\infty}^{\infty} F\left(t+\frac{u}{2}\right) \overline{G}\left(t-\frac{u}{2}\right) dt = \mathcal{A}_{f,g}(0,-u),
\end{aligned}$$

$$\mathcal{A}_f(0,0) = \|f\|_2^2 = \mathcal{A}_F(0,0) = \|F\|_2^2 \tag{31.15}$$

and similarly

$$\begin{aligned}
W_{f,g}(u,0) &= \int_{-\infty}^{\infty} f\left(u+\frac{x}{2}\right) \overline{g}\left(u-\frac{x}{2}\right) dx \\
W_{f,g}(0,0) &= 2\int_{-\infty}^{\infty} f(x)\overline{g}(-x)\, dx = 2\int_{-\infty}^{\infty} F(x)\overline{G}(-x)\, dx \\
&= W_{F,G}(0,0), \qquad (31.16) \\
W_{F,G}(u,0) &= \int_{-\infty}^{\infty} F\left(u+\frac{x}{2}\right) \overline{G}\left(u-\frac{x}{2}\right) dx = W_{f,g}(0,-u),
\end{aligned}$$

It also follows from (31.13) that

$$\frac{1}{2\pi}\int_{-\infty}^{\infty} \mathcal{A}_{f,g}(v,-u)e^{itv}dv = F\left(t+\frac{u}{2}\right)\overline{G}\left(t-\frac{u}{2}\right),$$

or

$$\frac{1}{2\pi}\int_{-\infty}^{\infty} \mathcal{A}_{f,g}\left(v,-(t_1-t_2)\right)e^{iv(t_1+t_2)/2}dv = F(t_1)\overline{G}(t_2),$$

where $t_1 = t+u/2$, $t_2 = t-u/2$. Similarly

$$\frac{1}{2\pi}\int_{-\infty}^{\infty} W_{f,g}\left(v,-\frac{(w_1+w_2)}{2}\right)e^{iv(w_1-w_2)}dv = F(w_1)\overline{G}(w_2).$$

As a special case, we obtain for $t_1 = t_2 = t$ and $w_1 = w_2 = w$,

$$\frac{1}{2\pi}\int_{-\infty}^{\infty} \mathcal{A}_{f,g}(v,0)e^{ivt}dv = F(t)\overline{G}(t),$$

$$\frac{1}{2\pi}\int_{-\infty}^{\infty}\int_{-\infty}^{\infty} \mathcal{A}_{f,g}(v,0)e^{ivt}dvdt = \int_{-\infty}^{\infty} F(t)\overline{G}(t)dt,$$

$$\frac{1}{2\pi}\int_{-\infty}^{\infty}\int_{-\infty}^{\infty} \mathcal{A}_{f}(v,0)e^{ivt}dvdt = \|F\|_2^2,$$

and

$$\frac{1}{2\pi}\int_{-\infty}^{\infty} W_{f,g}(v,-w)dv = F(w)\overline{G}(w)$$

$$\frac{1}{2\pi}\int_{-\infty}^{\infty}\int_{-\infty}^{\infty} W_{f,g}(v,-w)dvdw = \int_{-\infty}^{\infty} F(w)\overline{G}(w)dw,$$

$$\frac{1}{2\pi}\int_{-\infty}^{\infty}\int_{-\infty}^{\infty} W_{f}(v,-w)dvdw = \|F\|_2^2.$$

31.4.2 The Zak Transform

Let us denote the Zak transform of a function $f(t)$ by $Z_f(t,w)$ where

$$Z_f(t,w) = \sum_{k=-\infty}^{\infty} f(t+k)e^{-2\pi ikw};$$

see Chapter 29. Recall from Equation (29.8) that

$$Z_f(t,w)\overline{Z}_g(t,w) = \sum_{m,n=-\infty}^{\infty} c_{m,n}e^{2\pi imt}e^{2\pi inw}, \quad 0 \le t,w \le 1 \tag{31.17}$$

where

$$c_{m,n} = \langle f, E_{2\pi m}T_{-n}g\rangle = \int_{-\infty}^{\infty} f(t)\overline{g}(t+n)e^{-2\pi imt}dt.$$

For the sake of symmetry in the formula we are about to derive, let us define $\tilde{\mathcal{A}}_{f,g}(u,v) = \mathcal{A}_{f,g}(-u, 2\pi v)$ so that

$$\begin{aligned}\tilde{\mathcal{A}}_{f,g}(u,v) &= \int_{-\infty}^{\infty} f\left(t-\frac{u}{2}\right)\overline{g}\left(t+\frac{u}{2}\right)e^{-2\pi ivt}dt \\ &= e^{-i\pi uv}\int_{-\infty}^{\infty} f(t)\overline{g}(t+u)e^{-2\pi ivt}dt.\end{aligned}$$

Thus,

$$c_{m,n} = (-1)^{mn}\tilde{\mathcal{A}}_{f,g}(n,m),$$

and it follows from (31.17) that

$$\begin{aligned}|Z_f(t,w)|^2 &= \sum_{m,n=-\infty}^{\infty} (-1)^{mn}\tilde{\mathcal{A}}_f(n,m)e^{2\pi imt}e^{2\pi inw}, \\ |Z_g(t,w)|^2 &= \sum_{k,l=-\infty}^{\infty} (-1)^{kl}\overline{\tilde{\mathcal{A}}_g}(l,k)e^{-2\pi ikt}e^{-2\pi ilw}.\end{aligned}$$

If we multiply these last two series and then integrate over the square $0 \le t \le 1, 0 \le w \le 1$, we find

$$\int_0^1\int_0^1 |Z_f(t,w)|^2\,|Z_g(t,w)|^2\,dtdw = \sum_{m,n=-\infty}^{\infty} \tilde{\mathcal{A}}_f(n,m)\overline{\tilde{\mathcal{A}}_g}(n,m),$$

but, on the other hand, we have by (31.17) and Parseval's equality for the Fourier series (cf. Example A, Section 1.5)

$$\int_0^1\int_0^1 |Z_f(t,w)|^2\,|Z_g(t,w)|^2\,dtdw = \sum_{m,n=-\infty}^{\infty} |c_{m,n}|^2 = \sum_{m,n=-\infty}^{\infty} \left|\tilde{\mathcal{A}}_{f,g}(n,m)\right|^2.$$

Therefore,

$$\begin{aligned}\sum_{m,n=-\infty}^{\infty} \left|\tilde{\mathcal{A}}_{f,g}(n,m)\right|^2 &= \sum_{n,m=-\infty}^{\infty} \tilde{\mathcal{A}}_f(n,m)\overline{\tilde{\mathcal{A}}_g}(n,m) \\ &= \int_0^1\int_0^1 |Z_f(t,w)Z_g(t,w)|^2\,dt\,dw\,. \qquad (31.18)\end{aligned}$$

31.5 Parseval's Relation

Parseval's relations for the cross–ambiguity and the cross–Wigner functions are

$$\frac{1}{2\pi}\int_{-\infty}^{\infty}\int_{-\infty}^{\infty} \mathcal{A}_{f_1,g_1}(u,v)\overline{\mathcal{A}}_{f_2,g_2}(u,v)\,du\,dv = \langle f_1, f_2\rangle\overline{\langle g_1, g_2\rangle}, \qquad (31.19)$$

and

$$\frac{1}{2\pi}\int_{-\infty}^{\infty}\int_{-\infty}^{\infty} W_{f_1,g_1}(u,v)\overline{W}_{f_2,g_2}(u,v)\,du\,dv = \langle f_1, f_2\rangle\overline{\langle g_1, g_2\rangle}. \qquad (31.20)$$

Formula (31.20) is also called *Moyal's formula.*

We only prove the first formula since the proof of the second is similar. From (31.3) it is clear that for fixed u the Fourier transform of $\mathcal{A}_{f,g}(u,v)$ taken with respect to v is $\sqrt{2\pi}f\left(t+\frac{u}{2}\right)\overline{g}\left(t-\frac{u}{2}\right)$. Therefore, from Parseval's relation for the Fourier transform, we have

$$\frac{1}{2\pi}\int_{-\infty}^{\infty} \mathcal{A}_{f_1,g_1}(u,v)\overline{\mathcal{A}}_{f_2,g_2}(u,v)dv =$$
$$\int_{-\infty}^{\infty} f_1\left(t+\frac{u}{2}\right)\overline{g}_1\left(t-\frac{u}{2}\right)\overline{f}_2\left(t+\frac{u}{2}\right)g_2\left(t-\frac{u}{2}\right)dt\,.$$

Integrating both sides with respect to u from $-\infty$ to ∞ yields the result.

As a special case, we have

$$\frac{1}{2\pi}\int_{-\infty}^{\infty}\int_{-\infty}^{\infty} \left|\mathcal{A}_{f,g}(u,v)\right|^2 du\,dv = \|f\|^2\,\|g\|^2,$$
$$\frac{1}{2\pi}\int_{-\infty}^{\infty}\int_{-\infty}^{\infty} \left|W_{f,g}(u,v)\right|^2 du\,dv = \|f\|^2\,\|g\|^2, \qquad (31.21)$$
$$\frac{1}{2\pi}\int_{-\infty}^{\infty}\int_{-\infty}^{\infty} \left|\mathcal{A}_f(u,v)\right|^2 du\,dv = \frac{1}{2\pi}\int_{-\infty}^{\infty}\int_{-\infty}^{\infty} \left|W_f(u,v)\right|^2 du\,dv$$
$$= \|f\|^4 = \mathcal{A}_f^2(0,0). \qquad (31.22)$$

Equation (31.22) is known as the *radar uncertainty principle.* Mathematically, it says that, regardless of which function f is chosen, the graph of the function $\left|\mathcal{A}_f(u,v)\right|^2$ cannot be concentrated arbitrarily close to the origin.

Since

$$\left|\mathcal{A}_{f,g}(u,v)\right| \le \int_{-\infty}^{\infty}\left|f\left(t+\frac{u}{2}\right)\right|\left|\overline{g}\left(t-\frac{u}{2}\right)\right|dt,$$

it follows from the Cauchy-Schwarz inequality and (31.15) that for any u and v,

$$\left|\mathcal{A}_{f,g}(u,v)\right| \le \sqrt{\mathcal{A}_f(0,0)\mathcal{A}_g(0,0)}$$

in particular,

$$\left|\mathcal{A}_f(u,v)\right| \le \mathcal{A}_f(0,0).$$

We also have an important inequality involving the second partial derivatives of $\mathcal{A}_f$ at the origin. If $f(t)$ represents a signal in the time domain, we use the quantities

$$T^2 = \frac{1}{E}\int_{-\infty}^{\infty} t^2\left|f(t)\right|^2 dt \quad \text{and} \quad W^2 = \frac{1}{E}\int_{-\infty}^{\infty} w^2\left|F(w)\right|^2 dw$$

as a measure of the signal duration in both the time and frequency domains, where $E = \|f\|^2$. Recall from the uncertainty principle for the Fourier transform (Theorem 12.9) that $\frac{1}{2} \leq TW$. But it is easy to see from (31.1) and (31.13) that

$$-\frac{\partial^2 \mathcal{A}_f(0,0)}{\partial v^2} = \int_{-\infty}^{\infty} t^2 \left|f(t)\right|^2 dt,$$

and

$$-\frac{\partial^2 \mathcal{A}_f(0,0)}{\partial u^2} = \int_{-\infty}^{\infty} w^2 \left|F(w)\right|^2 dw.$$

Therefore, under the same assumptions on f as in Theorem 12.9, we have

$$\frac{1}{2}\|f\|^2 \leq \sqrt{\frac{\partial^2 A(0,0)}{\partial u^2} \, \frac{\partial^2 A(0,0)}{\partial v^2}}.$$

31.6 The Ambiguity Transformation

In this section we will work with the cross–ambiguity function $\mathcal{B}_{f,g}(u,v)$ given by (31.2). The following properties of $\mathcal{B}_{f,g}$ are easy to establish:

$$\frac{1}{2\pi}\int_{-\infty}^{\infty} \mathcal{B}_{f,g}(t-\tau, v)e^{itv}dv = f(t)\overline{g}(\tau), \tag{31.23}$$

$$\mathcal{B}_{f,g}(u,v) = e^{-iuv}\mathcal{B}_{F,G}(-v,u), \tag{31.24}$$

$$\mathcal{B}_{g,f}(u,v) = e^{-iuv}\overline{\mathcal{B}_{f,g}}(-u,-v), \tag{31.25}$$

and

$$\frac{1}{2\pi}\int_{-\infty}^{\infty}\int_{-\infty}^{\infty} \mathcal{B}_{f_1,g_1}(u,v)\overline{\mathcal{B}}_{f_2,g_2}(u,v)dudv = \langle f_1, f_2\rangle\overline{\langle g_1, g_2\rangle}. \tag{31.26}$$

The relation (31.23) follows from Definition 31.2 and the inversion formula for the Fourier transform, (31.24) follows from (31.14), (31.25) from property 5 of Section 31.3, and (31.26) from (31.19).

Let $g = f$ in (31.23) and set

$$H(t,\tau) = \frac{1}{2\pi}\int_{-\infty}^{\infty} \mathcal{B}_f(t-\tau, v)e^{itv}dv = f(t)\overline{f}(\tau). \tag{31.27}$$

It is easy to see that H satisfies the following properties:

1. $\overline{H}(t,\tau) = H(\tau,t)$
2. $H(t,t) \geq 0$

3. $H(x,x)H(t,\tau) = H(t,x)H(x,\tau)$.

Let $\tilde{L}^2(\mathcal{R}^2)$ denote that subset of $L^2(\mathcal{R}^2)$ consisting of all functions $H(t,\tau)$ satisfying conditions (1)–(3). Consider the mapping $U[\mathcal{B}_f] = H$ defined by (31.27). It is easy to see that $\|\mathcal{B}_f\| = \sqrt{2\pi}\|H\| = \sqrt{2\pi}\|f\|^2$, where $\|H\|$ and $\|\mathcal{B}_f\|$ are defined in the usual way, e.g.,

$$\|H\|^2 = \int_{-\infty}^{\infty}\int_{-\infty}^{\infty} |H(t,\tau)|^2\, dt d\tau.$$

For any $F(t,\tau) \in \tilde{L}^2(\mathcal{R}^2)$, there exists $f \in L^2(\mathcal{R})$ such that $U[\mathcal{B}_f] = F$. In fact, if F is given, we can choose $f(t) = F(t,\tau_0)/\sqrt{F(\tau_0,\tau_0)}$ where τ_0 is chosen so that $F(\tau_0,\tau_0) > 0$.

Recall from Definition 31.2 that the cross–ambiguity transformation $\mathcal{B}$ is defined as a mapping from $L^2(\mathcal{R}) \times L^2(\mathcal{R}) \to L^2(\mathcal{R}^2)$, given by $\mathcal{B}(f,g) = \mathcal{B}_{f,g}(u,v)$. Clearly, this mapping is bilinear. If $\{f_n\}_{n=1}^{\infty}$ and $\{g_n\}_{n=1}^{\infty}$ are two sequences in $L^2(\mathcal{R})$ such that $f_n \to f$ and $g_n \to g$ in $L^2(\mathcal{R})$, then by writing $f_n g_n - fg$ as $f_n g_n - f_n g + f_n g - fg$ and using the Cauchy-Schwarz inequality, we conclude that $\mathcal{B}$ is continuous and $\mathcal{B}_{f_n,g_n}(u,v) \to \mathcal{B}_{f,g}(u,v)$ uniformly. In fact, we have the following stronger result.

THEOREM 31.1 **(Auslander–Tolimieri** [5])

The cross–ambiguity transformation $\mathcal{B}$ is a continuous transformation from $L^2(\mathcal{R}) \times L^2(\mathcal{R})$ into $L^2(\mathcal{R}^2)$ and the image of $\mathcal{B}$ spans a dense subspace of $L^2(\mathcal{R}^2)$.

Let $\{\psi_n(t)\}_{n=1}^{\infty}$ be an orthonormal basis of $L^2(\mathcal{R})$ and set $\psi_{m,n}(u,v) = \mathcal{B}_{\psi_m,\psi_n}(u,v)$. From (31.26), we have

$$\frac{1}{2\pi}\langle \psi_{k,l}, \psi_{m,n}\rangle = \langle \psi_k, \psi_m\rangle \overline{\langle \psi_l, \psi_n\rangle} = \delta_{k,m}\delta_{l,n}$$

which shows that $\{\psi_{m,n}(u,v)\}_{m,n=1}^{\infty}$ is an orthonormal set in $L^2(\mathcal{R}^2)$ and by Theorem 31.1 that set is a basis. Now let $f \in L^2(\mathcal{R})$. Then

$$f(t) = \sum_{n=1}^{\infty} \hat{f}_n \psi_n(t) \quad \text{with} \quad \|f\|^2 = \sum_{n=1}^{\infty} \left|\hat{f}_n\right|^2.$$

Similarly, if $F \in L^2(\mathcal{R}^2)$, we have

$$F(u,v) = \sum_{m,n=1}^{\infty} \hat{F}_{m,n}\psi_{m,n}(u,v), \quad \text{with} \quad \sum_{m,n=1}^{\infty} \left|\hat{F}_{m,n}\right|^2 < \infty.$$

Therefore, if $\mathcal{B}_f \in L^2(\mathcal{R}^2)$, then

$$\mathcal{B}_f(u,v) = \sum_{m,n=1}^{\infty} c_{m,n}\psi_{m,n},$$

with

$$c_{m,n} = \langle \mathcal{B}_f, \psi_{m,n} \rangle = \langle \mathcal{B}_{f,f}, \mathcal{B}_{\psi_m,\psi_n} \rangle = \langle f, \psi_m \rangle \overline{\langle f, \psi_n \rangle} = \hat{f}_m \overline{\hat{f}_n}.$$

Conversely, if $c_{m,n} = \hat{f}_m \overline{\hat{f}_n}$, then $f = \sum \hat{f}_m \psi_m \in L^2(\mathcal{R})$ and $H(t,\tau) = f(t)\overline{f}(\tau) \in \tilde{L}^2(\mathcal{R}^2)$. Thus, a function $F(u,v) \in L^2(\mathcal{R}^2)$ is an ambiguity function if and only if $\hat{F}_{m,n} = \hat{f}_m \overline{\hat{f}_n}$. Consequently, $F \in L^2(\mathcal{R}^2)$ is an ambiguity function if and only if (i) $\overline{c}_{m,n} = c_{n,m}$, (ii) $c_{m,n} \geq 0$ and (iii) $c_{k,k} c_{m,n} = c_{m,k} c_{k,n}$. This leads to the following theorem.

THEOREM 31.2
The set of ambiguity functions $\mathcal{B}_f(u,v)$ for all $f \in L^2(\mathcal{R})$ is a closed subset of $L^2(\mathcal{R}^2)$.

Proof: Let $\{\mathcal{B}_{f_k}(u,v)\}_{n=1}^{\infty}$ be a sequence of ambiguity functions such that $\lim_{k\to\infty} \mathcal{B}_{f_k} = F$ in $L^2(\mathcal{R}^2)$, where $f_k \in L^2(\mathcal{R}), k = 1, 2, \ldots$. Let

$$c_{m,n}(k) = \langle \mathcal{B}_{f_k}, \psi_{m,n} \rangle \quad \text{and} \quad c_{m,n} = \langle F, \psi_{m,n} \rangle.$$

The Cauchy-Schwarz inequality implies that $\lim_{k\to\infty} c_{m,n}(k) = c_{m,n}$ for each $k = 1, 2, \ldots$. But since for each k, the coefficients $c_{m,n}(k)$ satisfy conditions (i)–(iii) above, it follows that $c_{m,n}$ also satisfy the same conditions, and therefore F is an ambiguity function. ∎

The next theorem gives a necessary and sufficient condition for the sum of two ambiguity functions to be an ambiguity function.

THEOREM 31.3
Let $f, g \in L^2(\mathcal{R})$, and $\mathcal{B}_f$ and $\mathcal{B}_g$ be their corresponding ambiguity functions. Then $\mathcal{B}_f + \mathcal{B}_g$ is an ambiguity function if and only if $f = \lambda g$, λ is a constant.

Thus, the sum of two ambiguity functions that are not proportional is never an ambiguity function.

31.7 The Wigner Distribution of Band-limited Functions

If $f(t)$ and $g(t)$ are functions band-limited to $[-\sigma_1, \sigma_2] (\sigma_1, \sigma_2 > 0)$, i.e., their Fourier transforms $F(w)$ and $G(w)$ vanish outside $[-\sigma_1, \sigma_2]$, then their

Wigner distribution $W_{f,g}(u,v)$ is also band-limited in v to $[-\sigma_2, \sigma_1]$, i.e., $W_{f,g}(u,v) = 0$ for $v < -\sigma_2$ or $v > \sigma_1$ and all u. For, by (31.14)

$$W_{f,g}(u,v) = W_{F,G}(-v,u) = 2e^{-2iuv} \int_{-\infty}^{\infty} F(w)\overline{G}(-2v-w)e^{-2iuw}dw$$
$$= 2e^{-2iuv} \int_{-\sigma_1}^{\sigma_2} F(w)\overline{G}(-2v-w)e^{-2iuw}dw.$$

But since $G(x) = 0$ for $-\sigma_1 \leq x \leq \sigma_2$, it follows by inspection that $W_{f,g}(u,v) = 0$ for $v < -\sigma_2$ or $v > \sigma_1$. In particular, if $\sigma_1 = \sigma_2 = \sigma$, then $W_{f,g}(u,v)$ is band-limited to $[-\sigma, \sigma]$ in v for all u if f and g are band-limited to $[-\sigma, \sigma]$. Under this condition the Whittaker-Shannon-Kotel'nikov sampling theorem (Theorem 12.23) asserts that

$$f(t) = \sum_{k=-\infty}^{\infty} f(t_k)\frac{\sin\sigma(t-t_k)}{\sigma(t-t_k)},$$

and

$$g(-t) = \sum_{m=-\infty}^{\infty} g(-t_m)\frac{\sin\sigma(t-t_m)}{\sigma(t-t_m)},$$

where $t_k = k\pi/\sigma$. Multiplying these two series together and using the fact that the set $\{\sin\pi(t-n)/\pi(t-n)\}_{n=-\infty}^{\infty}$ is orthonormal on $(-\infty, \infty)$, we find from (31.16)

$$W_{f,g}(0,0) = 2\int_{-\infty}^{\infty} f(t)g(-t)dt = \frac{2\pi}{\sigma}\sum_{k=-\infty}^{\infty} f(t_k)g(-t_k).$$

31.8 Convolution

From the definition of the Wigner distribution, it follows that for fixed u, $W_f(u,v)$ and $W_g(\alpha-u,v)$ are the inverse Fourier transforms of $\sqrt{2\pi}\, f\left(u+\frac{x}{2}\right)\overline{f}\left(u-\frac{x}{2}\right)$ and $\sqrt{2\pi}\, g\left(\alpha-u+\frac{x}{2}\right)\overline{g}\left(\alpha-u-\frac{x}{2}\right)$. Thus, by the convolution theorem for the Fourier transform (Theorem 12.14), we have

$$\int_{-\infty}^{\infty} W_f(u,v)W_g(\alpha-u,\beta-v)\,dv =$$
$$(2\pi)\int_{-\infty}^{\infty} f\left(u+\frac{x}{2}\right)\overline{f}\left(u-\frac{x}{2}\right)g\left(\alpha-u+\frac{x}{2}\right)\overline{g}\left(\alpha-u-\frac{x}{2}\right)e^{i\beta x}dx$$
$$= 4\pi\int_{-\infty}^{\infty} f(x)\overline{f}(2u-x)g(\alpha+x-2u)\overline{g}(\alpha-x)e^{2i\beta(x-u)}dx,$$

therefore,

$$\int_{-\infty}^{\infty}\int_{-\infty}^{\infty} W_f(u,v)W_g(\alpha-u,\beta-v)du\,dv$$
$$= 4\pi\int_{-\infty}^{\infty} f(x)\overline{g}(\alpha-x)e^{2i\beta x}dx\int_{-\infty}^{\infty}\overline{f}(2u-x)g(\alpha+x-2u)e^{-2i\beta u}du\,.$$

Using the substitution $2u = y + x$ in the last integral yields

$$\int_{-\infty}^{\infty}\int_{-\infty}^{\infty} W_f(u,v)W_g(\alpha-u,\beta-v)du\,dv$$
$$= (2\pi)\int_{-\infty}^{\infty} f(x)\overline{g}(\alpha-x)e^{i\beta x}dx\int_{-\infty}^{\infty}\overline{f}(y)g(\alpha-y)e^{-i\beta y}dy$$
$$= (2\pi)\left|\int_{-\infty}^{\infty} f(x)\overline{g}(\alpha-x)e^{i\beta x}dx\right|^2$$
$$= 2\pi\left|\mathcal{B}_{f,g^-}(\alpha\,,\,-\beta)\right|^2 = 2\pi\left|\langle f(x),E_{-\beta}T_\alpha g^-(x)\rangle\right|^2, \tag{31.28}$$

where $g^-(x) = g(-x)$.

In particular, if g is even, we have

$$\frac{1}{2\pi}\int_{-\infty}^{\infty}\int_{-\infty}^{\infty} W_f(u,v)W_g(\alpha-u,\beta-v)\,du\,dv = |\mathcal{B}_{f,g}(\alpha,-\beta)|^2 = |\langle f(x),E_{-\beta}T_\alpha g(x)\rangle|^2$$

31.9 Generalized Functions

To the best of our knowledge, N. G. de Bruijn was the first to define the Wigner distribution of generalized functions, but his definition of a generalized function was non-standard. He defined a generalized function as a mapping F from the set of all positive numbers into a testing-function space $\mathbf{S}$ such that $N_\alpha F(\beta) = F(\alpha+\beta)$, where $\alpha,\beta > 0$ and N_α is a smoothing operator defined on $\mathbf{S}$.

In this section we will define the Wigner distribution of generalized function using De Bruijn's idea. Our definition may not be as general as his but is simpler and consistent with our definition of generalized functions.

Let $S_{A,B}$ be the linear space of all entire functions f that satisfy

$$|f(x+iy)| \le Me^{-Ax^2+By^2}, \tag{31.29}$$

for all real x and y, where $A,B,M > 0$. Let

$$\mathbf{S} = \bigcup_{A>0}\bigcup_{B>0} S_{A,B}.$$

We define the inner product

$$\langle f, g\rangle = \int_{-\infty}^{\infty} f(x)\overline{g}(x)dx$$

on $\mathbf{S}$ so that $\|f\| = \sqrt{\langle f, f\rangle}$. Clearly, if $f \in \mathbf{S}$, then $f \in L^1(\mathcal{R})\bigcap L^2(\mathcal{R})$. We say that a sequence $\{f_n\}_{n=1}^{\infty}$ in $\mathbf{S}$ converges to zero in $\mathbf{S}$ if there are positive numbers, A and B, such that

$$f_n(x+iy)e^{Ax^2-By^2} \to 0$$

uniformly for all x and y. The space $\mathbf{S}$ is complete with respect to this notion of convergence. Convergence in $\mathbf{S}$ is linear, that is if $f_n \to f$ and $g_n \to g$ in $\mathbf{S}$ then $\alpha f_n + \beta g_n \to \alpha f + \beta g$ since if

$$|f_n(x+iy)| \le M_1 e^{-A_1x^2+B_1y^2} \quad \text{and} \quad |g_n(x+iy)| \le M_2 e^{-A_2x^2+B_2y^2},$$

then

$$|\alpha f_n(x+iy) + \beta g_n(x+iy)| \le M_3 e^{-A_3x^2+B_3y^2},$$

where

$$A_3 = \min(A_1, A_2), \quad B_3 = \max(B_1, B_2), \quad M_3 = \max\left(|\alpha|M_1 + |\beta|M_2\right).$$

We define $\mathbf{S}^2 = \mathbf{S}\times\mathbf{S}$ as the linear space of all entire functions $f(z_1, z_2)$ of two complex variables z_1, z_2 that satisfy

$$|f(x_1+iy_1, x_2+iy_2)| \le M\exp\left(-A(x_1^2+x_2^2) + B(y_1^2+y_2^2)\right), \qquad (31.30)$$

for some A, B, and $M > 0$. Convergence in $\mathbf{S}^2$ is defined as that in $\mathbf{S}$.

Let $\mathbf{S}^*$ denote the dual space of $\mathbf{S}$. Thus, if $\{f_n\}_{n=1}^{\infty}$ is a sequence in $\mathbf{S}$ converging to $f \in \mathbf{S}$ and $F \in \mathbf{S}^*$, then $\lim_{n\to\infty}\langle F, f_n\rangle = \langle F, f\rangle$. Moreover, if $\{F_n\}_{n=1}^{\infty}$ is a sequence in $\mathbf{S}^*$, then $\lim_{n\to\infty} F_n = F \in \mathbf{S}^*$ if and only if

$$\lim_{n\to\infty}\langle F_n, f\rangle = \langle F, f\rangle \quad \text{for all} \quad f \in \mathbf{S}.$$

THEOREM 31.4

Let $f, g \in \mathbf{S}$. Then their Wigner distribution $W_{f,g}(u,v) \in \mathbf{S}^2$.

Proof: From the representation

$$W_{f,g}(u,v) = 2e^{-2iuv}\int_{-\infty}^{\infty} f(2u-x)\overline{g}(x)e^{2ivx}dx$$

and the fact that f and g are entire functions satisfying (31.29), it follows that the integral converges absolutely and uniformly on any compact subset of the complex u and v planes; hence $W_{f,g}$ is entire in u and v.

We only need to show that $W_{f,g}$ satisfies an estimate of type (31.30). Let $u = u_1 + iu_2, v = v_1 + iv_2$ and $x = t + ib$, where b is some fixed number to be determined later. Without loss of generality we may assume that f and g satisfy (31.30) for the same constants A, B and M, because otherwise we can replace A_f and A_g by their minimum, B_f, B_g and M_f, M_g by their maximum. Therefore, since

$$\left|f\left(u + \frac{x}{2}\right)\right| \le M \exp\left[-Au_1^2 + Bu_2^2 + Bbu_2 + \frac{Bb^2}{4} - Au_1 t - \frac{At^2}{4}\right],$$

and

$$\left|g\left(u - \frac{x}{2}\right)\right| \le M \exp\left[-Au_1^2 + Bu_2^2 - Bbu_2 + \frac{Bb^2}{4} + Au_1 t - \frac{At^2}{4}\right]$$

it follows that

$$|W_{f,g}(u,v)| \le M^2 \exp\left[-2Au_1^2 + 2Bu_2^2 + \frac{Bb^2}{2} + bv_1\right] \int_{-\infty}^{\infty} \exp\left[-\frac{At^2}{2} + v_2 t\right] dt,$$

which, if we employ (6.9), gives us

$$|W_{f,g}(u,v)| \le \sqrt{\frac{2\pi}{A}}\, M^2 \exp\left[-2Au_1^2 + 2Bu_2^2 + \frac{Bb^2}{2} + bv_1 + \frac{v_2^2}{2A}\right].$$

By minimizing over b, we obtain

$$|W_{f,g}(u,v)| \le \sqrt{\frac{2\pi}{A}}\, M^2 \exp\left[-2Au_1^2 - \frac{v_1^2}{2B} + 2Bu_2^2 + \frac{v_2^2}{2A}\right],$$

which can be reduced to the desired form

$$|W_{f,g}(u,v)| \le C \exp\left[-\alpha(u_1^2 + v_1^2) + \beta(u_2^2 + v_2^2)\right],$$

where $\alpha = \min(2A, 1/2B), \beta = \max(2B, 1/2A)$ and $C = \sqrt{2\pi/A}\, M^2$. ∎

Now we can define the Wigner distribution of generalized functions.

DEFINITION 31.5 *Let $F, G \in \mathbf{S}^*$ and $f, g \in \mathbf{S}$. The Wigner distribution $W_{F,G}$ of F and G is defined as an element of $(\mathbf{S}^2)^*$ given by*

$$\langle W_{F,G}, W_{f,g}\rangle = \langle F, g\rangle\langle G, f\rangle.$$

If we denote by $\tilde{\mathbf{S}}^2$ that subspace of $\mathbf{S}^2$ consisting of functions of the form $W_{f,g}$ for some f and $g \in \mathbf{S}$, then $W_{F,G} \in (\tilde{\mathbf{S}}^2)^*$. The definition makes sense by Theorem 31.4. If $F, G \in \mathbf{S}$, then this definition agrees with Definition 31.3.

N. De Bruijn has shown [31] that if H is a generalized function in $(\mathbf{S}^2)^*$, then there exists a unique linear mapping ω_H of $\mathbf{S}$ into $\mathbf{S}^*$ such that

$$\langle H, W_{f,g} \rangle = \langle \omega_H g, f \rangle$$

for all $f, g \in \mathbf{S}$. The correspondence between H and ω_H is called *Weyl's correspondence.*

31.10 Applications

The idea of Wigner distribution started in quantum physics and later spread to optics, signal analysis, and radar theory in the form of ambiguity function. Therefore, it is needless to say that the Wigner distribution and the ambiguity function have too many physical applications to be listed here. In this section we will discuss only one application and refer the interested reader to [31, 184, 185, 205, 249] for more examples. The example we are about to discuss is from radar theory and it has been chosen not only because it is interesting, but also because it explains how the phrase "*radar ambiguity function*" came about.

The main purpose of a radar is to detect the presence of an object of interest and provide other useful information about it, such as its velocity and size. Determining the object's location is called the *problem of target detection.* The radar accomplishes its task by transmitting an electromagnetic wave $f(t)$ or a pulse which is reflected by the target. The echoed signal $f_e(t)$ is then picked up by the receiver and analyzed in order to determine the target location.

In practice, echoes of different strengths are received, but here it is assumed in effect that there is only a single echo of known strength. Since system and background noise is unavoidable, for a mathematical model to be realistic, it must account for the noise which will be assumed to be Gaussian and uniformly distributed in time and frequency. If we further assume that the noise is additive to the input signal, then we may write the echoed signal in the form

$$f_e(t) = f(t - \tau_0) + n(t) \tag{31.31}$$

where $n(t)$ is a function representing the Gaussian noise and $f(t - \tau_0)$ is the transmitted signal after suffering a time-delay τ_0. The object of the receiver is to determine τ_0 by analysis of the received signal $f_e(t)$. For simplicity, we assume that the target is stationary; hence τ_0 is independent of time t.

The transmitted signal f is, of course, known and a copy of it is available at the receiver for comparison with f_e. Through probabilistic and statistical arguments, P. Woodward [249] showed that if f_e is known, then the probability of determining τ is given by

$$q(\tau) = \frac{2}{N_0} \int_T f_e(t) f(t-\tau) dt, \tag{31.32}$$

where T is the time interval for which f_e is available and N_0 is the mean noise power per unit bandwidth. Substituting (31.31) into (31.32) and separating f_e into its signal and noise components, we find

$$q(\tau) = g(\tau) + h(\tau),$$

where

$$g(\tau) = \frac{2}{N_0} \int_T f(t-\tau_0) f(t-\tau) dt,$$

and

$$h(\tau) = \frac{2}{N_0} \int_T n(t) f(t-\tau) dt.$$

The functions g and h are called the signal and noise functions, respectively. The structure of q can now be analyzed. It can be shown that g has a maximum at τ_0 and the value of the maximum is

$$g(\tau_0) = g_{\max} = \frac{2E}{N_0},$$

where E is the energy of the signal. Thus, if the noise were removed, $q(\tau)$ would have a maximum at τ_0. Therefore, determining τ_0 can be accomplished by searching through the maxima (peaks) of $q(\tau)$.

If the signal is suitably chosen, g will be very small for τ sufficiently far from τ_0 and even though g may have other unavoidable maxima besides the one at τ_0, the one at τ_0 will be a global maximum, i.e., $g(\tau) \leq g(\tau_0)$ for all τ. Having a pronounced global maximum at τ_0 makes it easier to circumvent ambiguities of range determination. The function $h(\tau)$ is a random function of τ whose variance can be shown to be equal to $g_{\max} = 2E/N_0 = R$.

The quantity R is the signal-to-noise ratio (S/N) in $f_e(\tau)$ and it is also the value of the highest peak of the signal function $g(\tau)$ and the variance of the noise function $h(\tau)$. The mathematical significance of R is that when R is less than unity, the variation of q with τ is due mainly to the noise, but as R increases the effect of the noise starts to diminish. As R is made larger than unity, the peaks of g start to separate into clearly distinguishable peaks separated by regions of almost zero probability. The true value of the time-delay τ_0 can be estimated with accuracy represented by the width of any one of these peaks, but which peak is located at the true value of τ_0 is a matter of guesswork. Even if there is only one peak, τ_0 cannot be determined with complete precision because the accuracy with which τ_0 can be determined is measured by the width of the peak which is located at τ_0.

Complete precision occurs when $R = \infty$, that is in the absence of noise, and in this case the peak at τ_0 becomes a delta-function at τ_0. This ambiguity in estimating τ_0 is what gave rise to the term ambiguity function.

In the model discussed above, the signal f was assumed to be real-valued; thus, its Fourier transform $\hat{f}$ satisfies $\overline{\hat{f}(w)} = \hat{f}(-w)$ and consequently, f is determined completely by its positive spectrum, $\hat{f}(w)(w \geq 0)$. If we introduce the function

$$\psi_f(t) = \sqrt{\frac{2}{\pi}} \int_0^\infty \hat{f}(w) e^{-itw} dw,$$

we find $f(t) = Re\{\psi_f(t)\}$. The waveform $u_f(t)$ of a signal f is defined by

$$u_f(t) = \psi_f(t) e^{-i\omega_0 t},$$

where ω_0 is the carrier frequency defined by $\omega_0 = \int_{-\infty}^{\infty} w \left|\hat{\psi}_f(w)\right|^2 dw$. By passing from the real-valued functions f, f_e and n to their complex-valued form, Equation (31.29) can be written as $\psi_{f_e}(t) = \psi_f(t - \tau_0) + \psi_n(t)$.

If different ranges are to be distinguishable at the receiver, the signal waveform must be different from its shifted version as much as possible. One way to express this is to require that

$$\int |\psi_f(t) - \psi_f(t + \tau)|^2 dt$$

be as large as possible except in a neighborhood of $\tau = 0$. This implies that

$$Re \int \psi_f(t) \overline{\psi}_f(t + \tau) dt = Re\, e^{-\omega_0 \tau} \int u_f(t) \overline{u}_f(t + \tau) dt$$

should be as small as possible, except in a neighborhood of $\tau = 0$. Let us consider the function

$$I(\tau) = \left| \int u_f(t) \overline{u}_f(t + \tau) dt \right|.$$

The waveform u_f must be chosen so that $I(\tau)$ is as close to zero as possible, except for a neighborhood of $\tau = 0$ where it has a global maximum so as to avoid any ambiguities in range determination. It can be shown that even in the absence of noise, two stationary targets at ranges represented by τ_0 and $\tau_0 + \tau$ cannot be distinguished if $I(\tau) = I(0)$, which again results in some ambiguities in target detection.

Just as the range is measured by time-delay, the radial velocity can be measured by the Doppler frequency shift. This is accomplished by introducing the correlation function for a combined time and frequency shift

$$\mathcal{A}(\tau, w) = \int_{-\infty}^{\infty} u_f(t) \overline{u}_f(t + \tau) e^{-i\omega t} dt$$

which plays the role of the function g in the theory of ideal reception discussed above. It can be shown that a target with range and velocity determined by (τ_0, w_0) cannot be distinguished from a target at $(\tau_0 + \tau, w_0 + w)$ if $\mathcal{A}(\tau, w) = \|f\|^2$. This is the radar uncertainty principle that we have alluded to in Section 31.5 of this chapter (cf. (31.23)). For more details on radar theory and applications, see [205, 249].

32

The Wavelet Transform

One of the vital ideas in mathematical analysis is the decomposition of a function as a linear combination of some basic functions of elementary form which we shall call building blocks. This idea of function decomposition, which dates back to the work of Fourier, facilitates the study of functions and the operators acting on them. Many building blocks have appeared over the years; the newest comer on the stage is wavelets. Wavelets are building blocks for a large number of function and generalized function spaces, but unlike other building blocks, they are generated from one single function by translation and dilation.

In the last few years the wavelet transform has emerged as one of the most promising function transforms of this century with a great potential for applications. Never has any transformation spurred so much interest and a great number of publications in such a short time as the wavelet transform. Its applications have almost eclipsed those of the Fourier transform and rivaled those of the Radon transform.

Wavelet theory provides a unified framework and coherent theory for a number of different ideas and techniques that have been independently developed in several fields since the beginning of the century. The word "wavelets" started to appear in the literature only in the early 1980s; nevertheless, wavelets themselves were recognized in one form or another much earlier than that by different scientists in various fields. The fundamentals of wavelet theory were known to mathematicians, in particular, to harmonic analysts, in the form of the Calderon-Zygmund operator, Littlewood-Paley analysis, Franklin basis and atomic decompositions of functions. In fact,the first orthonormal wavelet basis was discovered by Haar in 1909 [82] and later was refined by Franklin in 1927 [63].

On the other hand, subband coding systems and tree structured digital filter banks, ideas that have been known to electric and communication engineers for a while, turned out to be related to multiresolution analysis which is a mathematical structure contrived to generate orthonormal wavelet basis. Filter banks had been used in speech processing long before wavelets were formally introduced.

Although the basic ideas of wavelet theory were known to many scientists in different fields, it was the work of the French School, led by a geophysicist, J. Morlet, a theoretical physicist, A. Grossmann, and a mathematician, Y. Meyer, that has attracted the attention of the scientific community. In a series of papers [156, 161, 162, 171, 172], they established the mathematical foundation of wavelet analysis and introduced the term "ondelettes" or "wavelets." Wavelet theory also owes a great deal to I. Daubechies [50] for constructing orthonormal wavelet bases consisting of compactly supported functions and to S. Mallat [153, 154] for introducing the concept of multiresolution analysis.

32.1 Definitions

DEFINITION 32.1 *Let $\psi \in L^2(\mathcal{R})$, a and b be real numbers with $a \neq 0$. We define the continuous wavelet transform of a function $f(t)$ with respect to ψ as*

$$W_\psi[f](a,b) = \int_{-\infty}^{\infty} f(x)\overline{\psi}_{a,b}(x)dx \; , \tag{32.1}$$

where

$$\psi_{a,b}(x) = \frac{1}{\sqrt{|a|}}\psi\left(\frac{x-b}{a}\right) . \tag{32.2}$$

The function ψ is called the mother wavelet or the analyzing wavelet. The mother wavelet is assumed to satisfy

$$C_\psi = 2\pi \int_{-\infty}^{\infty} \frac{|\hat{\psi}(w)|^2}{|w|} dw < \infty \; , \tag{32.3}$$

where $\hat{\psi}$ is the Fourier transform of ψ. Condition (32.3) is called the admissibility condition. If there is only one mother wavelet involved in a formula, we shall write $W[f]$ instead of $W_\psi[f]$.

The constant $1/\sqrt{|a|}$ in (32.2) is a normalizing constant introduced so that

$$||\psi_{a,b}||^2 = \int_{-\infty}^{\infty} |\psi_{a,b}(x)|^2 dx = \int_{-\infty}^{\infty} |\psi(x)|^2 dx = ||\psi||^2 \, .$$

If $\psi \in L^1(\mathcal{R})$, then the admissibility condition is satisfied only if

$$\hat{\psi}(0) = 0 \quad \text{or} \quad \int_{-\infty}^{\infty} \psi(x)dx = 0.$$

In many cases of interest it is more convenient to discretize the parameters a and b. The standard choice is $a = a_0^m$, $a_0 \neq 0, 1$, and $b = nb_0 a_0^m$, $b_0 \neq 0$, where m and n are integers. Since m and n take all positive and negative values, we choose, without loss of generality, $a_0 > 1$ and $b_0 > 0$.

DEFINITION 32.2 *We define the discrete wavelet transform of f as the doubly indexed sequence*

$$\tilde{f}(m,n) = W[f](m,n) = \langle f, \psi_{m,n} \rangle = \int_{-\infty}^{\infty} f(x) \overline{\psi}_{m,n}(x) dx ,$$

where

$$\psi_{m,n}(x) = a_0^{-\frac{m}{2}} \psi(a_0^{-m} x - nb_0) .$$

The series $\sum_{m,n=-\infty}^{\infty} \tilde{f}(m,n)\psi_{m,n}(x)$ is called the wavelet series of f and the functions $\{\psi_{m,n}(x)\}$ are called wavelets.

It should be emphasized that some authors use the term "discrete wavelet transform" to indicate a wavelet transformation of a sequence of numbers rather than of a function. Such a transformation of sequences is very useful in applications, but unfortunately is beyond the scope of this book. The discrete wavelet transform can be viewed as the values of the continuous wavelet transform at the lattice of sample points $\{(a_0^m, nb_0 a_0^m)\}_{m,n=-\infty}^{\infty}$. As with the Gabor transform (see Chapter 30), the following questions naturally arise:

1) Is the continuous wavelet transform, $W[f]$, of f and consequently f itself completely determined by the sequence $\{\tilde{f}(m,n)\}$?
2) And if so, can we recover f from $\{\tilde{f}(m,n)\}$?
3) Does the wavelet series of f converge to f and in what sense?

If $f \in L^2(\mathcal{R})$, then the answer to the first question is related to whether the system of wavelets $\{\psi_{m,n}\}$ is complete in $L^2(\mathcal{R})$. The answers to the other questions are obvious if the wavelets form a frame or an orthonormal basis for $L^2(\mathcal{R})$. We shall discuss this in more detail later in this chapter.

32.2 The Continuous Wavelet Transform

32.2.1 Existence

The continuous wavelet transform of $f \in L^2(\mathcal{R})$ exists as can be seen by applying the Cauchy–Schwarz inequality to (32.1). In most practical applications, the mother wavelet is assumed to be continuous and to vanish at $\pm\infty$; therefore, the continuous wavelet transform exists for a larger class of functions than $L^2(\mathcal{R})$. If the mother wavelet is bounded, then for the wavelet transform to exist, it suffices that $f \in L^1(\mathcal{R})$.

32.2.2 Elementary Properties

1. **Translation:**

$$W_\psi[f(x+c)](a,b) = W_\psi[f(x)](a,b+c)$$

2. **Dilation:**

$$W_\psi\left[f\left(\frac{x}{c}\right)\right](a,b) = \sqrt{c}\, W_\psi[f(x)]\left(\frac{a}{c},\frac{b}{c}\right), \qquad c > 0 .$$

3. **Interchanging of mother wavelets:** If ϕ and ψ are admissible mother wavelets, then

$$W_\psi[\phi](a,b) = \overline{W}_\phi[\psi]\left(\frac{1}{a},-\frac{b}{a}\right) .$$

32.2.3 Relationship with the Fourier Transform

Upon observing that

$$\mathcal{F}(\psi_{a,b}(x))(w) = \sqrt{|a|}\; e^{ibw}\hat{\psi}(aw) ,$$

where $\mathcal{F}$ denotes the Fourier transformation, and using the Parseval relation for the Fourier transform, we obtain another version of the continuous wavelet transform

$$\begin{aligned} W_\psi[f](a,b) &= \langle f, \psi_{a,b}\rangle = \langle \hat{f}, \hat{\psi}_{a,b}\rangle \\ &= \sqrt{|a|}\int_{-\infty}^{\infty} \hat{f}(w)e^{-ibw}\overline{\hat{\psi}}(aw)dw , \end{aligned} \tag{32.4}$$

which implies

$$\sqrt{|a|}\; \hat{f}(w)\overline{\hat{\psi}}(aw) = \frac{1}{2\pi}\int_{-\infty}^{\infty} W_\psi[f](a,b)e^{iwb}db . \tag{32.5}$$

32.2.4 Parseval's Relation

We shall derive three different versions of Parseval's relation for the continuous wavelet transform. But first, let us remark that the continuous wavelet transform is a continuous linear transformation from $L^2(\mathcal{R})$ into the space

$$L^2(\mathcal{R}^2, a^{-2}dadb) = \left\{ F(a,b) : \int_{-\infty}^{\infty}\int_{-\infty}^{\infty} |F(a,b)|^2 \frac{dadb}{a^2} < \infty \right\},$$

which is a Hilbert space with inner product

$$\langle F, G\rangle = \int_{-\infty}^{\infty}\int_{-\infty}^{\infty} F(a,b)\overline{G}(a,b)d\mu(a,b)$$

where $d\mu(a,b) = da\,db/a^2$.

Let $f, g \in L^2(\mathcal{R})$ and denote their continuous wavelet transforms by F and G, respectively. In view of (32.4) and (32.5), we obtain

$$\begin{aligned}
\langle F, G\rangle_{L^2(\mathcal{R}^2,\mu)} &= \int_{-\infty}^{\infty}\int_{-\infty}^{\infty} \frac{da\,db}{a^2}\left(F(a,b)\sqrt{|a|}\int_{-\infty}^{\infty} \overline{\hat{g}}(w)e^{ibw}\hat{\psi}(aw)\,dw\right) \\
&= \int_{-\infty}^{\infty} \frac{da}{a^2}\int_{-\infty}^{\infty} \sqrt{|a|}\,\overline{\hat{g}}(w)\hat{\psi}(aw)dw \int_{-\infty}^{\infty} F(a,b)e^{ibw}db \\
&= 2\pi \int_{-\infty}^{\infty} \frac{da}{|a|}\int_{-\infty}^{\infty} \hat{f}(w)\overline{\hat{g}}(w)|\hat{\psi}(aw)|^2 dw \\
&= 2\pi \int_{-\infty}^{\infty} \hat{f}(w)\overline{\hat{g}}(w)\left(\int_{-\infty}^{\infty} \frac{|\hat{\psi}(aw)|^2}{|a|}da\right)dw\,.
\end{aligned}$$

But

$$\int_{-\infty}^{\infty} \hat{f}(w)\overline{\hat{g}}(w)\left(\int_{-\infty}^{\infty} \frac{|\hat{\psi}(aw)|^2}{|a|}da\right)dw =$$

$$\left\{\int_{-\infty}^{0} \hat{f}(w)\overline{\hat{g}}(w)\left(\int_{-\infty}^{\infty} \frac{|\hat{\psi}(aw)|^2}{|a|}da\right)dw + \int_{0}^{\infty} \hat{f}(w)\overline{\hat{g}}(w)\left(\int_{-\infty}^{\infty} \frac{|\hat{\psi}(aw)|^2}{|a|}da\right)dw\right\}.$$

If we make the change of variable, $y = aw$, in the inside integrals, we obtain the first version of Parseval's relation

$$\langle W_\psi[f], W_\psi[g]\rangle_{L^2(\mathcal{R}^2,\mu)} = \langle F, G\rangle_{L^2(\mathcal{R}^2,\mu)} = C_\psi\langle \hat{f}, \hat{g}\rangle_{L^2(\mathcal{R})} = C_\psi\langle f, g\rangle_{L^2(\mathcal{R})}\,, \tag{32.6}$$

where C_ψ is given by (32.3). In particular,

$$\|F\|^2_{L^2(\mathcal{R}^2,\mu)} = C_\psi\|f\|^2_{L^2(\mathcal{R})}\,.$$

If ψ is normalized so that $C_\psi = 1$, it will follow that the continuous wavelet transform is an isometry from $L^2(\mathcal{R})$ into $L^2(\mathcal{R}^2,\mu)$. To simplify the notation, we will from now on drop the subscripts from the inner product.

Another version of Parseval's relation can be obtained if we restrict a to be non-negative. Repeating the above calculation with $a > 0$ leads to

$$\langle F, G\rangle = 2\pi \int_{-\infty}^{\infty} \hat{f}(w)\bar{\hat{g}}(w) \left(\int_{0}^{\infty} \frac{|\hat{\psi}(aw)|^2}{a} da \right) dw \; .$$

Under the more stringent admissibility condition

$$C_\psi = 2\pi \int_{0}^{\infty} \frac{|\hat{\psi}(w)|^2}{w} dw = 2\pi \int_{-\infty}^{0} \frac{|\hat{\psi}(w)|^2}{|w|} dw < \infty \tag{32.7}$$

we obtain

$$\langle F, G\rangle = C_\psi \langle f, g\rangle \; .$$

The admissibility condition (32.7) is satisfied if for example ψ is a real-valued function because under this assumption $\overline{\hat{\psi}}(w) = \hat{\psi}(-w)$.

A third version of Parseval's relation, which is more general than the ones given above, can be obtained as follows. Let ψ and ϕ be two mother wavelets such that

$$C_{\psi,\phi} = 2\pi \int_{-\infty}^{\infty} \frac{\hat{\phi}(w)\overline{\hat{\psi}}(w)}{|w|} dw < \infty \, ,$$

and denote the continuous wavelet transform of $f, g \in L^2(\mathcal{R})$ with respect to ψ and ϕ by $W_\psi[f]$ and $W_\phi[g]$, respectively. Then, by (32.4) and (32.5), we have

$$\begin{aligned}
\langle W_\psi[f], W_\phi[g]\rangle &= \int_{-\infty}^{\infty}\int_{-\infty}^{\infty} \frac{dadb}{a^2} W_\psi[f](a,b)\sqrt{|a|} \int_{-\infty}^{\infty} \bar{\hat{g}}(w) e^{ibw}\hat{\phi}(aw)dw \\
&= \int_{-\infty}^{\infty} \frac{da}{|a|^{\frac{3}{2}}} \left(\int_{-\infty}^{\infty} \bar{\hat{g}}(w)\hat{\phi}(aw)dw \right) \left(\int_{-\infty}^{\infty} W_\psi[f](a,b) e^{ibw} db \right) \\
&= 2\pi \int_{-\infty}^{\infty} \frac{da}{|a|} \left(\int_{-\infty}^{\infty} \hat{f}(w)\bar{\hat{g}}(w)\hat{\phi}(aw)\overline{\hat{\psi}}(aw)dw \right) \\
&= 2\pi \int_{-\infty}^{\infty} \hat{f}(w)\bar{\hat{g}}(w) \left(\int_{-\infty}^{\infty} \frac{\hat{\phi}(aw)\overline{\hat{\psi}}(aw)}{|a|} da \right) dw \\
&= C_{\psi,\phi}\langle \hat{f}, \hat{g}\rangle = C_{\psi,\phi}\langle f, g\rangle \; .
\end{aligned}$$

Thus,

$$\langle W_\psi[f], W_\phi[g]\rangle = c_{\psi,\phi}\langle f, g\rangle \; . \tag{32.8}$$

If $\psi = \phi$, Formula (32.8) reduces to (32.6).

32.2.5 Inversion Formula

The following inversion formula, which is a consequence of (32.8), is rather surprising because it shows that if the continuous wavelet transform is taken with respect to a mother wavelet ψ, the inversion formula may be taken with respect to a different mother wavelet ϕ. On one hand, we have

$$\langle W_\psi[f], W_\phi[g]\rangle = \int_{-\infty}^{\infty}\int_{-\infty}^{\infty} \frac{dadb}{a^2} W_\psi[f](a,b) \int_{-\infty}^{\infty} \overline{g}(x)\phi_{a,b}(x)dx$$

$$= \int_{-\infty}^{\infty} \left(\int_{-\infty}^{\infty}\int_{-\infty}^{\infty} \frac{dadb}{a^2} W_\psi[f](a,b)\phi_{a,b}(x)\right) \overline{g}(x)dx. \tag{32.9}$$

On the other hand, we have from (32.8)

$$\langle W_\psi[f], W_\phi[g]\rangle = C_{\psi,\phi} \int_{-\infty}^{\infty} f(x)\overline{g}(x)dx\,. \tag{32.10}$$

Since (32.9) and (32.10) are valid for all $g \in L^2(\mathcal{R})$, we conclude that

$$f(x) = \frac{1}{C_{\psi,\phi}} \int_{-\infty}^{\infty}\int_{-\infty}^{\infty} W_\psi[f](a,b)\phi_{a,b}(x)\frac{da\,db}{a^2}\,. \tag{32.11}$$

In particular if $\phi = \psi$, we obtain

$$f(x) = \frac{1}{C_\psi} \int_{-\infty}^{\infty}\int_{-\infty}^{\infty} W_\psi[f](a,b)\psi_{a,b}(x)\frac{da\,db}{a^2}\,. \tag{32.12}$$

32.2.6 Reproducing-Kernel-Hilbert Space Associated with the Continuous Wavelet Transform

The continuous wavelet transform is a continuous transformation from $L^2(\mathcal{R})$ into $L^2\left(\mathcal{R}^2,\mu\right)$ that preserves the inner product up to a constant. It is actually an isometry if ψ is normalized so that $C_\psi = 1$. The transformation, however, is not onto. Let us denote the image of $L^2(\mathcal{R})$ under the continuous wavelet transformation by $\mathbf{H}$. We will show that $\mathbf{H}$ is a reproducing-kernel Hilbert space. Let $W_\psi[f]$ be the continuous wavelet transform of $f \in L^2(\mathcal{R})$, then by (32.8) we have

$$W_\psi[f](a,b) = \langle f, \psi_{a,b}\rangle = \frac{1}{C_{\psi,\phi}}\langle W_\psi[f], W_\phi[\psi_{a,b}]\rangle$$

$$= \frac{1}{C_{\psi,\phi}} \int_{-\infty}^{\infty}\int_{-\infty}^{\infty} \frac{da'db'}{a'^2} W_\psi[f](a',b')\overline{W}_\phi[\psi_{a,b}](a',b')$$

$$= \int_{-\infty}^{\infty}\int_{-\infty}^{\infty} W_\psi[f](a',b')k(a,b,a',b')\frac{da'\,db'}{a'^2}\,,$$

where

$$k(a,b,a',b') = \frac{1}{C_{\psi,\phi}} \overline{W}_{\phi}[\psi_{a,b}](a',b') = \frac{1}{C_{\psi,\phi}} \langle \phi_{a',b'}, \psi_{a,b} \rangle$$

is the reproducing kernel. In particular, if $\psi = \phi$

$$k(a,b,a',b') = \frac{1}{C_{\psi}} \langle \psi_{a',b'}, \psi_{a,b} \rangle \,.$$

32.2.7 Local Properties of the Continuous Wavelet Transform

THEOREM 32.1

Let ψ be an admissible mother wavelet satisfying $\int (1+|x|)|\psi(x)|dx < \infty$.

1. *If f is a bounded function that satisfies Lipschitz condition of order $\alpha, 0 < \alpha \leq 1$, then*

$$|W_{\psi}[f](a,b)| \leq c|a|^{\alpha+\frac{1}{2}} \quad \textit{for some constant} \quad c > 0\,.$$

2. *If f is bounded and Hölder continuous at x_0, with exponent $0 < \alpha \leq 1$, i.e.,*

$$|f(x_0+h) - f(x_0)| \leq a|h|^{\alpha} \qquad \textit{for some } a > 0\,,$$

then

$$|W_{\psi}[f](a,b)| \leq c|a|^{\frac{1}{2}}(|a|^{\alpha} + |b|^{\alpha}), \quad \textit{for some constant} \quad c > 0\,.$$

Proof: 1) Since $\int_{-\infty}^{\infty} \psi(x)dx = 0$, it follows that $\langle 1, \psi_{a,b} \rangle = 0$; hence

$$W_{\psi}[f](a,b) = \langle f, \psi_{a,b} \rangle - f(b)\langle 1, \psi_{a,b} \rangle.$$

Thus,

$$\begin{aligned} |W_{\psi}[f](a,b)| &\leq \frac{1}{\sqrt{|a|}} \int_{-\infty}^{\infty} |f(x) - f(b)| \left| \psi\left(\frac{x-b}{a}\right) \right| dx \\ &\leq \frac{M}{\sqrt{|a|}} \int_{-\infty}^{\infty} |x-b|^{\alpha} \left| \psi\left(\frac{x-b}{a}\right) \right| dx \\ &= M|a|^{\alpha+\frac{1}{2}} \int_{-\infty}^{\infty} |y|^{\alpha} |\psi(y)| dy = c|a|^{\alpha+\frac{1}{2}}\,. \end{aligned}$$

2) Without loss of generality, we may take $x_0 = 0$. As in part (1)

$$|W_{\psi}[f](a,b)| \leq \frac{1}{\sqrt{|a|}} \int_{-\infty}^{\infty} |f(x) - f(0)| \left| \psi\left(\frac{x-b}{a}\right) \right| dx$$

$$= \frac{M}{\sqrt{|a|}} \int_{-\infty}^{\infty} |x|^{\alpha} \left| \psi\left(\frac{x-b}{a}\right) \right| dx$$
$$= M|a|^{\alpha+\frac{1}{2}} \int_{-\infty}^{\infty} \left| y + \frac{b}{a} \right|^{\alpha} |\psi(y)| dy$$
$$\leq c|a|^{\frac{1}{2}} (|a|^{\alpha} + |b|^{\alpha}) .$$

The last inequality follows from the inequality $(1+c)^{\alpha} \leq 2(1+c^{\alpha})$, for $c \geq 0$. ∎

More general results on the local behavior of the continuous wavelet transform can be found in [49, 160].

32.3 The Discrete Wavelet Transform

The question of whether a function $f \in L^2(\mathcal{R})$ is completely determined by its discrete wavelet transform (wavelet coefficients) is answered in the affirmative if the wavelets form a complete system in $L^2(\mathcal{R})$. For, if there exists another function $g \in L^2(\mathcal{R})$ such that

$$\tilde{f}(m,n) = \langle f, \psi_{m,n} \rangle = \tilde{g}(m,n) = \langle g, \psi_{m,n} \rangle ,$$

then from the completeness of the wavelet system it will follow that $f = g$.

If the wavelets form a frame or an orthonormal basis of $L^2(\mathcal{R})$, then they are complete and in addition f can be reconstructed from its discrete wavelet transform $\{\tilde{f}(m,n) = \langle f, \psi_{m,n} \rangle\}$ by means of the formula

$$f = \sum_{m,n=-\infty}^{\infty} \tilde{f}(m,n) \psi_{m,n} ,$$

if the wavelets constitute an orthonormal basis, and by the formula

$$f = \sum_{m,n=-\infty}^{\infty} \tilde{f}(m,n) \tilde{\psi}_{m,n} ,$$

if the wavelets constitute a frame, where $\{\tilde{\psi}_{m,n}\}$ is the dual frame. Actually, it turns out that wavelets can form a basis not only for $L^2(\mathcal{R})$, but also for a large class of function spaces [49, Ch. 9], [160], including spaces of analytic functions [273].

32.3.1 Orthonormal Basis and Frames of Wavelets

The purpose of this section is to report on some necessary and sufficient conditions for a system of wavelets to form a frame or an orthonormal basis. Most of the material in this section can be found in [49].

The first theorem gives a necessary condition for $\{\psi_{m,n}\}$ to be a frame.

THEOREM 32.2
If $\psi_{m,n}(x) = a_0^{-\frac{m}{2}}\psi(a_0^{-m}x - nb_0), m, n = 0, \pm1, \pm2...$ constitute a frame for $L^2(\mathcal{R})$ with frame bounds A and B, then

$$\frac{b_0 \ln a_0}{2\pi}A \le \int_0^\infty \frac{|\hat{\psi}(w)|^2}{w}dw \le \frac{b_0 \ln a_0}{2\pi}B \,,$$

and

$$\frac{b_0 \ln a_0}{2\pi}A \le \int_{-\infty}^0 \frac{|\hat{\psi}(w)|^2}{|w|}dw \le \frac{b_0 \ln a_0}{2\pi}B \,.$$

In particular, if $\{\psi_{m,n}\}$ is a tight frame, then

$$A = B = \frac{2\pi}{b_0 \ln a_0}\int_0^\infty \frac{|\hat{\psi}(w)|^2}{w}dw = \frac{2\pi}{b_0 \ln a_0}\int_{-\infty}^0 \frac{|\hat{\psi}(w)|^2}{|w|}dw \,,$$

and if they constitute an orthonormal basis,then

$$\frac{b_0 \ln a_0}{2\pi} = \int_0^\infty \frac{|\hat{\psi}(w)|^2}{w}dw = \int_{-\infty}^0 \frac{|\hat{\psi}(w)|^2}{|w|}dw \,.$$

The next theorem gives a sufficient condition for $\psi \in L^2(\mathcal{R})$ to generate a frame of wavelets.

THEOREM 32.3 [49, p. 69]
Let ψ and a_0 be such that

$$\inf_{1\le|w|\le a_0} \sum_{m=-\infty}^{\infty} |\hat{\psi}(a_0^m w)|^2 > 0 \,,$$

$$\sup_{1\le|w|\le a_0} \sum_{m=-\infty}^{\infty} |\hat{\psi}(a_0^m w)|^2 < \infty \,.$$

If $\beta(s) = \sup_w \sum_{m=-\infty}^{\infty} |\hat{\psi}(a_0^m w)||\hat{\psi}(a_0^m w + s)|$ decays at least as fast as $(1+|s|)^{-(1+\epsilon)}$ for some $\epsilon > 0$, then there exists $\tilde{b}_0 > 0$ such that $\{\psi_{m,n}(x) = a_0^{-\frac{m}{2}}\psi(a_0^{-m}x - nb_0)\}$ form a frame for any $b_0 < \tilde{b}_0$.

The above assumptions are satisfied if $|\hat{\psi}(w)| \le C|w|^\alpha(1 + |w|)^{-\gamma}$ with $\alpha > 0$ and $\gamma > \alpha + 1$.

To reconstruct f from its wavelet coefficients when $\{\psi_{m,n}\}$ constitute a frame, we need to calculate the dual frame $\tilde{\psi}_{m,n}$. To this end, let us write the wavelets in terms of the translation and dilation operators, T_a and D_a, respectively, so that

$$\psi_{m,n}(x) = D_{a^m} T_{nb} \psi(x) .$$

The frame operator is given by

$$Sf = \sum_{m,n} \langle f, D_{a^m} T_{nb} \psi \rangle D_{a^m} T_{nb} \psi , \quad f \in L^2(\mathcal{R}) .$$

Since

$$\begin{aligned} \langle f, D_{a^k} T_{nb} \psi \rangle &= a^{-\frac{k}{2}} \int_{-\infty}^{\infty} f(x) \overline{\psi}(a^{-k}x - nb) dx \\ &= a^{\frac{k}{2}} \int_{-\infty}^{\infty} f(a^k x) \overline{\psi}(x - nb) dx \\ &= \langle D_{a^{-k}} f, T_{nb} \psi \rangle , \end{aligned}$$

it follows that

$$\begin{aligned} D_{a^k}(Sf) &= \sum_{m,n} \langle f, D_{a^m} T_{nb} \psi \rangle D_{a^{k+m}} T_{nb} \psi \\ &= \sum_{m,n} \langle f, D_{a^{m-k}} T_{nb} \psi \rangle D_{a^m} T_{nb} \psi \\ &= \sum_{m,n} \langle D_{a^k} f, D_{a^m} T_{nb} \psi \rangle D_{a^m} T_{nb} \psi \\ &= S(D_{a^k} f) . \end{aligned}$$

That is, the frame operator S commutes with the dilation operator D_{a^k}; hence

$$S^{-1}(D_{a^k} f) = D_{a^k}(S^{-1} f) .$$

From the definition of the dual frame (Theorem 1.7), we have

$$\begin{aligned} \tilde{\psi}_{m,n}(x) &= S^{-1}(\psi_{m,n}(x)) = S^{-1}(D_{a^m} T_{nb} \psi(x)) \\ &= D_{a^m} S^{-1}(T_{nb} \psi(x)) = D_{a^m} \tilde{\psi}_{0,n}(x) = a^{-\frac{m}{2}} \tilde{\psi}_{0,n}(a^{-m} x) , \end{aligned}$$

where $\tilde{\psi}_{0,n}(x) = S^{-1}(\psi(x - nb))$.

Unlike the dual Gabor frame (see Definition 30.3 and Theorem 29.4), the dual wavelet frame is not generated by one single function. The reason is that the wavelet frame operator does not commute with the translation operator.

32.3.2 Multiresolution Analysis

The concept of a multiresolution analysis of a Hilbert space of functions was formulated by S. Mallet and Y. Meyer in 1986 and was eloquently presented by Mallat in [153]. It is a mathematical structure that is contrived to produce orthonormal wavelet basis using a set of rules and procedures. Without loss of generality, we shall take $a = 2$ and $b = 1$.

DEFINITION 32.3 *A multiresolution analysis of $L^2(\mathcal{R})$ consists of a nested sequence of closed subspaces $\{V_n\}_{n=-\infty}^{\infty}$ of $L^2(\mathcal{R})$ and a function $\phi \in V_0$, called the scaling function or the father wavelet, such that*

1) $\ldots \subset V_1 \subset V_0 \subset V_{-1} \subset V_{-2} \subset \cdots,$

2) $\bigcup_{i=-\infty}^{\infty} V_i$ *is dense in* $L^2(\mathcal{R})$,

3) $\bigcap_{i=-\infty}^{\infty} V_i = \{0\}$,

4) $V_{n+1} = D_2 V_n = \{f(2^{-1}x) : f(x) \in V_n\}$,

5) $\{T_n\phi(x) = \phi(x-n)\}_{n=-\infty}^{\infty}$ *is an orthonormal basis for* V_0.

From the definition, it readily follows that

i) $f(x) \in V_0$ if and only if $f(2^{-m}x) \in V_m$,

ii) V_0 is invariant under translation by an integer, i.e., $T_k V_0 = V_0$,

iii) $V_m = D_{2^m} T_k V_0 = \{f(2^{-m}x - k) : f(x) \in V_0\}$,

iv) for each m, $\{\phi_{m,n}(x) = 2^{-\frac{m}{2}}\phi(2^{-m}x - n)\}_{n=-\infty}^{\infty}$ is an orthonormal basis for V_m.

It will be shown that given a multiresolution analysis, one can construct an orthonormal wavelet basis of $L^2(\mathcal{R})$. Since $V_n \subset V_{n-1}$, let us define the orthogonal complement of V_n in V_{n-1} by W_n so that

$$V_{n-1} = V_n \oplus W_n \quad \text{for all} \quad n\ ,$$

where $\oplus$ denotes the direct sum; see Section 1.5. By induction, we can show that

$$L^2(\mathcal{R}) = \bigoplus_{i=-\infty}^{\infty} W_i \tag{32.13}$$

where

$$\bigoplus_{i=-k}^{k} W_i = W_{-k} \bigoplus \ldots \bigoplus W_0 \bigoplus \ldots \bigoplus W_k\ ,$$

and $f(x) \in W_0$ if and only if $f(2^{-n}x) \in W_n$. Unlike the spaces V_ns, the W_ns are disjoint; a property that will play a crucial role later on.

LEMMA 32.1

Let $\phi, \psi \in L^2(\mathcal{R})$. Then

1) The set $\{\phi(x-n)\}_{n=-\infty}^{\infty}$ is orthonormal if and only if

$$\sum_{k=-\infty}^{\infty} |\hat{\phi}(w+2\pi k)|^2 = 1 \quad a.e.$$

2) The sets $\{\phi(x-n)\}_{n=-\infty}^{\infty}$ and $\{\psi(x-m)\}_{m=-\infty}^{\infty}$ are biorthogonal, i.e., $\langle \psi_m, \phi_n \rangle = 0$ for all m and n if and only if

$$\sum_{k=-\infty}^{\infty} \hat{\psi}(w+2\pi k)\overline{\hat{\phi}}(w+2\pi k) = 0 \quad a.e.$$

where $\phi_n(x) = \phi(x-n)$ and $\psi_m(x) = \psi(x-m)$.

Proof: 1) Since $\hat{\phi}_n(w) = e^{inw}\hat{\phi}(w)$, we have from Parseval's relation for the Fourier transform

$$\begin{aligned}
\langle \phi_n, \phi_m \rangle &= \langle \phi_0, \phi_{m-n} \rangle = \langle \hat{\phi}_0, \hat{\phi}_{m-n} \rangle \\
&= \int_{-\infty}^{\infty} |\hat{\phi}(w)|^2 e^{i(n-m)w} dw = \sum_{k=-\infty}^{\infty} \int_{2\pi k}^{2\pi(k+1)} |\hat{\phi}(w)|^2 e^{i(n-m)w} dw \\
&= \sum_{k=-\infty}^{\infty} \int_0^{2\pi} |\hat{\phi}(w+2\pi k)|^2 e^{i(n-m)w} dw \\
&= \int_0^{2\pi} e^{i(n-m)w} \sum_{k=-\infty}^{\infty} |\hat{\phi}(w+2\pi k)|^2 dw\,.
\end{aligned}$$

Thus, from the completeness of $\{e^{inw}\}_{n=-\infty}^{\infty}$ in $L^2[0, 2\pi]$, it follows that $\langle \phi_n, \phi_m \rangle = \delta_{n,m}$ if and only if $\sum_{k=-\infty}^{\infty} |\hat{\phi}(w+2\pi k)|^2 = 1$ a.e.

2) Similarly,

$$\begin{aligned}
\langle \psi_n, \phi_m \rangle &= \langle \psi_0, \phi_{m-n} \rangle = \langle \hat{\psi}_0, \hat{\phi}_{m-n} \rangle \\
&= \int_{-\infty}^{\infty} \hat{\psi}(w)\overline{\hat{\phi}}(w) e^{i(n-m)w} dw \\
&= \sum_{k=-\infty}^{\infty} \int_0^{2\pi} \hat{\psi}(w+2\pi k)\overline{\hat{\phi}}(w+2\pi k) e^{i(n-m)w} dw \\
&= \int_0^{2\pi} e^{i(n-m)w} \left(\sum_{k=-\infty}^{\infty} \hat{\psi}(w+2\pi k)\overline{\hat{\phi}}(w+2\pi k) \right) dw
\end{aligned}$$

Therefore, $\langle \psi_n, \phi_m \rangle = 0$ for all m and n if and only if

$$\sum_{k=-\infty}^{\infty} \hat{\psi}(w+2\pi k)\overline{\hat{\phi}}(w+2\pi k) = 0 .$$

∎

LEMMA 32.2

The scaling function ϕ satisfies the following conditions:

1.

$$\sum_{k=-\infty}^{\infty} |\hat{\phi}(w+2\pi k)|^2 = 1 \quad a.e. \tag{32.14}$$

2.

$$\hat{\phi}(w) = M\left(\frac{w}{2}\right)\hat{\phi}\left(\frac{w}{2}\right) , \tag{32.15}$$

where $M(w)$ is a periodic function with period 2π that belongs to $L^2[0,2\pi]$ and satisfies

$$|M(w)|^2 + |M(w+\pi)|^2 = 1 \quad a.e. \tag{32.16}$$

Proof: Part (1) follows from Lemma 32.1 and the fact that $\{\phi(x-n)\}_{n=-\infty}^{\infty}$ is orthonormal.

2) Since $\phi \in V_0 \subset V_{-1}$ and $\{\phi_{-1,n}(x) = 2^{\frac{1}{2}}\phi(2x-n)\}_{n=-\infty}^{\infty}$ is an orthonormal basis for V_{-1}, we have

$$\phi(x) = \sqrt{2}\sum_{n=-\infty}^{\infty} \alpha_n \phi(2x-n) , \quad \text{with} \quad \sum_{n=-\infty}^{\infty} |\alpha_n|^2 < \infty , \tag{32.17}$$

where $\alpha_n = \langle \phi, \phi_{-1,n} \rangle$. By taking the Fourier transform of both sides of (32.17) we obtain (32.15), where

$$M\left(\frac{w}{2}\right) = \frac{1}{\sqrt{2}}\sum_{n=-\infty}^{\infty} \alpha_n e^{inw/2} . \tag{32.18}$$

Clearly $M(w) \in L^2[0,2\pi]$. By substituting (32.15) into (32.14) and replacing w by $2w$, we have

$$1 = \sum_{k=-\infty}^{\infty} |M(w+\pi k)|^2 \, |\hat{\phi}(w+\pi k)|^2 .$$

By writing the series on the right-hand side as a sum of two series, one taken over k even and the other over k odd we find

$$1 = \sum_{m=-\infty}^{\infty} |M(w+2m\pi)|^2 |\hat{\phi}(w+2\pi m)|^2$$

$$+ \sum_{m=-\infty}^{\infty} |M(w+2m\pi+\pi)|^2 |\hat{\phi}(w+2m\pi+\pi)|^2$$
$$= |M(w)|^2 \sum_{m=-\infty}^{\infty} |\hat{\phi}(w+2\pi m)|^2$$
$$+ |M(w+\pi)|^2 \sum_{m=-\infty}^{\infty} |\hat{\phi}(w+\pi+2\pi m)|^2$$
$$= |M(w)|^2 + |M(w+\pi)|^2,$$

which is (32.16). ∎

LEMMA 32.3
The Fourier transform of any function $f \in W_0$ can be written in the form

$$\hat{f}(w) = \lambda_f(w)\hat{\psi}(w) \,, \tag{32.19}$$

where $\lambda_f(w)$ is a periodic function with period 2π and $\hat{\psi}$ is independent of f. Moreover, $\lambda_f \in L^2[0,\pi]$ and

$$\|f\|_{L^2(\mathcal{R})} = \|\lambda_f\|_{L^2[0,\pi]} \,. \tag{32.20}$$

Proof: Since $f \in W_0$, it follows that f is in V_{-1} and is orthogonal to V_0. That $f \in V_{-1}$ implies

$$f(x) = \sqrt{2} \sum_{n=-\infty}^{\infty} f_n \phi(2x-n) \,, \text{with} \quad \sum_{n=-\infty}^{\infty} |f_n|^2 < \infty \,,$$

where $f_n = \langle f, \phi_{-1,n} \rangle$. By repeating the same argument as in Lemma 32.2, we obtain

$$\hat{f}(w) = M_f\left(\frac{w}{2}\right) \hat{\phi}\left(\frac{w}{2}\right) \,, \tag{32.21}$$

in which

$$M_f(w) = \frac{1}{\sqrt{2}} \sum_{n=-\infty}^{\infty} f_n e^{inw} \,.$$

Clearly M_f is periodic with period 2π and belongs to $L^2[0,2\pi]$. On the other hand, since f is orthogonal to V_0, it is orthogonal to $\{\phi(x-n)\}_{n=-\infty}^{\infty}$ and by Lemma 32.1

$$\sum_{k=-\infty}^{\infty} \hat{f}(w+2\pi k)\overline{\hat{\phi}}(w+2\pi k) = 0 \quad \text{a.e.} \tag{32.22}$$

Substituting (32.21) and (32.15) into (32.22) and repeating the same argument used in the proof of part (2) of Lemma 32.2 , we obtain

$$M_f(w)\overline{M}(w) + M_f(w+\pi)\overline{M}(w+\pi) = 0 \quad \text{a.e.}$$

Since $M(w)$ and $M(w+\pi)$ cannot vanish simultaneously on a set of positive measure, we may set $M_f(w) = K_f(w)\overline{M}(w+\pi)$ to obtain $K_f(w) + K_f(w+\pi) = 0$ a.e. One possible solution of this equation is $K_f(w) = e^{iw}\lambda_f(2w)$, where $\lambda_f(2w) = \lambda_f(2w+2\pi)$. Therefore, by combining this with (32.21) and (32.15), we have

$$\hat{f}(w) = \lambda_f(w)\hat{\psi}(w) ,$$

with

$$\hat{\psi}(w) = e^{i\frac{w}{2}}\overline{M}\left(\frac{w}{2}+\pi\right)\hat{\phi}\left(\frac{w}{2}\right) . \tag{32.23}$$

In view of (32.23), (32.16) and (32.14), it is easy to verify that

$$\sum_{k=-\infty}^{\infty} |\hat{\psi}(w+2\pi k)|^2 = 1 \quad \text{a.e.} \tag{32.24}$$

Finally, to prove (32.20), we recall that λ_f is periodic with period 2π and use (32.24) to conclude

$$\begin{aligned} \|\hat{f}\|^2 &= \int_{-\infty}^{\infty} |\hat{f}(w)|^2 dw = \sum_{k=-\infty}^{\infty} \int_{2\pi k}^{2\pi(k+1)} |\lambda_f(w)|^2 |\hat{\psi}(w)|^2 dw \\ &= \int_0^{2\pi} |\lambda_f(w)|^2 \sum_{k=-\infty}^{\infty} |\hat{\psi}(w+2\pi k)|^2 dw = \int_0^{2\pi} |\lambda_f(w)|^2 dw , \end{aligned}$$

which implies (32.20). ∎

THEOREM 32.4

For a given multiresolution analysis, there exists an orthonormal wavelet basis for $L^2(\mathcal{R})$.

Proof: We will show that the function ψ constructed in Lemma 32.3 is the mother wavelet. It suffices to show that $\{\psi(x-n)\}_{n=-\infty}^{\infty}$ is an orthonormal basis for W_0 since it will follow that $\{2^{-\frac{m}{2}}\psi(2^{-m}x-n)\}_{n=-\infty}^{\infty}$ is an orthonormal basis for W_m and because the W_m's are disjoint and satisfy (32.13) it will follow that $\{2^{-\frac{m}{2}}\psi(2^{-m}x-n)\}_{m,n=-\infty}^{\infty}$ is an orthonormal basis of $L^2(\mathcal{R})$.

To show that $\{\psi(x-n)\}_{n=-\infty}^{\infty}$ is an orthonormal basis of W_0 we need to show that

i) $\psi \in W_0$,

ii) $\{\psi(x-n)\}_{n=-\infty}^{\infty}$ is an orthonormal set,

iii) the set $\{\psi(x-n)\}_{n=-\infty}^{\infty}$ spans W_0.

(ii) is a consequence of (32.24) and Lemma 32.1 . To show (i) we write (32.23) in the form $\hat{\psi}(w) = G\left(\frac{w}{2}\right)\hat{\phi}\left(\frac{w}{2}\right)$, where $G(w/2) = e^{iw/2}\overline{M}(w/2+\pi)$. Since $G(w)$ is a periodic function with period 2π and belongs to $L^2[0,2\pi]$,

$$G(w) = \frac{1}{\sqrt{2}} \sum_{n=-\infty}^{\infty} g_n e^{inw}\,, \qquad \|G\|^2 = \pi \sum_{n=-\infty}^{\infty} |g_n|^2 < \infty\,.$$

Hence,

$$\hat{\psi}(w) = \frac{1}{\sqrt{2}} \sum_{n=-\infty}^{\infty} g_n e^{in\frac{w}{2}} \hat{\phi}\left(\frac{w}{2}\right)\,,$$

and if we take the inverse Fourier transform of the last equation, we find

$$\psi(x) = \sqrt{2} \sum_{n=-\infty}^{\infty} g_n \phi(2x-n) = \sum_{n=-\infty}^{\infty} g_n \phi_{-1,n}(x)\,;$$

thus $\psi \in V_{-1}$.

The proof of (i) will be complete if we show that ψ is orthogonal to V_0. To this end it suffices to show that $\langle \psi, \phi_{0,n}\rangle = 0$ for all n, which in view of Lemma 32.1 is equivalent to showing that

$$\sum_{k=-\infty}^{\infty} \hat{\psi}(w+2\pi k)\overline{\hat{\phi}}(w+2\pi k) = 0\,. \tag{32.25}$$

By substituting (32.23) and (32.15) into the left-hand side of (32.25) and using the periodicity of $M(w)$, we obtain

$$\begin{aligned}
&\sum_{k=-\infty}^{\infty} \hat{\psi}(w+2\pi k)\overline{\hat{\phi}}(w+2\pi k)\\
&= \sum_{k=-\infty}^{\infty} e^{i(w+2\pi k)/2}\overline{M}\left(\frac{w}{2}+\pi k+\pi\right)\overline{M}\left(\frac{w}{2}+\pi k\right)\left|\hat{\phi}\left(\frac{w}{2}+\pi k\right)\right|^2\\
&= \overline{M}\left(\frac{w}{2}+\pi\right)\overline{M}\left(\frac{w}{2}\right)\sum_{m=-\infty}^{\infty} e^{i\frac{w}{2}}\left|\hat{\phi}\left(\frac{w}{2}+2\pi m\right)\right|^2\\
&-\overline{M}\left(\frac{w}{2}+\pi\right)\overline{M}\left(\frac{w}{2}\right)\sum_{m=-\infty}^{\infty} e^{i\frac{w}{2}}\left|\hat{\phi}\left(\frac{w}{2}+\pi+2m\right)\right|^2\\
&= e^{i\frac{w}{2}}\overline{M}\left(\frac{w}{2}+\pi\right)\overline{M}\left(\frac{w}{2}\right) - e^{i\frac{w}{2}}\overline{M}\left(\frac{w}{2}+\pi\right)\overline{M}\left(\frac{w}{2}\right) = 0.
\end{aligned}$$

Now we prove (iii). From (32.19), we have for any $f \in W_0$,

$$\hat{f}(w) = \left(\sum_{n=-\infty}^{\infty} \lambda_{f,n} e^{inw}\right)\hat{\psi}(w)\,,$$

where

$$\lambda_f(w) = \sum_{n=-\infty}^{\infty} \lambda_{f,n} e^{inw}, \quad \sum |\lambda_{f,n}|^2 < \infty .$$

Therefore, by taking the inverse Fourier transform we obtain

$$f(x) = \sum_{n=-\infty}^{\infty} \lambda_{f,n} \psi(x-n),$$

which shows that $\{\psi(x-n)\}_{n=-\infty}^{\infty}$ indeed spans W_0, and this completes the proof. ∎

It should be noted that the mother wavelet associated with a multiresolution analysis is not unique. In addition, the mother wavelet in Theorem 32.4 is defined indirectly via its Fourier transform as shown in (32.23). A more direct approach to define it is to use (32.23) and (32.18) to obtain

$$\hat{\psi}(w) = e^{i\frac{w}{2}} \left(\frac{1}{\sqrt{2}} \sum_{n=-\infty}^{\infty} (-1)^n \alpha_n e^{-inw/2} \right) \hat{\phi}\left(\frac{w}{2}\right) .$$

If we apply the inverse Fourier transform to this equation, we obtain

$$\begin{aligned} \psi(x) &= \sqrt{2} \sum_{n=-\infty}^{\infty} (-1)^{1-n} \alpha_{1-n} \phi(2x-n) \\ &= \sum_{n=-\infty}^{\infty} (-1)^{1-n} \alpha_{1-n} \phi_{-1,n}(x) . \end{aligned} \tag{32.26}$$

If ϕ has compact support, the series (32.26) terminates and consequently ψ becomes a finite linear combination of translated versions of $\phi(2x)$.

32.3.3 Another Approach to Multiresolution Analysis

If $\{\phi(x-n)\}_{n=-\infty}^{\infty}$ is not an orthonormal basis of V_0, but rather is a Reisz basis, we can use the following orthogonalization process to generate an orthonormal wavelet basis. First, recall that if $\{\phi(x-n)\}_{n=-\infty}^{\infty}$ is a Riesz basis then there exists $A, B > 0$ such that

$$0 < A \leq \Phi(w) = \sum_{k=-\infty}^{\infty} |\hat{\phi}(w+2\pi k)|^2 \leq B < \infty .$$

Therefore, we can define

$$\hat{\phi}(w) = \frac{\hat{\phi}(w)}{\sqrt{\Phi(w)}} .$$

It follows immediately that $\sum_{k=-\infty}^{\infty} |\tilde{\hat{\phi}}(w+2\pi k)|^2 = 1$, and as a consequence of Lemma 32.1 $\{\tilde{\phi}(x-n)\}_{n=-\infty}^{\infty}$ is an orthonormal set. It is routine to show that the span of $\{\tilde{\phi}(x-n)\}_{n=-\infty}^{\infty} = \tilde{V}_0$ is the same as the span of $\{\phi(x-n)\}_{n=-\infty}^{\infty} = V_0$. Therefore, the multiresolution analysis is preserved under this orthogonalization process. This orthogonalization process, which is due to Lemarié and Meyer [143] is different from the familiar Gram-Schmidt orthogonalization process which is not preserved under translation.

Another approach to construct a multiresolution analysis is to start with a function $\phi \in L^2(\mathcal{R})$ that satisfies the relations

$$\phi(x) = \sum_{n=-\infty}^{\infty} c_n \phi(2x-n)\,, \qquad \sum_{n=-\infty}^{\infty} |c_n|^2 < \infty\,, \tag{32.27}$$

and

$$0 < A \leq \sum_{k=-\infty}^{\infty} |\hat{\phi}(w+2\pi k)|^2 \leq B < \infty\,. \tag{32.28}$$

A function ϕ that satisfies Equation (32.27) is said to satisfy the self-symmetry property. We define V_0 as the closed span of $\{\phi(x-n)\}_{n=-\infty}^{\infty}$ and V_m as the span of $\{\phi_{m,n}(x)\}_{n=-\infty}^{\infty}$. Condition (32.28) ensures that $\{\phi_{m,n}(x)\}$ is a Riesz basis for V_m for each m and condition (32.27) ensures that the V_m's satisfy the nested property of the spaces of the multiresolution analysis, property (1) in Definition 32.3. To ensure that properties (2) and (3) of the multiresolution analysis are satisfied, we further assume that $\hat{\phi}(w)$ is bounded, continuous and $\hat{\phi}(0) \neq 0$; see [49, pp. 141-142].

Taking the Fourier transform of (32.27) leads to

$$\hat{\phi}(w) = M\left(\frac{w}{2}\right)\hat{\phi}\left(\frac{w}{2}\right)\,,$$

where $M(w) = \frac{1}{2}\sum_{n=-\infty}^{\infty} c_n e^{inw}$. Since $\hat{\phi}(0) \neq 0$, $M(0) = 1$, which implies that

$$\sum_{n=-\infty}^{\infty} c_n = 2\,. \tag{32.29}$$

If $|\,\hat{\phi}(w)| \leq C(1+|w|)^{-\frac{1}{2}-\epsilon}$, $\epsilon > 0$, then $\Phi(w) = \sum_{k=-\infty}^{\infty} |\hat{\phi}(w+2\pi k)|^2$ is continuous, and consequently we can use the aforementioned orthogonalization trick. So, let

$$\tilde{\hat{\phi}}(w) = \frac{\hat{\phi}(w)}{\sqrt{\Phi(w)}} \quad \text{and} \quad \tilde{M}\left(\frac{w}{2}\right) = \frac{\tilde{\hat{\phi}}(w)}{\tilde{\hat{\phi}}\left(\frac{w}{2}\right)}\,, \tag{32.30}$$

and because of (32.15) it follows that

$$\tilde{M}\left(\frac{w}{2}\right) = \sqrt{\Phi\left(\frac{w}{2}\right)/\Phi(w)}\, M\left(\frac{w}{2}\right)\,. \tag{32.31}$$

Similarly, since

$$\hat{\tilde{\psi}}(w) = e^{iw/2}\overline{\tilde{M}}\left(\frac{w}{2}+\pi\right)\hat{\tilde{\phi}}(w/2),$$

we see from (32.30) and (32.31) that

$$\hat{\tilde{\psi}}(w) = e^{iw/2}\sqrt{\frac{\Phi\left(w/2+\pi\right)}{\Phi(w)\Phi(w/2)}}\left(\frac{\hat{\phi}(w+2\pi)\,\hat{\phi}(w/2)}{\hat{\phi}(w/2+\pi)}\right). \tag{32.32}$$

Clearly $\tilde{M}(0) = 1$ and $|\tilde{M}(0)|^2 + |\tilde{M}(\pi)|^2 = 1$. Thus, $\tilde{M}(\pi) = 0$ and consequently $M(\pi) = 0$, which implies $\sum_{n=-\infty}^{\infty}(-1)^n c_n = 0$. This and (32.29), imply that $\sum_{n=-\infty}^{\infty} c_{2n} = 1 = \sum_{n=-\infty}^{\infty} c_{2n+1}$.

32.4 Orthonormal Wavelets with Compact Support

Except for the Haar wavelets, the first examples of orthonormal wavelet basis consisted of infinitely supported functions. It was I. Daubechies [50] who discovered orthonormal wavelet basis consisting of compactly supported smooth functions. In this section we outline the main ideas of Daubechies' construction. The interested reader can find the details of her work in [49, 50].

If the scaling function ϕ of the multiresolution analysis has compact support, then the series in (32.17) has only finitely many terms and so do the series in (32.18) and (32.26). That is, the function

$$M(w) = \frac{1}{\sqrt{2}}\sum_{n=-\infty}^{\infty} \alpha_n e^{inw}$$

is actually a trigonometric polynomial and the mother wavelet ψ is also compactly supported. We assume that the mother wavelet ψ is such that $\psi \in C^m$ and all its first m derivatives are bounded. In addition, $|\psi(x)| \leq C(1+|x|)^{-m-1-\epsilon}$, $\epsilon > 0$. Similar conditions are also imposed on ϕ.

Under these conditions, it can be shown that $\int_{-\infty}^{\infty} x^k\psi(x)dx = 0$ and hence $\hat{\psi}^{(k)}(0) = 0$, $k = 0, 1, ..., m$. From (32.23), it follows that $M(w)$ has a zero of order $m+1$ at $w = \pi$. Therefore, we may write $M(w) = \left[(1+e^{iw})/2\right]^{m+1}\tilde{L}(w)$, where $\tilde{L}(w)$ is a trigonometric polynomial. But since $|M(w)|^2 + |M(w+\pi)|^2 = 1$, we must have $M_0(w) = |M(w)|^2 = \left(\cos^2 \frac{w}{2}\right)^{m+1} L(w)$ with

$$M_0(w) + M_0(w+\pi) = 1, \tag{32.33}$$

where $L(w) = |\tilde{L}(w)|^2$ is also a polynomial in $\cos w = 1 - 2\sin^2 \frac{w}{2}$.

Thus, we may write $M_0(w) = \left(\cos^2 \frac{w}{2}\right)^N P\left(\sin^2 \frac{w}{2}\right)$, where P is a polynomial and $N = m + 1$. Because of (32.33), it is easy to see that P satisfies

$$(1-y)^N P(y) + y^N P(1-y) = 1 \,, \quad y \in [0,1].$$

This equation is then solved by using Bezout's theorem [49, p. 168] to give the solution

$$P(y) = P_N(y) + y^N R\left(\frac{1}{2} - y\right) , \tag{32.34}$$

where

$$P_N(y) = \sum_{k=0}^{N-1} \binom{N-1+k}{k} y^k ,$$

and R is an odd polynomial chosen such that $P(y) \geq 0$ for $y \in [0,1]$. The following theorem summarizes Daubechies' construction.

THEOREM 32.5

Let $M(w) = (1/\sqrt{2}) \sum_n \alpha_n e^{inw}$ be a trigonometric polynomial such that $|M(w)|^2 + |M(w+\pi)|^2 = 1$ and $M(0) = 1$. Define ϕ and ψ by

$$\hat{\phi}(w) = \frac{1}{\sqrt{2\pi}} \prod_{k=1}^{\infty} M(2^{-k} w) ,$$

$$\hat{\psi}(w) = -e^{-i\frac{w}{2}} \overline{M}\left(\frac{w}{2} + \pi\right) \hat{\phi}\left(\frac{w}{2}\right) .$$

Then ϕ and ψ are compactly supported L^2-functions, satisfying

$$\phi(x) = \sqrt{2} \sum_n \alpha_n \phi(2x - n) ,$$

$$\psi(x) = \sqrt{2} \sum_n (-1)^n \alpha_{1-n} \phi(2x - n) .$$

Moreover, $\{\psi_{n,k}(x)\}$ is a tight frame for $L^2(\mathcal{R})$ with frame bound 1. This tight frame is an orthonormal basis if and only if there exists a compact set K congruent to $[-\pi, \pi]$ modulo 2π, containing a neighborhood of 0 such that

$$\inf_j \inf_{w \in K} |M(2^{-j} w)| > 0 .$$

By spectral factorization of (32.34), with different values of N and R, Daubechies used the above procedure to construct a family of smooth, compactly supported mother wavelets. None of these wavelets, which are known as Daubechies' wavelets, can be easily given in a closed form.

32.5 Examples of Wavelets

1. **The Haar wavelets:** The simplest and the oldest system of orthonormal wavelet basis is the Haar wavelets. The mother wavelet is defined by

$$\psi(x) = \begin{cases} 1 \;, 0 \le x \le \frac{1}{2} \\ -1 \;, \frac{1}{2} \le x < 1 \\ 0 \;, \text{otherwise} \;. \end{cases}$$

The scaling function $\phi(x) = \chi_{[0,1)}(x)$.

2. **The B-spline based wavelets:** They are also called the Battle-Lemarié wavelets [9, 142]. The scaling function is taken as a B-spline with knots at the integers. The simplest of these is the piecewise linear spline

$$\phi(x) = \begin{cases} 1 - |x| \;, 0 \le |x| \le 1 \\ 0 \quad\;\; , \text{otherwise} \;. \end{cases}$$

It is easy to see that $\phi(x) = \frac{1}{2}\phi(2x+1) + \phi(2x) + \frac{1}{2}\phi(2x-1)$ and

$$\hat{\phi}(w) = \frac{1}{\sqrt{2\pi}} \left(\frac{\sin\left(\frac{w}{2}\right)}{\left(\frac{w}{2}\right)} \right)^2 .$$

This choice of ϕ does not lead to orthonormal wavelet basis; therefore, we need to use the Lemarié–Meyer orthogonalization procedure described in Section 32.3.2. To calculate $\Phi(w) = \sum_{k=-\infty}^{\infty} |\hat{\phi}(w+2\pi k)|^2$, we differentiate the relation

$$\csc^2 \pi w = \frac{1}{\pi^2} \sum_{k=-\infty}^{\infty} \frac{1}{(w+k)^2}$$

twice to conclude that

$$\Phi(w) = \frac{1}{6\pi}\left(1 + 2\cos^2 \frac{w}{2}\right)$$

Using Equations (32.30)–(32.32), we find

$$\tilde{M}\left(\frac{w}{2}\right) = \sqrt{\frac{1+2\cos^2 \frac{w}{4}}{1+2\cos^2 \frac{w}{2}}} \cos^2 \frac{w}{4} \,,$$

and

$$\hat{\tilde{\psi}}(w) = 16\sqrt{3}e^{iw/2} \sqrt{\frac{1+2\sin^2 \frac{w}{4}}{\left(1+2\cos^2 \frac{w}{2}\right)\left(1+2\cos^2 \frac{w}{4}\right)}} \left(\frac{\sin^4 \frac{w}{4}}{w^2}\right) .$$

3. **Shannon Wavelet:** The scaling function ϕ for the underlying multiresolution analysis is $\phi(x) = \sin \pi x / \pi x$ or $\hat{\phi}(w) = \frac{1}{\sqrt{2\pi}} \chi_{[-\pi,\pi]}(w)$. It can be easily verified that

$$\phi(x) = \sum_{n=-\infty}^{\infty} \frac{\sin(n\pi/2)}{(n\pi/2)} \phi(2x - n)$$

which in virtue of (32.26) leads to

$$\psi(x) = \sum_{n=-\infty}^{\infty} \frac{(-1)^{n+1} \cos(n\pi/2)}{(n-1)\pi} \phi(2x - n)$$

One can also show directly that

$$\psi(x) = \frac{\sin 2\pi x - \sin \pi x}{\pi x} .$$

is a mother wavelet. To this end, notice that $\hat{\psi}(w) = (1/\sqrt{2\pi})\chi_J(w)$ where $J = [-2\pi, -\pi) \cup (\pi, 2\pi]$.

4. **Meyer Wavelet:** It is an example of a mother wavelet that is infinitely differentiable with rapid decay at $\pm\infty$. It is defined in terms of its Fourier transform

$$\hat{\psi}(w) = \begin{cases} (1/\sqrt{2\pi}) e^{iw/2} \sin\left[\frac{\pi}{2} \nu\left(\frac{3}{2\pi}|w| - 1\right)\right] , & \frac{2\pi}{3} \le |w| \le \frac{4\pi}{3} \\ (1/\sqrt{2\pi}) e^{iw/2} \cos\left[\frac{\pi}{2} \nu\left(\frac{3}{4\pi}|w| - 1\right)\right] , & \frac{4\pi}{3} \le |w| \le \frac{8\pi}{3} , \\ 0 & \text{otherwise} , \end{cases}$$

where ν is a C^∞-function satisfying

$$\nu(x) = \begin{cases} 0 \text{ if } x \le 0 \\ 1 \text{ if } x \ge 1 , \end{cases}$$

and $\nu(x) + \nu(1 - x) = 1$.

5. **A Generalized Meyer Wavelet:** Meyer's wavelet can be generalized as follows (see [239]): Let P be a probability measure with support in $[-\epsilon, \epsilon] \subset \left[-\frac{\pi}{3}, \frac{\pi}{3}\right]$ and define ϕ by its Fourier transform as follows;

$$\hat{\phi}(w) = \left[\int_{w-\pi}^{w+\pi} dP\right]^{\frac{1}{2}} .$$

Then $\hat{\phi}$ has support in $\left[-\frac{4\pi}{3}, \frac{4\pi}{3}\right]$ and $\hat{\phi}(w) = 1$ for $|w| < \frac{2\pi}{3}$. One can verify that

$$\sum_{k=-\infty}^{\infty} |\hat{\phi}(w + 2\pi k)|^2 = \int_{-\infty}^{\infty} dP = 1 ,$$

and $\hat{\phi}(w) = M\left(\frac{w}{2}\right)\hat{\phi}\left(\frac{w}{2}\right)$ with

$$M\left(\frac{w}{2}\right) = \begin{cases} \hat{\phi}(w) \ , & |w| \leq \frac{4\pi}{3} \\ 0 \ , & \frac{4\pi}{3} < |w| \leq 2\pi \ , \end{cases}$$

extended periodically with period 4π to all w. The mother wavelet is given by

$$\hat{\psi}(w) = e^{-i\frac{w}{2}} \left[\int_{\left|\frac{w}{2}\right|-\pi}^{|w|-\pi} dP \right]^{\frac{1}{2}} .$$

32.6 Generalized Functions

Wavelets constitute a basis not only for $L^2(\mathcal{R})$, but also for a large number of function spaces including the $L^p(\mathcal{R})$ spaces. The wavelet expansions of generalized functions have been studied by several people. A. Zayed [256] and M. Holschneider [109] investigated the wavelet expansions of periodic generalized functions. The expansion of analytic hyperfunctions on the real line in wavelet series has recently been derived by A. Zayed [255].

G. Walter [239] studied the wavelet expansion of tempered distributions and tried to extend the concept of multiresolution analysis to the space of tempered distributions. One of the central ideas in Walter's work is to seek generalized function solutions of Equation (32.27). This approach is more elegant and fruitful than the direct approach of the embedding method, which he has also used.

To use the embedding method, both the scaling function and the mother wavelet need to be smooth but not necessarily infinitely differentiable. A prototype condition, for example, is that ϕ and $\psi \in S_r$ (r positive integer), the space of r-differentiable functions f such that, $f^{(k)}(x)$, $k = 0, 1, ..., r$ are rapidly decreasing at $\pm\infty$. The dual space S_r^* is characterized by the fact that if $F \in S_r^*$, then

$$F(x) = D^r\left[(x^2+1)^k \mu(x)\right] , \tag{32.35}$$

where μ is a bounded measure on $\mathcal{R}$, k is a non-negative integer and $D = d/dx$. The wavelet coefficients of $F \in S_r^*$ are defined as $F_{m,n} = \langle F, \psi_{m,n}\rangle$, which is a well-defined expression since $\psi \in S_r$. When (32.35) is substituted into $F_{m,n}$ and integration by parts is carried out, an estimate of the growth rate of $F_{m,n}$ can be obtained; see [239, Ch. 5].

32.7 Applications

One of the main advantages of using wavelets over other techniques such as the Gabor transform (see Chapter 30) is that wavelets possess the so-called "zoom-in" property, i.e; they can zoom in on very short-lived high-frequency phenomena. The graph of each wavelet $\psi_{m,n}$ is a scaled version of the graph of the mother wavelet ψ and is also translated to the proper time location. But the shape of the graph of each wavelet is adapted to the wavelet frequency that is represented by the integer m. For positive m, the graph is wider and flatter, whereas for negative m it is narrower and sharper. This is in contrast with the windowed-Fourier transform, where all the analyzing functions have the same envelope as the mother function, translated to the proper time location and filled in with high-frequency oscillations.

The applications of wavelets have exceeded any expectations. Their scope ranges from physics and engineering applications to abstract mathematics and partial differential equations. Many books, such as [39, 10, 160, 244] and whole issues of scientific journals, such as *IEEE Transactions on Information Theory,* Vol. 38 (1992), have been dedicated to wavelets and their applications. Areas in which wavelet analysis has proved to be useful include image and speech analysis, data compression, optics, radar, fractals, geophysics, robotics, astronomy, turbulence, mathematical statistics and partial differential equations.

33

The Radon Transform

As is often the case with many beautiful and significant mathematical discoveries, the Radon transform was discovered and went unnoticed for very many years. And when it was rediscovered, it was rediscovered independently by many people in different fields. The Radon transformation without doubt is one of the most versatile function transformations. Its applications are numerous and its scope is so immense. Chief among its applications are computer assisted tomography (CAT), in particular, what is known to the layman as CAT scan, and nuclear magnetic resonance (NMR). Not only the transform, but also its history deserves a great deal of attention.

The saga of the Radon transform began in 1917 with the publication of Johann Radon's seminal paper [197] "*Über die Bestimmung Von Funcktionen durch ihre Integralwerte länge gewisser Manningfaltigkeiten*," or "On the determination of functions from their integrals along certain manifolds." At that time, Radon was an assistant to Professor Emanuel Czuber at the University of Technology of Vienna. Later he became a full professor and the head of the mathematical institute of the University of Vienna.

Although his paper had some direct ramifications on solutions of hyperbolic partial differential equations with constant coefficients, it did not receive much attention even from Radon's colleagues at the University of Vienna. This may be attributed to the turmoil that permeated the political atmosphere in Europe during that period.

In that paper Radon demonstrated how one could reconstruct a function of two variables from its integrals over all straight lines in the plane. He also discussed other generalizations of this problem, for example, reconstructing a function from its integrals over other smooth curves as well as reconstructing a function of n variables from its integrals over all hyperplanes .

More than forty years later and thousands of miles away, Allen Cormack, a young South African physicist, was appointed in 1956 at the Radiology Department at the teaching hospital for the University of Cape Town's medical school. Cormack took on himself, as one of his first duties at the new job, the task of finding a set of maps of absorption coefficients for different sections of the human body. The results of the task would make X-ray radiotherapy treatments more efficient. He soon realized that what he needed to complete his task was measurements of the absorption of X-rays along lines in thin sections of the body. Since the logarithm of the ratio of incident to emergent X-ray intensities along a given line is just the line integral of the absorption coefficient along that line, the problem mathematically was equivalent to finding a function f from the values of its integrals along all or some lines in the plane.

> "This struck me as a typical nineteenth century piece of mathematics which a Cauchy or a Riemann might have dashed off in a light moment, but a diligent search of standard texts on analysis failed to reveal it, so I had to solve the problem myself," says Cormack. "I still felt that the problem must have been solved, so I contacted mathematicians on three continents to see if they knew about it, but to no avail" adds Cormack [48].

A few years later, Cormack immigrated to the United States and became a naturalized citizen. Because of the demands of his new position, he had to pursue his problem part time as a hobby. But by 1963 he had already found three alternative forms of solutions to the problem. Cormack continued working on some generalizations of his problem, such as recovering a function from its line integrals along circles through the origin. Because there was almost no response to his publications, or at least that was what he thought, Cormack felt somewhat disappointed and forgot about the problem for a while.

In the late 1960s, at the Central Research Laboratories of a company called Electrical and Musical Industries (EMI), Godfrey Hounsfield, a British engineer, used some of Cormack's ideas to develop a new X-ray machine that revolutionized the field of medical imaging. Soon after that Cormack and Hounsfield joined forces and collaborated in refining the invention and developing the CT-scanning technique. Their work culminated in their receiving the 1979 Nobel Prize in Psychology or Medicine. In their Nobel Prize addresses they acknowledged the work of other pioneers in the field, in particular, the work of Radon in 1917.

33.1 Definitions

Before we give a general definition of the Radon transform, we begin with its simplest version. Let L be any straight line in the plane and ds be the

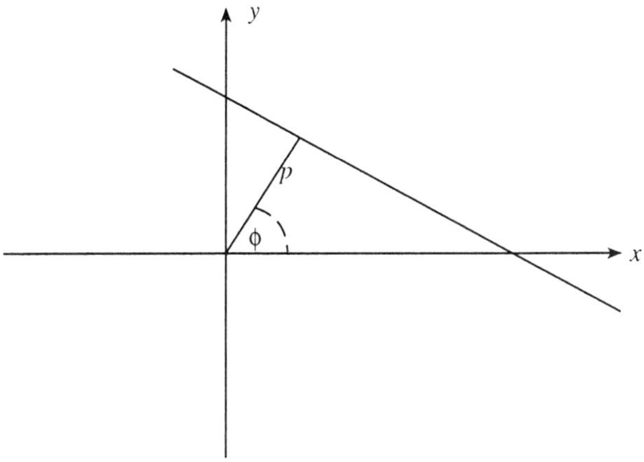

FIGURE 33.1
Vector representation of a straight line

arc length along L. Given a measurable function f, we denote its integral along L as usual by

$$\int_L f(x, y)ds \, .$$

The totality of all these line integrals constitutes the Radon transform of f and each line integral is called a sample of the Radon transform of f. Therefore, the Radon transform $f^\dagger$ of f can be viewed as a function defined on all straight lines in the plane and the value of $f^\dagger$ at a given line is the integral of f over that line.

The inversion of the Radon transform is clearly equivalent to the problem of reconstructing f from the values of its line integrals. If we write the equation of L in the form $p = x\cos\phi + y\sin\phi$, where p is the length of the normal from the origin to L and ϕ is the positive angle that the normal makes with the positive x-axis (see Figure 33.1), then the Radon transform of f can be written in the form

$$R[f](p, \phi) = f^\dagger(p, \phi) = \int_L f(x, y)ds \, .$$

If we rotate the coordinate system by an angle ϕ, and label the new axes by p and s, then $x = p\cos\phi - s\sin\phi$, $y = p\sin\phi + s\cos\phi$ and

$$f^\dagger(p, \phi) = \int_{-\infty}^{\infty} f(p\cos\phi - s\sin\phi, p\sin\phi + s\cos\phi)ds \, . \tag{33.1}$$

Example A: Let $f(x,y) = e^{-a(x^2+y^2)}, a > 0$. Then

$$f^\dagger(p,\phi) = \int_{-\infty}^{\infty} e^{-a(p^2+s^2)}ds = e^{-ap^2}\int_{-\infty}^{\infty} e^{-as^2}ds = \sqrt{\pi/a}\, e^{-ap^2} .$$

Example B: Let $f(x,y) = xe^{-(x^2+y^2)}$ and $g(x,y) = ye^{-(x^2+y^2)}$. Then

$$f^\dagger(p,\phi) = \int_{-\infty}^{\infty} (p\cos\phi - s\sin\phi)e^{-(p^2+s^2)}ds = \sqrt{\pi}\, pe^{-p^2}\cos\phi ,$$

$$g^\dagger(p,\phi) = \sqrt{\pi}\, p\sin\phi e^{-p^2} .$$

Hence

$$f^\dagger(p,\phi) + ig^\dagger(p,\phi) = \sqrt{\pi}\, pe^{-p^2}e^{i\phi} .$$

Formula (33.1) is practical to use in two dimensions; however, it does not lend itself easily to higher dimensions. To generalize it to higher dimensions, let us introduce the unit vectors $\boldsymbol{\xi} = (\cos\phi, \sin\phi)$ and $\boldsymbol{\xi}^\perp = (-\sin\phi, \cos\phi)$, so that $\mathbf{x} = (x,y) = (r,\theta) = p\boldsymbol{\xi} + t\boldsymbol{\xi}^\perp$ for some scalar parameter t, where r and θ are the polar coordinates. The equation of the line L can now be written in terms of the unit vector $\boldsymbol{\xi}$ as $p = \boldsymbol{\xi}.\mathbf{x}$. Using the definition of the δ-function introduced in Section 3.3.2 , we may write

$$R[f](p,\phi) = \int_{-\infty}^{\infty} f(p\boldsymbol{\xi} + t\boldsymbol{\xi}^\perp)dt = \iint_{\mathcal{R}^2} f(\mathbf{x})\delta(p - \boldsymbol{\xi}.\mathbf{x})d\mathbf{x} ,$$

where $d\mathbf{x} = dxdy$. Note that the integral is taken over a line orthogonal to the line $\theta = \phi$ and that $f^\dagger(-p,\phi) = f^\dagger(p,\phi+\pi)$; therefore, we may assign negative values for p and restrict ϕ to $[0,\pi]$. To simplify the notation, we may write the integral on the right-hand side as a single integral. Now we are ready to give a general definition of the Radon transform.

DEFINITION 33.1 *Let* $\mathbf{x} = (x_1, ..., x_n) \in \mathcal{R}^n$, $n \geq 2$, $d\mathbf{x} = dx_1, ..., dx_n$, $f(\mathbf{x}) = f(x_1, ..., x_n)$ *and* $\boldsymbol{\xi}$ *be a unit vector in* $\mathcal{R}^n$. *The Radon transform of a function* $f(\mathbf{x})$ *is defined by*

$$R[f](p,\boldsymbol{\xi}) = f^\dagger(p,\boldsymbol{\xi}) = \int f(\mathbf{x})\delta(p - \boldsymbol{\xi}.\mathbf{x})d\mathbf{x} ,$$

where $p = \mathbf{x}\,.\,\boldsymbol{\xi} = x_1\xi_1 + ... + x_n\xi_n$ *is a hyperplane.*

If S^{n-1} is the generalized unit sphere $\sum_{i=1}^n x_i^2 = 1$, then $R[f]$ may be viewed as a function defined on $\mathcal{R} \times S^{n-1}$. Thus, the Radon transform of a function of n variables f is the totality of all integrals of f over all hyperplanes in $\mathcal{R}^n$. In other words, the Radon transform $f^\dagger$ of f is a function defined on all hyperplanes in $\mathcal{R}^n$ and the value of that function at any hyperplane is the integral of f over that hyperplane. The inversion of the Radon transform is equivalent to finding f from the values of its integrals over all hyperplanes.

The idea of the Radon transform has been generalized further. First, let us observe that in $\mathcal{R}^2$, if we integrate a function $f(x, y)$ over a straight line L_θ with direction θ, and then move this line parallel to itself, we obtain a function $L_\theta f = P^1 f$ defined on a line $L_{\theta^\perp}$ orthogonal to L_θ. The value of this function at any point $\mathbf{x}$ on $L_{\theta^\perp}$ is equal to the integral of f over the line with direction θ which intersects $L_{\theta^\perp}$ at the point $\mathbf{x}$. Similarly, in $\mathcal{R}^3$ if we integrate a function f over a plane P^2, and then move this plane parallel to itself, we obtain a function $P^2 f$ defined on a line orthogonal to P^2. The value of this function at any point $\mathbf{x}$ on this line is equal to the integral of f over the plane passing through $\mathbf{x}$ and parallel to P^2.

But another possibility emerges in $\mathcal{R}^3$. What happens if we integrate f over a straight line L_θ with direction θ in $\mathcal{R}^3$, and then move this line parallel to itself? Clearly, we obtain a function that is defined on a plane orthogonal to that line. The value of that function at any point $\mathbf{x}$ on the plane is equal to the integral of f over the line with direction θ which intersects the plane at $\mathbf{x}$.

This is essentially the idea of the X-ray transform. The terminology came from the field of medical imaging, where the integral along a line represents a measurement of the intensity of the X-ray beam at the detector after passing thorough the object to be radiographed. If $f(\mathbf{x})$ represents the density at the point $\mathbf{x}$, then an X-ray picture taken in the θ direction provides a function $L_\theta f = P^1 f$ defined on the plane orthogonal to θ. The value of the function $L_\theta f$ at a point $\mathbf{x}$ on this plane is equal to the integral of f along a line through $\mathbf{x}$ in the direction θ, i.e.,

$$[L_\theta f](\mathbf{x}) = \int_{-\infty}^{\infty} f(\mathbf{x} + t\theta)dt \;, \quad x \in \theta^\perp$$

where $\theta^\perp$ denotes the plane orthogonal to θ.

DEFINITION 33.2 *Let $S^{n-1} = \{\mathbf{x} \in \mathcal{R}^n, \|\mathbf{x}\| = 1\}$ be the unit sphere in $\mathcal{R}^n$ or equivalently the set of all unit vectors in $\mathcal{R}^n$. Let $\theta = (\theta_1, ..., \theta_k)$, where $1 \leq k \leq n$, $\theta_k \in S^{n-1}$ with $\theta_i.\theta_j = \delta_{i,j}$ and $\theta^\perp$ be the perpendicular subspace*

$$\theta^\perp = \{\mathbf{x} \in \mathcal{R}^n, \mathbf{x} \textit{ is orthogonal to } \theta_i, i = 1, ..., k\} \;.$$

Let π be a k-dimensional subspace of $\mathcal{R}^n$ determined by θ. Then the k-plane transform of $f(\mathbf{x}), \mathbf{x} \in \mathcal{R}^n$ in the direction θ at the point $\mathbf{x}'' \in \pi^\perp$ is defined by

$$(P^k f)(\pi, \mathbf{x}'') = \int_\pi f(\mathbf{x}'', \mathbf{x}')d\mathbf{x}' \;,$$

where $\mathbf{x} = (\mathbf{x}'', \mathbf{x}'), \mathbf{x}' \in \pi, \mathbf{x}'' \in \pi^{\perp}$. *Or, equivalently,*

$$(P^k f)(\theta, \mathbf{x}'') = \int f(\mathbf{x}'' + u_1\theta_1 + ... + u_k\theta_k)d\mathbf{u} ,$$

where $\mathbf{u} = (u_1, ..., u_k), \mathbf{x}'' \in \theta^{\perp}$.

The 1-plane transform P^1 *is called the X-ray transform and the* $(n-1)$*-plane transform is the Radon transform.*

33.2 Existence and Uniqueness

If $f \in L^1(\mathcal{R}^n)$, then its k-plane transform exists almost everywhere. In most applications, however, f is assumed to be locally integrable with compact support or infinitely differentiable with rapid decay at infinity, which is sufficient for the k-plane transform to exist.

As for the uniqueness of the Radon transform it will be shown later that if $f \in \mathcal{S}$ and its Radon transform $f^{\dagger}$ vanishes identically, then f must be the zero function. If $f \in \mathcal{S}$ has compact support, then so does $f^{\dagger}$.

A question of great importance is whether or not a function $f(\mathbf{x})$ can be uniquely determined by any infinite set of k-plane transforms, or in other words, is f uniquely determined by its integrals over an infinite number of subspaces of dimension k? It has been shown [217] that if $f \in L^1(\mathcal{R}^n)$ has compact support, then f is uniquely determined by any infinite set of X-ray transforms. Likewise, if f is infinitely differentiable with compact support, it is uniquely determined by any infinite set of Radon transforms. A more important question from a practical point of view is whether a nice function can be completely determined by a finite set of its k-plane transforms, and if not, how close can a finite set of k-plane transforms approximate the function? A complete answer is not yet known.

33.3 Elementary Properties

1. **Linearity:** The Radon transformation, as well as the k-plane transformation, is a linear transformation.

2. **Homogeneity:** The Radon transform of a function f is an even homogeneous function of degree -1, i.e.,

$$R[f](sp, s\boldsymbol{\xi}) = f^{\dagger}(sp, s\boldsymbol{\xi}) = |s|^{-1} f^{\dagger}(p, \boldsymbol{\xi}) , \quad s \neq 0$$

in particular if $s = -1$, we have

$$f^\dagger(-p, -\boldsymbol{\xi}) = f^\dagger(p, s) .$$

If $\boldsymbol{\eta} = s\boldsymbol{\xi}$, then $f^\dagger(p, \eta) = |\eta|^{-1} f^\dagger(|\eta|^{-1}p, |\eta|^{-1}\eta)$.

3. **Translation:**

$$R[f(\mathbf{x} - \mathbf{a})](p, \boldsymbol{\xi}) = R[f(\mathbf{x})](p - \mathbf{a}.\boldsymbol{\xi}, \boldsymbol{\xi})$$

4. **Change of Variables:** Let A be a non-singular matrix. If $\mathbf{y} = A\mathbf{x}$, then

$$R[f(A\mathbf{x})](p, \boldsymbol{\xi}) = |\det B| R[f(\mathbf{x})](p, B^T\boldsymbol{\xi}) ,$$

where $B = A^{-1}$ and B^T is the transpose of B.

In particular, if B is orthogonal, i.e., $B^{-1} = B^T = A$, and $|\det B| = 1$; then

$$R[f(A\mathbf{x})](p, \boldsymbol{\xi}) = R[f(\mathbf{x})](p, A\boldsymbol{\xi}) .$$

5. **Dilation:**

$$\begin{aligned} R[f(\lambda\mathbf{x})](p, \boldsymbol{\xi}) &= \frac{1}{\lambda^n} R[f(\mathbf{x})](p, \lambda^{-1}\boldsymbol{\xi}) , \quad \lambda(\text{ real }) \neq 0 \\ &= \frac{1}{\lambda^{n-1}} R[f(\mathbf{x})](\lambda p, \boldsymbol{\xi}) \end{aligned}$$

6. **Transform of the First Derivatives:**

$$R\left[\sum_{k=1}^{n} a_k \frac{\partial f}{\partial x_k}\right](p, \boldsymbol{\xi}) = (\mathbf{a}\,.\,\boldsymbol{\xi}) \frac{\partial}{\partial p} R[f(\mathbf{x})](p, \boldsymbol{\xi}) ,$$

where $\mathbf{a} = (a_1, ..., a_n)$, or

$$R\left[(\mathbf{a}.\nabla)f\right](p, \boldsymbol{\xi}) = (\mathbf{a}\,.\,\boldsymbol{\xi}) \frac{\partial}{\partial p} f^\dagger(p, \boldsymbol{\xi}) ,$$

where $\nabla = \left(\frac{\partial}{\partial x_1}, \cdots, \frac{\partial}{\partial x_n}\right)$, is the gradient operator. In particular,

$$R\left[\frac{\partial f}{\partial x_k}\right](p, \boldsymbol{\xi}) = \xi_k \frac{\partial}{\partial p} f^\dagger(p, \boldsymbol{\xi}) ,$$

7. **Transform of Higher Order Derivatives:** If L is a linear differential operator with constant coefficients, $L\left(\frac{\partial}{\partial x_1}, \cdots, \frac{\partial}{\partial x_n}\right)$, then

$$R\left[L\left(\frac{\partial}{\partial x_1}, ..., \frac{\partial}{\partial x_n}\right) f(\mathbf{x})\right](p, \boldsymbol{\xi}) = L\left(\xi_1 \frac{\partial}{\partial p}, ..., \xi_n \frac{\partial}{\partial p}\right) f^\dagger(p, \boldsymbol{\xi}) .$$

In particular,

$$R\left[\sum_{k=1}^{n}\sum_{l=1}^{n} a_k b_l \frac{\partial^2 f}{\partial x_k \partial x_l}\right] = (\mathbf{a}.\boldsymbol{\xi})(\mathbf{b}.\boldsymbol{\xi})\frac{\partial^2 f^\dagger(p,\boldsymbol{\xi})}{\partial p^2} ,$$

and

$$R\left[\frac{\partial^2 f(\mathbf{x})}{\partial x_k \partial x_l}\right](p,\boldsymbol{\xi}) = \xi_k \xi_l \frac{\partial^2 f^\dagger(p,\boldsymbol{\xi})}{\partial p^2} .$$

8. **Derivatives of the Transform:**

$$\left(\mathbf{a}.\frac{\partial}{\partial \boldsymbol{\xi}}\right) R[f(\mathbf{x})](p,\boldsymbol{\xi}) = -\frac{\partial}{\partial p} R[(\mathbf{a}.\mathbf{x})f(\mathbf{x})](p,\boldsymbol{\xi})$$

where $\mathbf{a}.\frac{\partial}{\partial \boldsymbol{\xi}} = \sum_{k=1}^{n} a_k \frac{\partial}{\partial \xi_k}$. In particular,

$$\frac{\partial}{\partial \xi_k} R[f(\mathbf{x})](p,\boldsymbol{\xi}) = -\frac{\partial}{\partial p} R[x_k f(\mathbf{x})](p,\boldsymbol{\xi}) .$$

For higher order derivatives, we have

$$\sum_{k,l=1}^{n} a_k b_l \frac{\partial^2}{\partial \xi_k \partial \xi_l} f^\dagger(p,\boldsymbol{\xi}) = \frac{\partial^2}{\partial p^2} R[(\mathbf{a}.\mathbf{x})(\mathbf{b}.\mathbf{x})f(\mathbf{x})](p,\boldsymbol{\xi}) .$$

In particular, if $\mathbf{x} = (x,y) \in \mathcal{R}^2$

$$\frac{\partial^{k+l}}{\partial \xi_1^k \partial \xi_2^l} f^\dagger(p,\boldsymbol{\xi}) = \left(-\frac{\partial}{\partial p}\right)^{k+l} R[x^k y^l f(x,y)](p,\boldsymbol{\xi}) .$$

9. **Integration:**

$$\int_{-\infty}^{\infty} f^\dagger(p,\boldsymbol{\xi})dp = \int f(\mathbf{x})d\mathbf{x} .$$

The proof of (1) and (3) is trivial and (2) follows from the observation that $\delta(ax) = |a|^{-1}\delta(x)$. The proof of (4) is straightforward if we note that $\mathbf{x}.\boldsymbol{\xi} = \langle A^{-1}\mathbf{y}, \boldsymbol{\xi}\rangle = \langle \mathbf{y}, (A^{-1})^T\boldsymbol{\xi}\rangle$, and (5) is a consequence of (4) if we set $A = \lambda I$. For (6), let $\boldsymbol{\epsilon}_k = (0,0,...,\epsilon,0,...,0)$ be the vector in which the only non-zero component is the kth component which is equal to $\epsilon \neq 0$, and note that

$$\frac{\partial f}{\partial x_k} = \lim_{\epsilon \to 0} \frac{f(\mathbf{x}+\boldsymbol{\epsilon}_k) - f(\mathbf{x})}{\epsilon} .$$

An application of (3) to this equation yields

$$R\left[\frac{\partial f}{\partial x_k}\right](p,\boldsymbol{\xi}) = \xi_k \frac{\partial}{\partial p} R[f](p,\boldsymbol{\xi}) ,$$

which in turn implies (6) and (7). Property (8) is a consequence of the relation

$$\frac{\partial}{\partial \xi_k}\delta(p - \mathbf{x}.\boldsymbol{\xi}) = -x_k\delta'(p - \mathbf{x}.\boldsymbol{\xi}) = -x_k\frac{\partial}{\partial p}\delta(p - \mathbf{x}.\boldsymbol{\xi}) \,.$$

Property (9) is a restatement of the fact that if we integrate an integrable function over a hyperplane P of dimension $(n-1)$, then integrate the result over the orthogonal complement of P, we obtain the integral of the function over the whole space.

33.4 Convolution

Let $f^\dagger$ and $g^\dagger$ be the Radon transforms of f and g, respectively. Let h be the convolution of f and g defined by

$$h(\mathbf{x}) = (f * g)(\mathbf{x}) = \int f(\mathbf{y})g(\mathbf{x} - \mathbf{y})d\mathbf{y} \,.$$

Then

$$h^\dagger(p, \boldsymbol{\xi}) = (f^\dagger \otimes g^\dagger)(p\,,\,\boldsymbol{\xi}) = \int f^\dagger(s, \boldsymbol{\xi})g^\dagger(p - s, \boldsymbol{\xi})ds,$$

where $\otimes$ denotes the convolution with respect to the radial distance.

That is the Radon transform of the convolution is the convolution of the Radon transforms taken with respect to the scalar parameter p. For,

$$\begin{aligned}
h^\dagger(p, \boldsymbol{\xi}) &= \int h(\mathbf{x})\delta(p - \boldsymbol{\xi}.\mathbf{x})d\mathbf{x} \\
&= \int f(\mathbf{y})d\mathbf{y} \int g(\mathbf{x} - \mathbf{y})\delta(p - \boldsymbol{\xi}.\mathbf{x})d\mathbf{x} \\
&= \int f(\mathbf{y})d\mathbf{y} \int g(\mathbf{z})\delta(p - \boldsymbol{\xi}.\mathbf{y} - \boldsymbol{\xi}.\mathbf{z})d\mathbf{z} \\
&= \int f(\mathbf{y})g^\dagger(p - \boldsymbol{\xi}.\mathbf{y}, \boldsymbol{\xi})d\mathbf{y} \\
&= \int f(\mathbf{y})d\mathbf{y} \int g^\dagger(p - s, \boldsymbol{\xi})\delta(s - \boldsymbol{\xi}.\mathbf{y})ds \\
&= \int g^\dagger(p - s, \boldsymbol{\xi})ds \int f(\mathbf{y})\delta(s - \boldsymbol{\xi}.\mathbf{y})d\mathbf{y} \\
&= \int f^\dagger(s, \boldsymbol{\xi})g^\dagger(p - s, \boldsymbol{\xi})ds \,.
\end{aligned}$$

33.5 Relationship With Other Transforms

First, let us introduce the hyperspherical polar coordinates $(\rho, \theta_1, \cdots, \theta_{n-2}, \phi)$ in $\mathcal{R}^n$:

$$\begin{aligned} x_1 &= \rho \cos\theta_1 \\ x_2 &= \rho \sin\theta_1 \cos\theta_2 \\ \vdots\ & \quad \vdots \\ x_{n-2} &= \rho \sin\theta_1 \sin\theta_2 \cdots \sin\theta_{n-3} \cos\theta_{n-2} \\ x_{n-1} &= \rho \sin\theta_1 \sin\theta_2 \cdots \sin\theta_{n-2} \cos\phi \\ x_n &= \rho \sin\theta_1 \sin\theta_2 \cdots \sin\theta_{n-2} \sin\phi, \end{aligned}$$

where $0 \le \rho, 0 \le \theta_j \le \pi, j = 1, 2, \cdots, n-2, 0 \le \phi \le 2\pi$. The unit volume element is given by

$$dV = \rho^{n-1} (\sin\theta_1)^{n-2} (\sin\theta_2)^{n-3} \cdots (\sin\theta_{n-2})\, d\rho d\theta_1 \cdots d\theta_{n-2} d\phi$$

and the unit surface element is given by

$$d\Omega = (\sin\theta_1)^{n-2} (\sin\theta_2)^{n-3} \cdots (\sin\theta_{n-2})\, d\theta_1 \cdots d\theta_{n-2} d\phi.$$

33.5.1 The X-ray and k-Plane Transforms

Obviously, a hyperplane perpendicular to $\boldsymbol{\xi}$ can be covered by parallel lines in any direction $\boldsymbol{\theta}$ perpendicular to $\boldsymbol{\xi}$, and since the x-ray transform of f in the direction of $\boldsymbol{\theta}$ is the line integral of f over a line with direction $\boldsymbol{\theta}$, it is possible to express the Radon transform as an integral of the X-ray transform. In fact, if $\boldsymbol{\theta}$ and $\boldsymbol{\xi}$ are two perpendicular unit vectors in $\mathcal{R}^n$ and $\mathbf{y} \in \theta^\perp$, then

$$R[f](p, \boldsymbol{\xi}) = \int_{\mathbf{y}.\boldsymbol{\xi}=p} \left(P^1 f\right)(\boldsymbol{\theta}, \mathbf{y})\, d\mathbf{y},$$

where $P^1 f$ is the X-ray transform of f.

More generally, if we denote the m-dimensional Radon transform by R^m, then the Radon and the k-plane transforms are related by

$$R^n[f](p, \boldsymbol{\xi}) = R^{n-k}\left[\left(P^k_{\boldsymbol{\theta}} f\right)\right](p, \boldsymbol{\xi}),$$

where $\boldsymbol{\xi} \in \theta^\perp$, $P^k_{\boldsymbol{\theta}} f = \left(P^k f\right)(\boldsymbol{\theta}, .)$ and R^{n-k} is taken over $\theta^\perp$.

33.5.2 The Fourier Transform

Consider the n-dimensional Fourier transform of $f(\mathbf{x})$,

$$\hat{f}(\mathbf{w}) = \frac{1}{(2\pi)^{n/2}} \int e^{i(\mathbf{w}.\mathbf{x})} f(\mathbf{x}) d\mathbf{x} .$$

By using the hyperspherical polar coordinates, we can write $\mathbf{w} = \rho\boldsymbol{\xi}$ where $\boldsymbol{\xi} \in S^{n-1}$. Therefore,

$$\hat{f}(\mathbf{w}) = \hat{f}(\rho\boldsymbol{\xi}) = \frac{1}{(2\pi)^{n/2}} \int e^{i\rho(\boldsymbol{\xi}.\mathbf{x})} f(\mathbf{x}) d\mathbf{x} .$$

Fix ρ and $\boldsymbol{\xi}$ and set $F(\mathbf{x}) = e^{i\rho(\boldsymbol{\xi}.\mathbf{x})} f(\mathbf{x})$; thus, $F^\dagger(p, \boldsymbol{\xi}') = \int_L e^{i\rho(\boldsymbol{\xi}.\mathbf{x})} f(\mathbf{x}) ds$, where L is the hyperplane $\boldsymbol{\xi}'.\mathbf{x} = p$ and ds is the $(n-1)$-dimensional surface area measure in $\mathcal{R}^n$. Hence,

$$F^\dagger(p, \boldsymbol{\xi}) = e^{i\rho p} \int_{\boldsymbol{\xi}.\mathbf{x}=p} f(\mathbf{x}) ds = e^{i\rho p} f^\dagger(p, \boldsymbol{\xi}) ,$$

which in view of property (9) of Section 33.3 leads to

$$\begin{aligned} \int_{-\infty}^{\infty} F^\dagger(p, \boldsymbol{\xi}) dp &= \int_{-\infty}^{\infty} e^{i\rho p} f^\dagger(p, \boldsymbol{\xi}) dp = \int_{-\infty}^{\infty} F(\mathbf{x}) d\mathbf{x} \\ &= \int_{-\infty}^{\infty} e^{i\rho(\boldsymbol{\xi}.\mathbf{x})} f(\mathbf{x}) d\mathbf{x} = (2\pi)^{n/2} \hat{f}(\rho\boldsymbol{\xi}) . \end{aligned}$$

Therefore,

$$\hat{f}(\rho\boldsymbol{\xi}) = \frac{1}{(2\pi)^{n/2}} \int_{-\infty}^{\infty} e^{i\rho p} f^\dagger(p, \boldsymbol{\xi}) dp . \tag{33.2}$$

If we denote the one-dimensional Fourier transform taken along the radial direction by $\mathcal{F}_r$, then Equation (33.2) can be written as

$$\mathcal{F}[f] = \frac{1}{(2\pi)^{(n-1)/2}} \mathcal{F}_r[f^\dagger] .$$

From (33.2), and the inversion formula of the Fourier transform, we obtain

$$f^\dagger(p, \boldsymbol{\xi}) = (2\pi)^{(n-2)/2} \int_{-\infty}^{\infty} \hat{f}(\rho\boldsymbol{\xi}) e^{-i\rho p} d\rho . \tag{33.3}$$

33.5.3 The Gegenbauer Transform

The *Gegenbauer transform* of a function $f(r)$ is defined by

$$F_l^\nu(x) = G_l^\nu[f](x) = \int_{|x|}^{\infty} r^{2\nu} f(r) C_l^\nu\left(\frac{x}{r}\right) \left(1 - \frac{x^2}{r^2}\right)^{\nu - \frac{1}{2}} dr ,$$

where $C_l^\nu(x)$ is the Gegenbauer polynomial of degree l. In preparation for what follows we shall normalize the Gegenbauer transform so that

$$F_l^\nu(x) = G_l^\nu[f](x) = k_l^\nu \int_{|x|}^\infty r^{2\nu} f(r) C_l^\nu\left(\frac{x}{r}\right)\left(1-\frac{x^2}{r^2}\right)^{\nu-\frac{1}{2}} dr \,,$$

where $k_l^\nu = \Omega_{n-1}/C_l^\nu(1)$, $\nu = (n-2)/2$ and Ω_n is the total surface area of the unit sphere in $\mathcal{R}^n$, i.e., $\Omega_n = (2\pi^{n/2})/\Gamma(n/2)$.

The inversion formula for the Gegenbauer transform is [51, p. 120]

$$f(r) = \frac{\gamma_l^\nu}{r}\int_r^\infty \left(\frac{d^{2\nu+1}}{dx^{2\nu+1}} F_l^\nu(x)\right) C_l^\nu\left(\frac{x}{r}\right)\left(\frac{x^2}{r^2}-1\right)^{\nu-\frac{1}{2}} dx \,,$$

where

$$\gamma_l^\nu = \frac{(-1)^{2\nu+1} k_l^\nu}{(2\pi)^{2\nu+1}} \,.$$

The Radon and Gegenbauer transforms are related when $f(\mathbf{x})$ can be written as a product of a function of r and a function of $\boldsymbol{\xi}$, where $\mathbf{x} = r\boldsymbol{\xi}$, $\boldsymbol{\xi} \in S^{n-1}$, or more precisely, when $f(\mathbf{x}) = g(r)S_l^m(\boldsymbol{\xi})$, where $S_l^m(\boldsymbol{\xi})$ is a real surface harmonic of degree l defined on $\mathcal{R}^n$ (see [60, Vol. 2, p. 242]).

By a real surface harmonic in $\boldsymbol{\xi}$ of degree l we mean a homogeneous polynomial in $\xi_1, \cdots, \xi_n$ of degree l that satisfies Laplace's equation. Spherical harmonics of degree l form a finite-dimensional vector space of functions on the unit sphere. The dimension of that vector space, $N(n,l)$ depends on the dimension of the space, n, and the degree of the polynomials l. It can be shown [60, Vol. 2, p. 237] that

$$N(n,l) = (2l+n-2)\frac{(l+n-3)!}{l!\,(n-2)!} .$$

Clearly, a basis for that vector space that is orthonormal on the unit sphere can be chosen. We denote the elements of this basis by S_l^m. Spherical harmonics with different degrees are orthogonal on the unit sphere. It is not difficult to show that if P_k is a polynomial in $\boldsymbol{\xi}$ of degree k, then it is a linear combination of spherical harmonics of degrees $\leq k$. Hence, $S_l^m(\boldsymbol{\xi})\,, m = 1, 2, \cdots, N(n,l)\,,\ l = 0, 1, \cdots$ is a complete orthonormal set of functions on the unit sphere.

A key formula in the derivation of the relation between the Radon and Gegenbauer transforms is the Funk-Hecke Theorem [60, Vol. 2, p. 247] which states that if $F(x)$ is a real-valued function which is continuous on $[-1,1]$ and $S_l^m(\boldsymbol{\xi})$ is a surface harmonic of degree l, then for any unit vector $\boldsymbol{\eta}$ in $\mathcal{R}^n$,

$$\int_{|\boldsymbol{\xi}|=1} F(\boldsymbol{\eta}.\boldsymbol{\xi}) S_l^m(\boldsymbol{\xi}) d\boldsymbol{\xi} = k_l^\nu S_l^m(\boldsymbol{\eta}) \int_{-1}^1 F(t) C_l^\nu(t)(1-t^2)^{\nu-\frac{1}{2}} dt \,. \quad (33.4)$$

Now let $\mathbf{x} = r\boldsymbol{\eta}$, where $\boldsymbol{\eta} \in S^{n-1}$. The Radon transform of $f(\mathbf{x}) = g(r)S_l^m(\boldsymbol{\eta})$ can be calculated as follows

$$\begin{aligned}
f^{\dagger}(p,\boldsymbol{\xi}) &= \int f(\mathbf{x})\delta(p-\boldsymbol{\xi}.\mathbf{x})d\mathbf{x} \\
&= \int_{|\boldsymbol{\eta}|=1} S_l^m(\boldsymbol{\eta})\,d\boldsymbol{\eta} \int_{-\infty}^{\infty} r^{n-1}g(r)H(r)\delta(p-r\boldsymbol{\xi}\,.\,\eta)dr \\
&= \int_{|\boldsymbol{\eta}|=1} S_l^m(\boldsymbol{\eta})\,d\boldsymbol{\eta} \int_{-\infty}^{\infty} r^{n-1}g(r)H(r)\frac{1}{|\boldsymbol{\xi}\,.\,\boldsymbol{\eta}|}\delta\left(\frac{p}{\boldsymbol{\xi}\,.\,\boldsymbol{\eta}}-r\right)dr \\
&= \int_{|\boldsymbol{\eta}|=1} \left(\frac{p}{\boldsymbol{\xi}\,.\,\boldsymbol{\eta}}\right)^{n-1} g\left(\frac{p}{\boldsymbol{\xi}\,.\,\boldsymbol{\eta}}\right) H\left(\frac{p}{\boldsymbol{\xi}\,.\,\boldsymbol{\eta}}\right)\frac{1}{|\boldsymbol{\xi}\,.\,\boldsymbol{\eta}|}S_l^m(\boldsymbol{\eta})\,d\boldsymbol{\eta} \\
&= \int_{|\boldsymbol{\eta}|=1} F(\boldsymbol{\xi}\,.\,\boldsymbol{\eta})S_l^m(\boldsymbol{\eta})\,d\boldsymbol{\eta}\,, \qquad (33.5)
\end{aligned}$$

where $H(r)$ is the Heaviside function and

$$F(t) = \left(\frac{p}{t}\right)^{n-1} g\left(\frac{p}{t}\right) H\left(\frac{p}{t}\right)\frac{1}{|t|}\,.$$

If we apply (33.4) to (33.5) we immediately obtain

$$\hat{f}(p,\boldsymbol{\xi}) = k_l^{\nu} S_l^m(\boldsymbol{\xi}) \int_{-1}^{1} \left(\frac{p}{t}\right)^{n-1} g\left(\frac{p}{t}\right) H\left(\frac{p}{t}\right)\frac{1}{|t|}C_l^{\nu}(t)(1-t^2)^{\nu-\frac{1}{2}}dt\,.$$

If $p > 0$, then

$$\begin{aligned}
\hat{f}(p,\boldsymbol{\xi}) &= k_l^{\nu} S_l^m(\boldsymbol{\xi}) \int_{0}^{1} \left(\frac{p}{t}\right)^{n-1} g\left(\frac{p}{t}\right) C_l^{\nu}(t)(1-t^2)^{\nu-\frac{1}{2}}\frac{1}{|t|}dt \\
&= k_l^{\nu} S_l^m(\boldsymbol{\xi}) \int_{p}^{\infty} r^{n-2}g(r)C_l^{\nu}\left(\frac{p}{r}\right)\left(1-\frac{p^2}{r^2}\right)^{\nu-\frac{1}{2}} dr\,,
\end{aligned}$$

or

$$R[f](p,\boldsymbol{\xi}) = G_l^{\nu}[g](p)S_l^m(\boldsymbol{\xi})\,. \qquad (33.6)$$

The analysis for $p < 0$ is similar.

Now suppose that $f(\mathbf{x})$ is given by

$$f(\mathbf{x}) = \sum_{l,m} a_{l,m}\, g_l(\rho)S_l^m(\boldsymbol{\xi})\,,$$

where $a_{l,m}$ are constants, $l = 0, 1, \ldots; m = 1, \ldots, N(n, l)$. Then

$$R[f](p,\boldsymbol{\xi}) = \sum_{l,m} a_{l,m}G_l^{\nu}[g_l](p)S_l^m(\boldsymbol{\xi})\,, \quad \nu = (n-2)/2\,.$$

Assuming that $R[f]$ is known, we may use the orthonormality of S_l^m to obtain

$$a_{l,m}G_l^\nu[g_l](p) = \int_{|\boldsymbol{\xi}|=1} R[f](p,\boldsymbol{\xi})S_l^m(\boldsymbol{\xi})d\boldsymbol{\xi} .$$

If the inverse Gegenbauer transform is invoked, the functions $a_{l,m}g_l(\rho)$ can be determined so that

$$a_{l,m}g_l(\rho) = \frac{\gamma_l^\nu}{\rho}\int_\rho^\infty \left(\frac{d^{2\nu+1}}{d\rho^{2\nu+1}}a_{l,m}G_l^\nu[g_l](p)\right) C_l^\nu\left(\frac{p}{\rho}\right)\left[\frac{p^2}{\rho^2}-1\right]^{\nu-\frac{1}{2}} dp .$$

33.6 Inversion Formula

The inversion formula for the Radon transform depends on n, the dimension of the space. Let us combine the inversion formula for the Fourier transform with (33.2), to obtain

$$\begin{aligned} f(\mathbf{x}) &= \frac{1}{(2\pi)^{n/2}}\int \hat{f}(\mathbf{w})e^{-i\mathbf{x}.\mathbf{w}}d\mathbf{w} \\ &= \frac{1}{(2\pi)^{n/2}}\int_0^\infty \int_{|\boldsymbol{\xi}|=1} \hat{f}(\rho\boldsymbol{\xi})e^{-i\rho(\mathbf{x}.\boldsymbol{\xi})}\rho^{n-1}d\rho d\boldsymbol{\xi} \\ &= \frac{1}{(2\pi)^n}\int_0^\infty \int_{|\boldsymbol{\xi}|=1}\int_{-\infty}^\infty e^{i\rho p}f^\dagger(p,\boldsymbol{\xi})e^{-i\rho(\mathbf{x}.\boldsymbol{\xi})}\rho^{n-1}d\rho d\boldsymbol{\xi}dp \\ &= \int_{|\boldsymbol{\xi}|=1} h(\mathbf{x}\,.\,\boldsymbol{\xi},\boldsymbol{\xi})\,d\boldsymbol{\xi} , \end{aligned}$$

where

$$h(t,\boldsymbol{\xi}) = \frac{1}{(2\pi)^n}\int_0^\infty \int_{-\infty}^\infty f^\dagger(p,\boldsymbol{\xi})e^{i\rho(p-t)}\rho^{n-1}d\rho dp . \qquad (33.7)$$

Since $h(t,\boldsymbol{\xi})$ is to be integrated over the unit sphere, it is sufficient to evaluate the integral of its even part in $\boldsymbol{\xi}$. Therefore, we replace h by

$$h_e(t,\boldsymbol{\xi}) = \frac{1}{2}[h(t,\boldsymbol{\xi}) + h(-t,-\boldsymbol{\xi})]$$

in (33.7) to obtain

$$\begin{aligned} h_e(t,\boldsymbol{\xi}) = \frac{1}{2(2\pi)^n}\bigg\{&\int_0^\infty \int_{-\infty}^\infty f^\dagger(p,\boldsymbol{\xi})e^{i\rho(p-t)}\rho^{n-1}d\rho dp \\ &+ \int_0^\infty \int_{-\infty}^\infty f^\dagger(p,-\boldsymbol{\xi})e^{i\rho(p+t)}\rho^{n-1}d\rho dp\bigg\} \end{aligned}$$

If we make the change of variables $p \to -p$ and $\rho \to -\rho$ in the second integral and use the fact that $f^\dagger(p, \boldsymbol{\xi}) = f^\dagger(-p, -\boldsymbol{\xi})$ (Property 2 of Section 33.3), the second integral becomes

$$\int_{-\infty}^{0}\int_{-\infty}^{\infty} f^\dagger(p,\boldsymbol{\xi})e^{i\rho(p-t)}|\rho|^{n-1}d\rho dp$$

and hence

$$\begin{aligned} h_e(t,\boldsymbol{\xi}) &= \frac{1}{2(2\pi)^n}\int_{-\infty}^{\infty}\int_{-\infty}^{\infty} f^\dagger(p,\boldsymbol{\xi})e^{i\rho(p-t)}|\rho|^{n-1}d\rho dp \\ &= \frac{1}{2(2\pi)^n}\int_{-\infty}^{\infty} e^{-i\rho t}|\rho|^{n-1}\left(\int_{-\infty}^{\infty} f^\dagger(p,\boldsymbol{\xi})e^{i\rho p}dp\right)d\rho\,. \end{aligned} \tag{33.8}$$

If we integrate the inside integral by parts $(n-1)$ times, we obtain

$$h_e(t,\boldsymbol{\xi}) = \frac{(i)^{n-1}}{2(2\pi)^n}\int_{-\infty}^{\infty}\left(\frac{|\rho|}{\rho}\right)^{n-1} e^{-i\rho t}\left(\int_{-\infty}^{\infty} e^{i\rho p}\left(\frac{\partial}{\partial p}\right)^{n-1} f^\dagger(p,\boldsymbol{\xi})dp\right)d\rho\,. \tag{33.9}$$

Thus, if n is odd, it will follow from the inversion formula for the Fourier transform that

$$\begin{aligned} h_e(t,\boldsymbol{\xi}) &= \frac{(i)^{n-1}}{2(2\pi)^n}\int_{-\infty}^{\infty} e^{-i\rho t}\left(\int_{-\infty}^{\infty} e^{i\rho p}\left(\frac{\partial}{\partial p}\right)^{n-1} f^\dagger(p,\boldsymbol{\xi})dp\right)d\rho \\ &= \frac{(i)^{n-1}}{2(2\pi)^{n-1}}\left(\frac{\partial}{\partial t}\right)^{n-1} f^\dagger(t,\boldsymbol{\xi})\,, \end{aligned}$$

whereas if n is even

$$h_e(t,\boldsymbol{\xi}) = \frac{(i)^{n-1}}{2(2\pi)^n}\int_{-\infty}^{\infty}(\text{sgn }\rho)e^{-i\rho t}\left(\int_{-\infty}^{\infty} e^{i\rho p}\left(\frac{\partial}{\partial p}\right)^{n-1} f^\dagger(p,\boldsymbol{\xi})dp\right)d\rho\,.$$

This equation, together with formula (14.6), implies

$$h_e(t,\boldsymbol{\xi}) = \frac{(i)^n}{2(2\pi)^{n-1}}\mathcal{H}\left[\frac{\partial^{n-1}}{\partial p^{n-1}} f^\dagger(p,\boldsymbol{\xi})\right](t)\,,$$

where $\mathcal{H}$ denotes the Hilbert transformation taken with respect to p. Putting these results together, we obtain the following inversion formula for the Radon transform: If $f^\dagger$ is the Radon transform of f, then

$$f(\mathbf{x}) = \int_{|\boldsymbol{\xi}|=1} h(\mathbf{x}\,.\,\boldsymbol{\xi}\,,\,\boldsymbol{\xi})\,d\boldsymbol{\xi}\,, \tag{33.10}$$

where

$$h(t,\boldsymbol{\xi}) = \begin{cases} c_n\frac{\partial^{n-1}}{\partial t^{n-1}} f^\dagger(t,\boldsymbol{\xi}) & \text{if } n \text{ is odd}\,, \\ c_n\mathcal{H}\left[\frac{\partial^{n-1}}{\partial p^{n-1}} f^\dagger(p,\boldsymbol{\xi})\right](t) & \text{if } n \text{ is even,} \end{cases} \tag{33.11}$$

and

$$c_n = \begin{cases} \frac{(i)^{n-1}}{2(2\pi)^{n-1}} \text{ if } n \text{ is odd} \\ \frac{(i)^n}{2(2\pi)^{n-1}} \text{ if } n \text{ is even.} \end{cases} \tag{33.12}$$

Example: For $n = 2$,

$$h(t, \boldsymbol{\xi}) = \frac{-1}{2(2\pi)} \frac{1}{\pi} \int_{-\infty}^{\infty} \frac{f_p^\dagger(p, \boldsymbol{\xi})}{p - t} dp$$

where $f_p^\dagger(p, \boldsymbol{\xi}) = \frac{\partial}{\partial p} f^\dagger(p, \boldsymbol{\xi})$. Hence, formula (33.10) takes the simpler form

$$f(\mathbf{x}) = \frac{-1}{4\pi^2} \int_{|\boldsymbol{\xi}|=1} d\boldsymbol{\xi} \int_{-\infty}^{\infty} \frac{f_p^\dagger(p, \boldsymbol{\xi})}{p - \mathbf{x}.\boldsymbol{\xi}} dp \,,$$

where $\boldsymbol{\xi} = (\cos\phi, \sin\phi)$. The change of variables $x = r\cos\phi$, $y = r\sin\phi$, leads to the inversion formula

$$f(r, \theta) = \frac{-1}{4\pi^2} \int_0^{2\pi} d\phi \int_{-\infty}^{\infty} \frac{f_p^\dagger(p, \phi)}{p - r\cos(\theta - \phi)} dp \,.$$

33.7 The Adjoint Radon Transformation

Let us define the operator K by

$$Kg(p, \boldsymbol{\xi}) = \begin{cases} c_n \frac{\partial^{n-1}}{\partial p^{n-1}} g(p, \boldsymbol{\xi}) & \text{if } n \text{ is odd,} \\ c_n \mathcal{H}\left(\frac{\partial^{n-1}}{\partial p^{n-1}} g(p, \boldsymbol{\xi})\right) & \text{if } n \text{ is even,} \end{cases} \tag{33.13}$$

where c_n is given by (33.12). Now we introduce the adjoint of the Radon transformation. Since

$$\begin{aligned} \langle \phi, R[f] \rangle &= \int_{-\infty}^{\infty} \int_{|\boldsymbol{\xi}|=1} \phi(p, \boldsymbol{\xi}) \overline{R[f]}(p, \boldsymbol{\xi}) d\boldsymbol{\xi} dp \\ &= \int_{-\infty}^{\infty} \int_{|\boldsymbol{\xi}|=1} \phi(p, \boldsymbol{\xi}) d\boldsymbol{\xi} dp \int \overline{f}(\mathbf{x}) \delta(p - \boldsymbol{\xi}.\mathbf{x}) d\mathbf{x} \\ &= \int \overline{f}(\mathbf{x}) \left(\int_{|\boldsymbol{\xi}|=1} \phi(\boldsymbol{\xi}.\mathbf{x}, \boldsymbol{\xi}) d\boldsymbol{\xi} \right) d\mathbf{x} = \langle R^*[\phi], f \rangle \,, \end{aligned} \tag{33.14}$$

we define the transformation R^*, given by

$$R^*[\phi](\mathbf{x}) = \int_{|\boldsymbol{\xi}|=1} \phi(\mathbf{x}.\boldsymbol{\xi}, \boldsymbol{\xi}) d\boldsymbol{\xi} \tag{33.15}$$

as the adjoint transformation of R. The application of the transformation R^* to a function ϕ corresponds to the integration of ϕ over all hyperplanes passing through a given point.

The inversion formula (33.10) can now be written in the form

$$f = R^* K[f^\dagger]\,. \tag{33.16}$$

A sufficient condition for this formula to hold is that f belongs to the Schwartz space of function $\mathcal{S}$.

Now we prove the uniqueness of the Radon transform that was alluded to in Section 33.2. Let $\phi \in \mathcal{S}$ be even on $\mathcal{R} \times S^{n-1}$. We need to show that if $R^* K\phi = f$, then $\phi = f^\dagger$. Thus, it suffices to show that if $R^* K\phi = 0$, then $\phi = 0$. From (33.13) and (33.11) it follows that $K\phi$ is equal to an expression similar to that in (33.8), but in which $f^\dagger$ is replaced by ϕ. Thus

$$(R^* K\phi)(\mathbf{x}) = \frac{1}{2(2\pi)^n} \int_{|\boldsymbol{\xi}|=1} d\boldsymbol{\xi} \int_{-\infty}^{\infty} e^{-i\rho(\mathbf{x}.\boldsymbol{\xi})} |\rho|^{n-1} \left(\int_{-\infty}^{\infty} \phi(p, \boldsymbol{\xi}) e^{i\rho p} dp \right) d\rho\,.$$

Let $\boldsymbol{\zeta} = \rho\boldsymbol{\xi}$, so that $d\boldsymbol{\zeta} = \rho^{n-1} d\rho d\boldsymbol{\xi}$ and consequently

$$(R^* K\phi)(\mathbf{x}) = \frac{1}{(2\pi)^n} \int e^{-i\mathbf{x}.\boldsymbol{\zeta}} d\boldsymbol{\zeta} \int_{-\infty}^{\infty} \phi\left(p, \frac{\boldsymbol{\zeta}}{|\boldsymbol{\zeta}|}\right) e^{i|\boldsymbol{\zeta}|p} dp\,.$$

If $R^* K\phi = 0$, then from the uniqueness of the n-dimensional Fourier transform it follows that the inside integral is zero, which in turn because of the uniqueness of the one-dimensional Fourier transform implies that $\phi = 0$, and this completes the proof.

33.8 Parseval's Relation

Let f and $g \in \mathcal{S}$ and denote their Radon transforms by $f^\dagger$ and $g^\dagger$, respectively. Then Parseval's relation takes the form

$$\begin{aligned}
&\int f(\mathbf{x})\overline{g}(\mathbf{x})\, d\mathbf{x} = \\
&\frac{(-1)^{(n-1)/2}}{2(2\pi)^{n-1}} \int_{|\boldsymbol{\xi}|=1} d\boldsymbol{\xi} \int_{-\infty}^{\infty} f^\dagger(p, \boldsymbol{\zeta}) \overline{g}_p^{\dagger(n-1)}(p, \boldsymbol{\xi})\, dp \\
&= \frac{1}{2(2\pi)^{n-1}} \int_{|\boldsymbol{\xi}|=1} d\boldsymbol{\xi} \int_{-\infty}^{\infty} f_p^{\dagger(m)}(p, \boldsymbol{\xi}) \overline{g}_p^{\dagger(m)}(p, \boldsymbol{\xi})\, dp
\end{aligned} \tag{33.17}$$

if n is odd , where $m = (n-1)/2$ and

$$\int f(\mathbf{x})\overline{g}(\mathbf{x})\, d\mathbf{x} =$$

$$c_n \int_{|\boldsymbol{\xi}|=1} d\boldsymbol{\xi} \int_{-\infty}^{\infty} \int_{-\infty}^{\infty} f^\dagger(p,\boldsymbol{\xi}) \overline{g^\dagger}(t,\boldsymbol{\xi})(p-t)^{-n}\, dp\, dt \tag{33.18}$$

if n is even, where

$$c_n = \frac{(-1)^{n/2}(n-1)!}{(2\pi)^n} .$$

We prove Parseval's relation for n odd and the proof for n even is left to the reader. Let $g^*(x) = \overline{g}(-x)$ and define

$$h(\mathbf{x}) = (f * g^*)(\mathbf{x}) = \int f(\mathbf{y})\overline{g}(\mathbf{y}-\mathbf{x})d\mathbf{y} . \tag{33.19}$$

By the convolution property of the Radon transform, we have

$$h^\dagger(p,\boldsymbol{\xi}) = \int f^\dagger(t,\boldsymbol{\xi})\overline{g^\dagger}(t-p,\boldsymbol{\xi})dt \tag{33.20}$$

since $R[g^*](p,\boldsymbol{\xi}) = \overline{R[g]}(-p,\boldsymbol{\xi})$. But from the inversion formula (33.10) we also have

$$h(\mathbf{x}) = \frac{(i)^{n-1}}{2(2\pi)^{n-1}} \int_{|\boldsymbol{\xi}|=1} h_p^{\dagger(n-1)}(\mathbf{x}\,.\,\boldsymbol{\xi},\boldsymbol{\xi})d\boldsymbol{\xi} ,$$

where

$$h_p^{\dagger(n-1)}(\mathbf{x}.\boldsymbol{\xi},\boldsymbol{\xi}) = \left.\frac{\partial^{n-1}}{\partial p^{n-1}} h^\dagger(p,\boldsymbol{\xi})\right|_{p=\mathbf{x}.\boldsymbol{\xi}} .$$

Thus, in view of (33.19) and (33.20), we have

$$\begin{aligned} h(0) &= \frac{(i)^{n-1}}{2(2\pi)^{n-1}} \int_{|\boldsymbol{\xi}|=1} h_p^{\dagger(n-1)}(0,\boldsymbol{\xi})d\boldsymbol{\xi} \\ &= \frac{(-i)^{n-1}}{2(2\pi)^{n-1}} \int_{|\boldsymbol{\xi}|=1} d\boldsymbol{\xi} \int_{-\infty}^{\infty} f^\dagger(t,\boldsymbol{\xi})\overline{g_p^{\dagger(n-1)}}(t,\boldsymbol{\xi})dt \\ &= \int f(\mathbf{y})\overline{g}(\mathbf{y})\, d\mathbf{y} \end{aligned} \tag{33.21}$$

which is the first equality in (33.17). The second equality in (33.17) is obtained from the first by integration by parts $m = (n-1)/2$ times.

If we define the operator H by

$$H[g](p) = \frac{1}{\sqrt{2}\,(2\pi)^m} \frac{\partial^m}{\partial p^m} g(p) ,$$

$m = (n-1)/2$, then (33.17) becomes

$$\langle f, g\rangle = \langle HRf, HRg\rangle . \tag{33.22}$$

D. Ludwig [151] showed that HR is a unitary transformation from $L^2(\mathcal{R}^n)$ onto $L^2(\mathcal{R} \times S^{n-1})$.

33.9 Generalized Functions

Different techniques have been used to extend the Radon transform to generalized functions. A new approach has recently been obtained by A. Zayed and P. Mikusinski in [271, 272]. But in this section we shall outline the classical approach which is due to I. Gel'fand, M. Graev, N. Vilenkin [69] and D. Ludvig [151].

Let $f \in \mathcal{S}(\mathcal{R}^n)$. Then since

$$I_k = \int_{-\infty}^{\infty} f^{\dagger}(p, \boldsymbol{\xi}) p^k dp = \int_{-\infty}^{\infty} p^k dp \int f(\mathbf{x}) \delta(p - \boldsymbol{\xi}.\mathbf{x}) d\mathbf{x}$$
$$= \int f(\mathbf{x}) (\boldsymbol{\xi}.\mathbf{x})^k d\mathbf{x} ,$$

it is easy to see that I_k is a polynomial in $\boldsymbol{\xi}$ that is homogeneous of degree k. Therefore, from the elementary properties of the Radon transform, we conclude that in order for a function $g(p, \boldsymbol{\xi})$ to be the Radon transform of a function $f \in \mathcal{S}(\mathcal{R}^n)$ it is necessary that

1. g be even in p and $\boldsymbol{\xi}$, homogeneous of degree -1,
2. belong to $S(\mathcal{R} \times S^{n-1})$, i.e., the space S on the cylinder, and
3. $\int_{-\infty}^{\infty} g(p, \boldsymbol{\xi}) p^k dp$ be a polynomial in $\boldsymbol{\xi}$ homogeneous of degree $= k$, where $k = 0, 1, \cdots$.

In fact, these conditions are also sufficient [69, p. 16]. Condition (3) is equivalent to saying that if S_l^m is a spherical harmonic of degree l and if $k < l$, then

$$\int_{|\boldsymbol{\xi}|=1} \int_{-\infty}^{\infty} g(p, \boldsymbol{\xi}) p^k S_l^m(\boldsymbol{\xi}) d\boldsymbol{\xi} dp = 0 .$$

Let

$$\psi(p, \boldsymbol{\xi}) = f_p^{\dagger(n-1)}(p, \boldsymbol{\xi}) \tag{33.23}$$

be the $(n-1)$st derivative of $f^{\dagger}$ with respect to p. If $f \in \mathcal{S}(\mathcal{R}^n)$, then $\psi(p, \boldsymbol{\xi})$ satisfies the following conditions:

i) ψ is even in p and $\boldsymbol{\xi}$, homogeneous of degree $-n$, i.e.,

$$\psi(\alpha p, \alpha \boldsymbol{\xi}) = |\alpha|^{-n} \psi(p, \boldsymbol{\xi}) , \quad \alpha \neq 0 ,$$

ii) $\psi \in S(\mathcal{R} \times S^{n-1})$, i.e., ψ is infinitely differentiable with respect to p and $\boldsymbol{\xi}$ and decreases rapidly, together with all its derivatives with respect to p as $|p| \to \infty$, uniformly in $\boldsymbol{\xi}$ on compact sets.

iii) For any integer $k \geq 0$, $\int_{-\infty}^{\infty} \psi(p, \boldsymbol{\xi}) p^k dp$ is a homogeneous polynomial in $\boldsymbol{\xi}$ of degree $k - n + 1$.

Let $\tilde{S}$ be the vector space of all functions ψ satisfying conditions (i)–(iii). For every $\psi \in \tilde{S}$, there exists $f \in S(\mathcal{R}^n)$ such that (33.23) holds. We provide $\tilde{S}$ with the following topology: A sequence $\{\psi_n\}_{n=0}^{\infty}$ converges to 0 in $\tilde{S}$ if and only if $\{f_n\}_{n=0}^{\infty}$ converges to 0 in S where ψ_n is related to f_n as in (33.23). Now we can define the Radon transform of generalized functions by using Parseval's relations (33.17) and (33.18).

DEFINITION 33.3 *Let $f \in \mathcal{S}(\mathcal{R}^n)$, $F \in \mathcal{S}^*(\mathcal{R}^n)$, and n be odd. Then the Radon transform, $F^\dagger$, of F is given by*

$$\langle F, f \rangle = \frac{(-1)^{(n-1)/2}}{2(2\pi)^{n-1}} \langle F^\dagger, \psi \rangle \ ,$$

where ψ is given by (33.23).

For n even, the Radon transform of $F \in \mathcal{S}^*(\mathcal{R}^n)$ can be defined similarly using (33.18). Another version of the Radon transform of generalized functions was considered in [151]. But before we introduce it, let us first extend the adjoint Radon transformation to generalized functions.

DEFINITION 33.4 *Let $f \in \mathcal{S}(\mathcal{R}^n)$ and $F \in \mathcal{S}^*(\mathcal{R} \times S^{n-1})$. The adjoint R^* of the Radon transformation is defined by*

$$\langle R^* F, f \rangle = \langle F, Rf \rangle \ .$$

To simplify the notation, we shall denote $\mathcal{S}(\mathcal{R}^n)$ by $\mathcal{S}$ and $\mathcal{S}(\mathcal{R} \times S^{n-1})$ by $\mathcal{S}_1$. Since R maps $\mathcal{S}$ continuously into $\mathcal{S}_1$, it follows that R^* also maps $(R[\mathcal{S}])^*$ continuously into $\mathcal{S}^*$. Here we have adopted the notation that if A is a set of functions, then $R[A]$ is its image under the transformation R. In fact, R^* is a one-to-one bicontinuous transformation from $(R[\mathcal{S}])^*$ onto $\mathcal{S}^*$; see Theorem 4.1 in [151]. The extension of R^* to all of $\mathcal{S}_1$ is more difficult.

The inversion formula, $R^*KR = I$, (cf. (33.16)) can now be used to extend the Radon transform to tempered distributions.

DEFINITION 33.5 *Let $f \in \mathcal{S}^*$ and $\psi \in \mathcal{S}$. The Radon transform of f is defined by*

$$\langle Rf, \phi \rangle = \langle f, R^* \phi \rangle = \langle f, \psi \rangle \ ,$$

where $\phi = KR\psi$.

Therefore, the Radon transform of $f \in \mathcal{S}^*$ is an element of $(KR[S])^*$. In fact, R is a one-to-one bicontinuous transformation from $\mathcal{S}^*$ onto $(KR[\mathcal{S}])^*$; see Theorem 4.6 in [151].

To extend the Radon transform to the space $\mathcal{D}^* = \mathcal{D}^*(\mathcal{R}^n)$ of Schwartz distributions, the following important result is needed.

THEOREM 33.1 [151]
Let $f \in S(\mathcal{R}^n)$. Then f has support in $|\mathbf{x}| \leq r$ if and only if its Radon transform $f^\dagger(p, \boldsymbol{\xi})$ has support in $|p| \leq r$. Thus, $R[\mathcal{D}(\mathcal{R}^n)] = (R[S(\mathcal{R}^n)]) \cap \mathcal{D}(\mathcal{R}^n)$. The Radon transformation is a continuous transformation from $\mathcal{D}(\mathcal{R}^n)$ into itself.

Theorem 33.1 and Definition 33.4 imply that R^* is a one-to-one bicontinuous transformation from $(R[\mathcal{D}])^*$ onto $\mathcal{D}^*$. Moreover, R is a one-to-one bicontinuous transformation from $\mathcal{D}^*$ onto $(KR[\mathcal{D}])^*$.

33.10 Applications

33.10.1 Physical Applications

The Radon transform has numerous applications in a large number of fields; to mention a few: computer assisted tomography (CAT), medical imaging, magnetic resonance (MR), radio and radar astronomy, electron microscopy, optics, geophysics, ultrasound and acoustics. A very rich source on the applications of the Radon transform can be found in a book by S. Dean [51]; see also [96].

The Radon transform arises naturally in various reconstruction problems. By a reconstruction problem we mean the problem of determining the internal structure of an object without having to cut it. The size of the object can be as small as a molecule or as large as a galaxy or somewhere in between like a human body.

There are two approaches to the reconstruction problem; one is called active and the other is called passive. In the active approach, a beam of X-rays, ultrasound waves, photons, protons, or a laser, is directed at the object and a probe is placed at the opposite side of the object to measure the intensity of the beam after passing through the object. Assuming, as usually is the case, that the probe's reading is proportional to the line integral of a function representing some physical property of the object and the beam, such as the attenuation coefficient, the probe's measurements can then be interpreted as samples of the Radon transform of the physical quantity to be measured. The beam and the probe are moved parallel

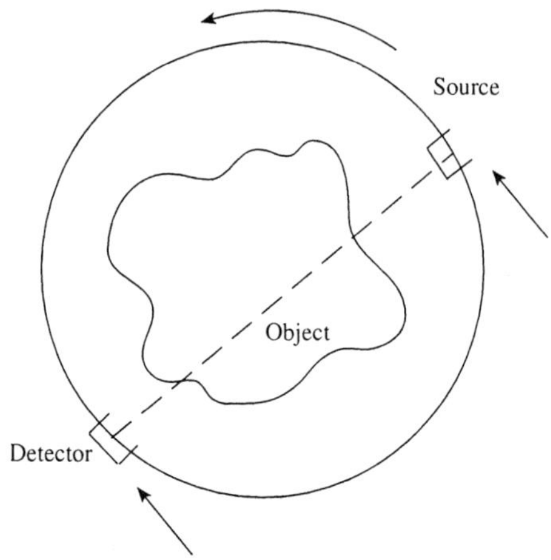

FIGURE 33.2
Parallel scanning

to each other to scan the object in one direction. This process is called parallel scanning. Then the beam and the probe are rotated and the parallel scanning process is repeated to take measurements from different angles; see Figure 33.2. Having collected enough measurements, one then invokes the inversion formula for the Radon transform to map the distribution of the physical quantity inside the object. In diagnostic radiology, for example, the data are recorded and processed by a computer to produce an image which can be displayed on a screen.

In the passive approach, the process is similar except that the source of the beam is placed inside and the probe is outside the object. The probe is then rotated around the object and measurements are recorded from different angles. The beam source may be a radioactive material injected into the human body or it may be a microwave or infrared emission from a far away celestial object.

In practice, however, the reconstruction problem is much more intricate than that, not only because of measurement errors akin to the probe accuracy, but also because in practice only a finite number of measurements are taken. This raises the question of whether a nice function can be determined by a finite number of samples of its Radon transform. Different techniques and algorithms have been developed to overcome these difficulties; for more details see [51, 176].

33.10.2 Mathematical Applications

The Radon transform can also be used to solve Cauchy's problem for hyperbolic partial differential equations with constant coefficients. Simply put, the Radon transform reduces problems with $n+1$ independent variables to problems with two independent variables. The link between the Radon transform and Cauchy's problem lies in property (7) in Section 33.3. Recall that if L is a linear differential operator with constant coefficients, then

$$R\left[L\left(\frac{\partial}{\partial x_1}, \ldots, \frac{\partial}{\partial x_n}; \frac{\partial}{\partial t}\right) f\right] = L\left(\xi_1 \frac{\partial}{\partial p}, \ldots, \xi_n \frac{\partial}{\partial p}; \frac{\partial}{\partial t}\right) R[f](p, \boldsymbol{\xi}; t)$$

For example consider the Cauchy problem

$$L\left(\frac{\partial}{\partial x_1}, \ldots, \frac{\partial}{\partial x_n}; \frac{\partial}{\partial t}\right) u(\mathbf{x}; t) = 0 , \tag{33.24}$$

subject to the initial conditions

$$\left.\frac{\partial^l u(\mathbf{x};t)}{\partial t^l}\right|_{t=0} = 0, \qquad 0 \leq l \leq m-2$$

$$\left.\frac{\partial^{m-1} u(\mathbf{x};t)}{\partial t^{m-1}}\right|_{t=0} = f(\mathbf{x}),$$

where m is the degree of the polynomial operator L. If we apply the Radon transform to (33.24) and use property (7) in Section 33.3, we find

$$L\left(\xi_1 \frac{\partial}{\partial p}, \ldots, \xi_n \frac{\partial}{\partial p}; \frac{\partial}{\partial t}\right) u^\dagger(p, \boldsymbol{\xi}; t) = 0 .$$

Since R and $\frac{\partial}{\partial t}$ commute, the initial conditions are converted into

$$\frac{\partial^l u^\dagger(p, \boldsymbol{\xi}; t)}{\partial t^l} = 0, \quad 0 \leq l \leq m-2,$$

$$\frac{\partial^{m-1} u^\dagger(p, \boldsymbol{\xi}; t)}{\partial t^{m-1}} = f^\dagger(p, \boldsymbol{\xi}).$$

The differential equation satisfied by $u^\dagger$ is one-dimensional in p and hence can be solved by using standard methods; for more details, see [51, Section 8.7].

34

Sequence Transformations

Although its main focus is function transformations, this book would not be complete without a discussion of an equally important concept, namely, sequence transformations. After all, the two are intimately related, as pointed out in Chapter 4; see also Appendix A. For example, in many practical situations, instead of applying a transformation directly to a function, a sequence of numbers is extracted first from the function, then an appropriate transformation is applied to that sequence. The transform of the sequence is then used to reconstruct the transform of the function. Speed and economics are usually the reason for favoring sequence transformations over function transformations.

With the advent of modern computers and digital processors, interest in sequence transformations has grown tremendously in the last thirty years. Fast algorithms to compute various sequence transformations more efficiently have been developed since the mid 1960s. The introduction of the fast Fourier transform by Cooley and Tukey in 1965 [47] was the beginning, but hardly the end, of a vast use of sequence transformations in a wide range of engineering applications. Soon after that came the fast Hartley, fast Walsh, and most recently the fast wavelets transforms [24, 55, 244]. Fast computers and algorithms have revolutionized numerical computation and made feasible some computing projects that not long ago could not have even been contemplated.

In this chapter we shall discuss briefly some of the most commonly used sequence transformations, the *discrete Fourier transform* (DFT) and its offspring, the *fast Fourier Transform* (FFT), the *Z-transform*, and the *Walsh transform*.

34.1 The Discrete Fourier Transform (DFT)

The essence of the discrete Fourier transform can be traced back to the work of Lagrange on the vibrating string problem [134], in which he discretized the problem by assuming that the string was made of a finite number of connected particles. The solution of the discretized problem involved formulas that unequivocally resembled the discrete Fourier transform. However, what has been described in [29, p. 4] as the first explicit formula for the discrete Fourier transform actually appeared in a paper by A. Claude Clairaut in 1754 on the determination of orbits of celestial bodies. The fact that the discrete Fourier transform can be traced to these two seemingly unrelated problems, i.e, the problem of a vibrating string and that of orbit determination, is no mystery to us now: both problems dealt with the reconstruction of a function from data with periodic pattern. Because of the periodicity of the data, it was natural for Lagrange and Clairaut to approximate their solutions by finite sums of sine and cosine functions, which is the quintessence of the discrete Fourier transform. For a lively discussion of the discrete Fourier transform and its history, we refer the reader to a recent book by W. Briggs and V. Henson [29].

34.1.1 Definition and Inversion Formula

DEFINITION 34.1 *The discrete Fourier transform of a sequence* $\{a_n\}_{n=0}^{N-1}$, *where* N *is a positive integer* ≥ 1, *is a sequence* $\{A_m\}_{m=0}^{N-1}$, *defined by*

$$A_m = \sum_{n=0}^{N-1} a_n (W_N)^{mn}, \quad m = 0, 1, \cdots, N-1, \tag{34.1}$$

where $W_N = e^{2\pi i/N}$.

Since

$$\sum_{m=0}^{N-1} W_N^{m(k-n)} = \frac{1 - W_N^{N(k-n)}}{1 - W_N^{(k-n)}}, \quad k \neq n,$$

it follows that

$$\sum_{m-0}^{N-1} W_N^{m(k-n)} = \begin{cases} 0 & \text{if } k \neq n \\ N & \text{if } k = n\,. \end{cases}$$

Using this relation, we obtain the *inversion formula* for the DFT.

Inversion Formula

$$a_n = \frac{1}{N}\sum_{m=0}^{N-1} A_m W_N^{-mn}\,, \quad n = 0, 1, \cdots, N-1\,. \tag{34.2}$$

Equations (34.1) and (34.2) are called a discrete Fourier transform (DFT) pair of order N. The factor $1/N$ and the negative sign in the exponent of W_N that appear in Equation (34.2) are sometimes introduced in (34.1) instead. We shall use the notation

$$\mathcal{F}_N[(a_n)] = A_m \quad , \quad \mathcal{F}_N^{-1}[(A_m)] = a_n,$$

to indicate that the discrete Fourier transform of order N of the sequence $\{a_n\}$ is $\{A_m\}$, and that the inverse transform of $\{A_m\}$, is $\{a_n\}$.

Example A: The DFT of the sequence $\{1, 0, 1, 1\}$ is $\{3, -i, 1, i\}$.
Because

$$W_N^{\pm(m+N)n} = W_N^{\pm mn}\,,$$

Equations (34.1) and (34.2) can be used to extend the sequences $\{a_n\}_{n=0}^{N-1}$ and $\{A_m\}_{m=0}^{N-1}$, as periodic sequences with period N. This means that

$$A_{m+N} = A_m\,, \quad a_{n+N} = a_n\,.$$

The summation limits, 0 and $N-1$, can now be replaced with n_1 and $n_1 + N - 1$, respectively, where n_1 is any integer. In the special case where $n_1 = -M$ and $N = 2M + 1$, Equations (34.1) and (34.2) become

$$A_m = \sum_{n=-M}^{M} a_n W_N^{mn}\,, \quad m = -M, -M+1, \cdots, M-1, M\,,$$

and

$$a_n = \frac{1}{2M+1}\sum_{m=-M}^{M} A_m W_N^{-mn}\,, \quad n = -M, -M+1, \cdots, M-1, M\,.$$

34.1.2 Elementary Properties

1. **Linearity:**

   The discrete Fourier transform is linear, that is

   $$\mathcal{F}_N[\alpha(a_n) + \beta(b_n)] = \alpha A_m + \beta B_m\,,$$

   for any complex numbers α and β, where the sum of two sequences is defined as

   $$(a_n) + (b_n) = (a_n + b_n)\,.$$

2. **Translation:**

$$\mathcal{F}_N[a_{n-k}] = W_N^{mk} A_m\,, \text{ or } e^{2\pi imk/N} A_m = \sum_{n=0}^{N-1} a_{n-k} W_N^{mn}\,.$$

3. **Modulation:**

$$\mathcal{F}_N[W_N^{nk} a_n] = A_{m+k}\,, \text{ or } A_{m+k} = \sum_{n=0}^{N-1} e^{2\pi ink/N} a_n W_N^{mn}\,.$$

4. **Complex Conjugation:**

$$\mathcal{F}_N[\bar{a}_{-n}] = \bar{A}_m\,, \text{ or } \bar{A}_m = \sum_{n=0}^{N-1} \bar{a}_{-n} W_N^{mn}\,.$$

5. **Symmetry:**

$$\mathcal{F}_N[a_{-n}] = A_{-m}\,, \text{ or } A_{-m} = \sum_{n=0}^{N-1} a_{-n} W_N^{mn}\,.$$

In (4) and (5), the fact that $\bar{W}_N = W_N^{-1}$ has been used. A sequence $\{a_n\}$ is said to be even if $\{a_{-n}\} = \{a_n\}$ and is said to be odd if $\{a_{-n}\} = \{-a_n\}$. The following are easy consequences of (4) and (5):

i) if $\{a_n\}$ is a sequence of real numbers, i.e., $\bar{a}_n = a_n$, then $\bar{A}_m = A_{-m}$,

ii) $\{a_n\}$ is real and even if and only if $\{A_m\}$ is real and even,

iii) $\{a_n\}$ is real and odd if and only if $\{A_m\}$ is pure imaginary and odd.

34.1.3 Convolution and Parseval's Relation

Convolution:
Let $\{a_n\}_{n=0}^{N-1}$, and $\{b_n\}_{n=0}^{N-1}$, be periodic sequences with period N. We define their convolution as a sequence $\{c_n\}_{n=0}^{N-1}$ given by

$$c_n = \sum_{k=0}^{N-1} a_k b_{n-k}\,.$$

The following relation is known as *the convolution relation of the DFT*:

$$\mathcal{F}_N[c_n] = \mathcal{F}_N[a_n]\mathcal{F}_N[b_n]\,, \text{ or } C_m = A_m B_m\,.$$

A consequence of this and (34.2) is the relation

$$\sum_{k=0}^{N-1} a_k b_{n-k} = \frac{1}{N} \sum_{m=0}^{N-1} A_m B_m W_N^{-mn} . \tag{34.3}$$

Parseval's Relation:
If we set $n = 0$ and $b_{-k} = \bar{d}_k$ in (34.3), we find that

$$\sum_{n=0}^{N-1} a_n \bar{d}_n = \frac{1}{N} \sum_{m=0}^{N-1} A_m \bar{D}_m ,$$

since $D_m = \bar{B}_m$. Setting $a_n = d_n$, leads to

$$\sum_{n=0}^{N-1} |a_n|^2 = \frac{1}{N} \sum_{m=0}^{N-1} |A_m|^2 .$$

34.2 The Fast Fourier Transform (FFT)

The fast Fourier transform was introduced in the realm of signal analysis in 1965 by Cooley and Tukey [47], and it was called for some time the Cooley–Tukey algorithm. But it turned out that the great mathematician C. F. Gauss actually used it in 1805 in computing the orbit of asteroid Pallas. To determine the orbit, Gauss used a set of a dozen regularly spaced data to fit into a trigonometric polynomial with a dozen unknown coefficients. This resulted in a system of twelve linear equations in twelve unknowns that he could have undoubtedly solved without difficulty. Inexplicably, instead of solving this system directly, he discovered a way to split the unknown coefficients and the equations into three subproblems that were easily solved. He then combined the solutions of the subproblems to obtain the solution of the original system. Gauss' idea of splitting the problem is the essence of the fast Fourier transform. This indisputable use of the fast Fourier transform by Gauss was discovered in one of his papers that appeared posthumously in 1866 as an unpublished paper and which went unnoticed for very many years.

A synopsis of the fast Fourier transform goes as follows. In order to determine A_m for $m = 0, 1, \cdots, M-1$ from (34.1), we must perform $M-1$ multiplications for each m; hence the total number of multiplications required to determine all the A_m's is $(M-1)^2$. This number can be reduced by using decimation.

Let us assume that M is even, so we may put $M = 2N$ and write

$$\mathcal{F}_{2N}[(a_n)] = A_m. \tag{34.4}$$

Let us split $\{a_n\}$ into two sequences, one consisting of terms with even subscripts and one with odd subscripts. That is

$$b_n = a_{2n} \quad , \quad c_n = a_{2n+1}.$$

Because

$$W_{2N}^{2km} = W_N^{km} \quad \text{and} \quad W_{2N}^{(2k+1)m} = W_N^{km} W_{2N}^m$$

we can show that

$$A_m = B_m + W_{2N}^m C_m \,. \tag{34.5}$$

For the evaluation of B_m and C_m, the total number of multiplications required is $2(N-1)^2$. To determine A_m from (34.5), we must calculate the product $W_{2N}^m C_m$, for each fixed m. Therefore, the total number of multiplications required to determine A_m from (34.5) is $2(N-1)^2+2N-1 = 2N^2 - 2N + 1$.

But had we determined A_m from (34.4), we would have performed $(2N-1)^2 = 4N^2 - 4N + 1$ multiplications. Thus, by splitting the sequence $\{a_n\}$ into two sequences and then applying the discrete Fourier transform, we reduce the number of multiplications required to evaluate A_m approximately by a factor of 2.

Once again, if N is even, we can repeat this process and split $\{b_n\}$ and $\{c_n\}$ into four sequences, each of length $N/2$; thus, B_m and C_m can be determined in terms of four discrete Fourier transforms each of order $N/2$. This process can be continued $k-1$ times if $M = 2^K$ for some positive integer K.

If we denote the required number of multiplications for the discrete Fourier transform of order $N = 2^K$ by $F(N)$, then

$$F(2N) = 2F(N) + N\,, F(2) = 1\,,$$

which leads to

$$F(N) = \frac{N}{2} \log_2 N.$$

34.3 The Z- Transform

34.3.1 Definitions

DEFINITION 34.2 *The $\mathcal{Z}$-transform of a sequence $\{f(n)\}_{-\infty}^{\infty}$ is defined by*

$$\mathcal{Z}[f(n)] = F(z) = \sum_{n=-\infty}^{\infty} f(n) z^{-n}\,,$$

for all complex numbers z for which the series converges.

The series converges at least in a ring of the form $0 \le r_1 < z < r_2 \le \infty$, whose radii, r_1 and r_2 depend on the behavior of $f(n)$ at $\pm\infty$. In fact,

$$r_1 = \limsup_{n\to\infty} \sqrt[n]{|f(n)|} \quad , \quad r_2 = \liminf_{n\to\infty} \frac{1}{\sqrt[n]{|f(-n)|}}.$$

If there is more than one sequence involved, we may denote r_1 and r_2 by $r_1(f)$ and $r_2(f)$, respectively. The function $F(z)$ is analytic in this ring, but it may be possible to continue it analytically beyond the boundaries of the ring. If $f(n) = 0$ for $n < 0$, then $r_2 = \infty$, and if $f(n) = 0$ for $n \ge 0$, then $r_1 = 0$.

Let $z = re^{i\theta}$. Then the $\mathcal{Z}$–transform evaluated at $r = 1$ is

$$\sum_{n=-\infty}^{\infty} f(n)e^{-in\theta},$$

which is called *the Fourier transform of the sequence* $\{f(n)\}_{-\infty}^{\infty}$.

Example A: Let a be a complex number and $f(n) = a^n$, for $n \ge 0$, and zero otherwise, then

$$\mathcal{Z}[f(n)] = \sum_{n=0}^{\infty} a^n z^{-n} = \frac{z}{z-a} \quad , \quad |z| > |a|, \tag{34.6}$$

and if $f(n) = na^n$, for $n \ge 0$, and zero otherwise, then

$$\sum_{n=0}^{\infty} na^n z^{-n} = \frac{az}{(z-a)^2} \quad , \quad |z| > |a|.$$

Example B: Let

$$u(n) = \begin{cases} 1, & n \ge 0 \\ 0, & n < 0. \end{cases}$$

Then

$$\mathcal{Z}[u(n)] = \frac{z}{z-1}.$$

Example C: Let

$$\delta(n) = \begin{cases} 1, & n = 0 \\ 0, & \text{otherwise.} \end{cases}$$

Then

$$\mathcal{Z}[\delta(n-k)] = z^{-k} \quad , \quad k = 0, \pm 1, \pm 2, \cdots.$$

34.3.2 Elementary Properties

Let the region of convergence of the $\mathcal{Z}$–transform of the sequence $\{f(n)\}$ be denoted by D_f.

1. **Linearity:**

$$\mathcal{Z}[af(n)+bg(n)] = a\mathcal{Z}[f(n)]+b\mathcal{Z}[g(n)] = aF(z)+bG(z), \quad z \in D_f \bigcap D_g.$$

The region $D_f \bigcap D_g$ contains the ring $r_1 < |z| < r_2$, where $r_1 =$ maximum $\{r_1(f), r_1(g)\}$ and $r_2 =$ minimum $\{r_2(f), r_2(g)\}$.

2. **Translation:**

$$\mathcal{Z}[f(n-k)] = z^{-k} F(z).$$

3. **Multiplication by Exponentials:**

$$\mathcal{Z}[(a^n f(n))] = F(z/a) \quad , \quad ar_1(f) < |z| < ar_2(f).$$

4. **Multiplication by powers of n:**

$$\mathcal{Z}[(nf(n))] = -z\frac{dF}{dz} \quad , \quad z \in D_f.$$

More generally,

$$\mathcal{Z}[(n^k f(n))] = (-1)^k \left(z\frac{d}{dz}\right)^k F(z), \quad ,k = 0, 1, 2, \cdots, \quad z \in D_f,$$

where

$$\left(z\frac{d}{dz}\right)^k F(z) = \left(z\frac{d}{dz}\right)^{(k-1)} \left(z\frac{dF}{dz}\right).$$

5. **Conjugation:**

$$\mathcal{Z}[\bar{f}(-n)] = \bar{F}(\frac{1}{\bar{z}}).$$

6. **Initial and Final Values:**

If $f(n) = 0$ for $n < 0$, then $\lim_{z\to\infty} F(z) = f(0)$, and conversely, if $F(z)$ is defined for $r_1 < |z|$ and for some integer m, $\lim_{z\to\infty} z^m F(z) = A$, then

$$f(m) = A \quad , \text{ and } \quad f(n) = 0 \quad \text{for } n < m.$$

34.3.3 Convolution, Product, and Parseval's Relation

Parseval's Relation:

Let $F(z)$ and $G(z)$ be the $\mathcal{Z}$–transforms of $\{f(n)\}$ and $\{g(n)\}$, respectively. Then

$$\sum_{n=-\infty}^{\infty} f(n)\bar{g}(n) = \frac{1}{2\pi}\int_{-\pi}^{\pi} F(e^{iw})\bar{G}(e^{iw})dw,$$

in particular,

$$\sum_{n=-\infty}^{\infty} |f(n)|^2 = \frac{1}{2\pi}\int_{-\pi}^{\pi} |F(e^{iw})|^2 dw.$$

Convolution and Product:

The convolution of two sequences $\{f(n)\}_{-\infty}^{\infty}$ and $\{g(n)\}_{-\infty}^{\infty}$ is a sequence $\{h(n)\}_{-\infty}^{\infty}$ defined by

$$h(n) = \sum_{k=-\infty}^{\infty} f(k)g(n-k).$$

The $\mathcal{Z}$–transform of the convolution of two sequences is the product of their $\mathcal{Z}$–transforms,

$$\mathcal{Z}[h(n)] = \mathcal{Z}[f(n)]\mathcal{Z}[g(n)] \quad , \quad z \in D_f \bigcap D_g,$$

or

$$H(z) = F(z)G(z).$$

The $\mathcal{Z}$–transform of the product of two sequences is given by

$$\mathcal{Z}[f(n)g(n)] = \frac{1}{2\pi i}\oint_{\gamma} F(w)G(z/w)\frac{dw}{w},$$

where γ is a closed contour surrounding the origin in the domain of convergence of $F(w)$ and $G(z/w)$.

34.3.4 Inversion Formulas

Suppose that we know the $\mathcal{Z}$–transform, $F(z)$, of a sequence, can we recover the sequence itself? To answer this question, let us first consider the sequences

$$f(n) = u(n) = \begin{cases} 1\,, & n \geq 0 \\ 0\,, & n < 0 \end{cases} \text{ and } g(n) = -u(-n-1) = \begin{cases} -1\,, & n \leq 0 \\ 0\,, & n \geq 0. \end{cases}$$

It is easy to see that

$$F(z) = \frac{z}{z-1}, \quad |z| > 1 \quad \text{and } G(z) = \frac{z}{z-1}, \quad |z| < 1.$$

This shows that the inverse $\mathcal{Z}$-transform of the function $z/(z-1)$ is not unique. In general, the inverse $\mathcal{Z}$-transform is not unique, unless its region of convergence is specified.

1. Inversion by Using Series Representation:
If $F(z)$ is given by its series

$$F(z) = \sum_{n=-\infty}^{\infty} a_n z^{-n}, \quad r_1 < z < r_2,$$

then its inverse $\mathcal{Z}$–transform is unique and is equal to $\{f(n) = a_n\}$ for all n.

2. Inversion by Using Complex Integration:
If $F(z)$ is given in a closed form as an algebraic expression and its domain of analyticity is known, then its inverse $\mathcal{Z}$-transform can be obtained by using the relation

$$f(n) = \frac{1}{2\pi i} \oint_\gamma F(z) z^{n-1} dz\,,$$

where γ is a closed contour surrounding the origin in the domain of analyticity of $F(z)$.

3. Inversion by Using Fourier Series:
If the domain of analyticity of F contains the unit circle, $|r| = 1$ and if F is single valued therein, then $F(e^{i\theta})$ is a periodic function with period 2π, and consequently it can be expanded in a Fourier series. The coefficients of the series form the inverse $\mathcal{Z}$-transform of F and they are given explicitly by

$$f(n) = \frac{1}{2\pi} \int_{-\pi}^{\pi} F(e^{i\theta}) e^{in\theta} d\theta.$$

4. Inversion by Using Partial Fractions:
By dividing Equation (34.6) by z and then differentiating both sides with respect to z, we obtain

$$\mathcal{Z}^{-1}\left[(z-a)^{-k}\right] = \frac{(n-k+1)_{k-1} a^{n-k}}{(k-1)!} u(n-k)\,, \quad k = 1, 2, \cdots, |z| > |a| > 0, \tag{34.7}$$

where

$$(a)_n = a(a+1)\cdots(a+n-1)\,, n = 1, 2, \cdots, \text{ and } (a)_0 = 1.$$

Moreover, from Example C,

$$\mathcal{Z}^{-1}\left[z^{-k}\right] = \delta(n-k). \tag{34.8}$$

Let $F(z)$ be a rational function of the form

$$F(z) = \frac{P(z)}{Q(z)} = \frac{a_N z^N + \cdots + a_1 z + a_0}{b_M z^M + \cdots + b_1 z + b_0}.$$

First, let $N < M$. The denominator $Q(z)$ can be factored over the field of complex numbers as

$$Q(z) = c(z - z_1)^{k_1} \cdots (z - z_m)^{k_m},$$

where c is a constant and $k_1 \cdots, k_m$ are positive integers satisfying $k_1 + \cdots + k_m = M$. Hence, F can be written in the form

$$F(z) = \sum_{i=1}^{m} \sum_{j=1}^{k_i} \frac{A_{i,j}}{(z - z_i)^j}, \tag{34.9}$$

where

$$A_{i,j} = \frac{1}{(k_i - j)!} \lim_{z \to z_i} \frac{d^{k_i - j}}{dz^{k_i - j}} (z - z_i)^{k_i} F(z).$$

The inverse $\mathcal{Z}$–transform of the fractional decomposition (34.9) in the region that is exterior to the smallest circle containing all the zeros of $Q(z)$ can be obtained by using (34.7).

If $N \geq M$, we must divide until F can be reduced to the form

$$F(z) = H(z) + \frac{R(z)}{Q(z)},$$

where the remainder polynomial, $R(z)$, has degree less than or equal to $M - 1$, and the quotient , $H(z)$, is a polynomial of degree at most $N - M$. The inverse $\mathcal{Z}$-transform of the quotient polynomial can be obtained by using (34.8) and that of $R(z)/Q(z)$ can be obtained as in the case $N < M$.

Example D:

Find the inverse $\mathcal{Z}$-transform of the function

$$F(z) = \frac{z^4 + 5}{(z-1)^2(z-2)}, \qquad |z| > 2.$$

Since

$$\frac{z^4 + 5}{(z-1)^2(z-2)} = z + 4 - \frac{1}{(z-1)^2} - \frac{5}{(z-1)} + \frac{16}{(z-2)}$$

we find with the aid of (34.7) and (34.8) that

$$\mathcal{Z}^{-1}\left[F(z)\right] = \delta(n+1) + 4\delta(n) - (n-1)u(n-2) - 5u(n-1) + (16)2^{n-1}u(n-1),$$

or

$$f(-1) = 1\,,\ f(0) = 4\,,\ f(1) = 11 \text{ and } f(n) = -n - 4 + (16)2^{n-1}\,,\ n \geq 2.$$

34.4 The Walsh Functions

In his 1923 seminal paper [238], J. Walsh introduced a complete system of orthonormal functions on the interval $I = [0, 1]$, which were later called the Walsh functions. Equally important to being a basis for $L^2(I)$ is their property that they assume only two values, ± 1, a property that has made them very valuable for digital circuits. The Walsh functions have been generalized in various directions, in one of which Chrestenson [38] constructed a complete orthonormal system of functions in $L^2(I)$ with the property that the functions take the values $\exp(2\pi i n/\alpha)$ for some integer α.

34.4.1 Definitions

The Walsh functions can be introduced in a number of different ways, the most recent of which is in terms of the Haar wavelets; see Example 1 in Section 32.5.

DEFINITION 34.3 *Let $\phi(t) = \chi_{[0,1]}(t)$ be the scalar function associated with the Haar mother wavelet, $\psi(t) = \phi(2t) - \phi(2t-1)$. The Walsh functions (see Figure 34.1) $W_n(t)$ are defined as $W_0(t) = \phi(t)$, $W_1(t) = \psi(t)$ and*

$$W_{2n}(t) = W_n(2t) + W_n(2t-1)\,,$$

and

$$W_{2n+1}(t) = W_n(2t) - W_n(2t-1)\,,$$

for $n = 1, 2, \cdots$.

Another definition of the Walsh functions can be given in terms of the Rademacher functions.

DEFINITION 34.4 *The Rademacher functions $\phi_n(t)$, $n = 0, 1, 2, \ldots$ are defined by*

$$\phi_0(t) = \chi_{[0,1]}(t) \quad \phi_n(t) = \phi_0(2^n t)\,.$$

The Rademacher functions form an orthonormal system in $L^2(I)$, but it is not a complete system.

DEFINITION 34.5 *The Walsh functions $W_n(t)$ can be defined as $W_0(t) = \phi_0(t)$ and*

$$W_n(t) = \prod_{k=1}^{m} a_k \phi_k(t)\,,$$

where $(a_1, a_2, \cdots, a_m)$ *is the binary representation of* n, *i.e.,*

$$n = \sum_{k=1}^{m} a_k 2^{k-1} = a_1 + a_2 2 + a_3 2^2 + \cdots + a_m 2^{m-1},$$

where $a_i = 0$ *or* 1 *for* $i = 1, 2, \cdots, m$.

For example,

$$W_7(t) = \phi_1(t)\phi_2(t)\phi_3(t) \quad , \qquad \text{and } W_{50}(t) = \phi_2(t)\phi_5(t)\phi_6(t)\,.$$

34.5 Basic Properties

The orthonormality of the Walsh functions can be easily established from Definition 34.3 by using induction on n and the fact that $\int_0^1 \psi(t)\,dt = 0$. It is also easy to see that if the Walsh functions are shifted so that they are defined on the interval $[-1/2\,,\,1/2]$, then W_{2n} is odd and W_{2n+1} is even ($n \geq 1$). This, together with other properties similar to those of the complex exponential functions e^{int}, $n = 0, \pm1, \pm2\ldots$, suggests the following definition

$$\text{cal}_n(t) = W_{2n+1}(t) \qquad \text{and} \qquad \text{sal}_n(t) = W_{2n}(t)\,,$$

where 'c' stands for cosine, 's' stands for sine, and 'al' stands for Walsh. In fact, we can reorder the Walsh functions so that the ones with even subscripts are even and the ones with odd subscripts are odd (cf. [92]).

The Walsh–Fourier series of a function $f(t)$ defined on $I = [0, 1]$ is given by

$$f(t) \sim \sum_{k=0}^{\infty} c_k W_k(t)\,,$$

where

$$c_k = \int_0^1 f(t) W_k(t)\,dt\,, \quad k = 0, 1, 2, \cdots.$$

Let us denote the nth partial sum of the *Walsh–Fourier series* of $f(t)$ by $S_n(t, f)$, i.e.,

$$S_n(x, f) = S_n(x) = \sum_{k=0}^{n} c_k W_k(x)\,.$$

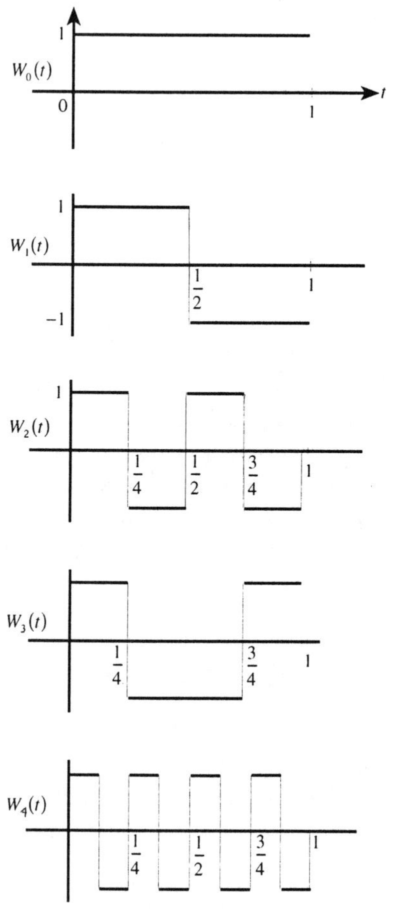

FIGURE 34.1
The first five Walsh functions

Thus, we have

$$S_n(x) = \int_0^1 f(t) D_n(x,t)\, dt\,, \tag{34.10}$$

where

$$D_n(x,t) = \sum_{k=0}^{n} W_k(x) W_k(t)\,.$$

Using Definition 34.5 and an induction argument, one can verify that

$$D_{2^n}(x,t) = \sum_{k=0}^{2^n-1} W_k(x) W_k(t) = \prod_{k=1}^{n} \left(1 + \phi_k(x)\phi_k(t)\right)\,.$$

For fixed $x \in [0,1]$ and a non-negative integer n, define $\alpha_n(x)$ and $\beta_n(x)$ by

$$\alpha_n(x) = \alpha_n = \frac{m}{2^n} \le x < \frac{m+1}{2^n} = \beta_n = \beta_n(x)\,.$$

If t is such that $|t-x| > 1/2^n$, then for some $1 \le k_0 \le n$, $\phi_{k_0}(x)$ and $\phi_{k_0}(t)$ have different signs; hence $(1 + \phi_{k_0}(x)\phi_{k_0}(t)) = 0$ and consequently $D_{2^n}(x,t) = 0$.

To this end, let $n = 2$ and assume for the moment that $x \in [0, 1/2]$. If $t \in (1/2, 1]$, then $\phi_1(x) = 1$ and $\phi_1(t) = -1$, but if $t \in [0, 1/2]$, then because $|t - x| > 1/4$, we must have either $x \in [0, 1/4]$, $t \in (1/4, 1/2]$, or $t \in [0, 1/4]$, $x \in (1/4, 1/2]$. In either case, $\phi_2(x)$ and $\phi_2(t)$ will have different signs, and hence $D_4(x,t) = 0$. The case in which $x \in (1/2, 1]$ can be treated similarly and the general case ($n \ge 2$) can be proved by induction.

It is also evident from the above argument that if $|t - x| \le 1/2^n$, then $D_{2^n}(x,t) = 2^n$. Thus, for sufficiently large n, we have

$$D_{2^n}(x,t) = 2^n \delta_{x,t}\,. \tag{34.11}$$

Let $F(x)$ be the anti-derivative of $f(x)$. Then, from (34.10) we have

$$\begin{aligned} S_{2^n}(x) &= 2^n \int_{\alpha_n}^{\beta_n} f(t)\, dt \\ &= 2^n \left(F(\beta_n) - F(\alpha_n)\right) = \frac{F(\beta_n) - F(\alpha_n)}{\beta_n - \alpha_n}\,, \end{aligned}$$

which implies that $S_{2^n}(x) \to f(x)$ almost everywhere. In particular, $S_{2^n}(x) \to f(x)$ at every point of continuity of $f(x)$. Ironically, the convergence behavior of $S_n(x)$, when n is not a power of 2, is more complicated. Walsh showed that if f is of bounded variation and x_0 is a point of continuity of f or a dyadic rational, then $S_n(x_0)$ converges, but if x_0 is neither, then $S_n(x_0)$ diverges.

34.6 The Discrete Walsh Transform

DEFINITION 34.6 *The discrete Walsh transform of a sequence* $\{f(k/2^n$ $k = 0, 1, 2, \cdots, 2^n - 1$ *is given by*

$$F_m = \frac{1}{2^n} \sum_{k=0}^{2^n-1} f\left(\frac{k}{2^n}\right) W_m\left(\frac{k}{2^n}\right), \quad m = 0, 1, 2, \cdots, \tag{34.12}$$

The discrete Walsh transform can be thought of as a mapping from a set of points in the time domain into a sequence of points in the so-called sequence domain.

By multiplying both sides of (34.12) by $W(j/2^n)$ and summing over m we obtain

$$\begin{aligned}\sum_{m=0}^{2^n-1} F_m W_m\left(\frac{j}{2^n}\right) &= \frac{1}{2^n} \sum_{m=0}^{2^n-1} \sum_{k=0}^{2^n-1} f\left(\frac{k}{2^n}\right) W_m\left(\frac{k}{2^n}\right) W_m\left(\frac{j}{2^n}\right) \\ &= \frac{1}{2^n} \sum_{k=0}^{2^n-1} f\left(\frac{k}{2^n}\right) \sum_{m=0}^{2^n-1} W_m\left(\frac{k}{2^n}\right) W_m\left(\frac{j}{2^n}\right),\end{aligned}$$

which, in view of (34.11), yields

The Inversion Formula of the Discrete Walsh Transform

$$f\left(\frac{j}{2^n}\right) = \sum_{m=0}^{2^n-1} F_m W_m\left(\frac{j}{2^n}\right).$$

Appendix A

Signal Analysis Terminology

In this appendix we give a mathematical description of the signal analysis terminology used throughout this book. By an analogue signal (or a signal for short), we mean a function, $f(t)$, $t \in \mathcal{R}$, whether real or complex-valued, and by a digital signal we mean a sequence of numbers, $\{f(n)\}$, where n runs over a set of integers whether finite or infinite. A signal $f(t)$ is said to be causal if $f(t) = 0$ for $t < 0$.

When a signal f, analogue or digital, is passed through a circuit or a system, certain operations or/and transformations are applied to it which result in producing an output signal g. Mathematically, this can be expressed as $g = L[f]$ for some operator L. Therefore, we may define a system as a transformation from a set $\mathcal{A}$ into a set $\mathcal{B}$. If the elements of $\mathcal{A}$ and $\mathcal{B}$ are functions, we say that the system is analogue or continuous, and if the elements of $\mathcal{A}$ and $\mathcal{B}$ are sequences, we say that the system is digital or discrete.

A system is said to be linear if

$$L\left[af_1 + bf_2\right] = aL[f_1] + bL[f_2],$$

and time-invariant if

$$L\left[f(t - t_0)\right] = g(t - t_0).$$

A distortionless transmission of a signal through a circuit or a system means that the exact shape of the input signal is produced at the output regardless of the changes in the amplitude (magnitude) and whether the signal is delayed in time (within reasonable limits). If the input and output signals are given in the time domain, that is they are functions in time, then a distortionless transmission can be described by the equation

$$g(t) = L\left[f(t)\right] = Af(t - t_0), \tag{1.1}$$

where A is a constant and L is a linear, time-invariant operator. The Fourier transform $\hat{f}(w)$ of a signal $f(t)$, which is called the amplitude spectrum of the signal, represents the signal's frequency content. Its importance lies in the fact that it allows engineers to view the signal as a sum of sinusoids of different frequencies via the inverse Fourier transform

$$f(t) = \frac{1}{\sqrt{2\pi}} \int_{-\infty}^{\infty} \hat{f}(\omega) e^{-it\omega} d\omega \,.$$

The energy of a signal is defined as

$$E = \int_{-\infty}^{\infty} |f(t)|^2 \, dt,$$

which, in view of Parseval's relation for the Fourier transform, can also be written as

$$E = \int_{-\infty}^{\infty} \left| \hat{f}(w) \right|^2 \, dw.$$

Thus, the space of finite-energy signals is nothing but the familiar space $L^2(\mathcal{R})$.

We may regard $f(t)$ and $\hat{f}(w)$ as two different representations of the same signal, one in the time domain and the other in the frequency domain. In cases where it is useful to view a signal in both time and frequency simultaneously, techniques other than the Fourier transform, such as the Zak, Gabor and wavelet transforms, are used.

A finite-energy signal is said to be band-limited if its amplitude spectrum (its Fourier transform) vanishes outside an interval of the form $(-\sigma, \sigma)$. The smallest such σ is called the band width of the signal. For a band-limited signal, we have

$$f(t) = \frac{1}{\sqrt{2\pi}} \int_{-\sigma}^{\sigma} \hat{f}(w) e^{-iwt} dw,$$

$\hat{f} \in L^2(-\sigma, \sigma)$.

Analogously, a signal is said to be time-limited if $f(t)$ vanishes outside some interval $(-T, T)$ $(T > 0)$. It is a fact that a signal cannot be both time-limited and band-limited simultaneously, unless it is identically zero.

The average power (energy) of a signal f over the interval (a, b) is defined as

$$\frac{1}{(b-a)} \int_{a}^{b} |f(t)|^2 \, dt,$$

and its average power (over the whole real line $\mathcal{R}$) is defined as

$$\tilde{E} = \lim_{T\to\infty} \frac{1}{2T} \int_{-T}^{T} |f(t)|^2 \, dt.$$

A signal is said to be of finite power if $\tilde{E}$ is finite.

A linear, time-invariant system can be described by a convolution in the form

$$g(t) = \int_{-\infty}^{\infty} f(x)h(t-x)\, dx\,,$$

where the function h is called the *impulse response* of the system. Its Fourier transform, $H(\omega)$, is called the *system transfer function*, or the system function for short. It is easy to see that h can also be defined as the response of the system to an input signal represented by the delta function, i.e., $h(t) = L[\delta(t)]$. In view of the convolution theorem for the Fourier transform, we obtain,

$$\hat{g}(w) = H(w)\hat{f}(w).$$

For the system described by Equation (1.1), $H(\omega) = Ae^{it_0\omega}$.

A system that converts an analogue signal into a digital signal is called an *analogue-digital* (A-D) converter and a *digital-analogue* (D-A) converter is the system that does the opposite. Mathematically, an analogue-digital converter can be represented by a discrete function transformation that maps a function into a sequence of numbers, and likewise a digital-analogue converter can be represented by a sequence transformation that maps a sequence of numbers into a function.

For an analogue system L_a with input $f(t)$, output $g(t)$ and impulse response $h(t)$, it is sometimes desirable to have a digital system L_d that simulates this analogue system. This means if the input signal of L_d is the sample values, $\{f(nT)\}$, of f at time $t = nT$ for some fixed number T, then its output is $\{g(nT)\}$, i.e.,

$$L_d\left[f(nT)\right] = g(nT)\,.$$

Such a system, if it exists, will be called a *digital simulator* of L_a.

A filter is a circuit or a system that has some frequency selective mechanism. Filters are usually classified according to their system transfer functions. The most common types are the low pass, high pass, band pass, and band stop. The system transfer functions of these filters are:

1. Low pass:

$$H(w) = \begin{cases} Ae^{it_0 w}, & \text{for } |w| \le w_1, \\ 0, & \text{otherwise,} \end{cases}$$

   where w_1 is the cut-off frequency

2. High pass:

$$H(w) = \begin{cases} Ae^{it_o w}, & \text{for } |w| \geq w_2, \\ 0, & \text{otherwise,} \end{cases}$$

3. Band pass:

$$H(w) = \begin{cases} Ae^{it_o w}, & \text{for } w_2 \leq |w| \leq w_1, \\ 0, & \text{otherwise,} \end{cases}$$

4. Band stop:

$$H(w) = \begin{cases} 0, & \text{for } w_2 \leq |w| \leq w_1, \\ Ae^{it_o w}, & \text{otherwise,} \end{cases}$$

Appendix B

Tables of Integral Transforms

The Fourier Transform

No.	$f(x)$	$F(w) = \int_{-\infty}^{\infty} f(x) e^{iwx} dx$
1)	$1, \quad \lvert x\rvert < a$ $0, \quad a < \lvert x\rvert$	$2 \sin aw/w$
2)	$a + x \quad -a \leq x \leq 0$ $a - x \quad 0 \leq x \leq a$ $0 \quad a \leq \lvert x\rvert,$	$\left[\frac{\sin(aw/2)}{(w/2)}\right]^2$
3)	$1,$	$(2\pi)\delta(w)$
4)	$x^n, \quad n = 0, 1, 2 \cdots$	$(2\pi)(-i)^n \delta^{(n)}(w)$
5)	$x^{-1},$	$i\pi \text{ sgn}(w)$
6)	$(a - ix)^{-\nu}, \text{Re}\, a > 0, \quad \text{Re}\, \nu > 0$	$2\pi w^{\nu-1} e^{-aw}/\Gamma(\nu) \quad y > 0$ $0 \quad y < 0$
7)	$\lvert x\rvert^{-\nu}, \quad 0 < \text{Re}\, \nu < 1$	$\pi[\Gamma(\nu)]^{-1} \sec(\pi\nu/2) \lvert w\rvert^{\nu-1}$
8)	$\text{sgn}(x),$	$2(i)/w$

The Fourier Transform

No.	$f(x)$		$F(w) = \int_{-\infty}^{\infty} f(x)e^{iwx}dx$
9)	$x\,\mathrm{sgn}(x)$,		$(-2)/w^2$
10)	e^{iax},	a real	$(2\pi)\delta(w+a)$
11)	$e^{-a\lvert x\rvert}$,	$\mathrm{Re}\,a > 0$	$(2a)(a^2+w^2)^{-1}$
12)	$e^{-a^2x^2}$,	$a > 0$	$\sqrt{\pi}a^{-1}e^{-w^2/4a^2}$
13)	$(1-x^2)P_n^{(\nu,\nu)}(x)\chi_{(-1,1)}(x)$,	$\mathrm{Re}\,\nu > -1$	$i^n 2^{\nu+\frac{1}{2}}(n!)^{-1}$ $\Gamma(n+\nu+1)J_{n+\nu+\frac{1}{2}}(w)$ $(-i)^m w^m$
14)	$\delta^{(m)}(x)$		$(-i)^m w^m$
15)	$\cos ax$		$\pi[\delta(w+a)+\delta(w-a)]$
16)	$\sin ax$		$-i\pi[\delta(w+a)-\delta(w-a)]$

The Fourier Cosine Transform

No.	$f(x)$		$F(w) = \int_0^{\infty} f(x)\cos(wx)dx, \quad w > 0$
1)	$x^{-1/2}$,		$\sqrt{\pi/(2w)}$
2)	$(x^2+a^2)^{-1}$,	$\mathrm{Re}\,a > 0$	$(\pi/2a)e^{-aw}$
3)	$x^{-\nu}$,	$0 < \mathrm{Re}\,\nu < 1$	$\pi\left[2\Gamma(\nu)\right]^{-1}\sec(\pi\nu/2)w^{\nu-1}$
4)	$x^{\nu-1}e^{-ax}$,	$\mathrm{Re}\,a > 0, \nu > 0$	$\Gamma(\nu)\left(a^2+w^2\right)^{-\nu/2}\cos[\nu\tan^{-1}(w/a)]$

The Fourier Cosine Transform

No.	$f(x)$		$F(w)=\int_0^\infty f(x)\cos(wx)dx, \quad w>0$
5)	$e^{-x}\,(\sin x/x)\,,$		$(1/2)\tan^{-1}(2/w^2)$
6)	$e^{-bx}\cos(ax)\,,$	$\mathrm{Re}\,b > \lvert\mathrm{Im}\,a\rvert$	$(b/2)\left[(b^2+(a+w)^2)^{-1}\right.$ $\left.+\,(b^2+(a-w)^2)^{-1}\right]$
7)	$e^{-bx}\sin(ax)\,,$	$a>0, \quad \mathrm{Re}\,b>0$	$(1/2)\left[(a+w)(b^2+(a+w)^2)^{-1}\right.$ $\left.+\,(a-w)(b^2+(a-w)^2)^{-1}\right]$
8)	$(\sin ax/x)\,,$	$a>0$	$\pi/2, \quad w<a$ $\pi/4, \quad w=a$ $0, \quad a<w$
9)	$\cos(ax^2)\,,$	$a>0$	$(1/4)\sqrt{2\pi/a}\left[\cos(w^2/4a)+\sin(w^2/4a)\right]$
10)	$\sin(ax^2)\,,$	$a>0$	$(1/4)\sqrt{2\pi/a}\left[\cos(w^2/4a)-\sin(w^2/4a)\right]$
11)	$e^{-ax^2}\,,$	$\mathrm{Re}\,a>0$	$\frac{1}{2}\sqrt{\pi}\,a^{-\frac{1}{2}}e^{-w^2/4a}$
12)	$x^\nu e^{-ax^2}\,,\ \mathrm{Re}\,a>0$	$\mathrm{Re}\,\nu>-1$	$\frac{1}{2}a^{-\frac{1}{2}(1+\nu)}\Gamma\left(\frac{1}{2}+\frac{\nu}{2}\right)\times$ ${}_1F_1\left(\frac{1}{2}+\frac{1}{2}\nu;\frac{1}{2};-y^2/4a\right)$
13)	$(1-\cos ax)/x^2\,,$	$\mathrm{Re}\,a>0$	$\pi(a-w)/2 \quad \text{if } y<a$ $0 \quad \text{if } a<y$
14)	$(\cosh x+\cos a)^{-\frac{1}{2}}\,,$	$-\pi<a<\pi$	$2^{-\frac{1}{2}}\pi\mathrm{sech}(\pi w)P_{-\frac{1}{2}+iw}(\cos a)$
15)	$e^{-a\cosh x}\,,$	$\mathrm{Re}\,a>0$	$K_{iw}(a)$
16)	$x^{-\nu}J_\nu(ax)\,,$	$\mathrm{Re}\,\nu>-\frac{1}{2}$	$\sqrt{\pi}(2a)^{-\nu}[\Gamma(\nu+\frac{1}{2})]^{-1}$ $\times(a^2-w^2)^{\nu-\frac{1}{2}}\chi_{(0,a)}(w)$

The Fourier Sine Transform

No.	$f(x)$	$F(w) = \int_0^\infty f(x)\sin(wx)dx, \quad w > 0$
1)	$x^{-\nu}$, $\quad 0 < \mathrm{Re}\,\nu < 2$	$w^{\nu-1}\Gamma(1-\nu)\cos(\nu\pi/2)$
2)	x^{-1},	$\pi/2$
3)	$x(x^2+a^2)^{-1}$, $\quad \mathrm{Re}\,a > 0$	$(\pi/2)e^{-aw}$
4)	$x^{-1}(a^2+x^2)^{-1}$, $\quad \mathrm{Re}\,a > 0$	$\frac{\pi}{2}a^{-2}(1-e^{-ay})$
5)	e^{-ax}, $\quad \mathrm{Re}\,a > 0$	$w(w^2+a^2)^{-1}$
6)	$x^{-1}e^{-ax}$, $\quad \mathrm{Re}\,a > 0$	$\tan^{-1}(w/a)$
7)	$x^{\nu-1}e^{-ax}$, $\quad \mathrm{Re}\,a > 0\,, \nu > -1$	$\Gamma(\nu)(w^2+a^2)^{-\nu/2}$ $\times \sin[\nu\tan^{-1}(w/a)]$
8)	xe^{-ax^2}, $\quad \lvert\arg\, a\rvert < \pi/2$	$(\sqrt{\pi}/4)a^{-3/2}we^{-w^2/(4a)}$
9)	$x^{-1}\sin(ax)$, $\quad a > 0$	$(1/2)\log\lvert(w+a)(w-a)^{-1}\rvert$
10)	$(1-x^2)^{-1}\sin(\pi x)$	$\sin w\,, \quad 0 \le w \le \pi$ $0\,, \quad \pi \le w.$
11)	$\tan^{-1}(x/a)$, $\quad a > 0$	$(\pi/2w)e^{-aw}$
12)	$x(a^2-x^2)^{\nu-\frac{1}{2}}\chi_{(0,a)}(x)$	$2^{\nu-1}\sqrt{\pi}a^{\nu+1}\Gamma\left(\nu+\frac{1}{2}\right)$ $\times w^{-\nu}J_{\nu+1}(aw)$
13)	$\sin ax/x^2$, $\quad \mathrm{Re}\,a > 0$	$\pi w/2\,, \quad$ if $0 < w < a$ $\pi a/2\,, \quad$ if $a < w < \infty$
14)	$\sinh(\pi x/2)\,K_{ix}(a)$, $\quad a > 0$	$\frac{\pi}{2}\sin(a\sinh w)$

Fractional Integrals (Riemann-Liouville)

No.	$f(x)$		$\frac{1}{\Gamma(\mu)}\int_0^y f(x)(y-x)^{\mu-1}dx$
)	$x^{\nu-1}$	Re $\nu>0$	$\Gamma(\nu)[\Gamma(\nu+\mu)]^{-1}y^{\mu+\nu-1}$
)	$x^{\nu-1}(x+a)^{\lambda}$,	Re $\nu>0$	$a^{\lambda}y^{\mu+\nu-1}\Gamma(\nu)[\Gamma(\mu+\nu)]^{-1}$ $\times{}_2F_1(-\lambda,\nu;\mu+\nu,-y/a)$
)	$x^{\nu-1}e^{ax}$,	Re $\nu>0$	$\Gamma(\nu)[\Gamma(\mu+\nu)]^{-1}y^{\mu+\nu-1}$ $\times{}_1F_1(\nu;\mu+\nu;ay)$
)	$x^{-\mu-1}\exp(-a/x)$	Re $\mu>0$	$a^{-\mu}y^{\mu-1}\exp(-a/y)$
)	$x^{-2\mu}\exp(-a/x)$	Re $\mu>0$	$\left(a^{\frac{1}{2}-\mu}/\sqrt{\pi y}\right)\exp\left(-a/(2y)\right)K_{\mu-\frac{1}{2}}\left(a/(2y)\right)$
)	$x^{\mu-1}\sin(ax)$	Re $\mu>0$	$\sqrt{\pi}(y/a)^{\mu-\frac{1}{2}}\sin(ay/2)J_{\mu-\frac{1}{2}}(ay/2)$
	$x^{\mu-1}\cos(ax)$	Re $\mu>0$	$\sqrt{\pi}(y/a)^{\mu-\frac{1}{2}}\cos(ay/2)J_{\mu-\frac{1}{2}}(ay/2)$

The Hankel Transform

o.	$f(x)$		$F(y)=\int_0^\infty f(x)\sqrt{yx}J_\nu(yx)dx,\ y>0$
	$x^{\nu+1/2}\chi_{(0,1)}(x)$	$\mathrm{Re}\,\nu>-1$	$y^{-1/2}J_{\nu+1}(y)$
	$x^{\nu+1/2}(a^2-x^2)^{\mu}\chi_{(0,a)}(x)$,	$\mathrm{Re}\,\nu,\mathrm{Re}\,\mu>-1$,	$2^{\mu}\Gamma(\mu+1)a^{\nu+\mu+1}y^{-\mu-1/2}$ $\times J_{\nu+\mu+1}(ay)$
	$x^{\nu+1/2}(x^2+a^2)^{-\nu-1/2}$,	$\mathrm{Re}\,\nu>-1/2$	$\sqrt{\pi}y^{\nu-1/2}2^{-\nu}e^{-ay}[\Gamma(\nu+1/2)]^{-1}$
	$x^{\nu+\frac{1}{2}}(x^2+a^2)^{-1}$,	$-1<$ Re $\nu<\frac{3}{2}$	$a^{\nu}y^{\frac{1}{2}}K_\nu(ay)$
	$x^{-\frac{1}{2}}e^{-ax^2}$,	Re $\nu>-1$	$2\sqrt{\pi y/a}\ \exp\left(-\frac{y^2}{8a}\right)I_{\nu/2}\left(\frac{y^2}{8a}\right)$

The Hankel Transform

No.	$f(x)$		$F(y) = \int_0^\infty f(x)\sqrt{yx}J_\nu(yx)dx,\ y>0$
6)	$x^{\nu+1/2}e^{-ax}$,	$\mathrm{Re}\,a>0$, $\mathrm{Re}\,\nu>-1$	$a(\pi)^{-1/2}2^{\nu+1}y^{\nu+1/2}\Gamma(\nu+3/2)$ $\times(a^2+y^2)^{-\nu-3/2}$
7)	$x^{\nu+1/2}e^{-ax^2}$,	$\mathrm{Re}\,a>0$, $\mathrm{Re}\,\nu>-1$	$y^{\nu+1/2}(2a)^{-\nu-1}\exp\left(-y^2/4a\right)$
8)	$e^{-ax}/\sqrt{x}$,	$\mathrm{Re}\,a>0$, $\mathrm{Re}\ \nu>-1$	$y^{-\nu+1/2}\left[\sqrt{(a^2+y^2)}-a\right]^\nu$ $\times(a^2+y^2)^{-1/2}$
9)	$x^{-\nu-1/2}\cos(ax)$,	$\mathrm{Re}\,\nu>-1/2$	$\sqrt{\pi}2^{-\nu}y^{-\nu+1/2}[\Gamma(\nu+1/2)]^{-1}$ $\times(y^2-a^2)^{\nu-1/2}\chi_{(a,\infty)}(y)$
10)	$x^{1/2-\nu}\sin(ax)$,	$\mathrm{Re}\,\nu>1/2$	$a2^{1-\nu}\sqrt{\pi}y^{\nu+1/2}[\Gamma(\nu-1/2)]^{-1}$ $\times(y^2-a^2)^{\nu-3/2}\chi_{(a,\infty)}(y)$
11)	$x^{-1/2}J_{\nu-1}(ax)$,	$\mathrm{Re}\,\nu>-1$	$a^{\nu-1}y^{-\nu+1/2}\chi_{(a,\infty)}(y)$
12)	$x^{-1/2}J_{\nu+1}(ax)$,	$\mathrm{Re}\,\nu>-3/2$	$a^{-\nu-1}y^{\nu+1/2}\chi_{(0.a)}(y)$
13)	$x^{\nu-\mu+\frac{1}{2}}J_\mu(ax)$,	$-1<\mathrm{Re}\ \nu<\mathrm{Re}\ \mu$	$2^{\nu-\mu+1}y^{\nu+\frac{1}{2}}\left[\Gamma(\mu-\nu)a^\mu\right]^{-1}$ $\times(a^2-y^2)^{\mu-\nu-1}\chi_{(0,a)}(y)$
14)	$\sqrt{x}\ e^{-\beta x^2}J_\nu(ax)$,	$\mathrm{Re}\ \beta>0$	$(\sqrt{y}/2\beta)\exp\left(-\frac{a^2+y^2}{4\beta}\right)I_\nu\left(ay/2\beta\right)$
15)	$x^{\mu+\nu+\frac{1}{2}}K_\mu(ax)$,	$\nu+1>\lvert\mathrm{Re}\ \mu\rvert$	$2^{\nu+\mu}a^\mu\Gamma(\mu+\nu+1)y^{\nu+\frac{1}{2}}$ $\times(y^2+a^2)^{-\mu-\nu-1}$

The Hilbert Transform

No.	$f(x)$		$F(y) = (\pi)^{-1} \int_{-\infty}^{\infty} f(x)(x-y)^{-1} dx$
1)	1,		0
2)	$\chi_{(a,b)}(x)$		$\frac{1}{\pi} \log \left\|(b-y)(a-y)^{-1}\right\|$
3)	$x^{-1}\chi_{(a,\infty)}(x)$		$(\pi y)^{-1} \log \left\|a(a-y)^{-1}\right\|$, $0 \neq y \neq a, \quad a > 0$
4)	$(x+a)^{-1}$,	$\operatorname{Im} a > 0$	$i(y+a)^{-1}$
5)	$\sin(ax)$,	$a > 0$	$\cos(ay)$
6)	$\sin(ax)/x$,	$a > 0$	$[\cos(ay) - 1]/y$
7)	$\cos(ax)$,	$a > 0$	$-\sin(ay)$
8)	$[1 - \cos(ax)]/x$,	$a > 0$	$[\sin(ay)]/y$
9)	$\operatorname{sgn}(x) \sin(a\|x\|^{1/2})$	$a > 0$	$\cos(a\|y\|^{1/2}) + \exp(-a\|y\|^{1/2})$
10)	e^{iax}	$a > 0$	ie^{iay}
11)	$(x^2 + a^2)^{-1}$		$-(y/a)(y^2 + a^2)^{-1}$
12)	$x(x^2 + a^2)^{-1}$		$a(y^2 + a^2)^{-1}$
13)	$(\lambda x + \mu a)(x^2 + a^2)^{-1}$		$(\lambda a - \mu y)(y^2 + a^2)^{-1}$
14)	$\ln \|(b-x)/(x-a)\|$,	$a < b$	$-\pi \chi_{(a,b)}(y)$
15)	$\sin(ax) J_n(bx)$,	$0 < b < a$	$\cos(ay) J_n(by)$
16)	$\cos(ax) J_n(bx)$	$0 < b < a$	$-\sin(ay) J_n(by)$
17)	$\|x\|^{\nu} Y_{\nu}(a\|x\|)$,	$-\frac{1}{2} < \operatorname{Re} \nu < \frac{3}{2}$	$\operatorname{sgn}(y) \|y\|^{\nu} J_{\nu}(a\|y\|)$
18)	$\operatorname{sgn}(x) \|x\|^{-\nu} \mathbf{H}_{\nu}(a\|x\|)$	$\operatorname{Re} \nu > -\frac{3}{2}$	$\|y\|^{-\nu} J_{\nu}(a\|y\|)$

The Kontorovich-Lebedev Transform

No.	$f(x)$		$\int_0^\infty f(x)K_{ix}(y)dx \quad y>0$
1)	$x\sin ax$	$\lvert\text{Im}\, a\rvert < \pi/2$	$\frac{\pi}{2}y\sinh a\exp(-y\cosh a)$
2)	$\cos ax$	$\lvert\text{Im}\, a\rvert < \pi/2$	$\frac{\pi}{2}\exp(-y\cosh a)$
3)	$x\tanh(\pi x)K_{ix}(\beta)$	$\lvert\arg\beta\rvert < \pi$	$\frac{\pi}{2}\sqrt{\beta y}e^{-\beta-y}(\beta+y)^{-1}$
4)	$x\sin\left(\frac{\pi x}{2}\right)K_{ix/2}(a)$	$\lvert\arg a\rvert < \pi/2$	$\pi^{\frac{3}{2}}y(2a)^{-1/2}\exp\left(-a-y^2/8a\right)$
5)	$\cosh(ax)K_{ix}(\beta)$		$\frac{\pi}{2}K_0\left[\sqrt{y^2+\beta^2+2\beta y\cos a}\right]$
6)	$x\tanh(\pi x)P_{-\frac{1}{2}+ix}(z)$		$\sqrt{\pi y/2}\, e^{-yz}$

The K Transform

No.	$f(x)$		$\int_0^\infty f(x)K_\nu(xy)\sqrt{xy}\,dx$
1)	$x^{\nu+\frac{1}{2}}\chi_{(a,\infty)}(x)$		$a^{\nu+1}y^{-\frac{1}{2}}K_{\nu+1}(ay)$
2)	$x^{-\frac{1}{2}}e^{-ax}$	$-1<\text{Re}\,\nu<1$	$(\pi/\sqrt{y})\sin(\nu\theta)[\sin(\nu\pi)\sin\theta]^{-1}$, $\cos\theta = a/y,\ \text{Re}\,(a+y)>0$
3)	$x^{-\frac{1}{2}}e^{-ax^2}$	$-1<\text{Re}\,\nu<1$	$\frac{1}{4}\sec\left(\frac{\nu\pi}{2}\right)\left(\frac{\pi y}{a}\right)^{\frac{1}{2}}\exp\left(\frac{y^2}{8a}\right)K_{\nu/2}\left(\frac{y^2}{8a}\right)$
4)	$x^{-\frac{1}{2}}K_\nu(a/x)$	$\text{Re}\,a>0$	$\frac{\pi}{\sqrt{y}}K_{2\nu}\left(2\sqrt{ay}\right)$
5)	$x^{\frac{1}{2}}\mathbf{H}_\nu(ax)$,	$\text{Re}\,\nu > -\frac{3}{2}$	$a^{\nu+1}y^{-\nu-\frac{1}{2}}(y^2+a^2)^{-1}$, $\quad\text{Re}\,y > \lvert\text{Im}\,a\rvert$
6)	$x^{-\frac{1}{2}}J_{2\nu}\left(a\sqrt{x}\right)$ $+x^{-\frac{1}{2}}I_{2\nu}\left(a\sqrt{x}\right)$		$(\pi/\sqrt{y})I_\nu\left(a^2/4y\right)$

The Laplace Transform

No.	$f(t)$		$F(s)=\int_0^\infty f(t)e^{-st}\,dt$, $\mathrm{Re}\, s>0$
1)	1 ,		$1/s$,
2)	t^ν ,	$\mathrm{Re}\,\nu>-1$	$\Gamma(\nu+1)/s^{\nu+1}$
3)	0 , 1 , 0 ,	$0<t<a$ $a<t<b$ $b<t$	$(e^{-as}-e^{-bs})/s$
4)	n ,	$nb<t<(n+1)b$	$s^{-1}(e^{bs}-1)^{-1}$
5)	$t^{\nu-1}e^{-at}$,	$\mathrm{Re}\,\nu>0$	$\Gamma(\nu)(s+a)^{-\nu}$
6)	$t^{-3/2}e^{-a/4t}$,	$\mathrm{Re}\,a>0$	$2\sqrt{\pi}a^{-1/2}e^{-\sqrt{as}}$
7)	$\sin(at)$,		$a(s^2+a^2)^{-1}$
8)	$t^{-1}\sin(at)$,		$\tan^{-1}(a/s)$
9)	$\cos(at)$,		$s(s^2+a^2)^{-1}$
10)	$\sinh(at)$,		$a(s^2-a^2)^{-1}$
11)	$\cosh(at)$,		$s(s^2-a^2)^{-1}$
12)	$\sin(2\sqrt{at})$		$\sqrt{a\pi}s^{-3/2}e^{-a/s}$
13)	$(1-\cos at)/t$		$\frac{1}{2}\ln(1+a^2/s^2)$, $\mathrm{Re}\,s>\lvert\mathrm{Im}\,a\rvert$
14)	$t^{-\frac{1}{2}}\cos\left(2\sqrt{at}\right)$		$\sqrt{\pi}\ s^{-\frac{1}{2}}e^{-a/s}$
15)	$\sinh\left(2\sqrt{at}\right)$		$\sqrt{\pi a}\ s^{-\frac{3}{2}}e^{a/s}$
16)	$\delta^{(n)}(t-a)$		$s^n e^{-as}$

The Mellin Transform

No.	$f(x)$		$\int_0^\infty f(x)x^{s-1}dx$	
1)	$(a+x)^{-1}$		$\pi a^{s-1}\csc(\pi s)$,	$0 < \operatorname{Re} s < 1$
2)	$[(a+x)(b+x)]^{-1}$		$\pi\csc(\pi s)(a^{s-1}-b^{s-1})/(b-a)$	$0 < \operatorname{Re} s < 2$
3)	$(1+ax)^{-\nu}$		$a^{-s}B(s,\nu-s)$,	$0 < \operatorname{Re} s < \operatorname{Re}\nu$
4)	e^{-ax},	$a>0$	$a^{-s}\Gamma(s)$	$\operatorname{Re} s > 0$
5)	$(e^{ax}+1)^{-1}$,	$a>0$	$a^{-s}\Gamma(s)(1-2^{1-s})\zeta(s)$,	$\operatorname{Re} s > 0$
6)	$(e^{ax}-1)^{-1}$,	$a>0$	$a^{-s}\Gamma(s)\zeta(s)$,	$\operatorname{Re} s > 1$
7)	$e^{-x}(\ln x)^n$		$\frac{d^n}{ds^n}\Gamma(s)$	$\operatorname{Re} s > 0$
8)	$\ln(1+ax)$		$\pi s^{-1}a^{-s}\csc(\pi s)$,	$-1 < \operatorname{Re} s < 0$
9)	$\ln\left\|\frac{1+x}{1-x}\right\|$		$\pi s^{-1}\tan(\pi s/2)$	$-1 < \operatorname{Re} s < 1$
10)	$\sin(ax)$		$a^{-s}\Gamma(s)\sin(\pi s/2)$,	$-1 < \operatorname{Re} s < 1$
11)	$\cos(ax)$		$a^{-s}\Gamma(s)\cos(\pi s/2)$,	$0 < \operatorname{Re} s < 1$
12)	$\operatorname{sech}^2(ax)$		$4a^{-s}(1-s^{2-s})\Gamma(s)2^{-s}\zeta(s-1)$,	$\operatorname{Re} s > 0$
13)	$e^{-x}\operatorname{sech} x$		$(1-2^{1-s})\Gamma(s)2^{1-s}\zeta(s)$,	$\operatorname{Re} s > 0$
14)	$J_\nu(ax)$	$a>0$	$2^{s-1}a^{-s}\Gamma(\nu/2+s/2)$ $\times[\Gamma(\nu/2-s/2+1)]^{-1}$	$-\nu < \operatorname{Re} s$ $\operatorname{Re} s < 3/2$
15)	$Y_\nu(ax)$	$a>0$	$-2^{s-1}\pi^{-1}a^{-s}\Gamma(s/2+\nu/2)$ $\times\Gamma(s/2-\nu/2)\cos[(s-\nu)\pi/2]$	$-\nu < \operatorname{Re} s$ $\operatorname{Re} s < 3/2$

The Stieltjes Transform

No.	$f(x)$		$\int_0^\infty f(x)/(x+y)dx$
1)	$(a+x)^{-1}$		$(a-y)^{-1}\ln(a/y)$
2)	$(a^2+x^2)^{-1}$		$(a^2+y^2)^{-1}\left[\pi y/2a - \ln(y/a)\right]$
3)	$x(a^2+x^2)^{-1}$		$(a^2+y^2)^{-1}\left[\pi a/2 + y\ln(y/a)\right]$
4)	$x^{\nu}(a+x)^{-1}$	$-1 < \mathrm{Re}\,\nu < 1$	$\pi(a^{\nu}-y^{\nu})/\left((a-y)\sin(\nu\pi)\right)$
5)	x^{ν}	$-1 < \mathrm{Re}\,\nu < 0$	$-\pi y^{\nu}\csc(\pi\nu)$
6)	$\sin(a\sqrt{x})$		$\pi\exp\left(-a\sqrt{y}\right)$
7)	$x^{-1}\sin(a\sqrt{x})$		$\pi y^{-1}\left[1-\exp\left(-a\sqrt{y}\right)\right]$
8)	$x^{\frac{\nu}{2}+k}J_{\nu}(a\sqrt{x})$,	$-k-1 < \mathrm{Re}\,\nu$ $\mathrm{Re}\,\nu < -2k+\frac{3}{2}$	$2(-1)^k y^{\frac{\nu}{2}+k}K_{\nu}\left(a\sqrt{y}\right)$
9)	$x^{\nu/2+1/2}Y_{\nu}(a\sqrt{x})$	$-\frac{3}{2} < \mathrm{Re}\,\nu < \frac{1}{2}$	$2y^{\nu/2+1/2}K_{\nu}(a\sqrt{y})$
10)	$x^{\frac{\nu}{2}-\frac{1}{2}}\sin(a\sqrt{x})\,J_{\nu}(b\sqrt{x})$,	$-1 < \mathrm{Re}\,\nu < \frac{3}{2}$	$2y^{\frac{\nu}{2}-\frac{1}{2}}\sinh\left(a\sqrt{y}\right)K_{\nu}\left(b\sqrt{y}\right)$
11)	$x^{-\frac{1}{2}}e^{-ax}K_{\nu}(ax)$,	$-\frac{1}{2} < \mathrm{Re}\,\nu < \frac{1}{2}$	$\pi\sec(\nu\pi)y^{-\frac{1}{2}}e^{ay}K_{\nu}(ay)$
12)	e^{-ax}	$\mathrm{Re}\,a > 0$	$-e^{-ay}\mathrm{Ei}(-ay)$
13)	$x^{-\nu}e^{-ax}$	$\mathrm{Re}\,\nu < 1$	$\Gamma(1-\nu)\Gamma(\nu,ay)y^{-\nu}e^{ay}$

The Generalized Stieltjes Transform

No.	$f(x)$	$\int_0^\infty f(x)(x+y)^{-\rho}dx$
1)	$x^{\nu-1}$	$\frac{\Gamma(\nu)\Gamma(\rho-\nu)}{\Gamma(\rho)}y^{\nu-\rho}, \quad \text{Re }\rho > \text{Re }\nu$
2)	$x^{\nu-1}(a+x)^{-\mu}$	$\frac{\Gamma(\nu)\Gamma(\mu-\nu+\rho)y^{\nu-\rho}}{\Gamma(\mu+\rho)a^{\mu}}{}_2F_1(\mu,\nu;\mu+\rho;1-\frac{y}{a})$ $\text{Re }\rho > \text{Re}(\nu-\mu)$
3)	$x^{-\rho}e^{-ax}$	$\frac{\Gamma(1-\rho)}{\sqrt{\pi}}(a/y)^{\rho-\frac{1}{2}}e^{ay/2}K_{\rho-\frac{1}{2}}(ay/2)$
4)	$x^{-\frac{1}{2}}\cos(a\sqrt{x})$	$\frac{2\sqrt{\pi}}{\Gamma(\rho)}\left(2\sqrt{y}/a\right)^{\frac{1}{2}-\rho}K_{\rho-\frac{1}{2}}\left(a\sqrt{y}\right)$
5)	$\sin(a\sqrt{x})$	$\frac{2\sqrt{\pi y}}{\Gamma(\rho)}\left(2\sqrt{y}/a\right)^{\frac{1}{2}-\rho}K_{\rho-\frac{3}{2}}\left(a\sqrt{y}\right)$
6)	$x^{\frac{\nu}{2}}J_\nu(a\sqrt{x})$	$\frac{a^{\rho-1}}{2^{\rho}\Gamma(\rho)}y^{\frac{\nu}{2}+\frac{1}{2}-\frac{\rho}{2}}K_{\nu-\rho+1}\left(a\sqrt{y}\right)$
7)	$x^{\lambda}e^{ax}K_\nu(ax), \text{Re}\,\lambda > \lvert\text{Re}\,\nu\rvert - 1$	$\cos(\nu\pi)y^{\lambda+1-\rho}[\sqrt{\pi}\,\Gamma(\rho)]^{-1}\times$ $G^{3,2}_{2,3}\left(2ay\,\middle\vert\,\begin{matrix}-\lambda,\frac{1}{2}\\ \rho-\lambda-1,\nu,-\nu\end{matrix}\right)$ $\text{Re}\,\rho > \text{Re}\,\lambda+\frac{1}{2}$

Appendix C

List of Symbols

Symbol	Description	Page
$(a)_n$	Pochhammer symbol	p. 49
$\mathcal{A}[f]$	The Bargmann transform of f	p.462
$\mathcal{A}_{\alpha,n}[f]$	The Abel transform of f, $n = 1, 2$	p. 338
$\mathcal{A}_{f,g}$	The cross-ambiguity function	p. 508
B_σ^p	The set of generalized band-limited functions	p. 7
$B(\alpha, \beta)$	The beta-function	p. 45
$\mathcal{B}_{f,g}(u, v)$	A modified cross-ambiguity function	p. 508
$B_{\alpha,\sigma}^{p,w}$	A class of entire functions	p. 306
$\mathrm{Cas}\ (x)$	$\cos x + \sin x$	p. 266
$\mathcal{C}$	The set of complex numbers	p. 1
$d(n)$	The number of divisors of n	p. 48
D_a	The dilation (scaling) operator	p. 2
$D_\nu(x)$	The parabolic cylinder function	p. 58
$\mathcal{D}_K(I)$	The space of infinitely differentiable functions with support in a compact subset K of an open set I	p. 80
$\mathcal{D}(I)$	The space of infinitely differentiable functions with support in an open set I	p. 84
$\mathcal{D}$	The space of infinitely differentiable functions with compact support	p. 83
E_a	The modulation (exponentiation) operator	p. 2
$Exp(\mathcal{C})$	The space of entire functions of exponential type	p. 171
$Exp(\mathcal{C}; K)$	A subspace of the space entire functions of exponential type	p. 172
$E_\alpha(z)$	The Mittag–Leffler function	p. 48
$E_{a,b}(z)$	The generalized Mittag–Leffler function	p. 48
$E(p; a_r : q; b_s : z)$	MacRobert's E function	p. 68
$E_{q,l+1}^{p,m}$	The generalized MacRobert E function	p.454

$e^{a(d/dx)}$	A generalized translation operator	p. 309
$\mathrm{Erf}(x)$	The error function	p. 57
$\mathcal{E}(I)$	The space of infinitely differentiable functions on an open set I	p. 80
$\mathcal{E}$	The space of infinitely differentiable functions	p. 80
$\mathcal{F}[f] = \hat{f}$	The Fourier transform of f	p. 222
$\mathcal{G}[f]$	The Gabor transform of f	p. 492
F_g	The Gabor transform of f with repect to g	p. 492
$G^{m,n}_{p,q}$	The G function	p. 69
${}_pF_q$	The generalized hypergeometric function	p. 49
$H(x)$	The Heaviside function	p. 2
$H_n(x)$	The Hermite polynomial of degree n	p. 62
$\mathbf{H}_\nu(x)$	Struve's function	p. 67
$\mathcal{H}_\mu[f]$	The Hankel transform of f	p. 372
$\mathcal{H}(G)$	The space of all holomorphic functions on an open set G	p. 81
$I_\mu(x)$	The modified Bessel function of the first kind	p.63
$\mathcal{I}_\mu[f]$	The I transform of f	p. 373
$I^{\alpha,\eta}[f]$	The Erdélyi–Kober fractional integral operator	p. 349
$I^{\alpha,\eta}_m[f]$	The Erdélyi–Kober-type fractional integral operator	p. 349
$J_\mu(x)$	The Bessel function of the first kind and order μ	p. 63
$K_\mu(x)$	The modified Bessel function of the third kind	p. 64
$\mathcal{K}_\mu[f]$	The K (Meijer) transform of f	p. 397
$K^{\alpha,\eta}[f]$	The second Erdélyi–Kober fractional integral operator	p. 349
$K^{\alpha,\eta}_m[f]$	The second Erdélyi–Kober-type fractional integral operator	p. 349
$L^\alpha_n(x)$	The generalized Laguerre polynomial of degree n	p. 61
$L^\alpha_t(x)$	The Laguerre functions of index t	p. 55
$\mathcal{L}[f] = (\mathcal{L}f)$	The Laplace transform of f	p. 161
$\mathcal{L}_{a,b}$	A testing-function space	p. 81
$\mathcal{L}^*_{a,b}$	A space of generalized functions	p. 81
$\mathcal{L}_(w,z)$	A testing-function space	p. 85
$\mathcal{L}_(w,z)^*$	A space of generalized functions	p. 85
$L^p(I,\mu)$	The space of p-integrable functions with respect to a measure $d\,\mu$	p. 2
$L^p_{loc}(I,\mu)$	The space of locally p-integrable functions with respect to a measure $d\,\mu$	p. 3
$M_{\mu,\nu}(x)$	A Whittaker's function	p. 56
$\mathcal{M}[f]$	The Mellin transform of f	p. 201

$P^k[f]$	The k-plane transform of f	p. 561
$P_t(x)$	The Legendre function	p. 52
$P_t^{\mu}(x)$	The associated Legendre function	p. 52
$P_t^{(\alpha,\beta)}(x)$	The Jacobi function	p. 52
$P_n^{(\alpha,\beta)}(x)$	The Jacobi polynomial of degree n	p. 59
$\mathcal{P}_T$	The space of all infinitely differentiable, periodic functions with period T	p. 80
$\mathcal{P}_T^*$	The space of all periodic distributions with period T	p. 89
$Q_t^{\mu}(x)$	The associated Legendre function of the second kind	p. 52
$\mathcal{R}$	The set of real numbers	p. 1
$R[f] = f^{\dagger}$	The Radon transform of f	p. 554
$S[f]$	The Stieltjes transform of f	p. 175
$S_{\rho}[f]$	The generalized Stieltjes transform of f	p. 175
$\mathcal{S}$	The Schwartz space of infinitely differentiable, rapidly decreasing functions	p. 80
$\mathbf{S}$	The Bargmann–Segal–Foch space	p. 462
$s_{\nu,\mu}$	Lommel's function	p. 67
T_a	A translation operator	p. 2
$W_n(t)$	Walsh functions	p. 592
$W_c[f]$	The Weierstrass transform of f	p. 330
$W_{\mu,\nu}(x)$	A Whittaker's function	p. 56
$W_{\psi}[f]$	The wavelet transform of f with respect to the mother wavelet ψ	p. 532
$W_{f,g}(u,v)$	The Wigner distribution of f and g	p. 509
$Y_{\mu}(x)$	The Bessel function of the second kind	p. 63
$\mathcal{Y}_{\mu}[f]$	The Y transform of f	p. 373
$\mathbf{Y}(x)$	Young's function	p. 393
$Z_a[f](t,w)$	The Zak transform of f	p. 474
$\delta_{m,n}$	The Kronecker delta	p. 2
$\delta(x)$	The Dirac-delta function	p. 87
$\Phi_m[f]$	The Mehler–Fock transform of f	p. 416
$\Phi(a;c;x)$	A confluent hypergeometric function	p. 54
γ	Euler's constant	p. 44
$\gamma(a,x)$	The incomplete gamma-function	p. 56
$\Gamma(x)$	The gamma-function	p. 43
$\mu(n)$	The Möbius function	p. 47
$\psi_{a,b}(x)$	Wavelets generated by a mother wavelet	p. 532
$\Psi(a;c;x)$	A confluent hypergeometric function	p. 54
$\sigma_k(n)$	The sum of the kth powers of the divisors of n	p. 48
$\chi_A(x)$	The characteristic function of a set A	p. 2
$\zeta(x)$	The Riemann-zeta function	p. 47
$\zeta(s,\nu)$	The generalized Riemann-zeta function	p. 46

References

[1] N. H. Abel, *Solution de quelques problémes á l'aide d'integrales définites, oeuvres complétes,* Vol. 1, Grondahl Chistiamia, Norway (1881), pp. 16—18.

[2] L. Andrews and B. Shivamoggi, *Integral Transforms for Engineers and Mathematicians,* Macmillan, New York (1988).

[3] N. Aronszajn, Theory of reproducing kernels, *Trans. Amer. Math. Soc.*, Vol. 68 (1950), pp. 337—404.

[4] S. C. Arya, Abelian theorems for a generalized Stieltjes transform, *Bull. Unione Mat. Ital.*, Vol 1(1958), pp. 497—504.

[5] L. Auslander and R. Tolimieri, Radar ambiguity functions and group theory, *SIAM J. Math. Anal.,* Vol. 16(1985), pp. 577—601.

[6] V. Bargmann, P. Butera, L. Girardello, and J. R. Klauder, On the completeness of the coherent states, *Rep. Math. Phys.*, Vol. 2 (1971), pp. 221—228.

[7] V. Bargmann, On a Hilbert space of analytic functions and an associated integral transform, Part II, A family of related function spaces application to distribution theory, *Comm. Pure Appl. Math.*, Vol.20 (1967), pp. 1—101.

[8] V. Bargmann, On a Hilbert space of analytic functions and an associated integral transform, Part I, *Comm. Pure Appl. Math.*, Vol.14 (1961), pp. 187—214.

[9] G. Battle, A block spin construction of ondelettes, Part I, Lemarié functions, *Comm. Math. Phys.,* Vol. 110 (1987), pp. 601—615.

[10] J. Benedetto and M. Frazier, *Waveleys: Mathematics and Applications,* CRC Press, Boca Raton, FL (1994).

[11] J. Benedetto and D. Walnut, Gabor frames for L^2 and related spaces, in *Wavelets: Mathematics and Applications*, J. Benedetto and M. Frazier, Editors, CRC Press, Boca Raton, FL (1993).

[12] J. J. Benedetto, Analytic representation of generalized functions, *Math. Zeitschr.*, 97 (1967), pp. 303—319.

[13] A. Bennett, Higher wave number spectrum of passive scalar in isotropic turbulence, *Phys. Fluids,* Vol. 29(1986), pp. 1734—1735.

[14] C. Berenstein and D. Walnut, Local inversion of the Radon transform in even dimensions using wavelets, *75 years of Radon Transform,* S. Gindikin and P. Michor, Editors, International Press, Cambridge, Mass. (1994), pp. 45—69.

[15] R. Boas, *Entire Functions*, Academic Press, New York (1954).

[16] R. Boas, Inversion of a generalized Laplace integral, *Proc. Natl. Acad. Sci.*, USA, 28 (1942), pp. 21—24.

[17] R. Boas, Generalized Laplace integrals, *Bull. Amer. Math. Soc.*, 48 (1942), pp. 286—294.

[18] R. Boas, Jr., Some theorems on Fourier transforms and conjugate trigonometric integrals, *Trans. Amer. Math. Soc.,* Vol. 40 (1936), pp. 287—308.

[19] S. Bochner, A generalization of Poisson's summation formula, *Duke Math. J.,* Vol. 6 (1940), pp. 229—234.

[20] T. Boehme, The support of Mikusinski operators, *Trans. Amer. Math. Soc.,* Vol. 176 (1973), pp. 319—334.

[21] E. Borel, *Lecons sur les functions entieres,* Deuxiéme édition revue et augmentee d'une note de G. Valiron, Gauthier-Villars, Paris (1921).

[22] E. Borel, *Lecons sur les Séries divergentes,* Gauthier-Villars, Paris (1901).

[23] R. Bracewell, *The Fourier Transform and Its Applications,* McGraw-Hill, New York (1965).

[24] R. Bracewell, *The Hartley Transform*, Oxford University Press, Oxford, U.K. (1986).

[25] R. Bracewell, H. Bartelt, A. Lohmann, and N. Streibl, Optical synthesis of Hartley transform, *Applied Optics,* Vol. 24 (1985), pp. 1401—1402.

[26] B. Braaksma and B. Meulenbeld, Integral transforms with generalized Legendre functions as kernels, *Compositio Math.,* Vol. 18 (1967), pp. 235—287.

[27] H. Bremermann, *Distribution, Complex Variables and Fourier Transforms,* Addison-Wesley, New York (1965).

[28] W. Brenke, An application of Abel's integral equation, *Amer. Math. Monthly,* Vol. 29 (1922), pp. 58—60.

[29] W. Briggs and V. Henson, *The DFT: An Owner's Manual for the Discrete Fourier Transform,* Soc. for Indust. Appl. Math., Philadelphia (1995).

[30] U. Brosa and H. J. Krappe, On the dynamics of colliding droplets, *Quart. J. Mech. Appl. Math.*, 33(1980), pp. 159—177.

[31] N.G. de Bruijn, A theory of generalized functions with applications to Wigner distribution and Weyl correspondence, *Nieuw. Archief voor Wiskunde,* Vol. 21 (1973), pp. 205—280.

[32] Yu. A. Brychkov, H. Glaeske, A. Prudnikov and Vu. K. Tuan, *Multidimensional Integral Transformations*, Gordon and Breach, New York (1992).

[33] Yu. A. Brychkov and A. Prudnikov, *Integral Transforms of Generalized Functions*, Gordon and Breach, New York (1989).

[34] G. Buggle, *Die Kontorovich–Lebedev-Transformation und die Mehler–Fock Transformation verallgemeinerten Funcktionen*, Dissertation, TH. Darmstadt 1977.

[35] E. Cahen, Sur la fonction $\zeta(s)$ de Riemann et sur des fonctions analogues, *Ann. Ec. Norm.,* Vol. 3 (1894), pp. 75—164.

[36] F. M. Cholewinski and D. T. Haimo, Inversion of the Hankel potential transform, *Pacific J. Math.*, Vol. 37 (1971), pp. 319—330.

[37] F. M. Cholewinski and D. T. Haimo, Weierstrass–Hankel convolution transform, *J. Anal. Math.*, Vol 17 (1966), pp. 1—58.

[38] H. Chrestenson, A class of generalized Walsh functions, *Pacific J. Math.,* Vol. 5 (1955), pp. 17—31.

[39] C. Chui, *An Introduction to Wavelets,* Academic Press, New York (1992).

[40] R. Churchill, *Operational Mathematics,* McGraw-Hill, New York (1972).

[41] P. Civin, Inequalities for trigonometric integrals, *Duke Math. J.*, Vol. 8 (1941), pp. 656—665.

[42] T. A. Claasen and W. F. Mecklenbräuker, The Wigner distribution-A tool in time-frequency analysis, Part I. Continuous time signals, *Philips J. Research*, Vol. 35 (1980), pp. 217—250.

[43] E. Coddington and N. Levinson, *Theory of Ordinary Differential Equations*, McGraw-Hill, New York (1955).

[44] T. Colombeau, *New Generalized Functions and Multiplications of Distributions*, North-Holland, New York (1984).

[45] J. B. Conway, *A Course in Functional Analysis*, Springer-Verlag, New York (1990).

[46] R.G. Cooke, The inversion formula of Hardy and Titchmarsh, *Proc. London Math. Soc.,* Vol 24 (1925), pp. 381—420.

[47] J. Cooley and J. Tukey, An algorithm for the machine calculation of the complex Fourier series, *Math. Comp.,* Vol. 19 (1965), pp. 297—301.

[48] A. Cormack, My connection with the Radon transform, *75 years of Radon Transform,* S. Gindikin and P. Michor, Editors, International Press, Cambridge, Mass. (1994), pp. 32—35.

[49] I. Daubechies, *Ten Lectures on Wavelets,* SIAM Publications, Soc. Indust. Appl. Math., Philadelphia (1992).

[50] I. Daubechies, Orthonormal bases of compactly supported wavelets, *Comm. Pure Appl. Math.,* 41(1988), pp. 909—996.

[51] S. Deans, *The Radon Transform and Some of its Applications,* Wiley Interscience, New york (1983).

[52] E. Deeba and E. Koh, the continuous Jacobi transform, *Internat. J. Math. Math. Sci.,* Vol. 6, (1983), pp. 145—160.

[53] M. Djrbashian, *Harmonic Analysis and Boundary-Value Problems in the Complex Domain*, Birkhause, Basel (1994).

[54] P. Duren, *Theory of H^p Spaces*, Academic Press, New York (1970).

[55] D. Elliott and K. Rao, *Fast Transforms: Algorithms, Analyses, Applications*, Academic Press, New York (1982).

[56] A. Erdélyi, "Axially symmetric potentials and fractional integrations," *SIAM J. Appl. Math.*, Vol. 13 (1965), pp. 216—228.

[57] A. Erdélyi, "An integral equation involving Legendre functions," *SIAM J. Appl. Math.*, Vol. 12 (1964), pp. 15—30.

[58] A. Erdélyi, On Some Transformation, *Rend. Sem. Mat. Univ. Torino,* 10 (1950), pp. 217—234.

[59] A. Erdélyi, On some functional transformations, *Univ. Politec. Torino, Rend. Sem. Math.*, Vol.10 (1950), pp. 217—234.

[60] A. Erdelyi, W. Magnus, F. Oberhettinger, and F. Tricomi, *Higher Transcendental Functions,* Vol. I, II, McGraw-Hill, New York (1953).

[61] A. Erdelyi, W. Magnus, F. Oberhettinger, and F. Tricomi, *Higher Transcendental Functions,* Vol. III, McGraw-Hill, New York (1955).

[62] A. Erdelyi, W. Magnus, F. Oberhettinger, and F. Tricomi, *Tables of Integral Transforms,* Vol. I, II, McGraw-Hill, New York (1954).

[63] P. Franklin, A set of continuous orthogonal functions, *Math. Ann.*, Vol. 100 (1928), pp. 522—529.

[64] V. Fock, On the representation of an arbitrary function by an integral involving Legendre's function with a complex index, *Dokl. Akad. Nauk., SSSR,* Vol. 39 (1943), pp. 253—256.

[65] V. Foch, Verallgemeinerung und Lösung der Diracschen statistischen Gleichung, *Z. Phys.*, Vol. 49 (1928), pp. 339—357.

[66] J. Fourier, *Théorie Analytique de la chaleur,* Paris (1822), *The Analytical Theory of Heat*, translated by A. Freeman, Cambridge, England (1878).

[67] D. Gabor, Theory of Communication, *J. Inst. Electr. Eng.*, 93 (1946), pp.429-457.

[68] C. Gauss, *Theoria interpolationis methodo nova tractata,* Carl Friedrich Gauss Werke, Band 3, Königlichen Gesellschaft der Wissenschaften, Göttingen (1866).

[69] I. M. Gel'fand, M. I. Graev and N. Ya. Vilenkin, *Generalized Functions,* Vol. 5, Academic Press (1966).

[70] I. Gel'fand and G. Shilov, *Generalized Functions*, Vols. 1-4, Academic Press, New York (1964).

[71] I. Gel'fand, Eigenfunction expansions for an equation with periodic coefficients, *Dokl. Akad. Nauk. SSR*, Vol. 76 (1950), pp. 1117—1120.

[72] H. Glaeske and A. Hess, On the convolution theorem of the Mehler–Fock transform for a class of generalized functions, *Math. Nachr.*, 136 (1988), pp. 119—129.

[73] H. Glaeske, A convolution connected with the Kontorovich–Lebedev transform of generalized functions, *Math. Z.*, Vol. 193 (1986), pp. 67—78.

[74] M. L. Glasser, Some Bessel function integrals, *Kyungpook Math. J.*, Vol. 13 (1973), pp. 171—174.

[75] R. J. Glauber, in *Quantum Optics and Electronics*, C. De Witt, A. Blandin, and C. Cohen-Tannoudji, Editors, Gordon and Breach, New York (1964).

[76] R.R. Goldberg, An integral transform related to the Hilbert transform, *J. London Math. Soc.,* Vol. 35 (1960), pp. 200—204.

[77] A. Gonzalez Dominguez, Contribucion a la teoria de las funciones de Hille, *Ciencia y Tecnica*, Vol. 475 (1941), pp. 475—521.

[78] V. Gorbačuk and M. Gorbačuk, Trigonometric series and generalized periodic functions, *Soviet Math. Dokl.*, Vol. 2 (1981), pp. 342—346.

[79] I. Gradshteyn and I. Ryzhik, *Tables of Integrals, Series and Products,* Academic Press, New York (1965).

[80] R. E. Greenwood, Hankel and other extensions of Dirichlet's Series, *Ann. Math.*, 42 (1941), pp. 778—805.

[81] A. Grossmann and J. Morlet, Decomposition of Hardy functions into square integrable wavelets of constant shape, *SIAM J. Math. Anal.,* Vol. 15 (1984), pp. 723—736.

[82] A. Haar, Zur Theorie der orthogonalen Funktionensysteme, *Math. Ann.,* Vol. 69 (1910), pp. 321—371.

[83] D. T. Haimo, The dual Weierstrass–Laguerre transform, *Trans. Amer. Math. Soc.*, Vol. 290 (1985), pp. 597—613.

[84] D. T. Haimo, The dual Poisson–Laguerre transform, *Trans. Amer. Math. Soc.*, Vol. 144 (1969), pp. 271-300.

[85] D. T. Haimo and F. M. Cholewinski, The Poisson–Laguerre transform, *Bull. Amer. Math. Soc.*, Vol. 74 (1968), pp. 137—139.

[86] D. T. Haimo, The Weierstrass–Laguerre transform, *J. Math. Anal. Appl.*, Vol. 23 (1968), pp. 41—57.

[87] D. T. Haimo, Inversion of the reduced Poisson–Hankel transform, *J. Math. Anal. Appl.*, Vol. 25 (1972), pp. 323—343.

[88] P. Halmos, *Introduction to Hilbert Spaces*, Chelsea Publishing Company, New York (1957).

[89] G. H. Hardy and J. F. Littlewood, "Some properties of fractional integrals," *Math. Zeit.*, Vol. 27 (1928), pp. 265—606.

[90] G.H. Hardy, Some formulae in the theory of Bessel functions, *Proc. London Math. Soc., Ser. 2,* Vol. 23 (1925), lxi-lxiii.

[91] G. Hardy, On Hilbert transforms, *Messenger of Math.*, Vol. 54 (1924), pp. 20-27, pp. 81—88.

[92] H. Harmuth, *Transmission of Information by Orthogonal Functions,* Springer-Verlag, New York (1972).

[93] R. V. Hartley, A more symmetrical Fourier analysis applied to transmission problems, *Proc. Inst. Radio Eng.,* Vol. 30 (1942), pp. 144—150.

[94] N. Hayek and B. J. Gonzalez, On the distributional index ${}_2F_1$-transform, *Math. Nach.*, Vol. 165(1994), pp. 15—24.

[95] C. Heil and D. Walnut, Continuous and discrete wavelet transforms, *SIAM Rev.,* Vol. 31 (1989), pp. 628—666.

[96] S. Helgason, *The Radon Transform,* Birkhäuser, Boston, (1980).

[97] P. Heywood, On a transform discussed by Goldberg, *J. London Math. Soc.,* Vol. 38 (1963), pp. 162—168.

[98] O. Heaviside, *Electromagnetic Theory,* Vol. I-III (1993-1912), reissued by Benn Bros., London (1922).

[99] O. Heaviside, Operators in mathematical physics, *Proc. Roy. Soc. London, Ser. A*, Vol. 52 (1893), pp. 504—529.

[100] D. Hilbert, *Grundzüge einer allgemeinen Theorie der linearen Integralgleichungen,* (1912), pp. 75—77.

[101] E. Hille and R. S. Philips, *Functional Analysis and semigroups*, Amer. Math Soc. Colloq. Publ., Vol. 31, revised ed., Providence, Rhode Island (1957).

[102] I. I. Hirschman, Variation diminishing Hankel Transforms, *J. Anal. Math.*, Vol. 8 (1960), pp. 307—336.

[103] I. I. Hirschman and D. V. Widder, *The Convolution Transform,* Princeton University Press, Princeton, New Jersey (1955).

[104] I. I. Hirschman and D. V. Widder, General inversion formulas for convolution transforms, II, *Duke Math. J.*, Vol. 17 (1950), pp. 391—402.

[105] I. I. Hirschman and D. V. Widder, The inversion of a general class of convolution transforms, *Trans. Amer. Math. Soc.*, Vol. 66 (1949), pp. 135—201.

[106] I. I. Hirschman and D. V. Widder, A representation theory for a general class of convolution transforms, *Trans. Amer. Math. Soc.*, Vol. 67 (1949), pp. 69—97.

[107] I. I. Hirschman and D. V. Widder, Generalized inversion formulas for convolution transforms, *Duke Math. J.*, Vol. 15 (1948), pp. 659—696.

[108] I. I. Hirschman and D. V. Widder, An inversion and representation theory for convolution transforms with totally positive kernels, *Proc. Nat. Acad. Sci. USA*, Vol. 34 (1948), pp. 152—156.

[109] M. Holschneider, Wavelet analysis on the unit circle, *J. Math. Phys.*, Vol. 31 (1990), pp. 39—44.

[110] A. J. Janssen, Spaces of type W, growth of Hermite coefficients, Wigner distribution, and Bargmann transform, *J. Math. Anal. Appl.*, Vol. 152 (1990), pp. 368—390.

[111] A. J. Janssen, The Zak transform: A signal transform for sampled time-continuous signals, *Philips J. Research*, 43 (1988), pp. 23–69.

[112] A. J. Janssen, Bargmann transform, Zak transform, and coherent states, *J. Math. Phys.*, Vol. 23 (1982), pp. 720—731.

[113] A. J. Janssen, On the locus and spread of pseudo-density functions in the time-frequency plane, *Philips J. Research*, 37 (1982), pp. 79—110.

[114] A. J. Janssen, Weighted Wigner Distributions vanishing on lattices, *J. Math. Anal. Appl.*, Vol. 80 (1981), pp. 156—167.

[115] A. J. E. M. Janssen, Gabor representation of generalized functions, *J. Math. Anal. Appl.*, Vol. 83 (1981), pp. 377—394.

[116] D. S. Jones, The Kontorovich–Lebedev transform, *J. Inst. Math. Appl.*, Vol. 26 (1980), pp. 133—141.

[117] D. Jones, *Generalized Functions*, McGraw-Hill, London, England (1966).

[118] Y. Katznelson, *An Introduction to Harmonic Analysis*, Dover Publications, New York (1976).

[119] O. Kellogg, Unstetigkeiten bei den linearen Integralgleichungen mit Anwendung auf ein Problem von Riemann, *Math. Ann.*, Vol. 60 (1905), pp. 424—433.

[120] O. Kellogg, Unstetigkeiten in den linearen Integralgleichungen, *Math. Ann.*, Vol. 58 (1904), pp. 441—456.

[121] J. Klauder and B. S. Skagerstam, *Coherent States*, World Scientific, Singapore (1985).

[122] J. Klauder and E. C. Sudarshan, *Fundamentals of Quantum Optics*, W. A. Benjamin, New York (1968).

[123] H. Kober, On fractional integrals and derivatives, *Quart. J. Math.*, Vol. 11 (1940), pp. 193—221.

[124] E. L. Koh, E. Y. Deeba and M. A. Ali, On the generalized Meijer transformation, *Generalized Functions, Convergence Structures, and their Applications*, B. Stankovic, E. Papand, S. Pilipovic, V. Vladimivov, Editors, Plenum Press, New York (1988), pp. 219—225.

[125] E. L. Koh, and E. Y. Deeba, The Meijer transformations of generalized Functions, *Internat. J. Math. Math. Sci. Sc.*,, Vol. 10 (1987), pp. 267—286.

[126] E. Koh and A. Zemanian, The Complex Hankel and I transformations of generalized functions, *SIAM J. Appl. Math.*, Vol. 16, (1968), pp. 945—57.

[127] A. Kolmogorov and S. Fomin, *Introductory Real Analysis*, Dover Publishing, New York (1975).

[128] M. Kontorovitch and N. Lebedev, One method of solution of some problems in diffraction theory and other related problems, *J. Experiment. Theoret. Phys.*, Vol. 8 (1938), pp. 1192—1206.

[129] T. Koornwinder and G. Walter, The finite continuous Jacobi transform and its inverse, *J. Approx. Theory,* Vol. 60 (1990), pp. 83—100.

[130] E. Kratzel, Integral transformation of Bessel-type, *Proc. International Conference Generalized Functions and Operational Calculus*, Varna, Bulgaria (1975), pp. 148—155.

[131] E. Kratzel, Bemerkunger zur Meijer-Transformation und Anwendungen, *Math. Nachr.*, 30 (1965), pp. 327—334.

[132] E. Kratzel, Eine Verallgemeinerung der Laplace und Meijer Transformation, *Wiss. Z. Univ. Jena. Math. - Naturw Reihe*, 5 (1965), pp. 369—381.

[133] A. M. Kuethe, and J. Schetzer, *Foundations of Aerodynamics,* John Wiley & sons, New York (1967).

[134] J. Lagrange, *Recherches sur la nature et la propagation du son*, Miscellanea Taurinensia (Mélanges de Turin), Vol. I (1759), pp. 1—112.

[135] J.H. Lambert, Anlage zur Architektonik oder Theorie des Einfachen und Ersten in der Philosophischen und Mathematischen Erkenntnis, *Riga* 2 (1771).

[136] P.S. Laplace, *Théorie Analytique des Probabilités,* Vol. I, Part 2, Lerch, Paris (1820).

[137] N. Lebedev, *Special Functions and their Applications,* Translated by R. Silverman, Dover Publishing, New York (1972).

[138] N. Lebedev, Some singular integral equations connected with integral representations of mathematical physics, *Dokl. Akad. Nauk. SSSR,* Vol. 65 (1949), pp. 621—624.

[139] N. Lebedev, Expansion of an arbitrary function in an integral with respect to cylinder functions of imaginary order and argument (in Russian), *Prikl. Mat. Mekh.,* Vol. 13 (1949), p. 465.

[140] N. Lebedev, Sur une formule d'inversion, *Dokl. Akad. Nauk. USSR,* Vol. 52 (1946), pp. 655—658.

[141] G. W. Leibniz, Letter from Hanover, Germany, September 30, 1695 to G. A. L'Hospital, *Leibnizen Mathematische Schriften*, Vol. 2, Olms Verlag, Hildesheim, Germany (1962). First published in 1849.

[142] P. Lemarié, Une nouvelle base d'ondelettes de $L^2(R^n)$, *J. Math. Pures et Appl.,* Vol. 67 (1988), pp. 227—236.

[143] P. Lemarié and Y. Meyer, Ondelettes et bases hilbertiennes, *Rev. Mat. Iberoamericana,* Vol. 2 (1986), pp. 1—18.

[144] W.R. LePage, *Complex Variables and the Laplace Transform for Engineers,* Dover Publ., New York (1980).

[145] N. Levinson, *Gap and Density Theorems*, Amer. Math. Soc. Colloq.. Publs. Series, Vol. 26, Rhode Island (1940).

[146] B. Ja. Levin, *Distribution of Zeros of Entire Functions,* Amer. Math. Soc. Transl. Math. Monographs, Vol. 5, Providence, Rhode Island (1964).

[147] B. Levitan and I. Sargsjan, *Introduction to Spectral Theory,* Amer. Math. Soc. Transls., Vol. 39 (1975), Providence, R.I.

[148] M. Lighthill, *Fourier Analysis and Generalized Functions*, Cambridge University Press, Cambridge, England (1958).

[149] S. Lojasiewcz , Sur la valeur et la limite d'une distribution dans um point, *Studia Math.,* Vol. 16 (1957), pp. 1—36.

[150] J. Liouville, Sur le calcul des differentielles à indices quelconques, *J. École Polytechn.*, Vol. 13 (1832), 1—69.

[151] D. Ludwig, The Radon transform on Euclidean spaces, *Comm. Pure Appl. Math.,* Vol. 19 (1966), pp. 49—81.

[152] T. M. MacRobert, Barnes integrals as a sum of E-functions, *Math. Annal.*, Vol. 147(1962), pp.240—243.

[153] S. Mallat, Multiresolution approximations and wavelet orthonormal bases in $L^2(R)$, *Trans. Amer. Math. Soc.* , Vol 315 (1989), pp. 69—87.

[154] S. Mallat, A theory for multiresolution signal decomposition: The wavelet representation, *IEEE Trans. Pattern Anal. Machine Intelligence,* Vol. 11 (1989), pp. 674—693.

[155] B. N. Mandal and Pallabi Guha Roy, A Mehler–Fock type integral transform, *Appl. Math. Lett.*, 4(1991), pp. 29—32.

[156] R. Martinet, J. Morlet, and A. Grossmann, Analysis of sound patterns through wavelet transform, *Internat. J. Pattern Recog. Artif. Intelligence,* 1(1987), pp. 273—302.

[157] A. C. McBride, A theory of fractional integration for generalized functions, *SIAM J. Math. Anal.*, Vol. 6 (1975), 583—599.

[158] C. S. Meijer, Uber eine Erweiterung der Laplace - Transformation, *Proc. Amsterdam Akad. Wet.*, 43 (1940), pp. 599-608, 702—711.

[159] H. Mellin, Über den Zusammenhang zwischen den Linearen Differential und Differenzengleichungen, *Acta Math.,* Vol. 25(1902), pp. 139—164.

[160] Y. Meyer *Ondelettes et Opérateurs I: Ondelettes,* Hermann, Paris (1990).

[161] Y. Meyer, *Ondelettes, fonctions splines et analyses graduées,* Lectures given at the University of Torino, Italy (1986).

[162] Y. Meyer, Principe d'incertitude bases hilbertiennes et algébres d' opérateurs, *Seminaire Bourbaki,* Vol. 662 (1985).

[163] J. Mikusinski,*Operational Calculus,* Warsaw, Poland (1967).

[164] J. Mikusinski and P. Mikusinski, Quotients de suites et leurs applications dans l'analyse functionnelle, *Comptes Rendus,* Vol. 293, Serie I (1981), pp. 463—464.

[165] O. P. Misra and J. L. Lavoine, *Transform Analysis of Generalized Functions*, North-Holland, New York (1986).

[166] O. P. Misra, Generalized convolution for G-transformation, *Bull. Calcutta Math. Soc.,* Vol. 64 (1972), pp. 137—142.

[167] O. P. Misra, Distributional G-transformation, *Bull. Calcutta Math. Soc.,* Vol. 73 (1981), pp. 247—255.

[168] G. Mittag-Leffler, Sur la representation analytique d'une branche uniforme d'une fonction monogene, *Acta Math.*, Vol. 29 (1905), pp. 101—182.

[169] G. Mittag-Leffler, Sur la nouvelle fonction E_α,*Comptes Rendus Hebdomadaires. Acad. Sci.*, Paris, 2, Vol. 137(1903), pp. 554—558.

[170] M. Morimoto, *An Introduction to Sato's Hyperfunctions,* Amer. Math. Soc. Transl. Math. Monographs, Vol. 129, Providence, Rhode Island (1993).

[171] J. Morlet, G. Arens, E. Fourgeau, and D. Giard, Wave propagation and sampling theory, Part I: Complex signal and scattering in multilayered media, *J. Geophys.,* Vol. 47 (1982), pp. 203—221.

[172] J. Morlet, G. Arens, E. Fourgeau, and D. Giard, Wave propagation and sampling theory, Part II: Sampling theory and complex waves, *J. Geophys.,* Vol. 47 (1982), pp. 222-236.

[173] M. Naimark, *Linear Differential Operators*, Vol. I, *Elementary Theory of Linear Differential Operators*, George Harrap & Co., London, England (1967).

[174] R. Narain, The G-functions as unsymmetrical Fourier kernels II, *Proc. Amer. Math. Soc.*, Vol 14 (1963), pp. 18—28.

[175] R. Narain, The G-functions as unsymmetrical Fourier kernels I, *Proc. Amer. Math. Soc.*, Vol 13 (1962), pp. 950—959.

[176] F. Natterer, *The Mathematics of Computerized Tomography,* John Wiley & Sons, New York (1986).

[177] D. Naylor, On a Mellin type integral transform, *J. Math. Mech.* 12(a)(1963), pp. 265—274.

[178] K. Nickel, Lösung eines Integralgleichungs Systems aus der Tragflügetheorie, *Math. Z.,* Vol. 54 (1951), pp. 81—96.

[179] K. Nishimoto, *Fractional Calculus and its Applications,* Proc. Intern. Conference, Nihon University, Japan (1990).

[180] K. B. Oldham and J. Spanier, *The Fractional Calculus*, Academic Press, New York (1974).

[181] T. J. Osler, Taylor's series generalized for fractional derivatives and applications, *SIAM J. Math. Anal.*, Vol. 2 (1971), pp. 37—48.

[182] R. Paley and N. Wiener, *Fourier Transforms in the Complex Domain,* Amer. Math. Soc. Colloq. Publ., Vol. 19, Providence, Rhode Island (1934).

[183] A. Papoulis, *The Fourier Integral and its Applications*, McGraw-Hill, New York (1987).

[184] A. Papoulis, *Signal Analysis*, McGraw-Hill, New York (1977).

[185] A. Papoulis, Ambiguity function in Fourier optics, *J. Optics Soc.,* Vol. 64 (1974), pp. 779—788.

[186] E. Parzen, A simple proof and some extensions of sampling theorem, *Tech. Report* **7**, Stanford University, Stanford, California (1956).

[187] R.S. Pathak and J.N. Pandey, The G-transform of generalized functions, *Rocky Mountain J. Math.*, Vol.9 (1979), pp. 307—325.

[188] R.S. Pathak and J.N. Pandey, A distributional Hardy transformation, *Internat. J. Math. and Math. Sci.*, Vol. 2 (1979), pp. 693—701.

[189] R.S. Pathak and J.N. Pandey, The Kontorovich-Lebedev transformation of distributions, *Math. Zeitschr.*, Vol. 165 (1979), pp. 29—51.

[190] R.S. Pathak and J.N. Pandey, A distributional Hardy transformation, *Proc. Camb. Philos. Soc.*, Vol. 76 (1974), pp. 247—262.

[191] W.B. Pennington, Widder's inversion formula for the Lambert transform, *Duke Math. J.* , Vol. 27 (1960), pp. 56—568.

[192] H. Pollard, The completely monotonic character of the Mittag-Leffler function $E_\alpha(-x)$, *Bull. Amer. Math. Soc.*, Vol. 54 (1948), pp. 1115—1116.

[193] G. Polya, Bemerkungen über unendliche Folgen und ganze Funktionen, *Math. Ann.*, Vol. 88 (1923), pp. 169—183.

[194] G. Polya, Untersuchungen über Lücken und Singularitäten von Potenzreihen, *Math. Z.* , Vol. 29 (1929), pp. 549—640.

[195] E.L. Post, Generalized differentiation, *Trans. Amer. Math. Soc.*, Vol. 32 (1930), pp. 723—781.

[196] R. Powell and S. Shah, *Summability Theory and Applications*, Van Nostrand Reinhold, London, England (1972).

[197] J. Radon, Über die Bestimmung von Funktionen durch ihre Integralwärte längs gewisser manningfaltigkeiten, *Ber. Verh. Sächs. Akad. Wiss. Math.-Nat.*, Leipzig, Vol. 69 (1917), pp. 262—277.

[198] F. M. Ragab, E-transforms, *J. Res. Nat. Bur. Stand.*, Vol. 71B (1967), pp. 23—37.

[199] G. L. Rao and L. Debnath, A generalized Meijer transformation, *Internat. J. Math. Math. Sci.*, 8 (1985), pp. 359—365.

[200] G. L. N. Rao, Abelian theorems for a distributional generalized Stieltjes transform, *Rev. Real. Acad. Cienc. Exact. fis. Natur. Madrid*, Vol. 70 (1976), pp. 97—108.

[201] G. L. N. Rao, The n-dimensional generalized Laplace transform of certain distributions,*J. Indian Math. Soc.*, Vol. 39 (1975), pp. 219—226.

[202] G. L. N. Rao, A generalized Laplace transform of generalized functions, *Ranchi Univ. Math. J.*, Vol. 5 (1974), pp. 76—88.

[203] B. Riemann, Versuch einer allgemeinen Auffassung der Integration und Differentiation, Werke, *The collected works of Bernhard Riemann*, H. Weber, ed., 2nd edition, Dover, New York (1953).

[204] B. Riemann, Über die Anzahl der Primzahlen unter einer gegenbener Grössen, *Gesammelte Mathematische Werke*, (Ed. 1876), Leipzig, pp. 136—144.

[205] A. Rihaczek, *Principles of High-Resolution Radar,* McGraw-Hill, New York (1969).

[206] P. G. Rooney, A generalization of an inversion formula for the Gauss transformation, *Cand. Math. Bull.*, Vol. 6 (1963), pp. 45—53.

[207] P. G. Rooney, On the inversion of the Gauss transformation, *Cand J. Math.*, Vol. 9 (1957), pp. 459—464.

[208] W. Rudin, *Real and Complex Analysis,* McGraw-Hill, New York (1974).

[209] W. Rudin, *Functional Analysis*, McGraw-Hill, New York (1973).

[210] V. Sastry, A note on Sommerfeld's differaction problem, *J. Math. Phys.,* Vol. 20 (1979), pp. 2123—2125.

[211] M. Sato, Theory of hyperfunctions, Vol. I, *J. Fac. Sci. Univ. Tokyo,* Vol. 8 (1959-1960), pp. 139—193.

[212] M. Sato, Theory of hyperfunctions, Vol. II, *J. Fac. Sci. Univ. Tokyo,* Vol. 8 (1959-1960), pp. 287—437.

[213] W. Schempp, Radar ambiguity functions, the Heisenberg group and holomorphic theta series, *Proc. Amer. Math. Soc.*, Vol. 92 (1984), pp. 103—110.

[214] L. Schwartz, *Theorie des Distributions,* Hermann, Paris, (1966).

[215] J.C. Slater, *Concepts and Developments of Quantum Physics,* Dover Publishing, New York (1955).

[216] I. Sneddon, *The Use of Integral Transforms,* McGraw-Hill, New York (1972).

[217] D. Solmon, The X-ray transform, *J. Math. Anal. Appl.,* Vol. 56 (1976), pp. 61—83.

[218] H. M. Srivastava, Certain properties of a generalized Whittaker transform, *Mathematica (Cluj)*, Vol. 10 (1968), pp. 385—390.

[219] T. S. Stieltjes, Recherches sur les fractions continues, *Annales de la Faculté des Sciences de Toulouse*, Vol. 8 (1894), pp. 1—123

[220] G. Szegö, *Orthogonal Polynomials,* Amer. Math. Soc. Colloq. Publ., Vol. 23, Providence, Rhode Island (1939).

[221] Vu-Kim-Tuan, and S. Yakubovich, The Kontorovich–Lebedev integral transformation in a new class of functions, *Dokl. Akad. Nauk. USSR,* Vol. 29(1985), pp. 11—14.

[222] G. Temple, The theory of generalized functions, *Proc. Roy. Soc. London, Series A,* Vol. 228 (1955), pp. 175—190.

[223] G. Temple, Theories and applications of generalized functions, *J. London Math. Soc.,* Vol 28 (1953), pp. 134—148.

[224] E.C. Titchmarsh, *The Theory of Functions*, Oxford University Press, Oxford, England (1978).

[225] E. Titchmarsh, *Introduction to the Theory of Fourier Integrals*, Oxford University Press, Oxford, England (1967).

[226] E. Titchmarsh, *Eigenfunction Expansions Associated with Second Order Differential Equations,* Part I, Oxford University Press, Oxford, England (1962).

[227] E. Titchmarsh, Additional note on conjugate functions, *J. London Math. Soc.,* Vol. 4 (1930), pp. 204—206.

[228] E. Titchmarsh, On conjugate functions, *Proc. London Math. Soc.,* Vol. 29 (1928), pp. 49—80.

[229] E. Titchmarsh, Conjugate trigonometrical integrals, *Proc. London Math. Soc.,* Vol.24 (1925), pp. 109—130.

[230] A. K. Tiwari and A. Ko, Certain properties of distributional generalized Whittaker transform, *Indian J. pure Appl. Math.*, Vol. 13 (1982), pp. 348—361.

[231] U. N. Tiwari and J. N. Pandey, the Mehler–Fock transform of distributions, *Rocky Mount. J. Math.*, 10 (1980), pp. 401—408.

[232] F. Tricomi, On the finite Hilbert transformation, *Quart. J. Math.,* Vol. 2 (1951), pp. 199—211.

[233] F. Tricomi, The airfoil equation for a double interval, *Z. Angew. Math. Phys.,* Vol. 2 (1951), pp. 402—406.

[234] B. Van Der Pol and H. Bremmer, *Operational Calculus Based on the Two-Sided Laplace Integral,* Cambridge University Press, Cambridge, England (1959).

[235] R. S. Varma, On a generalization of Laplace integral, *Proc. Nat. Acad. Sci. India*, A20 (1951), pp. 209—216.

[236] J. Ville, Theorie et applications de la notion de signal analytique, *Cables et Transmissions,* Vol. 2A (1948), pp.61—74.

[237] R. C. Vyas and R. K. Saxena, Kummer's transform of two variables, *Riv. Math. Univ. Parma*, Vol. 10 (1969), pp. 23—32.

[238] J. Walsh, A closed set of normal orthogonal functions, *Amer. J. Math.*, Vol. 55 (1923), pp. 5—24.

[239] G. Walter, *Wavelets and Other Orthogonal Systems with Applications,* CRC Press, Boca Raton, FL (1994).

[240] G. Walter, Series de Fourier, Tipo de convergencia, *Bol. Ciencias Univ. Cat. Peru,* Vol. 2 (1969), pp. 40—53.

[241] G. Walter and A. Zayed, The continuous (α, β) Jacobi transform and its inverse when $\alpha + \beta + 1$ is a positive integer, *Trans. Amer. Math. Soc.*, Vol. 305 (1988), pp. 653—664.

[242] G. N. Watson, *A Treatise on the Theory of Bessel Functions*, Cambridge University Press, Cambridge, England (1966).

[243] H. Weyl, Bemerkungen zum Begriff des Differentialquotienten gebrochener Ordnung, *Viertel Jarhsschrifft Naturforsch. Gesellsch.*, Zürich, Vol. 62 (1917), pp. 296—302.

[244] M. Wickerhauser, *Adapted Wavelet Analysis from Theory to Software,* AK Peters, Wellesley, Mass. (1994).

[245] D.V. Widder, *The Laplace Transform*, Princeton University Press, Princeton (1966).

[246] D.V. Widder, An inversion of the Lambert transform, *Math. Mag.* Vol. 23 (1950), pp. 171—182.

[247] E. Wigner, On the quantum correction for thermodynamic equilibrium, *Phys. Rev.*, Vol. 40 (1932), pp. 749—759.

[248] C. Wilcox, The synthesis problem for radar ambiguity functions, *MRC Technical Report,* 157, Math. Research Center, U.S. Army, Univ. Wisconsin, Madison (1960).

[249] P. Woodward, *Probability and Information Theory with Applications to Radar,* McGraw-Hill, New York (1953).

[250] R. Wong, Asymptotic expansion of the Kontorovich–Lebedev transform, *J. Appl. Anal.*, Vol. 12(1981), pp. 161—172.

[251] R. Young, *An Introduction to Nonharmonic Fourier Series,* Academic Press, New York (1980).

[252] W.H. Young, On infinite integrals involving a generalisation of the sine and cosine functions, *Quart. J. Math.*, Vol. 4 (1912), pp. 161—177.

[253] J. Zak, Dynamics of electrons in solids in external fields, *Phys. Rev.*, Vol. 168 (1968), pp. 686—695.

[254] J. Zak, Finite translation in solid state physics, *Phys. Rev. Lett.*, Vol. 19 (1967), pp. 1385—1397.

[255] A. I. Zayed, Wavelet expansions of analytic hyperfunctions, *J. Integral Transforms, Special Functions,* Vol. 3 (1995), pp. 305—320.

[256] A. I. Zayed, Wavelet transform of periodic generalized functions, *J. Math. Anal. Appl.*, Vol. 183 (1994), pp. 391—412.

[257] A. I. Zayed, *Advances in Shannon's Sampling Theory,* CRC Press, Boca Raton, Fl (1993).

[258] A. I. Zayed, Kramer's sampling theorem for multidimensional signals and its relationship with Lagrange-type interpolation, *J. Multidim. Systems, Signal Processing,* Vol. 3 (1992), pp. 323-340.

[259] A. I. Zayed, On Kramer's sampling theorem associated with general Sturm-Liouville boundary-value problems and Lagrange interpolation, *SIAM J. Appl. Math.*, Vol. 51 (1991), pp. 575—604.

[260] A. I. Zayed, Generalized Faber expansions of hyperfunctions on analytic curves, *J. Math. Soc. Japan,* Vol. 42 (1990), pp. 155—170.

[261] A. I. Zayed, A generalized inversion formula for the continuous Jacobi transform, *Internat. J. Math. Math. Sci.*, Vol. 10, (1987), pp. 671—692.

[262] A. I. Zayed, Generalized functions and orthogonal polynomials on the unit circle, *Proc. Amer. Math. Soc.*, Vol. 88 (1983), pp. 407—415.

[263] A. I. Zayed, Asymptotic expansions of some integral transforms by using generalized functions, *Trans. Amer. Math. Soc.*, Vol. 272 (1982), pp. 785—802.

[264] A. I. Zayed, Laguerre series as boundary values, *SIAM J. Math. Anal,* Vol. 13, No. 2 (1982), pp. 263—279.

[265] A. I. Zayed, Hyperfunctions as boundary values of generalized axially symmetric potentials, *Ill. J. Math.*, Vol. 25 (1981), pp. 306—317.

[266] A. I. Zayed, P. Butzer and G. Hinsen, On Lagrange interpolation and Kramer-type sampling theorems associated with regular Sturm–Liouville problems, *SIAM J. Appl. Math.*, Vol. 50 (1990), pp. 893—909.

[267] A. I. Zayed and E. Deeba, Generalized continuous Jacobi transform, *J. Appl. Anal.*, Vol. 48 (1993), pp. 63—79.

[268] A. I. Zayed and D. T. Haimo, Inversion of integral transforms associated with a class of perturbed heat equations, *J. Math. Anal. Appl.*, Vol. 163 (1992), pp. 113—135.

[269] A. I. Zayed and D. T. Haimo, Inversion of integral transforms related a general form of heat equations, *J. Math. Anal. Appls.*, Vol. 131 (1988), pp. 246—264.

[270] A. I. Zayed and P. Mikusinski, On the extension of the Zak transform, *J. Methods Appl. Anal.*, Vol. 2 (1995), pp. 160—172.

[271] A. I. Zayed and P. Mikusinski, An extension of the Radon Transform, *Generalized Functions and Their Applications*, R.S. Pathak, Editor, Plenum Press, New York (1993), pp. 141—147.

[272] A. I. Zayed and P. Mikusinski, Radon transform of Boehmians, *Proc. Amer. Math. Soc.*, Vol. 118 (1993), pp. 561—570.

[273] A. I. Zayed and G. Walter, Characterization of analytic functions in terms of their wavelet coefficients, to appear in *J. Complex Variables: Theory, Appl.*

[274] A. I. Zayed and G. Walter, On the inversion of integral transforms associated with Sturm–Liouville problems, *J. Math. Anal. Appl.*, Vol. 164 (1992), pp. 285—306.

[275] A. I. Zayed and G. Walter, Series of orthogonal polynomials as hyperfunctions,*SIAM J. Math. Anal.*, Vol. 13 (1982), pp. 664—675.

[276] A. Zemanian, *Generalized Integral Transformations,* Dover Publishing, New York (1987).

[277] A. Zemanian, The Kontorovich-Lebedev transformation on distributions of compact support and its inversion, *Math. Proc. Camb. Philos. Soc.,* Vol. 77 (1975), pp. 139—143.

[278] A. Zemanian, *Distribution Theory and Transform Analysis*, McGraw-Hill, New York (1965).

[279] A. Zygmund, *Trigonometric Series*, Vol. 1, Cambridge University Press, Cambridge, England (1959).

[280] A. Zygmund, *Trigonometric Series*, Vol. II, Cambridge University Press, Cambridge, England (1959).

Index